VOLUME NINETY FOUR

ADVANCES IN GENETICS

Genetics and Molecular Biology of Entomopathogenic Fungi

ADVANCES IN GENETICS, VOLUME 94

Serial Editors

Theodore Friedmann
Department of Pediatrics, University of California at San Diego, School of Medicine, CA, USA

Jay C. Dunlap
Department of Genetics, The Geisel School of Medicine at Dartmouth, Hanover, NH, USA

Stephen F. Goodwin
Department of Physiology, Anatomy and Genetics, University of Oxford, Oxford, UK

VOLUME NINETY FOUR

Advances in GENETICS

Genetics and Molecular Biology of Entomopathogenic Fungi

Edited by

BRIAN LOVETT
University of Maryland, College Park, MD, United States

RAYMOND J. St. LEGER
University of Maryland, College Park, MD, United States

ELSEVIER

AMSTERDAM • BOSTON • HEIDELBERG • LONDON
NEW YORK • OXFORD • PARIS • SAN DIEGO
SAN FRANCISCO • SINGAPORE • SYDNEY • TOKYO

Academic Press is an imprint of Elsevier

Academic Press is an imprint of Elsevier
50 Hampshire Street, 5th Floor, Cambridge, MA 02139, USA
525 B Street, Suite 1800, San Diego, CA 92101-4495, USA
125 London Wall, London EC2Y 5AS, UK
The Boulevard, Langford Lane, Kidlington, Oxford OX5 1GB, UK

First edition 2016

ISBN: 978-0-12-804694-4
ISSN: 0065-2660

Publisher: Zoe Kruze
Acquisition Editor: Zoe Kruze
Editorial Project Manager: Helene Kabes
Production Project Manager: Vignesh Tamil
Designer: Greg Harris

Typeset by TNQ Books and Journals

DEDICATION

The editors dedicate this book to the doyen of insect pathology, Donald Roberts, and to his versatile little friend *Metarhizium robertsii*.

CONTENTS

CONTRIBUTORS

J.P.M. Araújo
Pennsylvania State University, University Park, PA, United States

L. Barelli
Brock University, St. Catharines, ON, Canada

M.J. Bidochka
Brock University, St. Catharines, ON, Canada

T.M. Butt
Swansea University, Swansea, Wales, United Kingdom

C.J. Coates
Swansea University, Swansea, Wales, United Kingdom

C. de Bekker
Ludwig-Maximilians-University Munich, Munich, Germany

H.H. De Fine Licht
University of Copenhagen, Frederiksberg, Denmark

B.G.G. Donzelli
Cornell University, Ithaca, NY, United States

I.M. Dubovskiy
SB Russian Academy of Sciences, Novosibirsk, Russia

J. Eilenberg
University of Copenhagen, Frederiksberg, Denmark

H.C. Evans
CAB International, Surrey, United Kingdom

W. Fang
Zhejiang University, Hangzhou, Zhejiang, China

A.E. Hajek
Cornell University, Ithaca, NY, United States

D.P. Hughes
Pennsylvania State University, University Park, PA, United States

A.B. Jensen
University of Copenhagen, Frederiksberg, Denmark

N.O. Keyhani
University of Florida, Gainesville, FL, United States

S.B. Krasnoff
USDA-ARS, Ithaca, NY, United States

R.G. Loreto
Pennsylvania State University, University Park, PA, United States; CAPES Foundation, Ministry of Education of Brazil, Brasília, DF, Brazil

B. Lovett
University of Maryland, College Park, MD, United States

H.-L. Lu
University of Maryland, College Park, MD, United States

S. Moonjely
Brock University, St. Catharines, ON, Canada

A. Ortiz-Urquiza
University of Florida, Gainesville, FL, United States

L. Quevillon
Pennsylvania State University, University Park, PA, United States

N.A. Ratcliffe
Swansea University, Swansea, Wales, United Kingdom; Universidade Federal Fluminense, Niteroi, Rio de Janeiro, Brazil

R.J. St. Leger
University of Maryland, College Park, MD, United States

C. Wang
Chinese Academy of Sciences, Shanghai, China

J.B. Wang
University of Maryland, College Park, MD, United States

H. Zhao
Zhejiang University, Hangzhou, Zhejiang, China

PREFACE

Insects are the most species-rich group of eukaryotes, and this diversity grants these animals profound power to influence ecosystems. The great research interest in insects arises from their ability to negatively impact human society: as vectors for disease in humans and our livestock and as destroyers of crops and stored products. It is estimated that insects destroy approximately 18% of the world annual crop production, and vector borne diseases kill millions every year. Thus, the diseases, pathogens, and immune responses of insects have been a long-standing research interest. Early interest grew mostly from economic concerns. For example, *Beauveria bassiana* was described by Agostino Bassi in 1835 as the cause of the devastating muscardine disease of silkworm, and it was instrumental in his development of the germ theory of disease (Steinhaus, 1956). In 1880, the pioneer immunologist Elie Metchnikoff was among the first to propose practical methods of microbial biological control of an insect crop pest, initiating trials of the fungus *Metarhizium anisopliae* against grain beetles (Lord, 2005; Mechnikoff, 1879, Fig. 1).

Such applied interest continues today, as we all still have a stake in either lengthening the life expectancy of useful insects or shortening the life expectancy of pestiferous ones. Since fungi are the commonest disease-causing agents in insects and populations of most insects are regulated by density-dependent factors involving pathogens and predators, we need to understand and be able to manipulate the interactions of pest insects with fungi and their other natural enemies in order to feed the world and prevent disease. Most research on fungal-insect pathogens has continued to focus on hypocrealean Ascomycetes from the genera *Beauveria* and *Metarhizium* (family Cordycipitaceae and Clavicipitaceae, respectively). These genera are tractable model species, are readily cultivatable, and have a particularly wide host range allowing them to be applied en masse against vectors of human disease and multifarious agricultural pests. Numerous registered mycoinsecticide formulations are based on *Beauveria* and *Metarhizium* spp. They also have a worldwide distribution with variants adapted from the arctic to the tropics and colonizing an impressive array of environments including forests, savannahs, swamps, coastal zones, and deserts.

The majority of chapters in this volume focus on *Metarhizium* and *Beauveria*, reflecting the preponderance of published data that has utilized

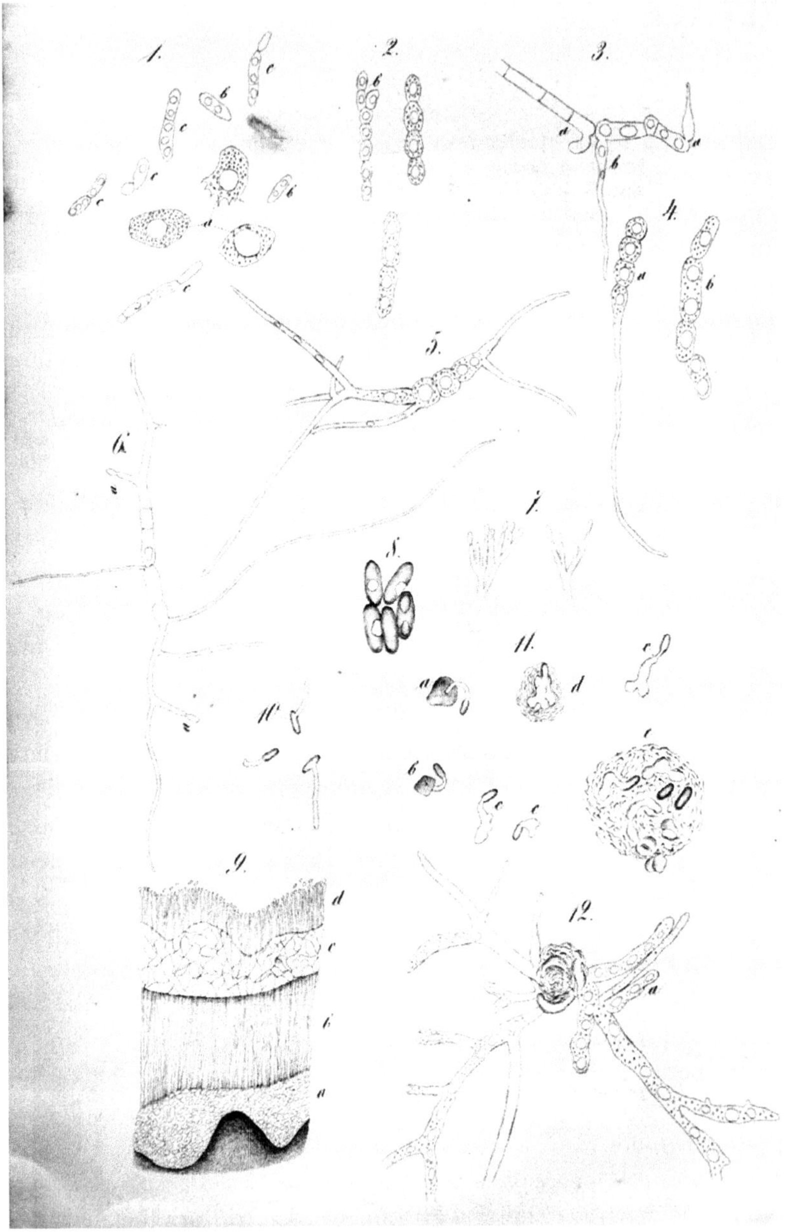

Figure 1 (1) Blood cells and fungal conidia from the blood of dying wheat chafer larvae. Magnified 550 times (Eyepiece 3 + System 9 alone picture). (2) Fungal conidia from the blood of living and apparently healthy larvae. Magnified 860 times (4 + 9). (3) Pieces of subcutaneous mycelial conidiophores formed on fresh cadavers. Magnified

these genera as models for entomopathogenicity. Yet, as clarified by Araújo and Hughes in Chapter 1, most major fungal taxonomic groups have members pathogenic to insects, so fungal-insect pathogens represent lifestyle adaptations that have likely evolved numerous times. Araújo and Hughes emphasize that, despite their ecological importance and potential applications, the entomopathogenic fungi are woefully understudied, especially regarding their biodiversity, which likely harbors one of the largest reservoirs of undocumented fungal species. Their chapter provides a broad overview of host–pathogen relationships by focusing on the impressive diversity of morphologies, ecologies, and interactions between insects and fungi. As host and pathogen are tied together in a very intimate relationship, the presented work strongly argues for empirical studies investigating both sides of the relationship in a much broader range of fungal-insect interactions than has hitherto been conducted. The authors reveal significant differences in host range and pathogenic strategies between the major groups of fungal pathogens. The basal groups, such as aquatic chytrids, infect mostly Diptera, while Microsporidia and Entomophthoromycota infect a wide range of hosts. Basidiomycota infects mostly Hemiptera, while Ascomycota, the most species-rich group, infects a vast number of insect groups. In terms of pathogenic strategies, the ascomycetes, for example, are all hemibiotrophic, switching from a biotrophic phase (parasitism) to a necrotrophic phase, growing on the dead host.

In contrast, the Entomophthoromycota are characteristically biotrophics with no somatic growth after death.

550 times. (4) Conidia from within the same larvae on the day after death. Magnified 550 times. (5) Overgrown conidia on the third day emerging from the larval cadaver in a humid chamber. Magnified 400 times (3 + 8). (6) Conidia and hyphae with two forming conidia (a) in a humidified chamber. Magnified 400 times. (7) Candelabra-shaped stalks with sterigmata forming conidia cultured in a humid chamber. Magnified 400 times. (8) Mature *Entomophthora anisopliae* spores. Magnified 1340 times (Eyepiece 3 + 14 System picture). (9) Cross section of larva covered with fungus. a-conidial layer, b-straight hyphal layer, c-felted layer, d-candelabra layer with sterigmata and conidia. Magnified 90 times (3 + 4). (10) Three conidia grown in sugar water. Magnified 550 times. (11) Germinating spores on larval cuticle. a, b- Spore germ tubes producing yellow spots. c- Bursiform germ linings on the cuticle. d- Further stage of germination. e- Spot on larval cuticle with ungerminated and germinating conidia. Magnified 550 times. (12) Spots on the cuticle with sprouts like rays of radiating hyphae; with notable formation of conidia. Magnified 400 times. *This figure is from Elie Metchnikoff's publication discovering Metarhizium anisopliae in Odessa in 1879; the legend is a translation of his original Russian legend for the figure.*

In their chapter (Chapter 2), De Fine Licht et al. focus on the Entomophthoromycota, as they are a large clade and particularly important natural regulators of insect populations. They relate how many genera in this group contain obligate insect pathogenic species with narrow host ranges, capable of producing epizootics in natural insect populations. The Entomophthoromycota are not easy to mass produce, and most studies have focused on the ecology of these organisms and their role in epizootics. The authors summarize and review the genetic information, albeit limited, which currently exists for Entomophthoromycota in order to provide a foundation that new genomic and transcriptomic information can build upon. Several genome-sequencing projects have been initiated among Entomophthoromycota species with diverse life histories and these will undoubtedly provide new insights into the biology of species and will aid phylogenetic analysis of this basal group within the fungal kingdom.

It is clear that understanding the genetics, biology, and ecology of entomopathogenic fungi is entering a new era. Wang et al. in Chapter 3 show how new insights into the ecological roles that these fungi occupy have been obtained by taking a genomic perspective on the evolution of insect pathogenicity and host range usage. They point out that from a phylogenetic and evolutionary perspective, *Beauveria* and *Metarhizium* are closely related to plant-associated fungi, and they retain many genes you would expect in a plant colonizer (or pathogen), consistent with their continuing abilities to form stable interactions with plants. The genus *Metarhizium* contains species with wide insect host ranges, for example, *Metarhizium robertsii*, and species such as *Metarhizium album*, *Metarhizium acridum* and *Metarhizium majus* show specificity for certain hemipterans, locusts, and beetles, respectively. Wang et al. dissect this observation by discussing the evolution of different components of infection processes. Looking at recognition, signal transduction, and effectors, they show how these differ in their evolutionary trajectories across broad host range and narrow host range species making use of nine whole *Metarhizium* genomes, as well as genomes from broad host range *B. bassiana* and narrow host range *Cordyceps militaris* and *Ophiocordyceps sinensis*. Interestingly, the components of pathogenicity differ not only in how fast they evolve but also in the way in which they change. In specialists, virulence factors show a high degree of amino acid divergence, whereas in generalists, there has been massive expansion of certain gene families. The authors show how the evolution and mechanism of pathogenic strategies and host range is closely linked with the diverse reproductive modes shown by entomopathogenic fungi that often determine the rates and patterns of genome evolution. These studies emphasize how genomic resources have helped make entomopathogenic fungi ideal model systems for answering basic questions in parasitology, entomology, and speciation.

Despite their long history of use and thousands of publications and patents, the important role of *Beauveria* and *Metarhizium* spp. in forming stable mutualistic interactions with plants, either as plant root colonizers or endophytes, has only become apparent in the last 20 years. Moonjely et al. (Chapter 4) emphasize the need to understand the multiple roles of these fungi as insect pathogens and plant growth promoters if we are to responsibly exploit them as biocontrol agents. The authors challenge the view that interkingdom host jumping occurred from ancestral fungi adapted to plants switching to arthropods and then back to plants. They suggest that many of these fungi evolved to infect insects while maintaining their mutualistic endosymbiosis with plants. They hypothesize that the driving force behind this evolution was the host plant demanding reciprocal nutrient exchange from the fungus in exchange for access to plant carbohydrates in the rhizosphere. Insect pathogenic fungi would be able to provide the plant with a source of nitrogen, or other growth limiting nutrients, derived from insect parasitism. The authors also review the other benefits that plants derive from *Beauveria* and *Metarhizium* endophytes including increases in plant biomass and productivity, alleviation of abiotic stresses (eg, drought, salinity, temperature fluctuations), and improved resistance to biotic stress (eg, herbivory, fungal disease).

Zhao et al. in Chapter 5 argue that as a natural control agent the "slow kill" characteristic of many entomopathogenic fungi allows them to replicate to large numbers, which is adaptive for the pathogen, but a severe limitation for application in modern agriculture or when attempting to control a disease vector. They describe how genetic engineering has proven an efficient tool to improve the efficacy of biofertilizers and mycoinsecticides by improving their tolerance to environmental stresses and increasing their virulence. The addition and expression of genes in ascomycete pathogens has become routine, and the authors describe several successful approaches including (1) inserting one or more foreign genes into the fungal genome whose product alters the pathogen's stress tolerance, alters the physiology of the target host insect or is toxic toward the target host and (2) altering expression of an endogenous gene so as to increase virulence or stress tolerance. Up until now, most "useful" genes for transfer have come from pathogens themselves, host insects, or from arthropods such as spiders that produce insect-specific toxins.

Each of these sources provides a vast array of biologically active metabolites. In addition, the authors describe how some methods appear to mitigate safety concerns regarding genetically engineered entomopathogenic fungi, and they discuss methods that have potential to ensure the success and safety of genetically modified strains.

Ortiz-Urquiza and Keyhani, in Chapter 6, show how research on *B. bassiana* has expanded greatly in the last 5 years due to a confluence of robust genetic tools and genomic resources for the fungus, and the recognition of its engagement in ecological interactions with plants and other organisms. Studies on *B. bassiana* currently range from physiological analyses related to its insect biological control potential to addressing fundamental questions regarding the molecular mechanisms of development, virulence, and fungal plant interactions. This work has established *B. bassiana* as an emerging model system, distinct from others in terms of the breadth of questions that can be addressed. The authors discuss generic ways of defining the concept of and genes involved in virulence and review examples of potential virulence determinants based on signaling molecules, effector genes encoding enzymes and toxins, and secondary metabolites. They convincingly show that *B. bassiana* is formidably well armed as an insect pathogen; although, the plethora of insecticidal gene products and their potential redundancy makes it hard to pinpoint genes crucial for pathogenicity that do not also have pleiotropic roles in fungal metabolism. In consequence, the underlying mechanisms of highly virulent *B. bassiana* strains remain elusive, but the functional outcomes are clearly worth studying and likely to lead to insights widely applicable to organisms, as well as those unique to this group of fungi.

The success of insect pathogenic fungi is in the context of insects having robust resistance to infection. The current great interest in insect immunity reflects applied consideration of their intrinsic importance, but, more significantly, insects are excellent models for studying host–pathogen interactions and immune responses. As detailed by Lu and St. Leger (Chapter 7), identification of host immune reactions or immune-related molecules in the best understood insect immune system, that of *Drosophila*, has already greatly benefited human health by assisting strategies for controlling human pathogens. The authors introduce early studies utilizing opportunistic human pathogens, as well as *Beauveria* and *Metarhizium*, which focused on molecules involved in sensing and signaling. They then describe how more recent studies have started to investigate immunity using a more holistic approach to provide a better understanding of behavioral avoidance of pathogens and how insect immunity integrates into an insect's general physiology to determine whether microbial infection leads to tolerance, resolution, or death. A key issue considered by the authors is that in order to understand the evolution of the insect immune system, with concomitant host–pathogen coevolution and trade-offs, we need to be able to quantify the strength of selection, and very little information is available on natural pathogens and their infection rates in wild populations of *Drosophila*. Similarly, the evolution of pathogen life histories in response to host genotypic variation is poorly understood. These challenges are being met in part by comparative genomics of insect immune defenses of different *Drosophila* species, mosquitoes, and social insects which are beginning to disentangle common themes from specific components of immunity. Likewise, the sequencing of entomopathogens is also beginning to play an important role in the development of insect-fungal model systems for studying host defense mechanisms. However, the authors point to the current focus on generalist pathogens, such as *Beauveria* and *Metarhizium*, as hindering any strong generalizations across the notoriously diverse array of host and parasite life histories. New data across a broad suite of study systems will be necessary to unravel the contributions of different host and pathogen life history traits to generating patterns of coevolution. This message resonates with the one by Araújo and Hughes (Chapter 1).

It also resonates with a review on social immunity in ants by Loreto and Hughes (Chapter 8). It is generally assumed that social life can lead to the rapid spread of infectious diseases and outbreaks. Social insects have limited genetic diversity in their crowded colonies, which should make them particularly prone to disease, as well as a potential model for human groups and their livestock. In practice, disease outbreaks are rare in ants and the expression of collective behaviors is invoked to explain the absence of epidemics in natural populations. For the sake of convenience, most studies on ants have been performed in the laboratory using *B. bassiana* and *Metarhizium* spp. to elicit and study social immunity. The authors discuss the limitations of the protocols in the published studies and argue that the use of generalist pathogens as a tool to understand infectious diseases in ants will better serve us when implemented in natural conditions. The generally held view that an ant nest is a "fortress," well defended against diseases, seems to be corroborated by the absence of outbreaks of *Beauveria* or *Metarhizium* in ant societies. However, as other fungi are commonly found infecting ants, the authors question if the absence of *Beauveria* or *Metarhizium* is, in fact, due to the efficient defenses. Additional studies on parasites known to be specialized on ants are encouraged as they would significantly complement our current knowledge. Along these lines, Lu and St. Leger (Chapter 7) discuss work with honey bee social immunity that exploits a natural honey bee pathogen, the Chalkbrood fungus.

The chapter by Butt et al. (Chapter 9) considers important questions relevant to the end user of entomopathogenic fungal technology including strain improvement, resistance monitoring, and risk assessment. As the authors note, fungal pathogens currently have a small share in the insect control market due to low virulence (slow kill and high inoculum load) compared to the chemical insecticides with which they compete and due to inconsistencies in their performance; however, as part of an integrated pest management system, entomopathogenic fungi have tremendous potential to impact human health and agriculture. Describing the route of a typical infection, they review the stratagems and tools employed by both the host and pathogen to endure stress, accumulate resources, and ultimately survive. They put these in the context of a microevolutionary arms race between host and pathogen that may work as a double-edged sword for practical purposes of pest control. For example, the known repellency of volatile organic compounds produced by some entomopathogens may reduce the risk of infection to natural predators and other nontarget hosts but interfere with "lure and kill" strategies for pest control.

Both *Beauveria* and *Metarhizium* are well known for producing a large array of biologically active secondary metabolites and secreted metabolites involved in pathogenesis and virulence that have potential or realized industrial, pharmaceutical, and agricultural uses. In Chapter 6, Ortiz-Urquiza and Keyhani include an overview of *Beauveria* secondary metabolites and their potential function in immunosuppression and as antimicrobial reagents. Donzelli and Krassnoff (Chapter 10) provide a comprehensive survey of the profusion of *Metarhizium* secondary metabolites. Notwithstanding the many known *Metarhizium* chemistries, the authors illustrate how the newly available *Metarhizium* genome sequences have revealed an abundance of biosynthetic pathways and a capacity for producing secondary metabolites that far exceeds the known chemistry. Some of the pathways characterized in their survey correspond to known products (eg, serinocyclins, destruxins), some are likely to be responsible for known chemistries (eg, cytochalasins, ovalicin), others are tantalizingly similar to pathways identified in other fungi (ergot, diketopiperazine, resorcylic acid lactones), but candidate products are still unknown from *Metarhizium*, and still others defy attempts to predict the type of molecule they produce. The authors' very reasonable conclusion is that our understanding of the services that secondary metabolites perform for the producing fungi, especially in interactions with other microbes, insect hosts, and plants is rudimentary at best.

With entomopathogenic fungi producing such complex chemistries, it is no surprise that certain insects have learned to recognize pathogen volatiles and avoid them. However, the hosts are not always the ones pulling the strings when it comes to their behavior; there are parasites that have evolved strategies to seize control of host behavior. In their chapter Hughes and colleagues (Chapter 11) discuss the role of behavior in driving disease in insect societies. With a nod to the chapter by Araújo and Hughes on the diversity of entomopathogens generally, they examine the diverse and stereotyped behaviors of arthropods (including familiar spiders and insects) under fungal entomopathogen manipulation using Niko Tinbergen's four-question paradigm: function, phylogeny, causation, and ontogeny. To explore the proximate causes of behavior manipulation from both host and parasite perspectives, they discuss case studies of a particularly dramatic example: the *Ophiocordyceps* fungi that produce the zombie ant phenomenon where ants are instructed to bite into leaves before dying in a "death grip." Other species induce less severe effects and some related pathogens of beetles do not manipulate at all. Hughes et al. describe how the mechanisms of behavioral control can be unraveled by comparing manipulators and non-manipulators, using tools such as phylogenomics, transcriptomics, and metabolomics. They also discuss how manipulation has evolved because ant societies establish complex conditions for parasite evolution and transmission. In this context, the usefulness of genetics algorithms to understand the evolution of complex behavior modification *in silico* is explored, especially where traditional study systems have not yet been established. The authors rightly describe the study of parasite manipulation of complex host behavior as entering "a golden age of discovery."

The range of articles in this volume covers many interrelated aspects of the genetics, biology, and ecology of these fascinating and useful insect-killing fungi. Although the individual chapters focus on specific issues, the reader should appreciate that each issue is multidimensional, being influenced by a milieu of fungal, insect, and environmental factors. It is the hope of the editors that the chapters untangle the complexity and illuminate the interactions between insects and fungi in a manner that shows the wide-ranging scope of the field and numerous applications of the knowledge it generates. Contained herein is evidence of the progress our field has made since Bassi first cautiously transferred a fungal hypha from a silkworm cadaver to a healthy insect, but this volume aims to serve as a foundation for future scientists hoping to shine light on our burgeoning field.

Raymond J. St. Leger and Brian Lovett
University of Maryland, College Park, MD, United States

REFERENCES

Lord, J. C. (2005). From Metchnikoff to Monsanto and beyond: the path of microbial control. *Journal of Invertebrate Pathology, 89*(1), 19–29.

Mechnikoff, I. (1879). *About harmful insects for agriculture. Issue III. Bread beetle. Diseases grain beetle larvae.* Research Mechnikov, professor of Novorossiysk University.

Steinhaus, E. A. (1956). Microbial control—the emergence of an idea. A brief history of insect pathology through the nineteenth century. *Hilgardia, 26*, 107–160.

CHAPTER ONE

Diversity of Entomopathogenic Fungi: Which Groups Conquered the Insect Body?

J.P.M. Araújo[1] and D.P. Hughes[1]
Pennsylvania State University, University Park, PA, United States
[1]Corresponding authors: E-mail: joaofungo@gmail.com; dph14@psu.edu

Contents

Advances in Genetics, Volume 94
ISSN 0065-2660
http://dx.doi.org/10.1016/bs.adgen.2016.01.001

Abstract

The entomopathogenic fungi are organisms that evolved to exploit insects. They comprise a wide range of morphologically, phylogenetically, and ecologically diverse fungal species. Entomopathogenic fungi can be found distributed among five of the eight fungal phyla. Entomopathogens are also present among the ecologically similar but phylogenetically distinct Oomycota or water molds, which belong to a different kingdom, the Stramenopila. As a group of parasites, the entomopathogenic fungi and water molds infect a wide range of insect hosts, from aquatic larvae to adult insects from high canopies in tropical forests or even deserts. Their hosts are spread among 20 of the 31 orders of insects, in all developmental stages: eggs, larvae, pupae, nymphs, and adults. Such assortment of niches has resulted in these parasites evolving a considerable morphological diversity, resulting in enormous biodiversity, the majority of which remains unknown. Here we undertake a comprehensive survey of records of these entomopathogens in order to compare and contrast both their morphologies and their ecological traits. Our findings highlight a wide range of adaptations that evolved following the evolutionary transition by the fungi and water molds to infect the most diverse and widespread animals on Earth, the insects.

1. INTRODUCTION

The kingdom Fungi is one of the major groups of eukaryotic microbes in terrestrial and aquatic ecosystems (Mueller & Schmit, 2007). There are approximately 100,000 described species of fungi (Kirk, Cannon, Minter, & Stalpers, 2008), which only represents a fraction of the estimated diversity, considered to be between 1.5 and 5 million species (Blackwell, 2011; Hawksworth & Rossman, 1997). Importantly, one of the hallmarks of Fungi is their propensity to form intimate interactions with other groups of life on Earth (Vega & Blackwell, 2005). According to (Hawksworth, 1988), 21% of all described species of fungi are associated with algae as lichens and 8% form intimate relationships with plants as mycorrhiza. Few if any organisms in terrestrial ecosystems exist in nature in the complete absence of fungi, and for this reason fungi are essential players in the maintenance of ecosystem health (Braga-Neto, Luizão, Magnusson, Zuquim, & Castilho, 2008). Another group often considered when discussing fungi are the Oomycota. These are colloquially known as water molds and belong to a very distant kingdom—Stramenopila—(Alexopoulos, Mims, & Blackwell, 1996), more closely related to brown algae (Kamoun, 2003). However, it is appropriate to discuss oomycetes with fungi as they were long considered to be in the same group and exhibit very similar ecologies, acting as parasites of both plants and animals.

The insects, with over 900,000 described species, represent the most species-rich groups of eukaryotes (Grimaldi & Engel, 2005, p. 12). They are known to form intimate relationships with many fungal groups that include: mutualistic endosymbionts that assist in nutrition (Suh, McHugh, Pollock, & Blackwell, 2005), fungi as food sources that are farmed as crops by leaf cutter ants (Currie et al., 2003), vertically transmitted parasites (Lucarotti & Klein, 1988), commensals (DeKesel, 1996), and pathogens with pronounced effects on host populations (Evans & Samson, 1982, 1984). However, even though we know that many different fungal–insect associations do exist, this subject remains among the most understudied fields in fungal biodiversity and likely harbors one of the largest reservoirs of undocumented fungal species (Vega & Blackwell, 2005).

A prominent characteristic of insects is their chitinous exoskeleton, which the great majority of entomopathogenic fungi and Oomycota need to penetrate (Evans, 1988). Following entry, some groups (ie, *Metarhizium* and *Beauveria* in the order Hypocreales, phylum Ascomycota) are known to grow inside the host as yeast-like hyphal bodies, multiplying by budding (Prasertphon & Tanada, 1968). Others, for example, some species within the Entomophthoromycota, produce protoplasts (cells without cell walls) instead (Butt, Hajek, & Humber, 1996). A third group encompassing some species within Oomycota, Chytridiomycota and species within the genus *Entomophthora* that infect aphids are known to grow directly as hyphal filaments inside the host's body (Lucarotti & Shoulkamy, 2000; Roberts & Humber, 1981; Samson, Ramakers, & Oswald, 1979; Zattau & McInnis, 1987). The majority of entomopathogenic fungi kill their hosts before the spore production starts (as such they are termed hemibiotrophic). A few of them, especially some in the phylum Entomophthoromycota, sporulate from the living body of their hosts (and as such are termed biotrophic) (Roy, Steinkraus, Eilenberg, Hajek, & Pell, 2006). All of entomopathogenic oomycetes kill the host before transmission.

All entomopathogenic fungi and water molds are transmitted via spores. There are two types to consider. The sexual spores are actively released into the environment. By definition, zoospores are motile spores that swim and, in the case of pathogenic fungi, reach their target host actively, via a flagellum attached to the spore. Such motile spores occur in the Chytridiomycota and Oomycota. In other groups, sexual spores are named to link them to their groups: (zygo)spores, (basidio)spores, and (asco)spores belonging to respectively, the "zygomycetes," Basidiomycota and Ascomycota. Each of the three types exhibits unique traits (Fig. 1). The asexual mitotic spores (always called conidia regardless of taxon) are often passively released (Roberts & Humber, 1981). Spore morphology and their germination behavior have been heavily relied upon in the classification and systematics of different groups of fungi (Alexopoulos et al., 1996). We will discuss, later, the diversity of these varied spores separately for each major group of entomopathogenic fungi.

This review has multiple aims. The first is to ask which groups of Fungi and Oomycota evolved the ability to exploit the insect body. We will then explore the strategies these organisms employ for both infection and subsequent transmission. We view each group of entomopathogens within the ecological framework that is its insect host, an approach that has surprisingly not been previously considered in a broad sense. Our overarching aim is to provide a clearer understanding of the diversity and ecology of this important group of parasites, highlight lacunae in our knowledge, and motivate other studies. Before proceeding further, however, it is necessary to introduce each of the fungal and oomycete groups that are known to infect insects. This is because many groups presented here are generally unfamiliar.

2. THE MAJOR GROUPS OF ENTOMOPATHOGENIC FUNGI AND OOMYCETES

2.1 Oomycota

The species belonging to Oomycota were in the past considered among Fungi due to multiple ecological and morphological similarities. However, phylogenetic studies (James et al., 2006) confirmed earlier suggestions by some authors (Kreisel, 1969; Pringsheim, Pfeffer, & Strasburger, 1858; Shaffer, 1975) that these organisms are not Fungi. They were therefore placed in the Stramenopila, a kingdom containing morphologically diverse organisms such as Hyphochytriomycota and Labyrinthulomycota (Alexopoulos et al., 1996; Beakes, Glockling, & Sekimoto, 2012). Despite having been previously considered in the same group, the phylum Oomycota has a number of biological characters that distinguish them from Fungi. The first one is reproduction by biflagellate zoospore with a longer tinsel flagellum

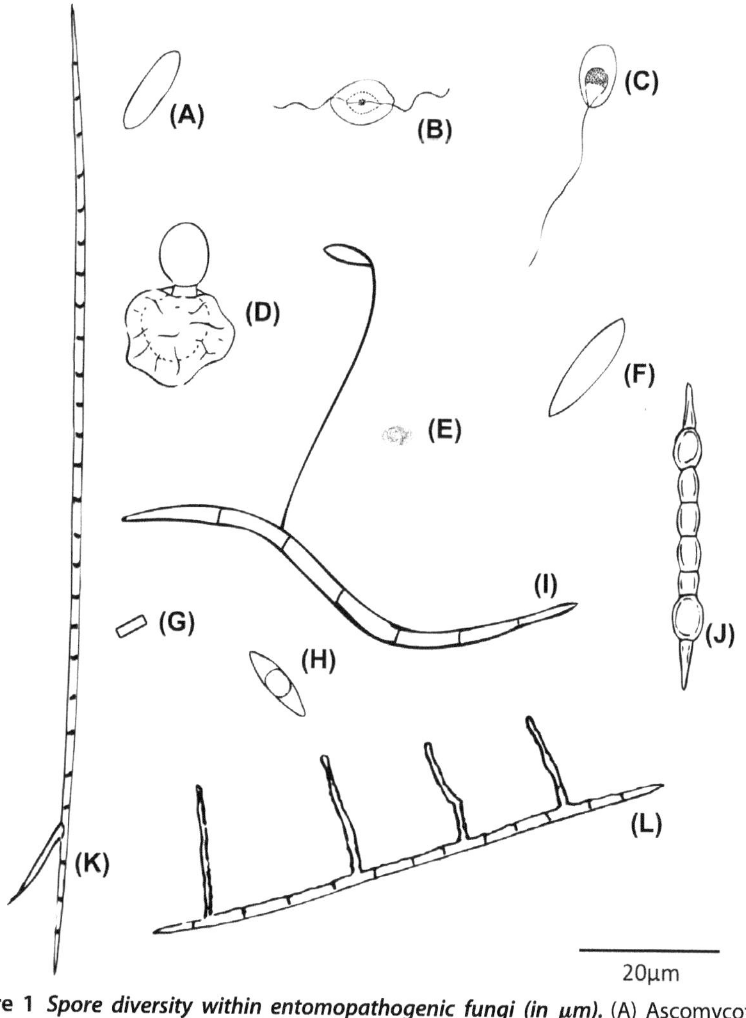

Figure 1 ***Spore diversity within entomopathogenic fungi (in μm).*** (A) Ascomycota—*Moelleriella sloaneae* (13–15 × 2.8–3 μm) (Chaverri, Liu, & Hodge, 2008); (B) Oomycota—*Lagenidium giganteum* (8–9 × 9–10 μm) (Couch, 1935); (C) Chytridiomycota—*Coelomomyces psophorae* (5 × 10) (Whisler, Zebold, & Shemanchuk, 1975); (D) Entomophthoromycota—*Entomophthora thripidum* (10–15 × 8–12 μm) (Samson et al., 1979); (E) Microsporidia—*Nosema hyperae* (3.1 × 1.7 μm) (Bulla & Cheng, 1977); (F) Basidiomycota—*Septobasidium maesae* (18–19.5 × 4–5 μm) (Lu & Guo, 2009); (G) Ascomycota—*Ophiocordyceps lloydii* (4 × 1 μm) (Kobayasi & Shimizu, 1978); (H) Ascomycota—*Hypocrella raciborskii* (10–16 × 2.5–4 μm) (Liu, Chaverri, & Hodge, 2006); (I) Ascomycota—*Ophiocordyceps camponoti-rufipedis* (80–95 × 2–3 μm) (Evans, Elliot, & Hughes, 2011); (J) Ascomycota—*Ophiocordyceps blattae* (40–60 × 4–6 μm) (Petch, 1924); (K) Ascomycota—*Ophiocordyceps camponoti-melanotici* (170–210 × 4–5 μm) (Evans et al., 2011); (L) Ascomycota—*Ophiocordyceps camponoti-novogranadensis* (75–95 × 2.5–3.5 μm) (Evans et al., 2011).

directed forward and a shorter whiplash flagellum directed backward (Fig. 1B) (Barr, 1992; Dick, 2001). They also reproduce by a thick-walled oospore, which is a structure not found in Fungi. At the cellular level, they possess mitochondria with tubular cristae, whereas the Fungi have mitochondria with flattened, plate-like cristae. Moreover, their cell walls contain cellulose, which is in contrast with Fungi that contain chitin as a cell wall component (Alexopoulos et al., 1996).

The Oomycota have evolved both parasitic and saprophytic (feeding from dead tissue and organic particles) lifestyles (Phillips, Anderson, Robertson, Secombes, & van West, 2008). As pathogens, oomycetes are able to infect a broad range of hosts such as algae, plants, protists, fungi, arthropods, and vertebrates, including humans (Kamoun, 2003). Certain genera are well-known plant pathogens, such as members of the genus *Phytophthora*, which was the causative agent of the Irish Potato Famine (Goss et al., 2014) and is currently causing Sudden Oak Death that affects millions of trees (Brasier, Denman, Brown, & Webber, 2004). Although better known as plant pathogens, the Oomycota do infect arthropods with records of infections on lobsters (Fisher, Nilson, & Shleser, 1975) and shrimps (Hatai, Rhoobunjongde, & Wada, 1992), as well as on insects (Pelizza, López Lastra, Maciá, Bisaro, & García, 2009; Samson, Evans, & Latgé, 1988; Seymour, 1984; Stephen & Kurtböke, 2011). For additional information about the entomopathogenic oomycetes, see Frances, Sweeney, and Humber (1989), Dick (1998), Scholte, Knols, Samson, and Takken (2004), Su, Zou, Guo, Huang, and Chen (2001) and Tiffney (1939).

2.2 Microsporidia

Traditionally, microsporidian species were classified within the phylum Apicomplexa as "sporozoan parasites" (Schwartz, 1998). However, an increasing number of studies are lending support to the hypothesis that Microsporidia are Fungi (Hibbett et al., 2007; Hirt et al., 1999; James et al., 2006). However, a conclusive resolution about microsporidians as an early lineage of Fungi will require further genetic studies from basal fungal taxa (James et al., 2006). Nevertheless, based on the studies mentioned earlier and the Microsporidia's ecological function as insect pathogens, we will include them among the entomopathogenic fungi in this study.

The most remarkable feature of this group is its unique spore—ranging from 1 to 40 μm (Wittner & Weiss, 1999, p. 8), which acts as a "syringe" injecting its protoplast material into the host (Keeling & Fast, 2002). Ohshima (1937) first suggested that the protoplasm is transmitted from the microsporidian spore to inside the host cell. We now know that this happens through the tube formed during adherence, which facilitates the subsequent discharge of the parasite's intracellular content to within the host's cell (Wittner & Weiss, 1999). The discharging of the polar tube occurs by breaking through the apex, which is the thinnest region of the spore wall. This event is compared by Keeling and Fast (2002) "to turning the finger of a glove inside-out."

The host range for most Microsporidia species is relatively restricted. They have been reported infecting a great number of domestic and wild animals such as fish (Kent, Shaw, & Sanders, 2014), amphibians, reptiles, birds (Kemp & Kluge, 1975), and mammals (Snowden & Shadduck, 1999), including some groups of humans, such as immunocompromised AIDS patients (Didier & Bessinger, 1999). Detailed studies on the biology and taxonomy of Microsporidia can be found in Bulla and Cheng (1977), Becnel and Andreadis (1999), Briano (2005), Lange (2010), Sokolova, Sokolov, and Carlton (2010), Kyei-Poku, Gauthier, Schwarz, and Frankenhuyzen (2011), Hossain, Gupta, Chakrabarty, Saha, and Bindroo (2012) and Vega and Kaya (2012).

2.3 Chytridiomycota

The Chytridiomycota is the phylum suggested to be the earliest diverging lineage of the Fungi (James et al., 2006). There are reports of them dating from Lower Devonian (about 400 million years ago (mya)) (Taylor, Remy, & Hass, 1992) and a parasitic chytrid-like fungus dates from the Antarctic Permian (about 250–300 mya) (Massini, 2007). Chytridiomycota is the only phylum among the kingdom Fungi that possesses motile cells at least once in its life cycle. These zoospores are equipped with a single posteriorly directed whiplash flagellum, which reflects their aquatic life cycle (for details, see Barr and Désaulniers (1988)). They respond to chemical gradients allowing them to actively locate their hosts, which is especially important for species pathogenic on aquatic organisms (Sparrow, 1960). They can also adaptively respond to environmental changes (eg, fluctuations in heat and humidity) in ways that reduce water loss or the collapse of the cell (Gleason & Lilje, 2009). The zoospores of chytrids are functionally equivalent to motile spores in the Oomycota, and so this is an example of convergent evolution, as both groups are aquatic.

The majority of chytrids are found as saprophytic organisms, especially in freshwaters and wet soils, but there are also some marine species (Gleason et al., 2011). However, a significant number of species are known to be parasites of plants, animals, rotifers, tardigrades, protists, and also other fungi (Dewel, Joines, & Bond, 1985; Karling, 1946; Martin, 1978; Sparrow, 1960). Diseases of insects caused by chytrids seem to be comparatively rare (Karling, 1948). For further reading see Voos (1969), Whisler et al. (1975), Millay and Taylor (1978), and Padua, Whisler, Gabriel, and Zebold (1986).

2.4 "Zygomycetes"

The phylum Zygomycota was traditionally organized as a single phylum and two classes, Zygomycetes and Trichomycetes (Alexopoulos, 1962; Alexopoulos et al., 1996). Both classes share common features like coenocytic mycelium (ie, lacking regular septation), asexual reproduction usually by sporangiospores and absence of flagellate cells and centrioles (Alexopoulos et al., 1996). Their main general characteristic is the production of a thick-walled resting spore (ie, zygospore) within a commonly ornamented zygosporangium, formed after fusion of two specialized hyphae called gametangia (Alexopoulos et al., 1996). The phylum is ecologically very diverse, widely distributed, and very common, with most species occurring as saprotrophs in both soil and dung. Some of them are fast growing and they are often found colonizing bread, fruits, and vegetables.

However, despite being placed in a single group, molecular phylogenetic studies validated the long-suggested hypothesis concerning the polyphyly of Zygomycota species and recognized five monophyletic taxa to replace the phylum. Thus, species that form arbuscular mycorrhizal associations with plants were accommodated within the phylum Glomeromycota and all other taxa distributed among four subphyla: Entomophthoromycotina, Kickxellomycotina, Mucoromycotina, and Zoopagomycotina without placement to any phylum (Hibbett et al., 2007; James et al., 2006). Thereafter, Humber (2012) proposed a detailed morphological and ecological description of a new phylum: Entomophthoromycota Humber, to accommodate species previously assigned to Entomophthoromycotina. This study was supported by a comprehensive phylogenetic study of this new phylum, which demonstrated its monophyletic nature (Gryganskyi et al., 2012). In our study, we will use this modern classification, which is supported by morphological, ecological, and phylogenetic data.

The Kickxellomycotina, Mucoromycotina, and Zoopagomycotina are composed mostly of saprobes. However, some families within Zoopagomycotina are known to predate nematodes (Zhang & Hyde, 2014). These are the relatively well-known "nematode trapping fungi." Within Zoopagomycotina mycoparasitic species are more common (Alexopoulos et al.,

1996). While Mucoromycotina is the largest and morphologically most diverse order within the zygomycetes, just one species of entomopathogenic fungi is assigned to the subphylum, ie, *Sporodiniella umbellata*, which occurs on various insects, notably membracids (plant-feeding insects in the order Hemiptera), in cocoa farms (Evans & Samson, 1977).

The trichomycetes, currently placed within subphylum Mucoromycotina, order Harpellales, are fungi that exclusively inhabit the guts of various arthropods (Horn & Lichtwardt, 1981). However, since the trichomycetes apparently do little, if any harm to their hosts under natural conditions (Horn & Lichtwardt, 1981) and the nature of their relationships is not fully understood, they will not be further discussed in this study.

One of the most important groups of all entomopathogens is Entomophthoromycota, which are mainly pathogens of insects. They exhibit specialized spore-producing cells (conidiophores) that have positive phototrophic growth. Their spores are usually discharged forcibly and this can occur by several different mechanisms, sometimes producing secondary and in some species tertiary conidia (ie, *Eryniopsis lampyridarum*, Fig. 2) (Humber, 2012; Thaxter, 1888). They frequently occur as epizootic events, killing a large number of insects in small patches of forest or agricultural systems (Roberts & Humber, 1981). For further reading see Nair and McEwen (1973), Humber (1976, 1981, 1982, 1984, 1989) and Scholte et al. (2004).

2.5 Basidiomycota

This group, together with Ascomycota, forms the subkingdom Dikarya, which exhibits a dikaryotic phase (Hibbett et al., 2007). They contain some of the most well-known fungi such as mushrooms, puffballs, earthstars, smuts, and rust fungi. The Basidiomycota are characterized by the formation of sexual spores called basidiospores, which are formed outside specialized reproductive cells called basidia. These spores are in most cases forcibly discharged by specialized structures (Pringle, Patek, Fischer, Stolze, & Money, 2005). Another important and unique trait for the group are clamp connections. Those are structures formed during the division of the nuclei on the tip of growing hyphae, which help to ensure the dikaryotic condition (Alexopoulos et al., 1996), and can be used to identify members of this phylum, even in fossil records (Krings, Dotzler, Galtier, & Taylor, 2011).

The basidiomycetes exhibit some important ecological traits. They colonize dead wood, decaying cellulose and lignin, also acting as leaf litter

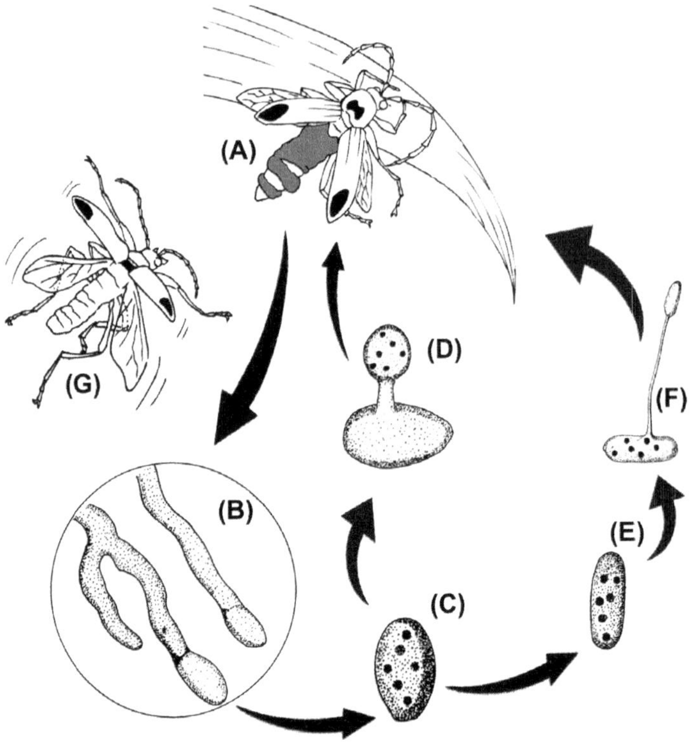

Figure 2 ***Entomophthoromycota*****—Eryniopsis lampyridarum.** (A) Cantharid beetle infected by *E. lampyridarum* died with its mandibles attached to flowering plants or grass. The elytra and wings gradually open as the fungus grows through the host's body; (B) Conidiophores emerges directly from the host's body; (C) Primary conidium; (D) Primary conidium bearing mature secondary conidium at the tip; (E) Secondary conidium; (F) Secondary conidia eventually will produce capilliconidia in absence of a suitable host; (G) Another cantharid beetle will get infected if it touches the exposed fungal hymenium. *References: (Humber, 1984; Roy et al., 2006; Thaxter, 1888).* (See color plate)

decomposers on the forest floor (Braga-Neto et al., 2008). Pathogenic basidiomycetes (ie, smut and rust fungi) are familiar scourges of plants, responsible for huge losses in agriculture. In addition, forest environments are also attacked by species like *Armillariella mellea*, which attack trees and *Heterobasidion annosum*, attacking specifically conifers (Kendrick, 2000). As animal pathogens, some species in the anamorphic genus *Nematoctonus* (linked to the teleomorphic genus *Hohenbuehelia*) are known to attack nematodes (Barron & Dierkes, 1977). A few genera are known as pathogens of insects, which infect scale insects (ie, *Septobasidium* and *Uredinella*, order

Septobasidiales) and termite eggs (ie, *Fibularhizoctonia*, order Atheliales, attacking eggs of the termite genus *Reticulitermes*).

The Septobasidiales exclusively attack scale insects (Hemiptera, Diaspididae) (Evans, 1989). The order includes two genera of entomopathogens: *Uredinella*, attacking single insects, and *Septobasidium*, attacking whole colonies of plant-feeding insects, with as many as 250 insects infected by one fungus (Couch, 1938). This character is one of the most remarkable differences between both genera, but morphological differences also exist. For example, the presence of a binucleate uredospore in *Uredinella* does not occur in *Septobasidium* (Couch, 1937). Due to this trait, *Uredinella* was described as "a new fungus intermediate between the rusts—a plant pathogen—and *Septobasidium*," exhibiting traits of both (Couch, 1937).

Another group within Basidiomycota was described on termite eggs (Matsuura, Tanaka, & Nishida, 2000). This fungus was found living inside the nest of termites, among their eggs, which they occasionally consume. The authors identified this fungus, based on molecular studies, within the order Atheliales, as being a species very close related to *Fibularhizoctonia* sp. (asexual state of the genus *Athelia*), however, not describing them formally (Matsuura et al., 2000). For more details and species descriptions see Couch (1938), Matsuura (2005, 2006), Yashiro and Matsuura (2007), Lu and Guo (2009), Matsuura, Yashiro, Shimizu, Tatsumi, and Tamura (2009) and Matsuura and Yashiro (2010).

2.6 Ascomycota

The phylum Ascomycota is the largest group in kingdom Fungi with about 64,000 species described (Kirk et al., 2008). The majority of them are filamentous, producing regularly septate hyphae. They are characterized by the formation of sexual spores (ie, ascospores) in sac-like structures, singularly called an ascus. As the most speciose group of fungi, it is not surprising that they also have diverse ecological breadth comprised by decomposers, plant pathogens, human and animal pathogens, as well as being known to form mutualistic relationships (ie, lichens) (Alexopoulos et al., 1996).

The majority of entomopathogenic species within Ascomycota have a well-developed parasitic phase that infects the host's body. Furthermore, after killing the insect, this group is able to colonize the cadaver switching to saprophytic nutrition (hemibiotrophic), maintaining hyphal growth, even after the host's death (Evans, 1988). According to the same author, these entomopathogens would have evolved and diversified in early, moist

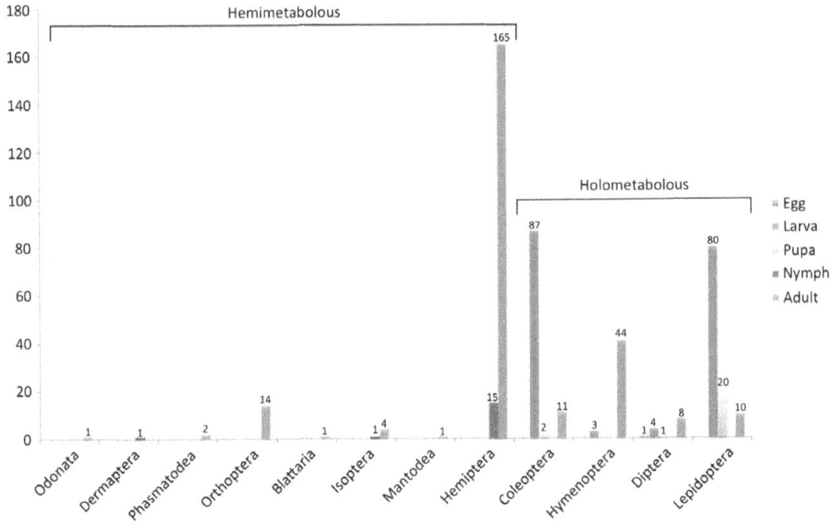

Figure 3 ***Hypocreales teleomorphic species.*** Hypocreales teleomorphic species numbers (y-axis) and their distribution across the different orders of insects, further divided into the stage of host development attacked. The Holometabolous orders have complete development with a larval stage, whereas the Hemimetabolous orders have an incomplete development with no distinct larval stage, but rather nymphs.

tropical forests, particularly rainforests. They are known to attack a wide range of different hosts (Fig. 3). The great majority of entomopathogenic ascomycetes form their spores inside structures called perithecia, a subglobose or flask-like ostiolate ascoma that contains many asci (Evans et al., 2011; Kirk et al., 2008; Kobayasi, 1941; Kobayasi & Shimizu, 1978). There is a wide diversity of spore types and shapes (Fig. 1). The phylum ranges from insect pathogens such as Pleosporales, Myriangiales, and Ascosphaerales, which have relatively few species, to the biggest group of entomopathogens, the hyperdiverse Hypocreales (Samson et al., 1988).

3. METHODS

3.1 Search Strategy

We are interested in determining which species of fungi and water molds successfully conquered the insect body. As such, our basic unit of analysis is the species name (binomial). The repositories for such names are electronic databases such as Mycobank (http://www.mycobank.org)

and Index Fungorum (http://www.indexfungorum.org) (Fig. 4). MycoBank is owned by the International Mycological Association and is an online database aimed as a service to the mycological and scientific community by documenting mycological nomenclatural novelties (new names and combinations) and associated data, for example, descriptions and illustrations. The Index Fungorum, another global fungal database, coordinated and supported by the Index Fungorum Partnerships, contains names of fungi (including yeasts, lichens, chromistan fungal analogues, protozoan fungal analogues, and fossil forms) at all ranks.

All groups (phyla) of fungi were investigated separately. Once we found a phylum containing at least one entomopathogenic species (six in total), a thorough search was made within such phylum, narrowing until the entomopathogenic genera and finally identifying all species recorded as insect pathogens in each phylum.

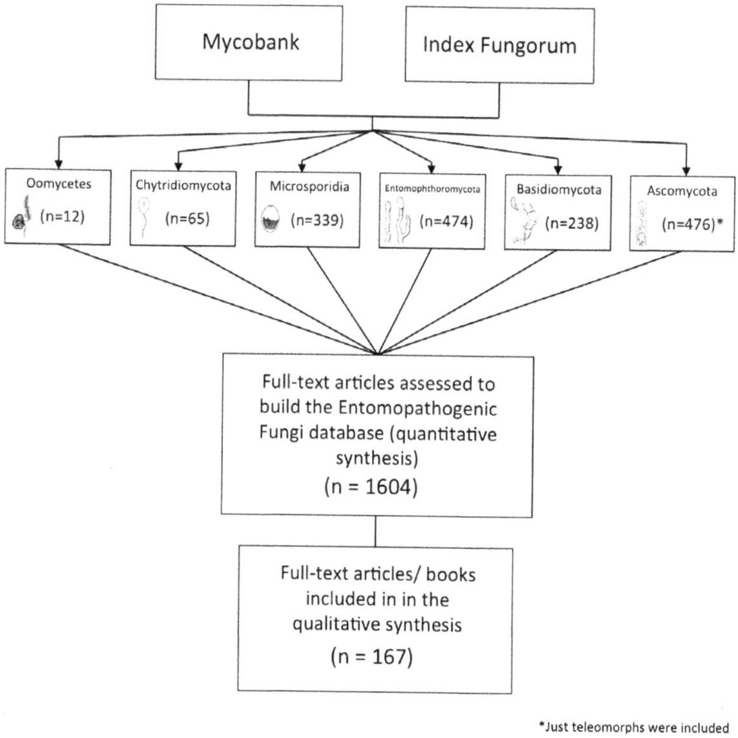

Figure 4 Flowchart shows the main sources consulted (Mycobank and Index Fungorum), the fungal/oomycetes phyla found infecting insects, and the number of species and sources consulted in this study.

3.2 Dealing With Name Changes

Species names are often not static and can change as taxonomists reorganize synonyms and as we advance with molecular phylogeny. We matched old records of entomopathogenic fungi with their current valid names, avoiding any duplicated record for the same organism.

3.3 Determining Host Associations

To determine the host association we first consulted the original formal description. This information is available on both Mycobank and Index Fungorum.

The complete list of original descriptions and references—that are not in the text—are listed within the tables organized by phyla, in the supplementary materials with species names, host association, and original reference(s).

3.4 Monographs and Atlases

We also consulted monographs and atlases of insect pathogenic fungi, eg, *Atlas of Entomopathogenic Fungi* (Samson et al., 1988), the monograph of hypocreloid fungi (Chaverri et al., 2008), *The Microsporidia and Microsporidiosis* (Wittner & Weiss, 1999), *The Genus Coelomomyces* (Couch & Bland, 1985), the *Genus Septobasidium* (Couch, 1938) among others. These were also useful for discovering both host associations and ecological aspects of the fungi reviewed herein.

4. RESULTS

We found 20 of 30 orders (sensu Grimaldi and Engel (2005)) of insects infected by fungi and Oomycetes (approximately 65% of insect orders are infected; Fig. 5). Microsporidia infects 14 orders of insects, Ascomycota (mostly Hypocreales) and Entomophthoromycota infect 13 and 10 orders of insects, respectively. Chytrids infect three, Basidiomycota and Oomycota, each infect two orders. There are several genera that are pathogenic to insects (Fig. 6). Later we examine incidences of parasitism for each of these six phylum focusing on the morphological and ecological traits they evolved to make them efficient and specialized parasites of insects.

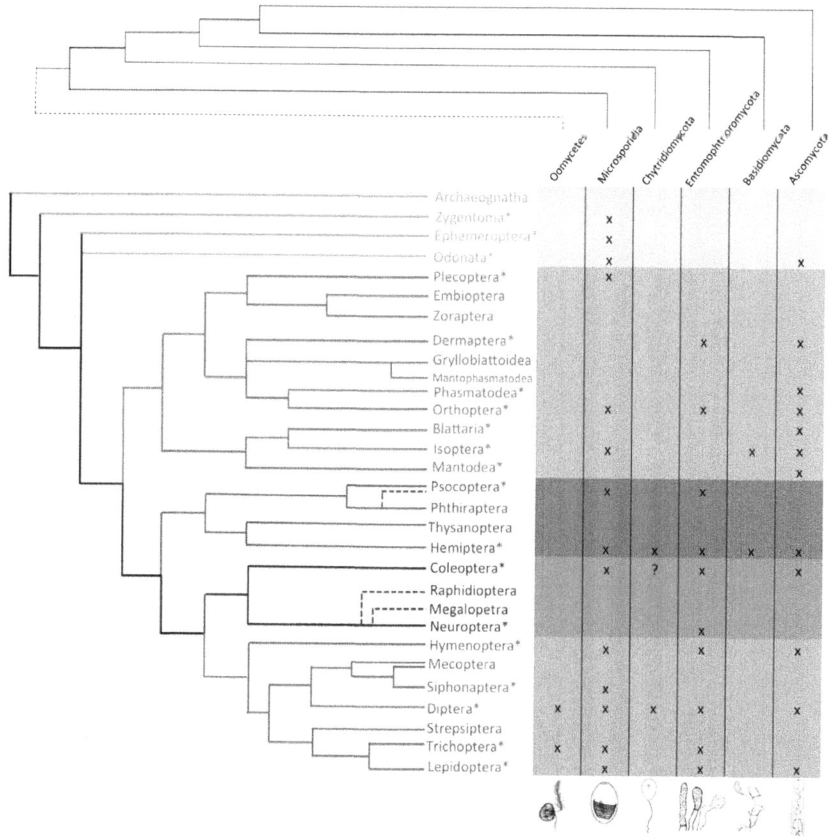

Figure 5 ***Insect orders × fungal phyla (and oomycetes): the parasitic relationship between entomopathogens and their hosts.*** On the left, the phylogeny of insect orders (*adapted from Grimaldi, D. & Engel, M. S. (2005). Evolution of the Insects. Cambridge University Press.*); on the top the phylogeny of fungal phyla and oomycetes (*adapted from James, T.Y., Kauff, F., Schoch, C.L., Matheny, P.B., Hofstetter, V., Cox, ... Miadlikowska, J. (2006). Reconstructing the early evolution of fungi using a six-gene phylogeny. Nature, 443(7113), 818–822.*); the table shows which fungal group infects each insect order. The uncertainty of a record is denoted with a question mark. (See color plate)

4.1 Incidence of Disease on Insects Caused by Fungus and Oomycetes

4.1.1 Oomycetes

The entomopathogenic oomycetes are comprised of 12 species distributed among six genera: *Lagenidium* (one species, *Lagenidium giganteum*), *Leptolegnia* (two species, *Leptolegnia caudata* and *Leptolegnia chapmanii*), *Pythium* (three

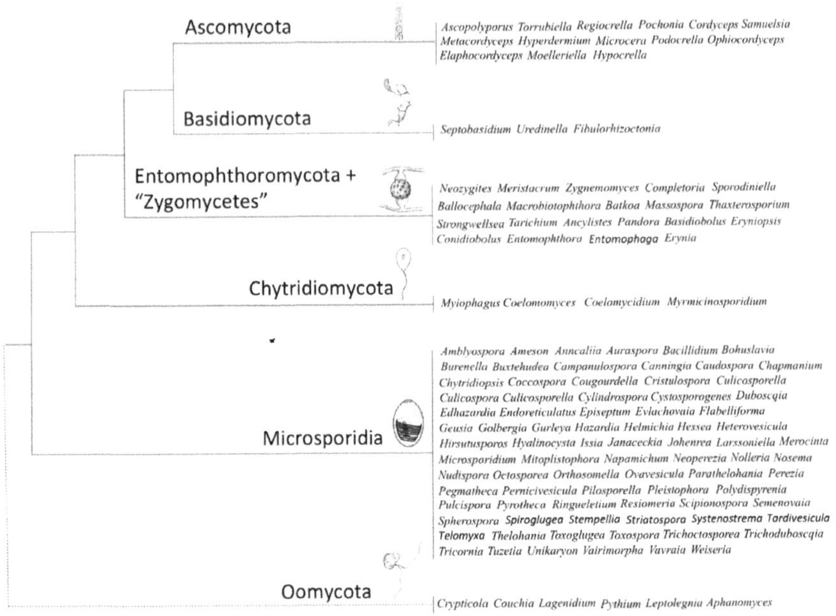

Figure 6 Diversity of genera of entomopathogens across Fungi and oomycetes.

species, *Pythium carolinianum*, *Pythium sierrensis*, and *Pythium flevoense*); *Crypticola* (two species, *Crypticola clavulifera* and *Crypticola entomophaga*); *Couchia* (three species, *Couchia amphora*, *Couchia linnophila*, and *Couchia circumplexa*), and *Aphanomyces* (one species, *Aphanomyces laevis*). They have been discovered attacking species of mosquito in the following genera: *Aedes*, *Anopheles*, *Chironomus*, *Culex*, *Forcipomyia*, *Glyptotendipes*, *Mansonia*, *Ochlerotatus*, *Pentaneura*, *Polypedilum*, *Tendipes*, and *Uranotaenia* (Martin, 1981, 2000; Scholte et al., 2004).

Oomycetes infections have been recorded from mosquito larvae in freshwater, primarily in well-aerated streams, rivers, ponds, lakes (Alexopoulos et al., 1996), and even treeholes (Saunders, Washburn, Egerter, & Anderson, 1988) and water that collects on leaf axils (Frances et al., 1989). A single example of oomycetes infecting a nondipteran was *Crypticola entomophaga*, which was described attacking caddis flies (Trichoptera), which are also aquatic (Dick, 1998).

Among the entomopathogenic oomycetes, the most well-known and broadly studied species is *L. giganteum* Couch, a facultative mosquito larvae parasite (see Fig. 7) (Scholte et al., 2004). In an experimental study researchers described the swimming behavior of zoospores towards the surface

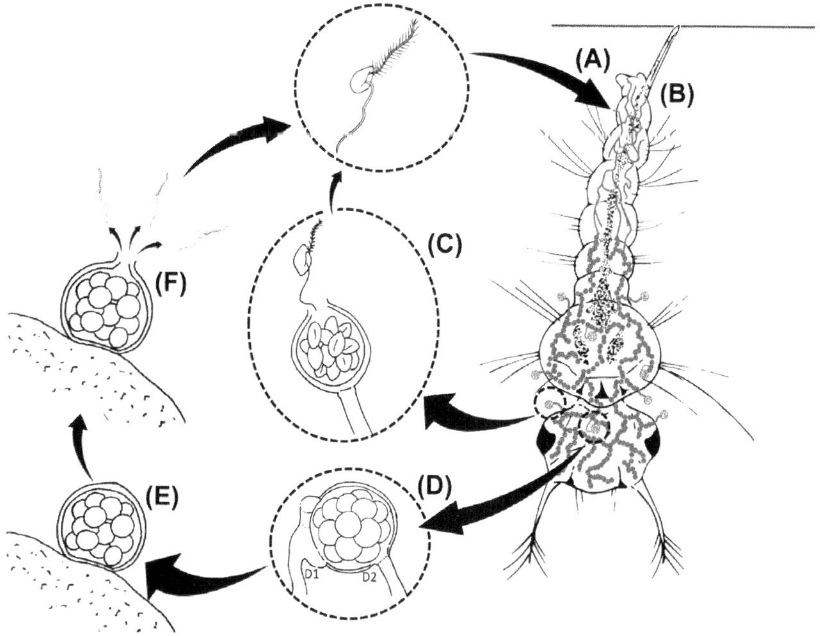

Figure 7 ***Oomycetes*—Lagenidium giganteum.** (A) Zoospore (n) adhere and penetrate the cuticle, starting the infection; (B) Mycelium starts to grow and proliferates within the larva's body; (C) Zoosporangium releasing asexual zoospores; (D) Oospore (Sexual part of the cycle)—(D1) Antheridium fertilizing (D2) oogonium; (E) Oospore detaches from the hyphae and settles down in the environment; (F) Releasing of sexual zoospores (2n). (See color plate)

of the water (Golkar et al., 1993). Their results suggested that spore-swimming behavior seems to be due to the cell shape and location of center of gravity rather than a sensory response.

Other oomycete genera like *Leptolegnia*, *Pythium*, *Crypticola* and *Aphanomyces* have received only limited attention (Scholte et al., 2004). Within the genus *Leptolegnia*, only *L. caudata* de Bary (Bisht, Joshi, & Khulbe, 1996) and *L. chapmanii* Seymour (Seymour, 1984) have been isolated from insects (Scholte et al., 2004). The life cycle of *L. chapmanii* was presented by Zattau and McInnis (1987), who reported that species infecting *Aedes aegypti*, the yellow fever and dengue disease vector. Effects of *L. chapmanii* on other aquatic invertebrates such as Odonata, Thichoptera, Coleoptera, Plecoptera, and Cladocera were tested, though no infections were observed, suggesting specificity (McInnis & Schimmel, 1985). Members of the genus *Pythium* are species spread across the world, occurring mostly as soil-inhabiting organisms or plant pathogens (Alexopoulos et al., 1996). Three species of *Pythium* are known to infect insect larvae (Phillips et al., 2008). *Aphanomyces* was recorded causing seasonal epizootic in insectaries (Seymour & Briggs, 1985), but few studies were published about this genus infecting insects.

4.1.2 Microsporidia

Microsporidia is a group of pathogens comprising 143 genera (Sprague & Becnel, 1999) with more than 1200 species (Wittner & Weiss, 1999). Among those, 69 genera were recorded infecting insects, attacking 12 orders (Fig. 5). According to Becnel and Andreadis (1999), the majority (42 of 69 genera) infect Diptera; five genera infect Ephemeroptera and Coleoptera; four genera infect Lepidoptera, followed by Trichoptera, infected by three, Orthoptera, Odonata, and Siphonaptera each infected by two genera; and Thysanura, Hymenoptera, and Isoptera with one genus of Microsporidia infecting each of them. These accounts are data based on described genera that possess an insect as type-host. Hence, this number certainly will increase in future publications.

As will be discussed further, the dipterans are the only insect group infected by five different groups of Fungi/oomycetes (only the Basidiomycota have not been recorded as pathogens of Diptera). Among the 42 microsporidian genera attacking Diptera, the largest, most widespread, and common is *Amblyospora* (Andreadis, 1985), which is known to infect 79 species of Diptera in 8 genera (Becnel & Andreadis, 1999). This genus of Microsporidia exhibits a complex life cycle, which requires an intermediate copepod host and two mosquito generations in order to complete its full life span (Sweeney & Becnel, 1991).

Another important group among the entomopathogenic Microsporidia is the genus *Nosema* (see Fig. 8). Some authors consider them the most important and widely distributed genus (Tsai, Lo, Soichi, & Wang, 2003), being responsible for the majority of microsporidian infections in Lepidoptera species (Tsai et al., 2003). A good example of their ecological and economical importance occurs with the species *Nosema bombycis* and *Nosema ceranae* that infect bees and are known to be responsible for great losses in apiculture (Higes, Martín, & Meana, 2006). These infections are restricted to the midgut epithelial cells of bees and occur by ingestion of spores by adults (Fig. 8). Once in the midgut, the spores are chemically stimulated to trigger the polar tube, which penetrates the host's cells, starting the infection processes (de Graaf, Raes, & Jacobs, 1994). Infectious spores are then

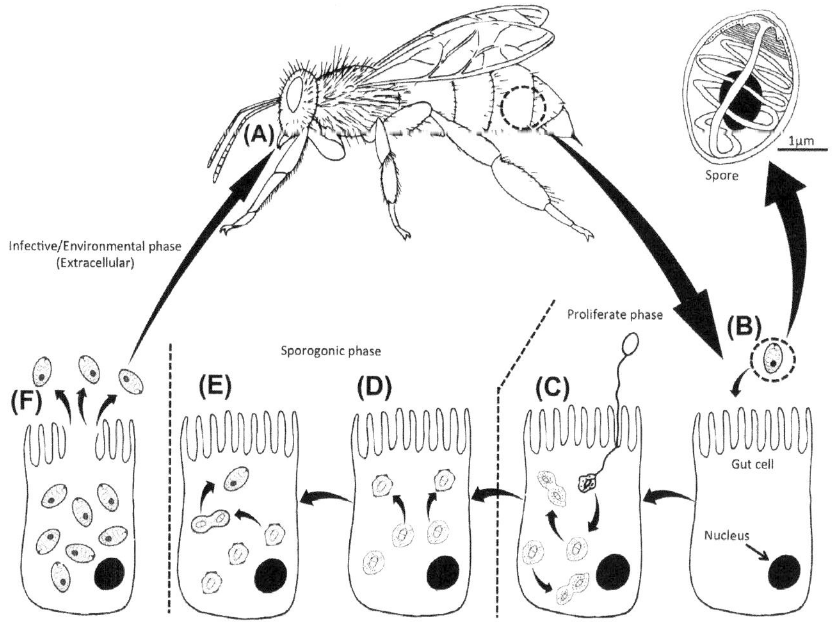

Figure 8 ***Microsporidia*—Nosema *sp.*** (A) Ingestion of spores; (B) Spore reaches the gut of the bee and is activated by its environment, triggering the polar tube to inject the sporoplasm into the host's cell; (C) Cellular multiplication (proliferate phase); (D and E) Transition from sporoplasm to spore; (F) Spores are released into the gut again, and will be spread in the bee's feces or will reinfect the same individual. (See color plate)

released with the feces and due to the characteristic thick three-layered wall structure, they are well adapted to resist in the environment until they are ingested by another adult bee (Wittner & Weiss, 1999).

4.1.3 Chytridiomycota

Among the chytrids, there are four genera that are entomopathogenic: *Myiophagus* (one species, *Myiophagus* cf. *ucrainicus*), *Coelomycidium* (one species, *Coelomycidium simulii*), *Myrmicinosporidium* (one species, *Myrmicinosporidium durum*), and the most diverse genus *Coelomomyces* (63 species). Most of the chytrid infections in insects have been recorded for Diptera.

The genus *Myiophagus* was described infecting dipteran pupae (Petch, 1948) and scale insects (Karling, 1948; Muma & Clancy, 1961). Doberski and Tribe (1978) reported *Catenaria auxiliaris* on coleopteran larvae, although, they are not sure if the colonization occurred after the larva's death (saprophytism) or if, in fact, parasitism occurred, leading to the death of the larvae (in Fig. 5 the uncertainty of this record is denoted with a question mark). Thus, since this relationship is not proven yet, we will not consider *C. auxiliaris* among the chytrids that parasitize insects.

The genus *Coelomycidium* is known to attack a specific group within Diptera order, the black flies (Simuliidae) (Jitklang, Ahantarig, Kuvangkadilok, Baimai, & Adler, 2012; McCreadie & Adler, 1999). This disease is identified by the observation of the larvae filled with spherical sporangia throughout the body cavity (Kim, 2011).

One group deserves special mention because of the effect they have on insect reproduction and behavior. The *Coelomomyces* species (Fig. 9) are relatively well known because their hosts are important human disease vectors (*Simulium* and the mosquitoes *Anopheles*, *Culex*, and *Aedes*). Species within this genus can infect eggs (Martin, 1978), larvae (the most common type of infection, see details in Travland (1979)), and adults (Lucarotti & Klein, 1988). In some species (ie, *Coelomomyces psophorae*, Fig. 8) a copepod is required to complete the whole life cycle (Whisler et al., 1975).

In some cases, the fungus does not kill the larvae. Rather, the chytrid remains inside the insect as it passes through the larval and pupal stages before maturing in the ovaries of adult females (Lucarotti, 1992). Once there and after the first mosquito's blood meal, the hypha matures to become sporangia, which is the fungal structure responsible for producing zoospores (Lucarotti & Shoulkamy, 2000). Thus, instead of laying eggs, the mosquito will 'lay' sporangia full of zoospores, ready to infect new larvae (Lucarotti & Klein, 1988). Fatefully, the fungus is reintroduced at the mosquito's breeding site by its own host.

4.1.4 Entomophthoromycota

The phylum Entomophthoromycota is composed mostly of pathogens of insects, with few pathogens of other invertebrates, desmid algae, and fern gametophytes, and some that live a saprophytic life (Humber, 2012). The entomopathogenic species are distributed among 19 genera: *Entomophthora, Conidiobolus, Entomophaga, Erynia, Meristacrum, Neozygites, Strongwellsea* and *Massospora, Pandora, Eryniopsis* (Fig. 2), *Batkoa, Tarichium, Completoria, Ballocephala, Zygnemomyces, Ancylistes, Macrobiotophthora, Thaxterosporium*, and *Basidiobolus*. It is difficult to say how many species of entomopathogens exist since many of these genera also infect different groups of hosts. In addition, the group is in constant taxonomic flux (Humber, 2012). However, since the scope of this work is not to provide a complete list of all species within

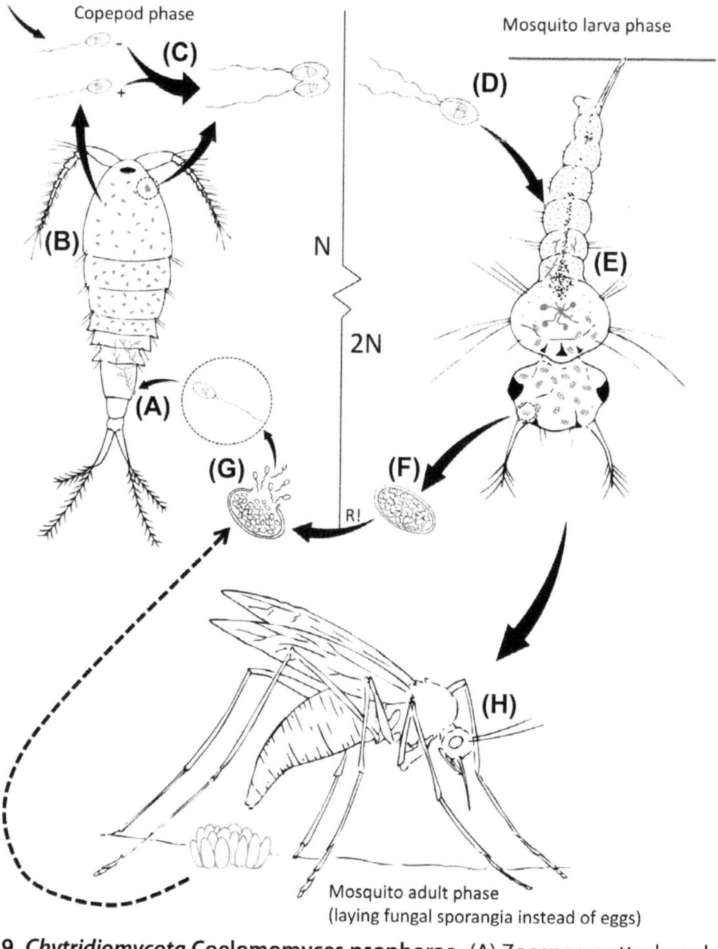

Figure 9 ***Chytridiomycota*** **Coelomomyces psophorae.** (A) Zoospores attach and penetrate the copepod cuticle; (B) Development of the gametophytic phase and dispersion of gametes into the environment; (C) Fusion of compatible gametes, inside the copepod or in the environment (plasmogamy); (D) Formation of zygote (kariogamy = 2n) and attachment to the cuticle of the mosquito larva; (E) Colonization and development of the sporophytic phase and formation of sporangium; (F) Resting sporangium released into the environment after the larva's death; (G) Meiosis and release of asexual zoospores; (H) If the larvae reach the adult stage, the fungus will migrate to the ovaries. Instead of laying eggs, the mosquito will lay fungal sporangia. (See color plate)

each group, but to present the diversity of morphologies and strategies to infect their hosts, we will provide a broad overview of entomopathogenic species among Entomophthoromycota, presenting some representative examples of their diversity.

The entomophthoroid fungi are well known as insect pathogens. This group attacks mainly adult insects, although two species of *Entomophthora* (*Entomophthora aquatica* and *Entomophthora conglomerata*) and *Erynia aquatica* are known to infect aquatic larval stages of mosquitoes (Scholte et al., 2004). Transmission within entomophthoroid fungi is via forcibly discharged spores into the environment, with the exception of one single genus (*Massospora*) that releases the spores passively, with the host still alive (Humber, 1981; Thaxter, 1888) (Fig. 10). In addition to *Massospora*, other groups like *Strongwellsea* and certain species of *Entomophthora*, *Erynia*, and *Entomophaga* (in addition, to the Ascomycete *Lecanicillium longisporum*) produce spores before the host death, in or on their living bodies (Roy et al., 2006). These fungi are characteristically biotrophics, consuming the host when they are still alive with no somatic growth after its death. This is one of the major differences when compared with hypocrealean fungi (discussed further in the Ascomycota section), which are all hemibiotrophic,

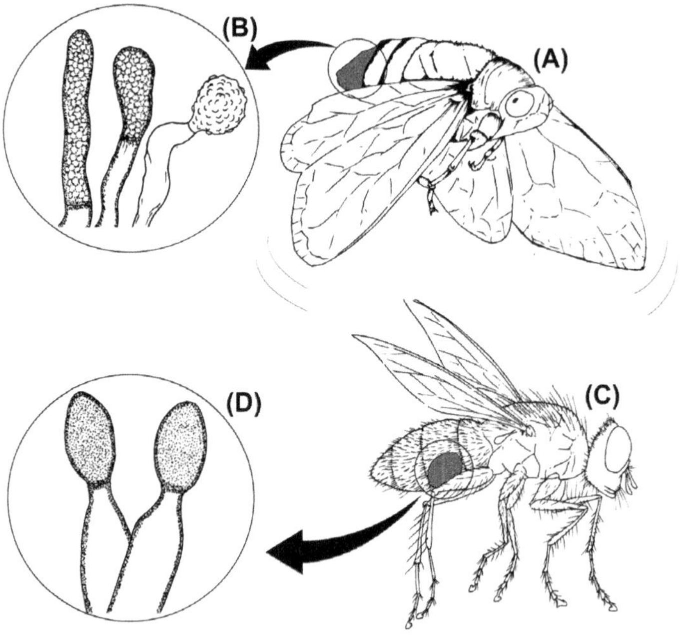

Figure 10 ***Entomophthoromycota*—Massospora cicadina *(A and B)*, Strongwellsea castrans *(C and D)*.** (A) A living cicada flying and dispersing spores while its body disintegrates due to fungal activity. (B) Spore-producing cells (Conidiophores) in different stages of development. (C) Fly exhibiting a hole on the abdomen caused by the fungal infection. (D) Conidiophores exhibiting a terminal spore. (See color plate)

switching from a biotrophic phase (parasitism) to a saprophytic phase, growing on or in the host's body, even after its death (Roy et al., 2006).

The infections caused by *Strongwellsea castrans* in *Hylemya brassicae* and *Hylemya platura* (Diptera) are classic examples of these peculiar situations where the sporulation occurs while the host is still alive (Nair & McEwen, 1973) (Fig. 10C and D). In this case, the infected fly is characterized by the presence of a large circular hole on the lateral side of the abdomen. However, surprisingly the infected insects can be observed acting normally, despite the big hole in its body, filled with fungal tissue and conidiophores (spore-producing cells). Both, males and females were described infected by *S. castrans*, causing castration and premature death (Nair & McEwen, 1973). Another similar case occurs with *Massospora cicadina*, which attack cicadas (Fig. 10A and B). This fungus also initiates sporulation when the host is still alive (Goldstein, 1929; Speare, 1921). Due to the pressure caused by the swelling mass of fungus, the collapse of its whole abdomen is inevitable, exposing the fungal tissue. Since the fungus maintains its growth inside the insect, over time the abdomen falls apart until just the head and thorax of the living insect remain (Speare, 1921). The ability to fly is retained increasing dispersion of spores in the environment.

Although Mucoromycotina is the largest and morphologically most diverse group of "zygomycetes," the subphylum has just one single entomopathogenic species, *S. umbellata* found attacking the hemipteran genus *Umbonia* in Ecuador (Evans & Samson, 1977; Samson et al., 1988) and the lepidopteran genus *Acraea* in Taiwan (Chien & Hwang, 1997).

4.1.5 Basidiomycota

Although the phylum exhibits great diversity of species—over 1500 genera and 31,000 species described (Kirk et al., 2008); just three genera are known to infect insects. Those are (1) *Fibularhizoctonia* spp. (an undescribed species, see Yashiro and Matsuura (2007)) infecting termite eggs, (2) *Uredinella* (two species, *Uredinella coccidiophaga* and *Uredinella spinulosa*) infecting scale insects, and (3) *Septobasidium* (c. 240 species attacking scale insects, Hemiptera).

The order Septobasidiales Couch (*Uredinella* and *Septobasidium*) exhibits a peculiar and complex relationship with their hosts, the Diaspididae (Hemiptera). Diaspididae are small, sedentary phythophagous insects, which spend their whole lives in one spot on a plant, a consequence of their sucking mouthpart structure (Grimaldi & Engel, 2005). To protect themselves, since they are not able to fly away from enemies (Heimpel & Rosenheim, 1995) and do not survive unprotected, juveniles start to secrete fine threads of white wax, which within the first 24 h after their hatching will form a complete covering over the insect's body (Couch, 1938).

This waxy protection is fragile but does afford some degree of defense; however, they are still exposed to external factors. An additional defense structure can be provided when a colony of such plant-feeding insects are infected by the fungus *Septobasidium*. The fungus can grow up to 20 cm and creates an elaborate system of tunnels and chambers inside its "body," which provide the Diaspididae with life-long protection (Couch, 1938). However, not all insects are protected as this fungus infects some members of the colony often causing dwarfism and castration. The atrophy is due to penetrant haustoria that drain plant sap and nutrients from the insect's body, resulting in undernourishment (Couch, 1931). Even uninfected adult insects are surrounded and held by hyphal threads, and so are unable to escape: providing an example of a fungus farming an insect (Couch, 1931, 1938). Juveniles (crawlers) may become infected as they attempt to move out of the parental chamber to establish a new colony (Couch, 1938).

With respect to the less speciose genus among Septobasidiales, *Uredinella*, there are only two described species: *U. coccidiophaga* and *U. spinulosa* (see Couch (1937, 1941)). They can be divided based on spore shape and the substratum in which they infect insects; leaf and trunk for *U. spinulosa* and just trunk for *U. coccidiophaga*. As the genus infects the body of single insect (unlike *Septobasidium*), the death of the insect means the death of the fungus also. Spores are produced in the spring and reach 0.2–1.5 mm in diameter (Couch, 1937). In contrast, *Septobasidium* exhibits an undefined lifetime, since its body is "renewed" each season, by the infection of the newborn crawlers.

Another case of Basidiomycota parasitic on insects can be found between *Fibularhizoctonia* spp. and some species of the subterranean termites *Reticulitermes*. These termite workers keep their eggs inside their nest in piles, taking care of them. Matsuura et al. (2000) found among these piles, some sclerotia, globose fungal structures, being cared for by the workers, as if they were eggs. The same study found that these sclerotia mimic the egg diameter and texture and because these traits are similar to those of the termite eggs themselves the worker termites mistake the sclerotium for a true egg and care for it. In nature it is rare to observe the fungus consuming the termite eggs, but there is the suggestion that the fungus becomes pathogenic and grows over the true termite eggs if the termites stop caring for the fungi (Matsuura, 2006).

4.1.6 Ascomycota

As mentioned, this diverse phylum comprises many entomopathogenic fungi: from the less speciose orders Pleosporales, Myriangiales, Ascosphaerales to hyperdiverse groups within the relatively well-known order Hypocreales. In each case the insect dies before the fungus begins its reproductive phase. Here, we describe each of these groups and their main characteristics.

Within the order Pleosporales, the entomopathogenic species belong to the genus *Podonectria* (Petch, 1921) that shares the unusual aspects of this group, such as bright coloration and fleshiness. All known species have been found infecting scale insects, covering the whole surface of the insects body with a cotton-like crust on which the perithecia is produced and later, multiseptated spores that do not disarticulate into part-spores (Kobayasi & Shimizu, 1977). The related anamorphs are the genus *Tetracrium* (Kobayasi & Shimizu, 1977; Petch, 1921) and the genus *Tetranacrium* (Roberts & Humber, 1981).

The Myriangiales includes a number of species associated with plants, resins, or scale insects on plants (Alexopoulos et al., 1996). The entomopathogenic species exhibit perennial growth for several years or at least until the scale-infested branch dies, probably because of decreased nourishment. The dead host insects can be found directly under each stroma, penetrated and covered by mycelium (Miller, 1938). The stroma are sometimes formed at the side of the insect (Petch, 1924). Growth commences when the rains begin with the fungus increasing in diameter, producing ascomata and then later, ascospores. Reproduction is entirely by ascospores, and no evidence of conidial (asexual spore) formation was found on stroma of any age or in culture (Miller, 1938). For taxonomic and additional discussions, see Miller (1938) and Petch (1924).

The order Ascosphaerales contains a unique group of bee pathogens within the genus *Ascosphaera*, which has approximately 30 species. These parasites are specialists that exploit the provisions of bees. Most species are exclusive saprophytes on honey, cocoons, larval feces, or nest materials such as leaf, mud, or wax of bees (Wynns, Jensen, Eilenberg, & James, 2012). However, some species are known as widespread fungal disease agents, attacking the brood of numerous species of solitary and social bees, causing a disease called "chalk-brood" (Klinger, James, Youssef, & Welker, 2013). The infection occurs when the larva ingests fungal spores. The fungus grows as hyphae within the body before killing the host and then developing spores on the cuticle of the dead larvae (Vojvodic, Boomsma, Eilenberg, &

Jensen, 2012). The morphology of *Ascosphaera* is very peculiar when compared to other fungal groups. The ascoma is a small brown to blackish brown spore cyst, which is a single enlarged cell containing ascospores (Wynns et al., 2012). Their spores also exhibit a curious similarity in appearance to pollen grains. For a detailed life cycle, see McManus and Youssef (1984).

The Hypocreales fungi encompass important genera of entomopathogenic fungi such as *Cordyceps, Tolypocladium, Hypocrella, Ophiocordyceps, Moelleriella, Samuelsia,* and *Torrubiella*. In addition to these species, there are many anamorphic species related to them, ie, *Hirsutella, Metarhizium, Hymenostilbe, Akanthomyces,* and many others (Roberts & Humber, 1981). In the past, these anamorphic species (which only produce asexual spores) were treated traditionally as a separate group within the now retired phylum Deuteromycota. However, with molecular techniques some of these species are now strongly supported or proven to be asexual stages of Ascomycota (Liu et al., 2001). Here we will address just the teleomorphic (ie, sexual stage) names since our goal is to provide the reader an overview of entomopathogenic groups and to avoid confusion between the same species, in which both teleomorphic and anamorphic (ie, asexual stage) phases are known.

Within the largest group of entomopathogens, the Hypocreales, we can highlight some genera that are notable due to their diversity and abundance in tropical forests worldwide. For example, *Hypocrella* (*Archersonia* is the anamorphic state) that was beautifully monographed by Chaverri et al. (2008). These fungi are known to infect whiteflies and scale insects in tropical forests, with few species recorded in the subtropics. They can cause epizootics in their host's population (but epizootics are by no means confined to this genus).

The genus *Ophiocordyceps*—especially species attacking ants—are known to cause huge infestations in small areas, called graveyards (Evans & Samson, 1982, 1984; Pontoppidan, Himaman, Hywel-Jones, Boomsma, & Hughes, 2009). Indeed, one of the most fascinating phenomena regarding entomopathogenic fungi is the zombie-ant behavior caused by *Ophiocordyceps unilateralis s.l.* (Andersen et al., 2009; Hughes et al., 2011). This species was originally described by Tulasne and Tulasne (1865) as *Torrubia unilateralis*. Species within this complex adaptively manipulate the behavior of worker ants, causing the insect to leave the colony to find an optimum microclimate site, which is required by the fungus to grow and produce ascospores (Fig. 11D). The ants die biting firmly on the underside or edge of a leaf, twig, branch, etc. (the death position is related to each species or group of species of the fungus).

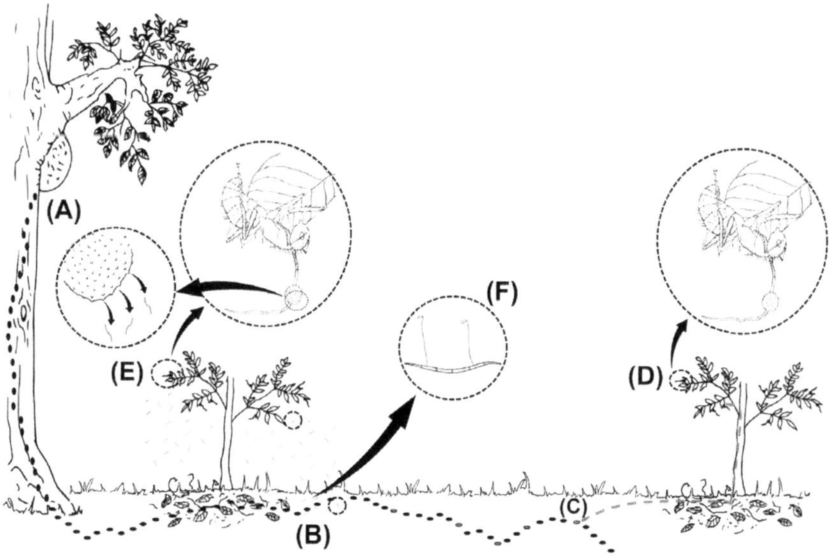

Figure 11 ***Ascomycota*****—Ophiocordyceps unilateralis *s.l.*** (A) Ants leave the nest to forage on the forest floor; (B) Eventually they get infected with *Ophiocordyceps* ascospores that were previously shot on the forest floor; (C and D) About 10 days after infection (depending on the species and the geographical location) the infected ant leaves the nest to die on an elevated position, biting the edge or the main vein of a leaf. The fungus places the ant on a precise location, which is optimal for fungal development and further dispersion of the spores; (E) Two to eight weeks after the ant's death, depending on the weather conditions, the fungus starts to produce spores and shoot them into the environment; (F) From 24 to 72 h after being shot, the spores will germinate and form a secondary spore, the capilliconidiospore. (See color plate)

Following the ant's death, the fungus grows a fruiting body from the back of the ant's head, which will spread the ascospores on the forest floor (Fig. 11E).

5. DISCUSSION

5.1 Factors Promoting Diversity Within Entomopathogenic Fungi and Oomycetes

5.1.1 Hemipterans as a Host Group Promoting Hyperdiversity of Entomopathogens

The host is the ecological niche for the fungus. Some ecological niches, ie, host groups, are notable because the abundance and diversity of entomopathogenic species infecting them are very high. For instance, the broad diversity found among entomopathogenic fungal species attacking sap-sucking

Hemiptera. Based on teleomorphic species (ie, species identified by their sexual stage) within Hypocreales, which is the most diverse group of entomopathogens in Ascomycota, we found 180 species of fungi exclusively parasitizing hemipteran insects (Fig. 3 and Supplementary table 5). Most cases are infections on adult stages (n = 165 records). The *Hypocrella–Aschersonia* species are responsible for the majority of infections, with 92 species infecting scale insects (Coccidae and Lecaniidae, Hemiptera) and whiteflies (Aleyrodidae, Hemiptera) (*Aschersonia* is the anamorphic stage and some species have no described teleomorphic stage). The other Hypocreales species attacking hemipterans are spread among the six other genera: *Moelleriella* (25 species), *Ophiocordyceps* (19 species), *Torrubiella* (18 species), *Cordyceps* (17 species), *Samuelsia* (6 species), and *Regiocrella* (2 species). These are all from the order Hypocreales. Other orders such as Myriangiales (3 species: *Myriangium asterinosporum*, *Myriangium curtisii*, and *Myriangium duriaei*, Ascomycota), Septobasidiales (approximately 240 species, Basidiomycota) and even the chytrid fungi *M. ucrainicus* (Chytridiomycota) in the order Chytridiales are known to infect exclusively hemipteran sap-sucking insects. No other group of insects is attacked either by so many different groups of fungi or so many distinct species of entomopathogens.

Why are hemipterans such common hosts? According to Spatafora, Sung, Sung, Hywel-Jones, and White (2007) the order Hypocreales evolved from an ancestor with a plant-based nutritional mode. They made a horizontal host-jump from plants to the insects feeding on plants. There are estimated to have been around five to eight interkingdom host-jumping events between Plant, Animal and Fungi within Hypocreales (Spatafora et al., 2007). In general terms the shifts between different host groups is suggested to be important for expanding the host range of the Hypocreales (Kepler et al., 2012; Nikoh & Fukatsu, 2000). But how did this happen?

Although the insects arose in the Devonian (earliest fossils from around 407 mya) their major diversification occurred after the seed plant (spermatophytes) radiation (Permian, 299–251 mya), when most of the modern orders of insects emerged, including small basal groups of Hemiptera (Grimaldi & Engel, 2005). Afterward, a new episode transformed all the ecosystems on Earth as the flowering plants (angiosperms) evolved and radiated (Lower Cretaceous, 130 mya) along with the diversification of insects (Gaunt & Miles, 2002; Grimaldi & Engel, 2005). A result of the codiversification of insects and flowering plants was the expansion of the different insect mouthparts (Labandeira, 1997). In the case of the hemipterans, the mouthparts evolved into two pairs of long and fine stylets, which are able to create strong suction in order to draw fluids from plant tissues (Grimaldi & Engel, 2005). This derived feature was essential for this group of insects, since it allowed them to exploit a new niche: living on the plants and feeding from their sap.

On the other hand, in the same way that hemipterans adapted their mouthparts into stylets, the fungi, in order to exploit another food source, switched from plant-based to insect-based ecology about 170–150 mya. This event would have been facilitated by the ecological proximity between hemipterans and a Hypocreales ancestors; one feeding from plant exudates, the other living inside the plant as endophytes, at least 190 mya (Sung, Poinar, & Spatafora, 2008). After the fungi adapted to their new ecological niche in insects they necessarily would have had to optimize horizontal transmission between insects. For instance, the hyperdiverse *Hypocrella–Aschersonia* group exhibit mitotic slime spores, adapted for short distance dispersal by rain-splash on leaf surfaces, which are a hot spot to find their hosts (Chaverri et al., 2008; Evans, 1989).

Afterward, the transition from growing within plants to infecting sap sucking insects would have provided a route to infect other insects that were not phytophagous. As a result, all these changes have given rise to three of the most important entomopathogenic families Clavicipitaceae (eg, *Hypocrella, Moelleriella, Samuelsia, Metacordyceps*), Cordycipitaceae (eg, *Cordyceps, Torrubiella*), and Ophiocordycipitaceae (*Ophiocordyceps, Tolypocladium*, formerly *Elaphocordyceps* (Quandt et al., 2014), *Polycephalomyces*) (Kepler et al., 2013; Sung et al., 2007, 2008). Those families have undoubtedly achieved great success with this host shifting from plant-based ecology to other groups, as illustrated by their species richness, ecological abundance, and worldwide distribution, especially in tropical forests.

5.1.2 Broad Range of Ecologies, High Diversity of Pathogens

Another interesting case of unusual diversity among fungal–insect infections can be found within the insect order Diptera (flies). Although the number of species is comparatively low (Fig. 3 and Supplementary table 5) it is notable that flies are the only order of insect with records of entomopathogens infecting all stages of development: eggs, larvae, pupae, and adults. Only the phylum—Basidiomycota—of the six phyla we recorded infecting insects do not infect dipterans. Why are flies such magnets for fungal infections? One reason is probably because they are the most ecologically diverse group of insects, found all over the world, occupying a broad range of niches: blood feeders, endoparasites and ectoparasites of vertebrates and invertebrates, gall makers, larval and adult predators, leaf miners, parasitoids, pollinators, saprophages, and wood borers (Grimaldi & Engel, 2005). In addition, their larvae can be found in many different breeding sites such as aquatic, semiaquatic (wet soils, stones on stream edges), or terrestrial (mushrooms, rotten wood, trunk). We suggest that by occupying such diverse niches, the flies have increased the opportunity for infection by fungal and oomycete pathogens, which like the insects, occupy diverse environments. It is still notable, however, that while flies are infected by five of the six different groups of entomopathogens and all developmental stages of flies are attacked, the overall species diversity of entomopathogens attacking them is still low given that flies are a very specious order (over 150,000 species). Flies would appear to be magnets for entomopathogens, but they are not cradles of diversity in the same way hemipterans have apparently been.

5.1.3 Susceptibility of Lepidoptera and Coleoptera Larval Stages to Fungal Infections

Beyond the insect orders discussed earlier (Hemiptera and Diptera), it is worthwhile to mention the prevalence of infections in larval stages of Lepidoptera and Coleoptera (Fig. 3) by hypocrealean fungi (Ascomycota). The larvae of both insect orders are the preferred host for two of the most diverse and ecologically abundant genera of entomopathogenic fungi, *Cordyceps* and *Ophiocordyceps* (Supplementary table 1). We found 80 and 87 species of teleomorphic Hypocreales infecting Lepidoptera and Coleoptera larvae, respectively. In contrast, we found just 10 and 11 records of teleomorphic Hypocreales species attacking adults of Lepidoptera and Coleoptera, respectively (Fig. 3). What factors promote such a predominance of infections on larval stages?

A number of biological traits that are different between the larvae and adults may be crucial to understanding this pattern: (1) partition of niches, (2) predictability in time–space scales, (3) feeding rate, and (4) protection (cuticle). (1) As holometabolous insects (which exhibit complete metamorphosis), the larval and adult stages are ecologically separated, occupying completely different microenvironments, thus avoiding competition between juveniles and adults (Gullan & Cranston, 2009). (2) Both coleopteran and lepidopteran larvae generally exhibit modest mobility compared to the wandering adults, tending to be closer to the breeding site and eating ferociously, hence being more predictable in time–space scales. (3) The larva needs to eat massive amounts of food and store that food, in order to grow as quickly as possible, making them a huge reservoir of energy. (4) Furthermore, larvae need to grow at a high rate and this would be impossible if they had the hard exoskeleton that adult coleopterans have. However, on the other hand, having such soft and thin skin would make these organisms much easier to be invaded by fungal spores equipped with their enzymatic and physical tools for infection. It is important to emphasize that the usual defenses that larvae exhibit—mimicry, aposematism, gregarious behavior, stinging hairs—that are very useful against predators are completely useless against the effective entomopathogenic fungi. These four ecological traits that distinguish larvae and adults from each other may explain why entomopathogenic fungi exhibit such a greater prevalence for infecting larvae rather than adults.

The other major holometabolous order infected by Hypocreales is Hymenoptera (wasps, bees, termites, and ants). Here, however, most infections are of the adult stages. There are few records (n = 47) so it is harder to contrast with Lepidoptera and Coleoptera. But it is noticeable that the hymenopterans build nests for their larvae, and in the case of ants, some wasps, and some bees these larvae are nursed and cleaned by their siblings, which is known to reduce fungal infections (Cremer, Armitage, & Schmid-Hempel, 2007).

6. CONCLUSION

This is the first time that an extensive review encompassing all entomopathogenic fungal phyla and oomycetes explored entomopathogenic fungi with a fungal–host approach. Despite the importance of insect–fungal associations, they have been overlooked and their diversity is poorly studied. The lack of interaction between mycologists and entomologists might play an important role in this gap of knowledge, and efforts to address this issue are crucial to better understand the parasitic relationship between insects and the multiple lineages of entomopathogenic fungi.

Fungi that are able to infect insects are not just comprised by a single monophyletic group. Different groups have arisen independently and repeatedly in many different lineages through fungal evolution (Humber, 2008). As presented here, they are spread from more basal to more complex Dikaria members. The basal groups, such as aquatic chytrids, infect mostly Diptera, while Microsporidia and Entomophthoromycota infect a wide range of hosts. Basidiomycota infects mostly Hemiptera, while Ascomycota, the most speciose group, infects a vast number of insect groups.

Insect pathologists, entomologists, and life scientists in general have traditionally seen entomopathogenic fungi as having a single role: to kill insect pests (Vega, 2008). But the coevolution of fungi and insects across hundreds of millions of years has resulted in a wide range of complex and intricate interactions. The purpose of this work is to provide a wide overview of these relationships by focusing on the impressive diversity of morphologies, ecologies, and interactions between insects and fungi. Our work also highlights the ways that biological and ecological aspects of the hosts likely played an important role to explain why and how some groups of insects are more susceptible to fungal infection than others.

ACKNOWLEDGMENTS

We are grateful to Harry Evans, Richard Humber, Ryan Kepler, Priscila Chaverri, Bhushan Shrestha, and James Becnel for the help and inputs to improve this work.

SUPPLEMENTARY DATA

Supplementary data related to this article can be found online at http://dx.doi.org/10.1016/bs.adgen.2016.01.001.

REFERENCES

Alexopoulos, C. J. (1962). *Introductory mycology*. Wiley.

Alexopoulos, C. J., Mims, C. W., & Blackwell, M. (1996). *Introductory mycology*. New York: John Wiley & Sons.

Andersen, S. B., Gerritsma, S., Yusah, K. M., Mayntz, D., Hywel-Jones, N. L., Billen, J. … Hughes, D. P. (2009). The life of a dead ant: the expression of an adaptive extended phenotype. *The American Naturalist, 174*(3), 424–433.

Andreadis, T. G. (1985). Experimental transmission of a microsporidian pathogen from mosquitoes to an alternate copepod host. *Proceedings of the National Academy of Sciences of the United States of America, 82*(16), 5574–5577.

Barr, D. J. S. (1992). Evolution and kingdoms of organisms from the perspective of a mycologist. *Mycologia*, 1–11.

Barr, D. J. S., & Désaulniers, N. L. (1988). Precise configuration of the chytrid zoospore. *Canadian Journal of Botany, 66*(5), 869–876.

Barron, G. L., & Dierkes, Y. (1977). Nematophagous fungi: Hohenbuehelia, the perfect state of *Nematoctonus*. *Canadian Journal of Botany, 55*(24), 3054–3062.

Beakes, G. W., Glockling, S. L., & Sekimoto, S. (2012). The evolutionary phylogeny of the oomycete "fungi". *Protoplasma, 249*(1), 3–19.

Becnel, J. J., & Andreadis, T. G. (1999). Microsporidia in insects. In M. Wittner, & L. M. Weiss (Eds.), *The microsporidia and microsporidiosis* (pp. 447–501). Washington: Am. Soc. Microbiol. Press.

Bisht, G. S., Joshi, C., & Khulbe, R. D. (1996). Watermolds: potential biological control agents of malaria vector *Anopheles culicifacies*. *Current Science Bangalore, 70*, 393–395.

Blackwell, M. (2011). The Fungi: 1, 2, 3 … 5.1 million species? *American Journal of Botany, 98*(3), 426–438.

Braga-Neto, R., Luizão, R. C. C., Magnusson, W. E., Zuquim, G., & Castilho, V. C. (2008). Leaf litter fungi in a Central Amazonian forest: the influence of rainfall, soil

and topography on the distribution of fruiting bodies. *Biodiversity and Conservation, 17*(11), 2701–2712.
Brasier, C., Denman, S., Brown, A., & Webber, J. (2004). Sudden oak death (*Phytophthora ramorum*) discovered on trees in Europe. *Mycological Research, 108*(10), 1108–1110.
Briano, J. A. (2005). Long-term studies of the red imported fire ant, *Solenopsis invicta*, infected with the microsporidia *Vairimorpha invictae* and *Thelohania solenopsae* in Argentina. *Environmental Entomology, 34*(1), 124–132.
Bulla, L. A., Jr., & Cheng, T. C. (1977). *Comparative pathobiology. Volume 2. Systematics of the microsporidia*. Plenum Press.
Butt, T. M., Hajek, A. E., & Humber, R. A. (1996). Gypsy moth immune defenses in response to hyphal bodies and natural protoplasts of entomophthoralean fungi. *Journal of Invertebrate Pathology, 68*(3), 278–285.
Chaverri, P., Liu, M., & Hodge, K. T. (2008). A monograph of the entomopathogenic genera *Hypocrella, Moelleriella*, and *Samuelsia* gen. nov. (Ascomycota, Hypocreales, Clavicipitaceae), and their aschersonia-like anamorphs in the Neotropics. *Studies in Mycology, 60*, 1–66.
Chien, C. Y., & Hwang, B. C. (1997). First record of the occurrence of *Sporodiniella umbellata* (Mucorales) in Taiwan. *Mycoscience, 38*(3), 343–346.
Couch, J. N. (1931). Memoirs: the biological relationship between *Septobasidium retiforme* (B. & C.) Pat. and *Aspidiotus osborni* New. and Ckll. *Quarterly Journal of Microscopical Science, 2*(295), 383–438.
Couch, J. N. (1935). A new saprophytic species of *Lagenidium*, with notes on other forms. *Mycologia, 27*(4), 376–387.
Couch, J. N. (1937). A new fungus intermediate between the rusts and *Septobasidium*. *Mycologia, 29*(6), 665–673.
Couch, J. N. (1938). *The genus* Septobasidium. The University of North Carolina Press.
Couch, J. N. (1941). A new *Uredinella* from Ceylon. *Mycologia*, 405–410.
Couch, J. N., & Bland, C. E. (1985). *The genus* Coelomomyces. Orlando: Academic Press.
Cremer, S., Armitage, S. A. O., & Schmid-Hempel, P. (2007). Social immunity. *Current Biology, 17*(16), R693–R702.
Currie, C. R., Wong, B., Stuart, A. E., Schultz, T. R., Rehner, S. A., Mueller, U. G.... Straus, N. A. (2003). Ancient tripartite coevolution in the attine ant-microbe symbiosis. *Science, 299*(5605), 386–388.
DeKesel, A. (1996). Host specificity and habitat preference of *Laboulbenia slackensis*. *Mycologia, 88*(4), 565–573.
Dewel, R. A., Joines, J. D., & Bond, J. J. (1985). A new chytridiomycete parasitizing the tardigrade *Milnesium tardigradum*. *Canadian Journal of Botany, 63*(9), 1525–1534.
Dick, M. W. (1998). The species and systematic position of Crypticola in the Peronosporomycetes, and new names for Halocrusticida and species therein. *Mycological Research, 102*(09), 1062–1066.
Dick, M. W. (2001). *Straminipilous fungi: Systematics of the peronosporomycetes including accounts of the marine straminipilous protists, the plasmodiophorids and similar organisms.*
Didier, E. S., & Bessinger, G. T. (1999). Host-parasite relationships in microsporidiosis: animal models and immunology. In M. Wittner, & L. M. Weiss (Eds.), *The microsporidia and microsporidiosis* (pp. 225–257). Washington, DC: ASM Press.
Doberski, J. W., & Tribe, H. T. (1978). *Catenaria auxiliaris* (Chytridiomycetes: Blastocladiales) identified in a larva of *Scolytus scolytus* (Coleoptera: Scolytidae). *Journal of Invertebrate Pathology, 32*(3), 392–393.
Evans, H. C. (1988). Coevolution of entomogenous fungi and their insect hosts. In K. A. Pirozynski, & D. L. Hawksworth (Eds.), *Coevolution of fungi with plants and animals.*
Evans, H. C. (1989). *Mycopathogens of insects of epigeal and aerial habitats*. London: Academic Press.

Evans, H. C., Elliot, S. L., & Hughes, D. P. (2011). Hidden diversity behind the zombie-ant fungus *Ophiocordyceps unilateralis*: four new species described from carpenter ants in Minas Gerais, Brazil. *PLoS One, 6*(3), e17024.
Evans, H. C., & Samson, R. A. (1977). *Sporodiniella umbellata*, an entomogenous fungus of the Mucorales from cocoa farms in Ecuador. *Canadian Journal of Botany, 55*(23), 2981–2984.
Evans, H. C., & Samson, R. A. (1982). *Cordyceps* species and their anamorphs pathogenic on ants (Formicidae) in tropical forest ecosystems I. The *Cephalotes* (Myrmicinae) complex. *Transactions of the British Mycological Society, 79*(3), 431–453.
Evans, H. C., & Samson, R. A. (1984). *Cordyceps* species and their anamorphs pathogenic on ants (Formicidae) in tropical forest ecosystems II. The *Camponotus* (Formicinae) complex. *Transactions of the British Mycological Society, 82*(1), 127–150.
Fisher, W. S., Nilson, E. H., & Shleser, R. A. (1975). Effect of the fungus *Haliphthoros milfordensis* on the juvenile stages of the American lobster *Homarus americanus*. *Journal of Invertebrate Pathology, 26*(1), 41–45.
Frances, S. P., Sweeney, A. W., & Humber, R. A. (1989). Crypticola clavulifera gen. et sp. nov. and *Lagenidium giganteum*: oomycetes pathogenic for dipterans infesting leaf axils in an Australian rain forest. *Journal of Invertebrate Pathology, 54*(1), 103–111.
Gaunt, M. W., & Miles, M. A. (2002). An insect molecular clock dates the origin of the insects and accords with palaeontological and biogeographic landmarks. *Molecular Biology and Evolution, 19*(5), 748–761.
Gleason, F. H., Küpper, F. C., Amon, J. P., Picard, K., Gachon, C. M. M., Marano, A. V. … Lilje, O. (2011). Zoosporic true fungi in marine ecosystems: a review. *Marine and Freshwater Research, 62*(4), 383–393.
Gleason, F. H., & Lilje, O. (2009). Structure and function of fungal zoospores: ecological implications. *Fungal Ecology, 2*(2), 53–59.
Goldstein, B. (1929). A cytological study of the fungus *Massospora cicadina*, parasitic on the 17-year cicada, *Magicicada septendecim*. *American Journal of Botany*, 394–401.
Golkar, L., LeBrun, R. A., Ohayon, H., Gounon, P., Papierok, B., & Brey, P. T. (1993). Variation of larval susceptibility to *Lagenidium giganteum* in three mosquito species. *Journal of Invertebrate Pathology, 62*(1), 1–8.
Goss, E. M., Tabima, J. F., Cooke, D. E. L., Restrepo, S., Fry, W. E., Forbes, M. … Grünwald, N. J. (2014). The Irish potato famine pathogen *Phytophthora infestans* originated in central Mexico rather than the Andes. *Proceedings of the National Academy of Sciences of the United States of America*, 201401884.
de Graaf, D. C., Raes, H., & Jacobs, F. J. (1994). Spore Dimorphism in *Nosema apis* (Microsporida, Nosematidae) developmental cycle. *Journal of Invertebrate Pathology, 63*(1), 92–94.
Grimaldi, D., & Engel, M. S. (2005). *Evolution of the insects*. Cambridge University Press.
Gryganskyi, A. P., Humber, R. A., Smith, M. E., Miadlikovska, J., Wu, S., Voigt, K. … Vilgalys, R. (2012). Molecular phylogeny of the Entomophthoromycota. *Molecular Phylogenetics and Evolution, 65*(2), 682–694.
Gullan, P. J., & Cranston, P. S. (2009). *The insects: An outline of entomology*. John Wiley & Sons.
Hatai, K., Rhoobunjongde, W., & Wada, S. (1992). *Haliphthoros milfordensis* isolated from gills of juvenile kuruma prawn (*Penaeus japonicus*) with black gill disease. *Transactions of the Mycological Society of Japan (Japan), 33*, 185–192.
Hawksworth, D. L. (1988). The variety of fungal-algal symbioses, their evolutionary significance, and the nature of lichens. *Botanical Journal of the Linnean Society, 96*(1), 3–20.
Hawksworth, D. L., & Rossman, A. Y. (1997). Where are all the undescribed fungi? *Phytopathology, 87*(9), 888–891.
Heimpel, G. E., & Rosenheim, J. A. (1995). Dynamic host feeding by the parasitoid *Aphytis melinus*: the balance between current and future reproduction. *Journal of Animal Ecology*, 153–167.

Hibbett, D. S., Binder, M., Bischoff, J. F., Blackwell, M., Cannon, P. F., Eriksson, O. E. ... Zhang, N. (2007). A higher-level phylogenetic classification of the Fungi. *Mycological Research, 111*(5), 509–547.

Higes, M., Martín, R., & Meana, A. (2006). *Nosema ceranae*, a new microsporidian parasite in honeybees in Europe. *Journal of Invertebrate Pathology, 92*(2), 93–95.

Hirt, R. P., Logsdon, J. M., Healy, B., Dorey, M. W., Doolittle, W. F., & Embley, T. M. (1999). Microsporidia are related to Fungi: evidence from the largest subunit of RNA polymerase II and other proteins. *Proceedings of the National Academy of Sciences of the United States of America, 96*(2), 580–585.

Horn, B. W., & Lichtwardt, R. W. (1981). Studies on the nutritional relationship of larval *Aedes aegypti* (Diptera: Culicidae) with *Smittium culisetae* (Trichomycetes). *Mycologia*, 724–740.

Hossain, Z., Gupta, S. K., Chakrabarty, S., Saha, A. K., & Bindroo, B. B. (2012). Studies on the life cycle of five microsporidian isolates and histopathology of the mid-gut of the silkworm *Bombyx mori* (Lepidoptera: Bombycidae). *International Journal of Tropical Insect Science, 32*(04), 203–209.

Hughes, D. P., Andersen, S. B., Hywel-Jones, N. L., Himaman, W., Billen, J., & Boomsma, J. J. (2011). Behavioral mechanisms and morphological symptoms of zombie ants dying from fungal infection. *BMC Ecology, 11*(1), 13.

Humber, R. A. (1976). The systematics of the genus *Strongwellsea* (Zygomycetes: Entomophthorales). *Mycologia, 68*, 1042–1060.

Humber, R. A. (1981). An alternative view of certain taxonomic criteria used in the Entomophthorales (Zygomycetes). *Mycotaxon, 13*, 191–240.

Humber, R. A. (1982). *Strongwellsea* vs. *Erynia*: the case for a phylogenetic classification of the Entomophthorales (Zygomycetes). *Mycotaxon, 15*, 167–184.

Humber, R. A. (1984). *Eryniopsis*, a new genus of the Entomophthoraceae (Entomophthorales). *Mycotaxon, 21*, 257–264.

Humber, R. A. (1989). Synopsis of a revised classification for the Entomophthorales (Zygomycotina). *Mycotaxon, 34*, 441–460.

Humber, R. A. (2008). Evolution of entomopathogenicity in fungi. *Journal of Invertebrate Pathology, 98*, 262–266.

Humber, R. A. (2012). Entomophthoromycota: a new phylum and reclassification for entomophthoroid fungi. *Mycotaxon, 120*, 477–492.

James, T. Y., Kauff, F., Schoch, C. L., Matheny, P. B., Hofstetter, V., Cox ... Miadlikowska, J. (2006). Reconstructing the early evolution of fungi using a six-gene phylogeny. *Nature, 443*(7113), 818–822.

Jitklang, S., Ahantarig, A., Kuvangkadilok, C., Baimai, V., & Adler, P. H. (2012). Parasites of larval black flies (Diptera: Simuliidae) in Thailand. *Songklanakarin Journal of Science & Technology, 34*(6).

Kamoun, S. (2003). Molecular genetics of pathogenic oomycetes. *Eukaryotic Cell, 2*(2), 191–199.

Karling, J. S. (1946). Brazilian Chytrids. VIII. Additional parasites of rotifers and nematodes. *Lloydia, 9*(1), 1–12.

Karling, J. S. (1948). Chytridiosis of scale insects. *American Journal of Botany*, 246–254.

Keeling, P. J., & Fast, N. M. (2002). Microsporidia: biology and evolution of highly reduced intracellular parasites. *Annual Reviews in Microbiology, 56*(1), 93–116.

Kemp, R. L., & Kluge, J. P. (1975). Encephalitozoon sp. in the blue-masked lovebird, *Agapornis personata* (Reichenow): first confirmed report of Microsporidan infection in birds. *The Journal of Protozoology, 22*(4), 489–491.

Kendrick, B. (2000). *The fifth kingdom* (3rd). Newburyport: Focus Publishing.

Kent, M. L., Shaw, R. W., & Sanders, J. L. (2014). *Microsporidia in fish* Microsporidia (pp. 493–520). John Wiley & Sons, Inc.

Kepler, R., Ban, S., Nakagiri, A., Bischoff, J., Hywel-Jones, N., Owensby, C. A., & Spatafora, J. W. (2013). The phylogenetic placement of hypocrealean insect pathogens in the genus *Polycephalomyces*: an application of one fungus one name. *Fungal Biology, 117*(9), 611–622.
Kepler, R. M., Sung, G.-H., Harada, Y., Tanaka, K., Tanaka, E., Hosoya, T.... Spatafora, J. W. (2012). Host jumping onto close relatives and across kingdoms by *Tyrannicordyceps* (Clavicipitaceae) gen. nov. and *Ustilaginoidea*_(Clavicipitaceae). *American Journal of Botany, 99*(3), 552–561.
Kim, S. K. (2011). Redescription of *Simulium (Simulium) japonicum* (Diptera: Simuliiae) and its entomopathogenic fungal symbionts. *Entomological Research, 41*(5), 208–210.
Kirk, P. M., Cannon, P. F., Minter, D. W., & Stalpers, J. A. (2008). *Dictionary of the fungi* (10th ed.). Wallingford, UK: CAB International.
Klinger, E. G., James, R. R., Youssef, N. N., & Welker, D. L. (2013). A multi-gene phylogeny provides additional insight into the relationships between several *Ascosphaera* species. *Journal of Invertebrate Pathology, 112*(1), 41–48.
Kobayasi, Y. (1941). *The genus* Cordyceps *and its allies* (pp. 53–260). Science Report of the Tokyo Bunrika Daigaku, Section B(84).
Kobayasi, Y., & Shimizu, D. (1977). Two new species of *Podonectria* (Clavicipitaceae). *Bulletin of National Science Museum Serie B Bot Kokuritsu Kagaku Hakubutsukan*, 017574592.
Kobayasi, Y., & Shimizu, D. (1978). *Cordyceps* species from Japan. *Bulletin of National Science Museum Tokyo, 4*, 43–63.
Kreisel, H. (1969). *Grundzüge eines natürlichen Systems der Pilze.* Cramer.
Krings, M., Dotzler, N., Galtier, J., & Taylor, T. N. (2011). Oldest fossil basidiomycete clamp connections. *Mycoscience, 52*(1), 18–23.
Kyei-Poku, G., Gauthier, D., Schwarz, R., & Frankenhuyzen, K. V. (2011). Morphology, molecular characteristics and prevalence of a *Cystosporogenes* species (Microsporidia) isolated from *Agrilus anxius* (Coleoptera: Buprestidae). *Journal of Invertebrate Pathology, 107*(1), 1–10.
Labandeira, C. C. (1997). Insect mouthparts: ascertaining the paleobiology of insect feeding strategies. *Annual Review of Ecology and Systematics*, 153–193.
Lange, C. E. (2010). *Paranosema locustae* (Microsporidia) in grasshoppers (Orthoptera: Acridoidea) of Argentina: field host range expanded. *Biocontrol Science and Technology, 20*(10), 1047–1054.
Liu, M., Chaverri, P., & Hodge, K. T. (2006). A taxonomic revision of the insect biocontrol fungus *Aschersonia aleyrodis*, its allies with white stromata and their *Hypocrella* sexual states. *Mycological Research, 110*(5), 537–554.
Liu, Z. Y., Yao, Y. J., Liang, Z. Q., Liu, A. Y., Pegler, D. N., & Chase, M. W. (2001). Molecular evidence for the anamorph–teleomorph connection in *Cordyceps sinensis*. *Mycological Research, 105*(07), 827–832.
Lu, C., & Guo, L. (2009). *Septobasidium maesae* sp. nov. (Septobasidiaceae) from China. *Mycotaxon, 109*, 103.
Lucarotti, C. J. (1992). Invasion of *Aedes aegypti* ovaries by *Coelomomyces stegomyiae*. *Journal of Invertebrate Pathology, 60*(2), 176–184.
Lucarotti, C. J., & Klein, M. B. (1988). Pathology of *Coelomomyces stegomyiae* in adult *Aedes aegypti* ovaries. *Canadian Journal of Botany, 66*(5), 877–884.
Lucarotti, C. J., & Shoulkamy, M. A. (2000). *Coelomomyces stegomyiae* infection in adult female *Aedes aegypti* following the first, second, and third host blood meals. *Journal of Invertebrate Pathology, 75*(4), 292–295.
Martin, W. W. (1978). Two additional species of *Catenaria* (Chytridiomycetes, Blastocladiales) parasitic in midge eggs. *Mycologia*, 461–467.
Martin, W. W. (1981). *Couchia circumplexa*, a water mold parasitic in midge eggs. *Mycologia*, 1143–1157.

Martin, W. W. (2000). Two new species of *Couchia* parasitic in midge eggs. *Mycologia*, 1149–1154.

Massini, J. L. G. (2007). A possible endoparasitic chytridiomycete fungus from the Permian of Antarctica. *Palaeontologia Electronica, 10*(3), 2493MB.

Matsuura, K. (2005). Distribution of termite egg-mimicking fungi ("termite balls") in *Reticulitermes* spp. (Isoptera: Rhinotermitidae) nests in Japan and the United States. *Applied Entomology and Zoology, 40*(1), 53–61.

Matsuura, K. (2006). Termite-egg mimicry by a sclerotium-forming fungus. *Proceedings of the Royal Society B: Biological Sciences, 273*(1591), 1203–1209.

Matsuura, K., Tanaka, C., & Nishida, T. (2000). Symbiosis of a termite and a sclerotium-forming fungus: sclerotia mimic termite eggs. *Ecological Research, 15*(4), 405–414.

Matsuura, K., & Yashiro, T. (2010). Parallel evolution of termite-egg mimicry by sclerotium-forming fungi in distant termite groups. *Biological Journal of the Linnean Society, 100*(3), 531–537.

Matsuura, K., Yashiro, T., Shimizu, K., Tatsumi, S., & Tamura, T. (2009). Cuckoo fungus mimics termite eggs by producing the cellulose-digesting enzyme β-glucosidase. *Current Biology, 19*(1), 30–36.

McCreadie, J. W., & Adler, P. H. (1999). Parasites of larval black flies (Diptera: Simuliidae) and environmental factors associated with their distributions. *Invertebrate Biology*, 310–318.

McInnis, T., Jr., & Schimmel, L. E. (March 22, 1985). Host range studies with the fungus *Leptolegnia*, a parasite of mosquito larvae (Diptera: Culicidae). *Noblet Journal of Medical Entomology, 22*(2), 226–227.

McManus, W. R., & Youssef, N. N. (1984). Life cycle of the chalk brood fungus, *Ascosphaera aggregata*, in the alfalfa leafcutting bee, *Megachile rotundata*, and its associated symptomatology. *Mycologia*, 830–842.

Millay, M. A., & Taylor, T. N. (1978). Chytrid-like fossils of Pennsylvanian age. *Science, 200*(4346), 1147–1149.

Miller, J. H. (1938). Studies in the development of two *Myriangium* species and the systematic position of the order Myriangiales. *Mycologia, 30*(2), 158–181.

Mueller, G. M., & Schmit, J. P. (2007). Fungal biodiversity: what do we know? what can we predict? *Biodiversity and Conservation, 16*(1), 1–5. http://dx.doi.org/10.1007/s10531-006-9117-7.

Muma, M. H., & Clancy, D. W. (1961). Parasitism of purple scale in Florida citrus groves. *Florida Entomologist*, 159–165.

Nair, K. S. S., & McEwen, F. L. (1973). *Strongwellsea castrans* (Phycomycetes: Entomophthoraceae), a fungal parasite of the adult cabbage maggot, *Hylemya brassicae* (Diptera: Anthomyiidae). *Journal of Invertebrate Pathology, 22*(3), 442–449.

Nikoh, N., & Fukatsu, T. (2000). Interkingdom host jumping underground: phylogenetic analysis of entomoparasitic fungi of the genus *Cordyceps*. *Molecular Biology and Evolution, 17*(4), 629–638.

Ohshima, K. (1937). On the function of the polar filament of *Nosema bombycis*. *Parasitology, 29*(02), 220–224.

Padua, L. E., Whisler, H. C., Gabriel, B. P., & Zebold, S. L. (1986). *In vivo* culture and life cycle of *Coelomomyces stegomyiae*. *Journal of Invertebrate Pathology, 48*(3), 284–288.

Pelizza, S. A., López Lastra, C. C., Maciá, A., Bisaro, V., & García, J. J. (2009). Efecto de la calidad del agua de criaderos de mosquitos (Diptera: Culicidae) sobre la patogenicidad e infectividad de las zoosporas del hongo *Leptolegnia chapmanii* (Straminipila: Peronosporomycetes). *Revista de Biologia Tropical, 57*(1–2), 371–380.

Petch, T. (1921). Fungi parasitic on scale insects. *Transaction of the British Mycological Society, 7*(1–2), 18–40.

Petch, T. (1924). Studies in entomogenous fungi V *Myriangium*. *Transaction of the British Mycological Society, 10*, 45–80.

Petch, T. (1948). A revised list of British entomogenous fungi. *Transactions of the British Mycological Society, 31*(3), 286–304.
Phillips, A. J., Anderson, V. L., Robertson, E. J., Secombes, C. J., & van West, P. (2008). New insights into animal pathogenic oomycetes. *Trends in Microbiology, 16*(1), 13–19.
Pontoppidan, M. B., Himaman, W., Hywel-Jones, N. L., Boomsma, J. J., & Hughes, D. P. (2009). Graveyards on the move: the spatio-temporal distribution of dead *Ophiocordyceps*-infected ants. *PLoS One, 4*(3), e4835.
Prasertphon, S., & Tanada, Y. (1968). The formation and circulation, in *Galleria*, of hyphal bodies of entomophtoraceous fungi. *Journal of Invertebrate Pathology, 11*(2), 260–280.
Pringle, A., Patek, S. N., Fischer, M., Stolze, J., & Money, N. P. (2005). The captured launch of a ballistospore. *Mycologia, 97*(4), 866–871.
Pringsheim, N., Pfeffer, W., & Strasburger, E. (1858). *Jahrbücher für wissenschaftliche Botanik* (Vol. 1). Wilh. Engelmann.
Quandt, C. A., Kepler, R. M., Gams, W., Araújo, J. P. M., Ban, S., Evans, H. C.… Spatafora, J. W. (2014). Phylogenetic-based nomenclatural proposals for Ophiocordycipitaceae (Hypocreales) with new combinations in *Tolypocladium*. *IMA Fungus, 5*(1), 121.
Roberts, D. W., & Humber, R. A. (1981). Entomogenous fungi. *Biology of Conidial Fungi, 2*, 201–236.
Roy, H. E., Steinkraus, D. C., Eilenberg, J., Hajek, A. E., & Pell, J. K. (2006). Bizarre interactions and endgames: entomopathogenic fungi and their arthropod hosts. *Annual Review of Entomology, 51*, 331–357.
Samson, R. A., Evans, H. C., & Latgé, J. P. (1988). *Atlas of entomopathogenic fungi*. Springer-Verlag GmbH & Co. KG.
Samson, R. A., Ramakers, P. M. J., & Oswald, T. (1979). *Entomophthora thripidum*, a new fungal pathogen of *Thrips tabaci*. *Canadian Journal of Botany, 57*(12), 1317–1323.
Saunders, G. A., Washburn, J. O., Egerter, D. E., & Anderson, J. R. (1988). Pathogenicity of fungi isolated from field-collected larvae of the western treehole mosquito, *Aedes sierrensis* (Diptera: Culicidae). *Journal of Invertebrate Pathology, 52*(2), 360–363.
Scholte, E. J., Knols, B. G. J., Samson, R. A., & Takken, W. (2004). Entomopathogenic fungi for mosquito control: a review. *Journal of Insect Science, 4*(19), 1–24.
Schwartz, K. V. (1998). *Five kingdoms: An illustrated guide to the phyla of life on earth*. New York: WH Freeman.
Seymour, R. L. (1984). *Leptolegnia chapmanii*, an oomycete pathogen of mosquito larvae. *Mycologia*, 670–674.
Seymour, R. L., & Briggs, J. D. (1985). Occurrence and control of *Aphanomyces* (Saprolegniales: Fungi) infections in laboratory colonies of larval *Anopheles*. *Journal of the American Mosquito Control Association, 1*, 100–102.
Shaffer, R. L. (1975). The major groups of Basidiomycetes. *Mycologia*, 1–18.
Snowden, K. F., & Shadduck, J. A. (1999). *Microsporidia in higher vertebrates. The microsporidia and microsporidiosis* (pp. 393–417). Washington, DC: ASM.
Sokolova, Y. Y., Sokolov, I. M., & Carlton, C. E. (2010). New microsporidia parasitizing bark lice (Insecta: Psocoptera). *Journal of Invertebrate Pathology, 104*(3), 186–194.
Sparrow, F. K., Jr. (1960). *Aquatic phycomycetes*. USA: Arbor.
Spatafora, J. W., Sung, G. H., Sung, J. M., Hywel-Jones, N. L., & White, J. F. (2007). Phylogenetic evidence for an animal pathogen origin of ergot and the grass endophytes. *Molecular Ecology, 16*(8), 1701–1711.
Speare, A. T. (1921). *Massospora cicadina* peck: a fungous parasite of the periodical cicada. *Mycologia, 13*(2), 72–82.
Sprague, V., & Becnel, J. J. (1999). Appendix: checklist of available generic names for microsporidia with type species and type hosts. In M. Wittner, & L. M. Weiss (Eds.), *The Microsporidia and microsporidiosis* (pp. 517–530). Washington, DC: ASM Press.

Stephen, K., & Kurtböke, D. I. (2011). Screening of Oomycete fungi for their potential role in reducing the biting midge (Diptera: Ceratopogonidae) larval populations in Hervey Bay, Queensland, Australia. *International Journal of Environmental Research and Public Health, 8*(5), 1560–1574.

Su, X., Zou, F., Guo, Q., Huang, J., & Chen, T. X. (2001). A report on a mosquito-killing fungus, *Pythium carolinianum*. *Fungal Diversity, 7*(129), 33.

Suh, S. O., McHugh, J. V., Pollock, D. D., & Blackwell, M. (2005). The beetle gut: a hyper-diverse source of novel yeasts. *Mycological Research, 109*, 261–265. http://dx.doi.org/10.1017/s0953756205002388.

Sung, G. H., Hywel-Jones, N. L., Sung, J. M., Luangsa-ard, J. J., Shrestha, B., & Spatafora, J. W. (2007). Phylogenetic classification of *Cordyceps* and the clavicipitaceous fungi. *Studies in Mycology, 57*, 5–59.

Sung, G. H., Poinar, G. O., Jr., & Spatafora, J. W. (2008). The oldest fossil evidence of animal parasitism by fungi supports a Cretaceous diversification of fungal–arthropod symbioses. *Molecular Phylogenetics and Evolution, 49*(2), 495–502.

Sweeney, A. W., & Becnel, J. J. (1991). Potential of microsporidia for the biological control of mosquitoes. *Parasitology Today, 7*(8), 217–220.

Taylor, T. N., Remy, W., & Hass, H. (1992). Fungi from the lower Devonian Rhynie chert: Chytridiomycetes. *American Journal of Botany*, 1233–1241.

Thaxter, R. (1888). *The Entomophthoreae of the United States*.

Tiffney, W. N. (1939). The host range of *Saprolegnia parasitica*. *Mycologia, 31*(3), 310–321.

Travland, L. B. (1979). Initiation of infection of mosquito larvae (*Culiseta inornata*) by *Coelomomyces psorophorae*. *Journal of Invertebrate Pathology, 33*(1), 95–105.

Tsai, S. J., Lo, C. F., Soichi, Y., & Wang, C. H. (2003). The characterization of microsporidian isolates (Nosematidae: *Nosema*) from five important lepidopteran pests in Taiwan. *Journal of Invertebrate Pathology, 83*(1), 51–59.

Tulasne, L. R., & Tulasne, C. (1865). *Selecta Fungorum Carpologia III* (p. 221). Paris Museum.

Vega, F. E. (2008). Insect pathology and fungal endophytes. *Journal of Invertebrate Pathology, 98*(3), 277–279.

Vega, F. E., & Blackwell, M. (2005). *Insect-fungal associations: Ecology and evolution*. Oxford: Oxford University Press.

Vega, F. E., & Kaya, H. K. (2012). *Insect pathology*. Academic Press.

Vojvodic, S., Boomsma, J. J., Eilenberg, J., & Jensen, A. B. (2012). Virulence of mixed fungal infections in honey bee brood. *Frontiers in Zoology, 9*(1), 5.

Voos, J. R. (1969). Morphology and life cycle of a new chytrid with aerial sporangia. *American Journal of Botany*, 898–909.

Whisler, H. C., Zebold, S. L., & Shemanchuk, J. A. (1975). Life history of *Coelomomyces psorophorae*. *Proceedings of the National Academy of Sciences of the United States of America, 72*(2), 693–696.

Wittner, M., & Weiss, L. M. (1999). *Microsporidia and microsporidiosis*.

Wynns, A. A., Jensen, A. B., Eilenberg, J., & James, R. (2012). *Ascosphaera subglobosa*, a new spore cyst fungus from North America associated with the solitary bee *Megachile rotundata*. *Mycologia, 104*(1), 108–114.

Yashiro, T., & Matsuura, K. (2007). Distribution and phylogenetic analysis of termite egg-mimicking fungi "termite balls" in *Reticulitermes* termites. *Annals of the Entomological Society of America, 100*(4), 532–538.

Zattau, W. C., & McInnis, T., Jr. (1987). Life cycle and mode of infection of *Leptolegnia chapmanii* (Oomycetes) parasitizing *Aedes aegypti*. *Journal of Invertebrate Pathology, 50*(2), 134–145.

Zhang, K. Q., & Hyde, K. D. (2014). *Nematode-trapping fungi* (Vol. 23). Springer Science & Business.

CHAPTER TWO

Utilizing Genomics to Study Entomopathogenicity in the Fungal Phylum *Entomophthoromycota*: A Review of Current Genetic Resources

H.H. De Fine Licht*[,1], A.E. Hajek[§], J. Eilenberg* and A.B. Jensen*
*University of Copenhagen, Frederiksberg, Denmark
[§]Cornell University, Ithaca, NY, United States
[1]Corresponding author: E-mail: hhdefinelicht@plen.ku.dk

Contents

Abstract

The order Entomophthorales, which formerly contained c. 280 species, has recently been recognized as a separate phylum, *Entomophthoromycota*, consisting of three recognized classes and six families. Many genera in this group contain obligate insect-pathogenic species with narrow host ranges, capable of producing epizootics in natural insect populations. Available sequence information from the phylum *Entomophthoromycota* can be classified into three main categories: first, partial gene regions (exons + introns) used for phylogenetic inference; second, protein coding gene regions obtained using degenerate primers, expressed sequence tag methodology or de novo transcriptome sequencing with molecular function inferred by homology analysis; and third, primarily forthcoming whole-genome sequencing data sets. Here we summarize the current genetic resources for *Entomophthoromycota* and identify research areas that are likely to be significantly advanced from the availability of new whole-genome resources.

Advances in Genetics, Volume 94
ISSN 0065-2660
http://dx.doi.org/10.1016/bs.adgen.2016.01.003

1. INTRODUCTION

The dissolution of the polyphyletic phylum *Zygomycota* (Hibbett et al., 2007; Humber, 2012; James et al., 2006) and recent reclassification, where the former order Entomophthorales was raised to its own phylum of *Entomophthoromycota* (Humber, 2012), resolves much previous uncertainty regarding the status of these phylogenetically basal zygosporic fungi (Blackwell, 2011; James et al., 2006). *Entomophthoromycota* contains c. 280 species (Blackwell, 2011; Gryganskyi, Humber, Smith, et al., 2013; Gryganskyi et al., 2012) and currently includes three classes and six families (Gryganskyi, Humber, Smith, et al., 2013; Gryganskyi et al., 2012; Humber, 1989) (Fig. 1). Although life histories of species within this phylum are variable, many of the best-known and most common genera contain species that are obligate insect or mite pathogens with relatively narrow host ranges (Humber, 2008). For example, the species *Pandora neoaphidis* and *Entomophthora planchoniana* only infect different aphid species (Jensen, Hansen, & Eilenberg, 2008) and *Entomophaga maimaiga* infects certain lepidopteran larvae (Hajek, 2007, Fig. 2). Other species are more specific and only infect a single or very few species of insect hosts (eg, Diptera-infecting species within the genus *Strongwellsea* and *Entomophthora* (Eilenberg & Michelsen, 1999)). Species in the *Entomophthoromycota* usually create two types of spores (Fig. 2): (1) zygospores or azygospores (commonly called resting spores) that are usually produced within the carcasses of hosts, have thick walls, and can persist in the environment for many years, eg, between cycles of 17-year cicadas (Roy, Steinkraus, Eilenberg, Hajek, & Pell, 2006); (2) conidia, which are relatively short-lived and in some species multinucleate, are produced externally on cadavers and are often actively ejected. Conidia that land on a host can germinate to produce a germ tube. In some cases conidia can also form an appressorium for infection, or if they do not land on a host, they can produce another, slightly smaller conidium that is actively ejected and allows the fungus another chance to randomly land on a host and infect it (Roy et al., 2006). Hosts of arthropod-pathogenic *Entomophthoromycota* are diverse, and some fungal species are known to manipulate the timing of host death and their resulting sporulation so that this occurs during the night when humidity is high (Krasnoff, Watson, Gibson, & Kwan, 1995). In numerous host/pathogen systems, the hosts die in elevated locations, which is believed to be driven by the pathogen. These manipulations of host behavior prior to

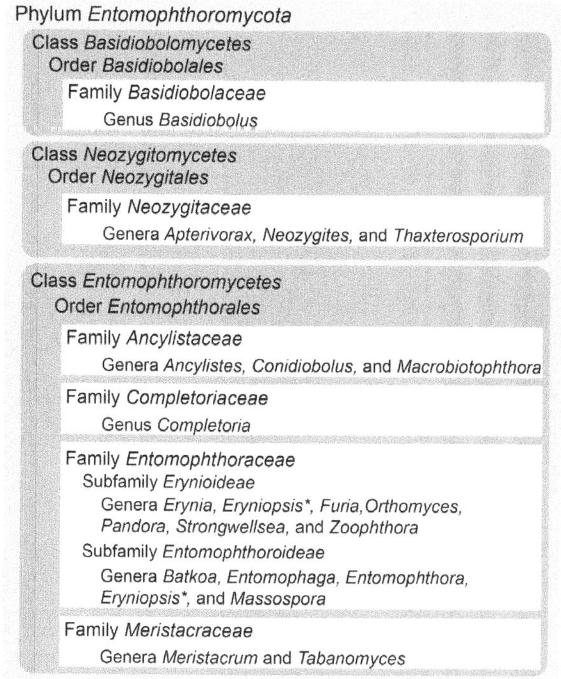

Figure 1 Classification of the phylum *Entomophthoromycota* with the c. 280 species distributed in three classes and six families (Gryganskyi, Humber, Smith, et al., 2013; Humber, 2012). The undescribed genus *Schizangiella* is not shown in this figure, but belongs to the *Basidiobolus* clade (Gryganskyi et al., 2012). * denotes the genus *Eryniopsis* which is paraphyletic between the two subfamilies *Erynioideae* and *Entomophthoroideae*, within *Entomophthoraceae*.

death improve the chance that discharged conidia will reach a healthy host to infect. The obligate pathogenic species do not grow outside of hosts in nature, yet some species are well known for their ability to cause epizootics in host populations.

Entomophthoromycota also contain primarily soil saprobic species in the genus *Conidiobolus* (eg, *Conidiobolus coronatus* and *Conidiobolus incongruus*) that can rarely infect mammals, including humans, and cause entomophthoromycosis (also called conidiobolomycosis) (Chayakulkeeree, Ghannoum, & Perfect, 2006; Jensen & Dromph, 2005; Prabhu & Patel, 2004; Ribes, Vanover-Sams, & Baker, 2000). Species in the genus *Basidiobolus* are soil saprobes as well but are also commonly known from the guts and feces of

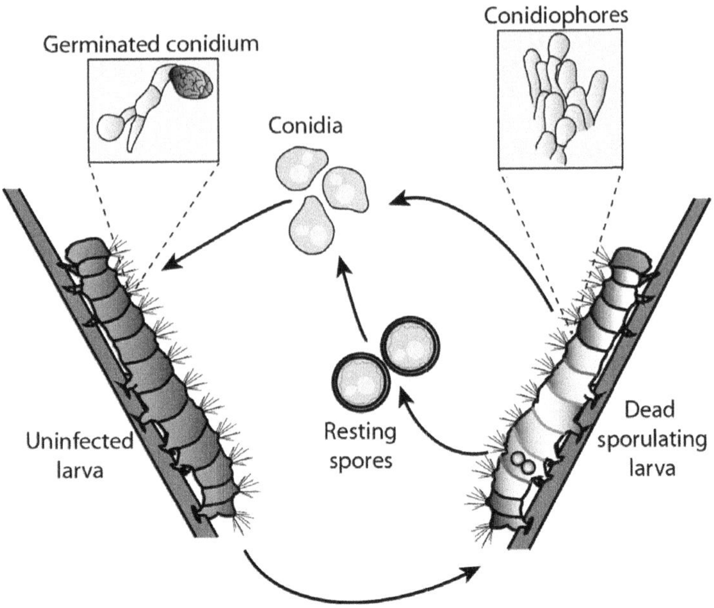

Figure 2 Drawing of the life cycle of *Entomophaga maimaiga*. Actively ejected conidia land on the uninfected host insect cuticle, germinate, and form an appressorium, which is a fungal cell enabling penetration of the cuticle. *Entomophaga maimaiga* grows inside the lepidopteran larva until it eventually kills the host, grows back outside of the body through the intersegmental membranes and the cuticle, and forms conidia-bearing conidiophores. Alternatively or in addition to conidial formation, *E. maimaiga* may develop thick-walled resting spores (azygospores) inside the dead larva, which often dies hanging onto a surface above ground level. The insect carcass containing resting spores eventually falls to the soil, the cuticle breaks, and resting spores are released onto and into the soil, where they may lie dormant from one to many years. (See color plate)

amphibians and reptiles (Manning, Waters, & Callaghan, 2007). *Basidiobolus ranarum* may occasionally infect humans and cause basidiobolomycosis (Jensen & Dromph, 2005; Vikram, Smilack, Leighton, Crowell, & De Petris, 2012). Additional genera contain species that are pathogens of freshwater green algae, fern prothallia, nematodes, and tardigrades (Bałazy, 1993). These examples together exemplify the diverse and highly specialized pathogenic niches occupied by species in the *Entomophthoromycota*, emphasizing that this group of fungi contains pathogenic members with varied hosts, which would require numerous methods of pathogenicity. Elucidation of the infection processes and pathogenicity by both classical biological methods and modern genomic tools is thus desirable, because it provides clues for understanding the main elements of host–pathogen evolution and the function of virulence genes in pathogenic fungi.

Several genome-sequencing projects have been initiated among the species with diverse life histories within *Entomophthoromycota*, and these will undoubtedly provide new insights into the biologies of species and will aid phylogenetic analysis of this basal group within the fungal kingdom. Here we summarize and review the genetic information, albeit limited, that currently exists for *Entomophthoromycota*, in order to provide a foundation that new genomic and transcriptomic information can build upon.

2. GENETIC TOOLS USED FOR PHYLOGENETIC INFERENCE, EVOLUTION, AND EPIZOOTIOLOGY

Early work examining evolution and host specificity within the *Entomophthoromycota* was based on restriction fragment length polymorphisms (RFLPs), allozymes, and DNA probes developed through RFLPs, to investigate the genus *Entomophaga*, with emphasis on the *Entomophaga aulicae* species complex associated with lepidopteran larvae (Hajek, Butler, et al., 1996; Hajek, Humber, Walsh, & Silver, 1991; Hajek, Walsh, Strong, & Silver, 1996; Walsh, 1996). *Zoophthora* spp. and *Entomophaga grylli* pathotypes from leafhopper, weevil, and grasshopper host species were identified and distinguished based on RAPD markers (Bidochka et al., 1995; Hajek, Hodge, Liebherr, Day, & Vandenberg, 1996; Hodge, Sawyer, & Humber, 1995). Questions about the origin of the emergent entomophthoralean *E. maimaiga* in North America were addressed using allozymes and RFLPs (Hajek et al., 1990). AFLP (amplified fragment length polymorphism) markers were used to suggest that *E. maimaiga* from Japan was responsible for 1989 epizootics in North American gypsy moth, *Lymantria dispar*, populations and the source for the *E. maimaiga* establishment in North America (Nielsen, Milgroom, & Hajek, 2005). These studies also demonstrated that strains of the introduced fungus are very homogeneous compared to native isolates from Japan, northeastern China, and Far Eastern Russia.

Today, the vast majority of available genomic information within the *Entomophthoromycota* is composed of partial gene and intron sequences developed for use in phylogenetic analyses. Most sequences deposited in the National Center for Biotechnology Information (NCBI) GenBank database are from the nuclear ribosomal DNA region, including the large (LSU) and small (SSU) subunits, 5.8S, and internal transcribed spacer (ITS) regions 1 and 2 (Table 1). The database also includes various gene regions that are frequently now used in fungal phylogenetics such as the RNA-polymerase

Table 1 The 964 *Entomophthoromycota* records in the National Center for Biotechnology Information (NCBI) nucleotide database (excluding sequence data from unknown genome regions)

Genus	SSU	LSU	5.8s-ITS1	ITS1-5.8s-ITS2	5.8s-ITS2	mt-SSU	alpha-tubulin	beta-tubulin	mt-atp6	MCM7	RPB1	RPB2	EF1-a	EF2-a	EFL	actin (ACT1)	G3P	GND1	TSR1	Proteases	Chitin synthase genes	Polyketide synthases	Fatty acid desaturases	EST-annotated genes	conserved DNA	repetitive DNA	SUM
Conidiobolus	49	71		16		22	3	3		1	2	19	30	1	1	1			1	3			2	1	13		239
Basidiobolus	18	30		30	6	7	2	5	21		4	20	28	2	28	29				1	1				2		234
Pandora	14	27		23		4		19			19	27	19									1		9			162
Entomophthora	25	24	25	2	22	1		2		1	1	2	1			2	10	1	1					11			131
Zoophthora	9	8	7	9	2	7						6								3		1					52
Entomophaga	4	3		4	11	3	2	2		1		1									2					6	39
Furia	7	6				5						6															24
Batkoa	7	7				5						4															23
Erynia	6	6				5						5															22
Neozygites	14																					1					15
Schizangiella	2	3		4								3															12
Eryniopsis	3	1				1						1															6
Massospora	2	1				1																					4
Strongwellsea	1																										1
Macrobiotophthora	1																										1

Data retrieved 28.07.15.

II genes (*RPB1* and *RPB2*), elongation factor 1-alpha (*EF1-alpha*), and others (Table 1). These regions have been used in single and multigene phylogenies of *Entomophthoromycota* (Einax & Voigt, 2003; Gryganskyi et al., 2012; James et al., 2006; Jensen, Gargas, Eilenberg, & Rosendahl, 1998; Keeling, 2003; Liu & Voigt, 2010; Nagahama, Sato, Shimazu, & Sugiyama, 1995; Schussler, Schwarzott, & Walker, 2001; Tanabe, 2000; Tanabe, Saikawa, Watanabe, & Sugiyama, 2004). The phylogenetic scope of currently sequenced *Entomophthoromycota* genes is also evident from the broad taxonomic spread of gene sequences in the NCBI GenBank, with many taxa represented. For some genera (eg, *Strongwellsea*, etc., Table 1), the only genetic information present are partially sequenced ITS, LSU, and/or SSU gene regions used for phylogenetic studies. Within the genus *Neozygites*, the only molecular data that have been reported are SSU-rDNA gene sequences that were used to delineate the closely related species *Neozygites tanajoae* from the common mite pathogen *Neozygites floridana* (Delalibera, Hajek, & Humber, 2004). The enigmatic form genus *Tarichium* is absent in our survey on available gene sequence data in the NCBI GenBank database (Table 1). This form genus contains species that are not known to produce conidia, which is a life stage required for traditional morphological and phylogenetic analyses. Therefore, because species in *Tarichium* are only known from environmentally sampled resting spores, their phylogenetic associations are not known and the form genus *Tarichium* has been used (Keller, 1991). Difficulty with DNA extraction from thick-walled resting spores has likely hindered studies of sampled *Tarichium* resting spores, although DNA has been successfully extracted from resting spores from other genera (Castrillo, Thomsen, Juneja, & Hajek, 2007). Many *Tarichium* resting spores may turn out to be the resting spore stage of new species that do not produce conidia or already recognized species within *Entomophthoromycota* for which resting spores have not been found (Gryganskyi, Humber, Smith, et al., 2013; Gryganskyi et al., 2012). However, genera such as *Conidiobolus* and *Basidiobolus* that contain species capable of infecting humans (and likely therefore are some of the best studied) have 70 and 30 LSU-rDNA sequences in the NCBI GenBank, respectively (Table 1).

Similarly the genera *Entomophthora* and *Pandora* in the large family *Entomophthoraceae* have many sequences deposited in GenBank due to several population genetic studies analyzing community structure and host specialization at varying geographical scales (Jensen & Eilenberg, 2001; Jensen, Eilenberg, & Lopez Lastra, 2009; Jensen et al., 2008; Jensen, Thomsen, & Eilenberg, 2001, 2006; Lihme, Jensen, & Rosendahl, 2009). One population biology study has intensively investigated the population structure of *Entomophthora muscae* in fly populations during springtime epizootics within one geographic area (Gryganskyi, Humber, Stajich, et al., 2013). Multilocus sequence typing demonstrated two clades of *E. muscae*, with several haplotypes in each. One clade principally infected *Delia radicum* during March and early April, and the second clade principally infected *Coenosia tigrina* during late April and May. These flies are in different dipteran families, yet members of each clade were found infecting both fly species, although with distinctly increased infection in only one species. Thus, two distinct subpopulations of *E. muscae* occurred, with little exchange of genes, in the same geographical location but with slightly different phenology (Gryganskyi, Humber, Stajich, et al., 2013). Similar cryptic host-specific variation has also been found among *E. aulicae* strains using RFLP-derived DNA probes (Walsh, 1996). From a sample of 28 *E. aulicae* isolates from *Choristoneura fumiferana* (eastern spruce budworm) 27 isolates were genetically distinct but closely related, whereas *E. aulicae* isolates from another host *Lambdina fiscellaria* (the hemlock looper) were much more genetically variable and perhaps consisted of several cryptic species (Walsh, 1996).

Gene fragments used in phylogenetic analysis have also been exploited in molecular diagnostic assays that have been developed to monitor human pathogenic *Entomophthoromycota* causing entomophthoromycosis. Species-specific primers within the LSU-rDNA region have been developed for the clinically relevant *C. coronatus*, *B. ranarum*, and *B. haptosporus* to aid in clinical diagnosis using a PCR-based assay (Voigt, Cigelnik, & O'Donnell, 1999). PCR-based assays have also been developed for entomophthororalean fungi with potential as biological control agents to monitor presence, prevalence, and spread in natural environments. For example, specific primers in two regions of ITS and SSU-rDNA were used in a cultivation-independent PCR-based assay to detect *P. neoaphidis* in soil samples from stinging nettle fields in Switzerland (Fournier, Enkerli, Keller, & Widmer, 2008), and this work was later extended to include 13 SNPs in six genomic regions (LSU, RPB1, RPB2, BTUB, EFL, SSU including ITS) (Fournier, Widmer, & Enkerli, 2010). Specific primers and RFLP analysis of ITS-rDNA have been used to investigate the relationship of species in the genera *Eryniopsis* and *Entomophaga* (Hajek, Jensen, Thomsen, Hodge, & Eilenberg, 2003) and to identify resting spores to link them with conidial phases, completing our understanding of the full life cycles of *Entomophthora* and *Pandora* species (Scorsetti, Jensen, Lopez Lastra, & Humber, 2012; Thomsen & Jensen, 2002). Specific primers were also used to verify in vitro isolates of the fastidious aphid pathogenic fungus *E. planchoniana*, obtained from protoplasts released from surface sterilized, living, infected aphids (Freimoser, Jensen, Tuor, Aebi, & Eilenberg, 2001). RFLP analysis of ITS-rDNA has also been used to detect dual infections of *Zoopthora radicans* and *Pandora blunckii* in the diamondback moth, *Plutella xylostella* (Guzmán-Franco, Atkins, Alderson, & Pell, 2008; Morales-Vidal, Alatorre-Rosas, Clark, Pell, & Guzmán-Franco, 2013) and, based on ITS-rDNA sequence information, a real-time PCR assay to quantitatively monitor infections in this system was developed and implemented (Guzmán-Franco, Atkins, Clark, Alderson, & Pell, 2011). ITS-rDNA sequence information has similarly been used to develop a real-time PCR assay based on fluorescence-labeled probes to detect and quantify *E. maimaiga* resting spores in the organic layer of forest soils (Castrillo et al., 2007).

3. HOST–PATHOGEN INTERACTIONS

Entomophthoralean fungi, like the majority of insect-pathogenic fungi (Boomsma, Jensen, Meyling, & Eilenberg, 2014), invade their hosts directly through the cuticle, propagate inside their hosts, and usually produce new infective conidia outside the cuticle upon host death. We are only recently beginning to understand the molecular mechanisms that come into play during this process. Scanning and transmission electron microscopy have shown the ultrastructure of conidial formation, discharge, and adhesion to host cuticle of species within the genus *Entomophthora* (Eilenberg, Bresciani, & Latgé, 1986; Eilenberg, Bresciani, Olesen, & Olson, 1995) and have also documented wide zones of complete histolysis in cuticle beneath penetrant hyphae of, for example, *Conidiobolus obscurus* infecting pea aphids, *Acyrthosiphon pisum* (Brey, Latgé, & Prevost, 1986). During the initial phases of infection the arthropod-pathogenic fungi utilize cuticle-degrading enzymes when infective spores germinate and penetrate the cuticle. Many cuticle-degrading enzymes are also active at the final stages of infection when the fungi need to grow out of the host to sporulate (Charnley, 2003).

Several proteases like the metalloprotease *ZrMEP1* and a trypsin-like serine protease *ZrSP1* that facilitate penetration of insect cuticle during infection have been cloned from *Z. radicans* (Xu, Baldwin, Kindrachuk, & Hegedus, 2006). Using degenerative primers trypsin-like serine protease-coding genes have also been found in *C. coronatus, Conidiobolus lamprauges, C. obscurus, Neozygites parvispora, Z. radicans*, and *B. ranarum* (Hu & St. Leger,

2004). With similar methods, two chitin synthase genes, *EaCHS1* and *EaCHS2*, which are active during hyphal growth, but not in protoplasts, have been characterized in *E. aulicae* (Thomsen & Beauvais, 1995). The biochemical properties, quantities, and types of the chitin, lipid, and protein molecules that make up insect cuticles vary substantially between insect species, tissues, and developmental stages, and differences in cuticular proteins may mediate insecticide resistance, drought tolerance, and differentiate sibling insect species (Willis, 2010). The generally narrow host ranges of species of insect-pathogenic *Entomophthoromycota*, coupled with the necessity for infective spores to penetrate the host cuticle, suggest that kinetic differences in cuticle-degrading enzymes could mediate host specificity. The metalloprotease *ZrMEP1* and trypsin-like serine protease *ZrSP1* from a *Z. radicans* strain highly infective towards cabbage butterfly (*Pieris brassicae*) were compared to enzymes produced by the *Z. radicans* strain after serial passages through a novel host, the diamondback moth (Xu et al., 2006). The coding regions of these enzymes were the same in the evolved and parent *Z. radicans* strains, possibly because experimental adaptation to the novel host only involved five serial passages and this, despite the strong selection pressure exerted by the novel host, was most likely insufficient to generate detectable differences. Likewise, expression levels of these proteases during infection of diamondback moths were unaltered in the evolved strain, leading the authors to conclude that these two proteases may not be major host specificity determinants (Xu et al., 2006). The cuticle can be viewed as the first line of disease defense in insects; thus, differences in cuticular components among host species are likely strong selective forces in host adaptation of obligate insect-pathogenic *Entomophthoromycota*. Future genomic studies, comparing entomophthoromycotan with hypocrealean insect-pathogenic fungi, could shed light on whether the fungal enzymatic machinery used to penetrate insect cuticles primarily consists of a conserved and widespread set of core enzymatic elements, which is similar among or within these very different groups of insect-pathogenic fungi, or, alternatively, is pathogen specific.

It is generally believed that the *Entomophthoromycota* are fast growing and overcome susceptible hosts by using all available nutrients in the hemolymph before beginning to colonize tissues and eventually the entire living host (Evans, 1989). Sporulation usually occurs relatively rapidly following host death and a saprophytic phase is minimal or completely absent, a strategy that may allow limited capacity for secondary metabolite production, such as toxins (Boomsma et al., 2014). In the ascomycete plant pathogen *Magnaporthe oryzae*, polyketides are involved in appressorium formation at the initial stages of infection (Möbius & Hertweck, 2009). Polyketides and many other biologically active secondary compounds are also produced by the polyketide biosynthetic pathway in ascomycete insect pathogens (see Donzelli & Krasnoff, 2016). Using degenerative primers targeting conserved domains, partial polyketide synthase (PKS)-encoding genes were identified in only *Pandora delphacis* and *Zoophthora lanceolata* from a set of 30 *Entomophthoromycota* species (Lee et al., 2001), suggesting that PKS genes are not widespread among species of *Entomophthoromycota*. *Entomophaga aulicae* produces a host cell-lytic factor with toxin-like effects just during the terminal stages of infection in *C. fumiferana* and *Trichoplusia ni* larvae, with a peak in activity around host death. The toxic activity is short-lived and has been suggested to be the causal agent for host death (Milne, Wright, Welton, & Budau, 1994), but whether this effect is due to secondary compounds or secreted enzymes remains unclear (also see discussion by Charnley (2003)). Secreted enzymes, and not toxins in the form of secondary compounds, may also be the source of the toxic effects previously reported in culture filtrates of *E. aulicae*, causing rapid paralysis and death of the host (Dunphy & Nolan, 1982), and in culture filtrates of *Batkoa apiculata* and *C. coronatus* (Yendol, Miller, & Behnke, 1968). Only species of some of the most basal lineages of *Entomophthoromycota*, eg, *Conidiobolus thromboides* and *C. coronatus* have unambiguously been shown to produce toxic secondary metabolites, as low-molecular weight azoxybenzenoid compounds: 4,4′-azoxybenzene dicarboxylic acid and 4,4′-hydroxymethyl azoxybenzene (Claydon, 1978). However, all of these studies investigating secondary compounds that are potential toxins rely on in vitro cultures. The production and function of these compounds during associations with hosts have not been studied, and there is thus no evidence for secondary metabolite production in the obligate insect-pathogenic lineages within *Entomophthoromycota*. Comprehensive expressed sequence tag (EST) analyses of *Z. radicans* have generated 899 unique sequences, of which 85% could be annotated (Xu, Baldwin, Kindrachuk, & Hegedus, 2009). A similar EST approach was used to obtain 1831 cDNA sequences from *C. coronatus*, which showed many transcripts involved in intermediate metabolism facilitating rapid growth (Freimoser, Screen, Hu, & St. Leger, 2003). Both of these EST gene expression analyses used laboratory systems where *Z. radicans* and *C. coronatus* were cultivated on specific hosts or insect cuticle, respectively. A later study by Grell, Jensen, Olsen, Eilenberg, and Lange (2011) used transposon-assisted signal sequence trapping (Becker et al., 2004), which excludes internal metabolites to exclusively look at secreted enzymes, to analyze field-collected aphids infected with either *P. neoaphidis, E. planchoniana,* or *C. obscurus.* Several transcripts were identified with sequence homology to extracellular secreted peptidases, glycoside hydrolases, lipases, glycosyl transferases, and a carbohydrate esterase and fatty acid desaturase, showing extensive capacity for penetrating the insect cuticle and rapid utilization of energy compounds in the insect hemolymph (Grell et al., 2011). A recent study characterized the interaction transcriptome of *Pandora formicae* in field-collected infected *Formica rufa* wood ants that were either sporulating or collected just prior to sporulation (Małagocka, Grell, Lange, Eilenberg, & Jensen, 2015) and found numerous subtilisin- and trypsin-like proteases and chitin metabolism-coding genes to be highly regulated during infection (Małagocka et al., 2015). Interestingly, specific lipase-coding genes with homology to known extracellular virulence factors involved with nutrient acquisition and disruption of host membranes in plant pathogenic fungi were also identified (Małagocka et al., 2015), which exemplifies the fundamental role of lipases for fungal virulence in pathogenic fungi irrespective of host kingdom.

Some insect-pathogenic *Entomophthoromycota*, including *E. maimaiga* and some Diptera-infecting members of the genus *Entomophthora*, proliferate within the host insect as protoplasts (which have also been documented in vitro), without producing a cell wall, and these structures are thought to functionally avoid detection by a host immune system that generally is triggered by cell wall epitopes (Beauvais, Latgé, Vey, & Prevost, 1989; Bidochka & Hajek, 1996; Latgé, Eilenberg, Beauvais, & Prevost, 1988). This intricate mechanism may have been a contributing factor for the evolution of a generally narrow host range within this group (Boomsma et al., 2014). Species that start infections as protoplasts generate cell walls later in infection and then grow as hyphal bodies and hyphae (Lopez Lastra, Gibson, & Hajek, 2001), before penetrating the intersegmental membranes from within and forming conidiophores externally. This transition is also evident from several EST and transcriptome data sets, where extensive regulation of transcripts involved with cell wall metabolism such as chitin synthases, chitinases, and chitin deacetylases have been documented in *P. neoaphidis, P. formicae, E. planchoniana,* and *C. obscurus* (Grell et al., 2011; Małagocka et al., 2015). It thus seems clear that comprehensive cell wall modulation takes place during infections by species of *Entomophthoromycota* to ensure correct development of morphological pathogenic structures, although how host cues may trigger fungal morphogenesis is unknown. Because the transition from protoplasts to cells with cell walls occurs toward the end of an infection, when most of the host tissues are utilized, an obvious possibility is that the depletion of certain nutritional substrates functions as a cue for the pathogenic *Entomophthoromycota* to initiate morphogenesis. But more intricate molecular interactions, perhaps involving osmotic pressure, the host immune response or host cell-lytic factors, may also be involved.

The *Entomophthoromycota* includes species that manipulate the behavior of the host so that the host dies at elevated positions (Roy et al., 2006; Hughes et al., 2016), the molecular mechanisms of fungal-induced "summit disease" in infected host insects are unknown but may further support the evolution of specialist pathogens with narrow host ranges within this group. However, similar extended phenotypes and changes in host behavior have been reported to occur in a number of insects infected with the ascomycete *Cordyceps*, viruses, and even a trematode (Roy et al., 2006; Van Houte, Ros, & van Oers, 2013), suggesting that these pathogens and parasites are triggering a preprogrammed insect behavior.

4. GENOME CHARACTERISTICS

Entomophthoromycota boasts one of the largest fungal genomes ever measured, at 8000 Mb in *E. aulicae*, which correlates with microscopic observations of extensive condensed chromatin in the nuclei (Murrin, Holtby, Noland, & Davidson, 1986). Also, *B. ranarum* has a large haploid genome of 350 Mb (Henk & Fisher, 2012), compared to an average genome size of around 40 Mb in the kingdom Fungi (Gregory et al., 2007; Henk & Fisher, 2012). Earlier microscopic analyses have found numerous chromosomes in *B. ranarum*, ranging from 60 to more than 500 (Olive, 1907; Robinow, 1963; Sun & Bowen, 1972), but it is unclear whether ploidy level or genome duplication governs genome size in *B. ranarum* (Henk & Fisher, 2012). In general, entomophthoromycotan genomes are considered to be haploid (Gryganskyi & Muszewska, 2014; Humber, 2012), and the basal chromosome count in *Entomophthoromycota* appears to be 8 (Humber, 1982), but 12, 16, and 32 have also been estimated (Riddle, 1906; Sawyer, 1933) and reviewed by Humber (1982). The large nuclei with numerous chromosomes and condensed chromatin seen in several species suggest that large genomes may be an ancestral trait within *Entomophthoromycota*, but less than 10 out of c. 280 species have been analyzed for either chromosome count or genome size, making it unclear whether having a large genome is unusual or the norm within the *Entomophthoromycota*. However, the first genome assembly (ver 1.0) within the *Entomophthoromycota* was the phylogenetically basal *C. coronatus*, which was sequenced with 454-technology by the US Department of Energy Joint Genome Initiative (JGI). *Conidiobolus coronatus* has 10,635 predicted genes and a genome size of 39.9 Mb (Chang et al., 2015), which is similar to the average fungal genome size of 40 Mb. The genus *Conidiobolus* is paraphyletic, consisting of one clade, exemplified by *C. coronatus*, including soil-living saprotrophs that appear to be facultative insect pathogens and another clade, exemplified by *C. thromboides*, which includes species that are primarily insect pathogens (Gryganskyi, Humber, Smith, et al., 2013; Gryganskyi et al., 2012). This may suggest that the larger genomes reported for *E. aulicae* (Murrin et al., 1986) and *B. ranarum* (Henk & Fisher, 2012) could be associated with specialization to obligate insect pathogenicity. This has been observed in some plant pathogenic fungi where the expansion of specific gene families and/or whole-genome duplication has been found to be associated with host specialization (Raffaele & Kamoun, 2012). If these inferences are correct it implies that the evolution of the large haploid genome of 350 Mb in *B. ranarum* is independent from potential genome expansion events within the order *Entomophthorales* and might be driven by adaptation to amphibian and reptile guts. This is supported by recent phylogenetic analyses where the genus *Basidiobolus* is the sister group to all other *Entomophthoromycota* and is placed somewhere between *Entomophthoromycota* and chytrid fungi (Gryganskyi, Humber, Smith, et al., 2013; Henk & Fisher, 2012). We therefore discuss the *Basidiobolus* genus in this chapter although it might not permanently be included in the *Entomophthoromycota*.

From the limited genomic information available for a handful of species within the *Entomophthoromycota*, a few interesting characteristics are emerging. Genome evolution responsible for host adaptation and specialization of the nonentomophthoroid mucoralean *Rhizopus delemar* that causes human mucormycosis is based on a species-specific whole-genome duplication event (Gryganskyi et al., 2010; Ma et al., 2009). Together with microscopic observations of enlarged nuclear contents in many *Entomophthoromycota*, this could suggest that genome duplications may be widespread and a common molecular evolutionary mechanism among basal fungi. However, comparisons with other mucoralean genomes showed that the whole-genome duplication in *R. delemar* is species specific (Ma et al., 2009). Sequencing of other basal fungi, including the entire genome of *C. coronatus* (Chang et al., 2015), suggests instead that among basal fungi, including *Entomophthoromycota*, individual gene duplication, such as tandem duplication, appears to be the main mechanism for gene expansion and evolutionary adaptation (Shelest & Voigt, 2014). The observed gene family expansions for serine proteases and chitinases in *P. formicae* is in concordance with individual gene duplication events (Małagocka et al., 2015), but whether these duplications are tandem duplications, as found in many other basal fungi (Shelest & Voigt, 2014), is not yet known.

5. INSIGHTS TO BE GAINED FROM *ENTOMOPHTHOROMYCOTA* GENOMIC RESOURCES

Forthcoming genomic sequence information for species of *Entomophthoromycota* will provide the potential for comparative analyses that undoubtedly will increase our knowledge on the biology of these fungi (Table 2). As of July 2015, one published (*C. coronatus* (NRRL 28638) (Chang et al., 2015)) and eight incomplete (*C. incongruus* (B7586), *C. thromboides* (FSU 785), *Basidiobolus meristosporus* (B9252 and CBS 931.73), *B. heterosporus* (B8920), *Z. radicans* (ARSEF 4784)), *E. maimaiga* (ARSEF 7190), and *E. muscae* (HHDFL130914-01) genome sequencing projects were listed in the Genomes OnLine Database (GOLD) (Reddy et al., 2015). This list is almost certainly incomplete, and more genome and transcriptome sequencing projects are likely being planned or are already under way, which will expand the phylogenetic breadth and include more entomophthoromycotan species of obligate insect-pathogenic fungi. This will allow comparative analysis of how natural selection on the genome influences gene family expansion and reduction, which may have governed transitions from soil saprobe to pathogenic lifestyles (Table 2).

Genomes from additional species of *Entomophthoromycota* will provide new information on the molecular details of host—pathogen interactions, both in obligate insect-pathogenic lineages and in the phylogenetically basal opportunistic human pathogens. Among insect-pathogenic fungi, the Hypocreales (Ascomycota) is by far the best-studied group and includes several strains (eg, *Metarhizium brunneum* F52) being used in biological control of insect pests today (Vega, Meyling, Luangsa-ard, & Blackwell, 2012). The use of new molecular methods and genomic data has provided considerable insight into the molecular details underlying hypocrealean infection processes and interactions with the host immune system (reviewed in Wang & St. Leger, 2014 and by Wang, St. Leger, & Wang, 2016). Many hypocrealean insect-pathogenic fungi are generalists attacking a wide range of insect hosts, and whole-genome sequencing of this group has, for

Table 2 Research areas that will benefit from genomic *Entomophthoromycota* resources

Scientific realm	Research question
Infection biology	Which virulence factors do pathogenic entomophthoromycotan species contain?
	How do secreted molecules interfere with hosts?
	To what extent do pathogenic species of *Entomophthoromycota* rely on toxins during infection?
	Is the absence of cell walls in pathogenic species of *Entomophthoromycota* enough to evade the host immune system?
	Do pathogenic species of *Entomophthoromycota* actively suppress the host immune response?
	What is the molecular mechanism governing behavioral manipulation of insect hosts and causing "summit disease"?
Life history adaptation	What is the enzymatic capacity to obtain nutrients from the hosts in the obligate biotrophic *Entomophthoromycota*?
	Which genes are involved in the morphogenesis from protoplasts to hyphal bodies or hyphae in the entomophthoromycotan species with this transition?
	How is the capacity for a species to grow saprotrophically in soil as well as pathogenically in insects and vertebrates reflected in the gene family content in the genus *Conidiobolus*?
	Can knowledge of the genome-wide inventory of enzyme-coding genes be used to formulate or improve artificial media for growth and conidial production?
Phylogeny	Is the evolution of *Entomophthoromycota* primarily driven by host-shift speciation or cospeciation events?
	How are the major taxonomic groups related at the deep nodes in the phylogeny?
	Are there gene families with significantly increased or decreased evolutionary mutation rates in species of *Entomophthoromycota*?
Ecology	Are significant changes in prevalence levels (endemic versus epidemic) in host—pathogen systems reflected in entomophthoromycotan gene expressions?
	How do two (or more) naturally occurring *Entomophthoromycota* infections in the same host species influence the genetic population structure of one another?

example, recently shown that a reduced total repertoire of genes and high expression of specific G protein—coupled receptors are associated with adaptation to a comparatively more narrow (locust-specific) host range in *Metarhizium acridum* when compared to the generalist *Metarhizium robertsii* (Gao et al., 2011; Hu et al., 2014). Similar insights into the infection processes of the generally more host-specific insect pathogens within the *Entomophthoromycota* are expected as comparative genomic analyses become possible (Table 2). The potential of species in the *Entomophthoromycota* as biological control agents is generally regarded to be highly questionable (ie, largely due to difficulty in mass production and storage), but a better understanding of the molecular mechanisms underlying host specificity may provide important information that could allow for use of species that can be mass produced, in cases where targeted pest species-specific control measures are desired (Table 2).

Fungi are osmotrophs and live by secreting enzymes into their immediate environment, and this implies that having an inventory of secreted proteins to a large extent might allow ecological niche(s) to be predicted for different species. For obligate pathogenic fungi, the ecological niche is composed of the host, and secreted proteins thus directly interact with host cells and proteins. Comparing all of the active genes (ie, transcriptomes) between pathogen growth stages or between fungi with different ecological niches (eg, pathogenic vs nonpathogenic or pathogenic fungi infecting different hosts) can therefore elucidate core pathogenicity genes in pathogenic fungi. This approach has been used to identify two expanded protein families of serine peptidases (S41) and fungalysin metallopeptidases (M36), where the latter is involved in keratin degradation and processing in the chytrid fungus *Batrachochytrium dendrobatidis* that has been decimating amphibian populations with alarming speed (Fisher et al., 2012; Rosenblum, Stajich, Maddox, & Eisen, 2008). Fungalysins are also expanded and expressed during pathogenic growth in the Dermatophyte fungi in the genera *Trichophyton* and *Arthroderma benhamiae*, which infect human nails, skin, and hair (Burmester et al., 2011; Martinez et al., 2012), highlighting the convergent use of fungalysins in fungi living in keratinaceous environments. More generally, expanded numbers of metalloprotease gene families appear not to be restricted to only human- and amphibian-associated pathogens but have also been discovered in the insect pathogens *Cordyceps militaris* (Zheng et al., 2011) and *Metarhizium* spp. (Gao et al., 2011; Hu et al., 2014), as well as in the nematode-trapping fungi *Arthrobotrys oligospora* (Yang et al., 2011) and *Monacrosporium haplotylum* (Meerupati et al., 2013). Interestingly, a recent study comparing three nematode-trapping fungi found that only a few subtilisin proteases (containing the peptidase_S8 domain) out of the up to 50 subtilisin gene copies in the *A. oligospora* genome were highly expressed during infections (Andersson et al., 2014). The entomophthoralean *P. formicae* has recently been found to have at least 26 putative subtilisin genes and several fungalysin metallopeptidases, and many are differentially expressed during the final stages of infection (Małagocka et al., 2015). These characteristics indicate that similar genomic evolutionary routes of pathogenicity exist toward animal hosts and may be replicated when more genomic information about species of *Entomophthoromycota* becomes available.

Species of *Entomophthoromycota* belong to some of the earliest diverging clades within the kingdom Fungi, which clearly sets them apart from the species-rich and well-studied Ascomycota and Basidiomycota in the subkingdom Dikarya (James et al., 2006). The basal fungi are highly underestimated in number of species and can contribute significantly to interspecific dynamics in all domains of life. A better understanding of the biology of basal fungi may also provide new targets for antifungal drug development, an increasing necessity because patients suffering from mucormycosis often respond poorly to standard treatments (Brown, Denning, et al., 2012; Brown, Cornforth, & Mideo, 2012). The recently increased focus on entomophthoromycotan genome characteristics will elucidate fundamental evolutionary mechanisms behind adaptation to selection (Shelest & Voigt, 2014), which will be important for interpreting general evolutionary insights obtained from fungal evolutionary genomics (Gladieux et al., 2014). However, the challenge of interpreting gene function predictions in *Entomophthoromycota* via comparisons of orthology with already sequenced organisms may be a daunting task. Even with more comprehensively developed molecular and analytical toolkits, the phylogenetic distance separating *Entomophthoromycota* and, for example, the Dikarya (Ascomycota and Basidiomycota) is so great that only the most conserved genes can be paired. This is likely to initially leave many of the important genes associated with the evolution of pathogenicity in *Entomophthoromycota* hanging as "orphans" without any functional annotation. However, as we learn more about entomophthoromycotan genomes and the genomes of basal fungi in general, we will refine our understanding of the evolution of pathogenicity and molecular biology of *Entomophthoromycota* insect pathogens.

ACKNOWLEDGMENTS

HHDFL was supported by an individual Carlsberg foundation fellowship (2012_01_0599) and a Villum Foundation Young Investigator Grant. The studies were in part also supported by Innovationsfonden (Denmark), project no. 1024151001.

REFERENCES

Andersson, K.-M., Kumar, D., Bentzer, J., Friman, E., Ahren, D., & Tunlid, A. (2014). Interspecific and host-related gene expression patterns in nematode-trapping fungi. *BMC Genomics, 15*, 968.

Bałazy, S. (1993). *Flora of Poland. Fungi (Mycota), vol XXIV, Entomophthorales*. Krakow, Poland: Polish Academy of Sciences.

Beauvais, A., Latgé, J.-P., Vey, A., & Prevost, M.-C. (1989). The role of surface components of the entomopathogenic fungus *Entomophaga aulicae* in the cellular immune response of *Galleria mellonella* (Lepidoptera). *Microbiology, 135*, 489–498.

Becker, F., Schnorr, K., Wilting, R., Tolstrup, N., Bendtsen, J. D., & Olsen, P. B. (2004). Development of in vitro transposon assisted signal sequence trapping and its use in screening *Bacillus halodurans* C125 and *Sulfolobus solfataricus* P2 gene libraries. *Journal of Microbiological Methods, 57*, 123–133.

Bidochka, M. J., & Hajek, A. E. (1996). Protoplast plasma membrane glycoproteins in two species. *Mycological Research, 100*, 1094–1098.

Bidochka, M. J., Walsh, S. R., Ramos, M. E., St. Leger, R. J., Silver, J. C., & Roberts, D. W. (1995). Pathotypes in the *Entomophaga grylli* species complex of grasshopper pathogens differentiated with random amplification of polymorphic DNA and cloned-DNA probes. *Applied and Environmental Microbiology, 61*, 556–560.

Blackwell, M. (2011). The fungi: 1, 2, 3... 5.1 million species? *American Journal of Botany, 98*, 426–438.

Boomsma, J. J., Jensen, A. B., Meyling, N. V., & Eilenberg, J. (2014). Evolutionary interaction networks of insect pathogenic fungi. *Annual Review of Entomology, 59*, 467–485.

Brey, P. T., Latgé, J. P., & Prevost, M. C. (1986). Integumental penetration of the pea aphid, *Acyrthosiphon pisum*, by *Conidiobolus obscurus* (Entomophthoraceae). *Journal of Invertebrate Pathology, 48*, 34–41.

Brown, G. D., Denning, D. W., Gow, N. A. R., Levitz, S. M., Netea, M. G., & White, T. C. (2012). Hidden killers: human fungal infections. *Science Translational Medicine, 4*, 165rv13.

Brown, S. P., Cornforth, D. M., & Mideo, N. (2012). Evolution of virulence in opportunistic pathogens: generalism, plasticity, and control. *Trends in Microbiology, 20*, 336–342.

Burmester, A., Shelest, E., Glöckner, G., Heddergott, C., Schindler, S., Staib, P.... Brakhage, A. A. (2011). Comparative and functional genomics provide insights into the pathogenicity of dermatophytic fungi. *Genome Biology, 12*, R7.

Castrillo, L. A., Thomsen, L., Juneja, P., & Hajek, A. E. (2007). Detection and quantification of *Entomophaga maimaiga* resting spores in forest soil using real-time PCR. *Mycological Research, 111*, 324–331.

Chang, Y., Wang, S., Sekimoto, S., Aerts, A. L., Choi, C., Clum, A. ... Berbee, M. L. (2015). Phylogenomic analyses indicate that early fungi evolved digesting cell walls of algal ancestors of land plants. *Genome Biology and Evolution, 7*, 1590–1601.

Charnley, A. K. (2003). Fungal pathogens of insects: cuticle degrading enzymes and toxins. *Advances in Botanical Research, 40*, 242–321.

Chayakulkeeree, M., Ghannoum, M. A., & Perfect, J. R. (2006). Zygomycosis: the re-emerging fungal infection. *European Journal of Clinical Microbiology & Infectious Diseases, 25*, 215–229.

Claydon, N. (1978). Insecticidal secondary metabolites from entomogenous fungi: *Entomophthora virulenta*. *Journal of Invertebrate Pathology, 32*, 319–324.

Delalibera, I., Hajek, A. E., & Humber, R. A. (2004). *Neozygites tanajoae* sp. nov., a pathogen of the cassava green mite. *Mycologia, 96*, 1002–1009.

Dunphy, G. B., & Nolan, R. A. (1982). Mycotoxin production by the protoplast stage of *Entomophthora egressa*. *Journal of Invertebrate Pathology, 39*, 261–263.

Donzelli, B. G. G., & Krasnoff, S. B. (2016). Molecular Genetics of Secondary Chemistry in Metarhizium Fungi. *Advances in Genetics, 94*, 365–436.

Eilenberg, J., Bresciani, J., & Latgé, J. P. (1986). Ultrastructural studies of primary spore formation and discharge in the genus *Entomophthora*. *Journal of Invertebrate Pathology, 48*, 318–324.

Eilenberg, J., Bresciani, J., Olesen, U., & Olson, L. (1995). Ultrastructural studies of secondary spore formation and discharge in the genus *Entomophthora*. *Journal of Invertebrate Pathology, 65*, 179–185.

Eilenberg, J., & Michelsen, V. (1999). Natural host range and prevalence of the genus *Strongwellsea* (Zygomycota: Entomophthorales) in Denmark. *Journal of Invertebrate Pathology, 73*, 189–198.

Einax, E., & Voigt, K. (2003). Oligonucleotide primers for the universal amplification of beta-tubulin genes facilitate phylogenetic analyses in the regnum fungi. *Organisms Diversity & Evolution, 3*, 185–194.

Evans, H. C. (1989). Mycopathogens of insects of epigeal and aerial habitats. In N. Wilding, N. M. Collins, P. M. Hammond, & J. F. Weber (Eds.), *Insect–fungus interactions* (pp. 205–238). London: Academic Press.

Fisher, M. C., Henk, D. A., Briggs, C. J., Brownstein, J. S., Madoff, L. C., McCraw, S. L., & Gurr, S. J. (2012). Emerging fungal threats to animal, plant and ecosystem health. *Nature, 484*, 186–194.

Fournier, A., Enkerli, J., Keller, S., & Widmer, F. (2008). A PCR-based tool for the cultivation-independent monitoring of *Pandora neoaphidis*. *Journal of Invertebrate Pathology, 99*, 49–56.

Fournier, A., Widmer, F., & Enkerli, J. (2010). Development of a single-nucleotide polymorphism (SNP) assay for genotyping of *Pandora neoaphidis*. *Fungal Biology, 114*, 498–506.

Freimoser, F. M., Jensen, A. B., Tuor, U., Aebi, M., & Eilenberg, J. (2001). Isolation and in vitro cultivation of the aphid pathogenic fungus *Entomophthora planchoniana*. *Canadian Journal of Microbiology, 47*, 1082–1087.

Freimoser, F. M., Screen, S., Hu, G., & St. Leger. (2003). EST analysis of genes expressed by the zygomycete pathogen *Conidiobolus coronatus* during growth on insect cuticle. *Microbiology, 149*, 1893–1900.

Gao, Q., Jin, K., Ying, S.-H., Zhang, Y., Xiao, G., Shang, Y.… Wang, C. (2011). Genome sequencing and comparative transcriptomics of the model entomopathogenic fungi *Metarhizium anisopliae* and *M. acridum*. *PLoS Genetics, 7*, e1001264.

Gladieux, P., Ropars, J., Badouin, H., Branca, A., Aguileta, G., de Vienne, D. M.… Giraud, T. (2014). Fungal evolutionary genomics provides insight into the mechanisms of adaptive divergence in eukaryotes. *Molecular Ecology, 23*, 753–773.

Gregory, T. R., Nicol, J. A., Tamm, H., Kullman, B., Kullman, K., Leitch, I. J.… Bennett, M. D. (2007). Eukaryotic genome size databases. *Nucleic Acids Research, 35*, D332–D338.

Grell, M. N., Jensen, A. B., Olsen, P. B., Eilenberg, J., & Lange, L. (2011). Secretome of fungus-infected aphids documents high pathogen activity and weak host response. *Fungal Genetics and Biology, 48*, 343–352.

Gryganskyi, A. P., Humber, R. A., Smith, M. E., Hodge, K., Huang, B., Voigt, K., & Vilgalys, R. (2013). Phylogenetic lineages in Entomophthoromycota. *Persoonia – Molecular Phylogeny and Evolution of Fungi, 30*, 94–105.

Gryganskyi, A. P., Humber, R. A., Smith, M. E., Miadlikovska, J., Wu, S., Voigt, K. … Vilgalys, R. (2012). Molecular phylogeny of the Entomophthoromycota. *Molecular Phylogenetics and Evolution, 65*, 682–694.

Gryganskyi, A. P., Humber, R. A., Stajich, J. E., Mullens, B., Anishchenko, I. M., & Vilgalys, R. (2013). Sequential utilization of hosts from different fly families by genetically distinct, sympatric populations within the *Entomophthora muscae* species complex. *PLoS One, 8*, e71168.

Gryganskyi, A. P., Lee, S. C., Litvintseva, A. P., Smith, M. E., Bonito, G., Porter, T. M. … Vilgalys, R. (2010). Structure, function, and phylogeny of the mating locus in the *Rhizopus oryzae* complex. *PLoS One, 5*, e15273.

Gryganskyi, A. P., & Muszewska, A. (2014). Whole genome sequencing and the Zygomycota. *Fungal Genomics & Biology, 04*, e116.

Guzmán-Franco, A. W., Atkins, S. D., Alderson, P. G., & Pell, J. K. (2008). Development of species-specific diagnostic primers for *Zoophthora radicans* and *Pandora blunckii*; two co-occurring fungal pathogens of the diamondback moth, *Plutella xylostella*. *Mycological Research, 112*, 1227–1240.

Guzmán-Franco, A. W., Atkins, S. D., Clark, S. J., Alderson, P. G., & Pell, J. K. (2011). Use of quantitative PCR to understand within-host competition between two entomopathogenic fungi. *Journal of Invertebrate Pathology, 107*, 155–158.

Hajek, A. E. (2007). Introduction of a fungus into North America for control of gypsy moth. In C. Vincent, M. Goettel, & G. Lazarovits (Eds.), *Biological control: International case studies* (pp. 53–62). UK: CABI Publishing.

Hajek, A. E., Butler, L., Walsh, S., Silver, J. C., Hain, F. P., Hastings, F. L. … Hain, F. P. (1996). Host range of the gypsy moth (Lepidoptera: Lymantriidae) pathogen *Entomophaga maimaiga* (Zygomycetes: Entomophthorales) in the field versus laboratory. *Environmental Entomology, 25*, 709–721.

Hajek, A. E., Hodge, K. T., Liebherr, J. K., Day, W. H., & Vandenberg, J. D. (1996). Use of RAPD analysis to trace the origin of the weevil pathogen *Zoophthora phytonomi* in North America. *Mycological Research, 100*, 349–355.

Hajek, A. E., Humber, R. A., Elkinton, J. S., May, B., Walsh, S., & Silver, J. C. (1990). Allozyme and restriction fragment length polymorphism analyses confirm *Entomophaga maimaiga* responsible for 1989 epizootics in North American gypsy moth populations. *Proceedings of the National Academy of Sciences of the United States of America, 87*, 6979–6982.

Hajek, A. E., Humber, R. A., Walsh, S., & Silver, J. C. (1991). Sympatric occurrence of two *Entomophaga aulicae* (Zygomycetes: Entomophthorales) complex species attacking forest Lepidoptera. *Journal of Invertebrate, 58*, 373–380.

Hajek, A. E., Jensen, A. B., Thomsen, L., Hodge, K. T., & Eilenberg, J. (2003). PCR-RFLP is used to investigate relations among species in the entomopathogenic genera *Eryniopsis* and *Entomophaga*. *Mycologia, 95*, 262–268.

Hajek, A. E., Walsh, S. R., Strong, D. R., & Silver, J. C. (1996). A disjunct Californian strain of *Entomophaga aulicae* infecting *Orgyia vetusta*. *Journal of Invertebrate Pathology, 68*, 260–268.

Henk, D. A., & Fisher, M. C. (2012). The gut fungus *Basidiobolus ranarum* has a large genome and different copy numbers of putatively functionally redundant elongation factor genes. *PLoS One, 7*, e31268.

Hibbett, D. S., Binder, M., Bischoff, J. F., Blackwell, M., Cannon, P. F., Eriksson, O. E. … Zhang, N. (2007). A higher-level phylogenetic classification of the fungi. *Mycological Research, 111*, 509–547.

Hodge, K. T., Sawyer, A. J., & Humber, R. A. (1995). RAPD-PCR for identification of *Zoophthora radicans* isolates in biological control of the potato leafhopper. *Journal of Invertebrate Pathology, 65*, 1–9.

Hu, G., & St. Leger, R. J. (2004). A phylogenomic approach to reconstructing the diversification of serine proteases in fungi. *Journal of Evolutionary Biology, 17*, 1204–1214.

Hu, X., Xiao, G., Zheng, P., Shang, Y., Su, Y., Zhang, X. … Wang, C. (2014). Trajectory and genomic determinants of fungal-pathogen speciation and host adaptation. *Proceedings of the National Academy of Sciences of the United States of America, 111*, 16796–16801.

Hughes, D. P., Araujo, J. P. M., Loreto, R. G., Quevillon, L., de Bekker, C., & Evans, H. C. (2016). From So Simple a Beginning: The Evolution of Behavioral Manipulation by Fungi. *Advances in Genetics, 94*, 437–469.

Humber, R. A. (1982). *Strongwellsea* vs. *Erynia*: the case for a phylogenetic classification of the Entomophthorales (Zygomycetes). *Mycotaxon, 15*, 167–184.

Humber, R. A. (1989). Synopsis of a revised classification for the Entomophthorales (Zygomycotina). *Mycotaxon, 34*, 441–460.

Humber, R. A. (2008). Evolution of entomopathogenicity in fungi. *Journal of Invertebrate Pathology, 98*, 262–266.

Humber, R. A. (2012). Entomophthoromycota: a new phylum and reclassification for entomophthoroid fungi. *Mycotaxon, 120*, 477–492.

James, T. Y., Kauff, F., Schoch, C. L., Matheny, P. B., Hofstetter, V., Cox, C. J. … Vilgalys, R. (2006). Reconstructing the early evolution of fungi using a six-gene phylogeny. *Nature, 443*, 818–822.

Jensen, A. B., & Dromph, K. M. (2005). The causal agents of "entomophthoramycosis" belong to two different orders: a suggestion for modification of the clinical nomenclature. *Clinical Microbiology and Infection, 11*, 249–250.

Jensen, A. B., & Eilenberg, J. (2001). Genetic variation within the insect-pathogenic genus *Entomophthora*, focusing on the *E. muscae* complex, using PCR—RFLP of the ITS II and the LSU rDNA. *Mycological Research, 105*, 307–312.

Jensen, A. B., Eilenberg, J. R., & Lopez Lastra, C. (2009). Differential divergences of obligately insect-pathogenic *Entomophthora* species from fly and aphid hosts. *FEMS Microbiology Letters, 300*, 180–187.

Jensen, A. B., Gargas, A., Eilenberg, J., & Rosendahl, S. (1998). Relationships of the insect-pathogenic order Entomophthorales (Zygomycota, Fungi) based on phylogenetic analyses of nuclear small subunit ribosomal DNA sequences (SSU rDNA). *Fungal Genetics and Biology, 24*, 325–334.

Jensen, A. B., Hansen, L. M., & Eilenberg, J. (2008). Grain aphid population structure: no effect of fungal infections in a 2-year field study in Denmark. *Agricultural and Forest Entomology, 10*, 279–290.

Jensen, A. B., Thomsen, L., & Eilenberg, J. (2001). Intraspecific variation and host specificity of *Entomophthora muscae* sensu stricto isolates revealed by random amplified polymorphic DNA, universal primed PCR, PCR-restriction fragment length polymorphism, and conidial morphology. *Journal of Invertebrate Pathology, 78*, 251–259.

Jensen, A. B., Thomsen, L., & Eilenberg, J. (2006). Value of host range, morphological, and genetic characteristics within the *Entomophthora muscae* species complex. *Mycological Research, 110*, 941–950.

Keeling, P. J. (2003). Congruent evidence from α-tubulin and β-tubulin gene phylogenies for a zygomycete origin of microsporidia. *Fungal Genetics and Biology, 38*, 298–309.

Keller, S. (1991). Arthropod-pathogenic Entomophthorales of Switzerland. II. *Erynia, Eryniopsis, Neozygites, Zoophthora* and *Tarichium*. *Sydowia, 43*, 39–122.

Krasnoff, S. B., Watson, D. W., Gibson, D. M., & Kwan, E. C. (1995). Behavioral effects of the entomopathogenic fungus, *Entomophthora muscae* on its host *Musca domestica*: postural changes in dying hosts and gated pattern of mortality. *Journal of Insect Physiology, 41*, 895–903.

Latgé, J. P., Eilenberg, J., Beauvais, A., & Prevost, M. C. (1988). Morphology of *Entomophthora muscae* protoplasts grown in vitro. *Protoplasma, 146*, 166–173.

Lee, T., Yun, S. H., Hodge, K. T., Humber, R. A., Krasnoff, S. B., Turgeon, G. B.... Gibson, D. M. (2001). Polyketide synthase genes in insect- and nematode-associated fungi. *Applied Microbiology and Biotechnology, 56*, 181–187.

Lihme, M., Jensen, A. B., & Rosendahl, S. (2009). Local scale population genetic structure of *Entomophthora muscae* epidemics. *Fungal Ecology, 2*, 81–86.

Liu, X.-Y., & Voigt, K. (2010). Molecular characters of zygomycetous fungi. In Y. Gherbawy, & K. Voigt (Eds.), *Molecular identification of fungi* (pp. 461–488). Berlin, Heidelberg: Springer Berlin Heidelberg.

Lopez Lastra, C. C., Gibson, D. M., & Hajek, A. E. (2001). Survival and differential development of *Entomophaga maimaiga* and *Entomophaga aulicae* (Zygomycetes: Entomophthorales) in *Lymantria dispar* hemolymph. *Journal of Invertebrate Pathology, 78*, 201–209.

Ma, L.-J., Ibrahim, A. S., Skory, C., Grabherr, M. G., Burger, G., Butler, M.... Wickes, B. L. (2009). Genomic analysis of the basal lineage fungus *Rhizopus oryzae* reveals a whole-genome duplication. *PLoS Genetics, 5*, e1000549.

Małagocka, J., Grell, M. N., Lange, L., Eilenberg, J., & Jensen, A. B. (2015). Transcriptome of an entomophthoralean fungus (*Pandora formicae*) shows molecular machinery adjusted for successful host exploitation and transmission. *Journal of Invertebrate Pathology, 128*, 47–56.

Manning, R. J., Waters, S. D., & Callaghan, A. A. (2007). Saprotrophy of *Conidiobolus* and *Basidiobolus* in leaf litter. *Mycological Research, 111*, 1437–1449.

Martinez, D. A., Oliver, B. G., Graser, Y., Goldberg, J. M., Li, W., Martinez-Rossi, N. M.... White, T. C. (2012). Comparative genome analysis of *Trichophyton rubrum* and related dermatophytes reveals candidate genes involved in infection. *mBio, 3*, e00259–e00312.

Meerupati, T., Andersson, K.-M., Friman, E., Kumar, D., Tunlid, A., & Ahrén, D. (2013). Genomic mechanisms accounting for the adaptation to parasitism in nematode-trapping fungi. *PLoS Genetics, 9*, e1003909.

Milne, R., Wright, T., Welton, M., & Budau, C. (1994). Identification and partial purification of a cell-lytic factor from *Entomophaga aulicae*. *Journal of Invertebrate Pathology, 54*, 253–259.

Möbius, N., & Hertweck, C. (2009). Fungal phytotoxins as mediators of virulence. *Current Opinion in Plant Biology, 12*, 390–398.

Morales-Vidal, S., Alatorre-Rosas, R., Clark, S. J., Pell, J. K., & Guzmán-Franco, A. W. (2013). Competition between isolates of *Zoophthora radicans* co-infecting *Plutella xylostella* populations. *Journal of Invertebrate Pathology, 113*, 137–145.

Murrin, F., Holtby, J., Noland, R. A., & Davidson, W. S. (1986). The genome of *Entomophaga aulicae* (Entomophthorales, Zygomycetes): base composition and size. *Experimental Mycology, 10*, 67–75.

Nagahama, T., Sato, H., Shimazu, M., & Sugiyama, J. (1995). Phylogenetic divergence of the entomophthoralean fungi – evidence from nuclear 18s ribosomal-RNA gene-sequences. *Mycologia, 87*, 203–209.

Nielsen, C., Milgroom, M. G., & Hajek, A. E. (2005). Genetic diversity in the gypsy moth fungal pathogen *Entomophaga maimaiga* from founder populations in North America and source populations in Asia. *Mycological Research, 109*, 941–950.

Olive, E. W. (1907). Cell and nuclear division in *Basidiobolus*. *Annales Mycologici, 5*, 404–418.

Prabhu, R. M., & Patel, R. (2004). Mucormycosis and entomophthoramycosis: a review of the clinical manifestations, diagnosis and treatment. *Clinical Microbiology and Infection, 10*, 31–47.

Raffaele, S., & Kamoun, S. (2012). Genome evolution in filamentous plant pathogens: why bigger can be better. *Nature Reviews Microbiology, 10*, 417–430.

Reddy, T. B. K., Thomas, A. D., Stamatis, D., Bertsch, J., Isbandi, M., Jansson, J.... Kyrpides, N. C. (2015). The Genomes OnLine Database (GOLD) v.5: a metadata management system based on a four level (meta)genome project classification. *Nucleic Acids Research, 43*, D1099–D1106.

Ribes, J. A., Vanover-Sams, C. L., & Baker, D. J. (2000). Zygomycetes in human disease. *Clinical Microbiology Reviews, 13*, 236–301.
Riddle, L. W. (1906). On the cytology of the Entomophthoraceae. *Proceedings of the American Academy of Arts and Sciences, 42*, 177–198.
Robinow, C. F. (1963). Observations on cell growth, mitosis, and division in the fungus *Basidiobolus ranarum*. *The Journal of Cell Biology, 17*, 123–152.
Rosenblum, E. B., Stajich, J. E., Maddox, N., & Eisen, M. B. (2008). Global gene expression profiles for life stages of the deadly amphibian pathogen *Batrachochytrium dendrobatidis*. *Proceedings of the National Academy of Sciences of the United States of America, 105*, 17034–17039.
Roy, H. E., Steinkraus, D. C., Eilenberg, J., Hajek, A. E., & Pell, J. K. (2006). Bizarre interactions and endgames: entomopathogenic fungi and their arthropod hosts. *Annual Review of Entomology, 51*, 331–357.
Sawyer, W. H. (1933). The development of *Entomophthora sphaerosperma* upon *Rhopobota vacciniana*. *Annals of Botany, 47*, 799–809.
Schussler, A., Schwarzott, D., & Walker, C. (2001). A new fungal phylum, the Glomeromycota: phylogeny and evolution. *Mycological Research, 105*, 1413–1421.
Scorsetti, A. C., Jensen, A. B., Lopez Lastra, C., & Humber, R. A. (2012). First report of *Pandora neoaphidis* resting spore formation in vivo in aphid hosts. *Fungal Biology, 116*, 196–203.
Shelest, E., & Voigt, K. (2014). Genomics to study basal lineage fungal biology: phylogenomics suggests a common origin. In M. Nowrousian (Ed.), *Fungal genomics* (2nd ed., Vol. 13, pp. 31–60). Berlin: Springer Verlag.
Sun, N. C., & Bowen, C. C. (1972). Ultrastructural studies of nuclear division in *Basidiobolus ranarum* Eidam. *Caryologia, 25*, 471–494.
Tanabe, Y. (2000). Molecular phylogeny of parasitic Zygomycota (Dimargaritales, Zoopagales) based on nuclear small subunit ribosomal DNA sequences. *Molecular Phylogenetics and Evolution, 16*, 253–262.
Tanabe, Y., Saikawa, M., Watanabe, M. M., & Sugiyama, J. (2004). Molecular phylogeny of Zygomycota based on EF-1α and RPB1 sequences: limitations and utility of alternative markers to rDNA. *Molecular Phylogenetics and Evolution, 30*, 438–449.
Thomsen, L., & Beauvais, A. (1995). Cloning of two chitin synthase gene fragments from a protoplastic entomophthorale. *FEMS Microbiology Letters, 129*, 115–120.
Thomsen, L., & Jensen, A. B. (2002). Application of nested-PCR technique to resting spores from the *Entomophthora muscae* species complex: implications for analyses of host-pathogen population interactions. *Mycologia, 94*, 794–802.
Van Houte, S., Ros, V. I. D., & van Oers, M. M. (2013). Walking with insects: molecular mechanisms behind parasitic manipulation of host behaviour. *Molecular Ecology, 22*, 3458–3475.
Vega, F. E., Meyling, N. V., Luangsa-ard, J. J., & Blackwell, M. (2012). Fungal entomopathogens. In F. E. Vega, & H. K. Kaya (Eds.), *Insect pathology* (pp. 171–220). London: Elsevier Inc.
Vikram, H. R., Smilack, J. D., Leighton, J. A., Crowell, M. D., & De Petris, G. (2012). Emergence of gastrointestinal basidiobolomycosis in the United States, with a review of worldwide cases. *Clinical Infectious Diseases, 54*, 1685–1691.
Voigt, K., Cigelnik, E., & O'Donnell, K. (1999). Phylogeny and PCR identification of clinically important zygomycetes based on nuclear ribosomal-DNA sequence data. *Journal of Clinical Microbiology, 37*, 3957–3964.
Walsh, S. R. A. (1996). *Development of molecular markers for the detection and differentiation of* Entomophaga *strains pathogenic for insects* (Ph.D. dissertation). Canada: University of Toronto.
Wang, C., & St. Leger, R. J. (2014). Genomics of entomopathogenic fungi. In F. Martin (Ed.), *The ecological genomics of fungi* (pp. 243–260). Hoboken: John Wiley & Sons, Inc.

Wang, J. B., St. Leger, R. J., & Wang, C. (2016). Advances in Genomics of Insect Pathogenic Fungi. *Advances in Genetics, 94,* 67–105.
Willis, J. H. (2010). Structural cuticular proteins from arthropods: annotation, nomenclature, and sequence characteristics in the genomics era. *Insect Biochemistry and Molecular Biology, 40,* 189–204.
Xu, J., Baldwin, D., Kindrachuk, C., & Hegedus, D. D. (2006). Serine proteases and metalloproteases associated with pathogenesis but not host specificity in the Entomophthoralean fungus *Zoophthora radicans. Canadian Journal of Microbiology, 52,* 550–559.
Xu, J., Baldwin, D., Kindrachuk, C., & Hegedus, D. D. (2009). Comparative EST analysis of a *Zoophthora radicans* isolate derived from *Pieris brassicae* and an isogenic strain adapted to *Plutella xylostella. Microbiology, 155,* 174–185.
Yang, J., Wang, L., Ji, X., Feng, Y., Li, X., Zou, C. … Zhang, K. Q. (2011). Genomic and proteomic analyses of the fungus *Arthrobotrys oligospora* provide insights into nematode-trap formation. *PLoS Pathogens, 7,* e1002179.
Yendol, W. G., Miller, E. M., & Behnke, C. N. (1968). Toxic substances from entomophthoraceous fungi. *Journal of Invertebrate Pathology, 10,* 313–319.
Zheng, P., Xia, Y., Xiao, G., Xiong, C., Hu, X., Zhang, S. … Wang, C. (2011). Genome sequence of the insect pathogenic fungus *Cordyceps militaris*, a valued traditional Chinese medicine. *Genome Biology, 12,* R116.

CHAPTER THREE

Advances in Genomics of Entomopathogenic Fungi

J.B. Wang*, R.J. St. Leger*,[1] and C. Wang[§,1]
*University of Maryland, College Park, MD, United States
[§]Chinese Academy of Sciences, Shanghai, China
[1]Corresponding authors: E-mail: stleger@umd.edu; cswang@sibs.ac.cn

Contents

Abstract

Fungi are the commonest pathogens of insects and crucial regulators of insect populations. The rapid advance of genome technologies has revolutionized our understanding of entomopathogenic fungi with multiple *Metarhizium* spp. sequenced, as well as *Beauveria bassiana, Cordyceps militaris*, and *Ophiocordyceps sinensis* among others. Phylogenomic analysis suggests that the ancestors of many of these fungi were plant endophytes or pathogens, with entomopathogenicity being an acquired characteristic. These fungi now occupy a wide range of habitats and hosts, and their genomes have provided a wealth of information on the evolution of virulence-related characteristics, as well as the protein families and genomic structure associated with ecological and econutritional heterogeneity, genome evolution, and host range diversification. In particular, their evolutionary transition from plant pathogens or endophytes to insect pathogens provides a novel perspective on how new functional mechanisms important for host switching and virulence are acquired. Importantly, genomic resources have helped make entomopathogenic fungi ideal model systems for answering basic questions in parasitology, entomology, and speciation. At the same time, identifying the selective

Advances in Genetics, Volume 94
ISSN 0065-2660
http://dx.doi.org/10.1016/bs.adgen.2016.01.002

forces that act upon entomopathogen fitness traits could underpin both the development of new mycoinsecticides and further our understanding of the natural roles of these fungi in nature. These roles frequently include mutualistic relationships with plants. Genomics has also facilitated the rapid identification of genes encoding biologically useful molecules, with implications for the development of pharmaceuticals and the use of these fungi as bioreactors.

1. INTRODUCTION

Entomopathogenic fungi are particularly well suited for development as biopesticides because unlike bacteria and viruses that have to be ingested to cause diseases, fungi typically infect insects by direct penetration of the cuticle followed by multiplication in the hemocoel (St. Leger, Wang, & Fang, 2011). However, entomopathogenic fungi are very heterogeneous; both they and their hosts have short generation times, and they occupy a wide range of habitats, with near ubiquity in the soil and on plants. The interactions between fungi, hosts, and the environment are therefore diverse and dynamic, which complicates comparisons between different fungi infecting different insects since their interactions may be necessarily disparate. The commonly accepted solution to this quandary was to pick a couple of related fungal species for a thorough study of host—pathogen interactions, and subsequently make comparisons with other distantly related species. Consequently, most of what we know about the biochemical and molecular basis of interactions between fungi and insects has been determined with the experimentally tractable hypocrealean ascomycete genera *Metarhizium* (family Clavicipitaceae) and *Beauveria* (family Cordycipitaceae). These fungi are able to degrade, penetrate, and assimilate the insect cuticle using a combination of cuticle-degrading enzymes and mechanical pressure while overcoming any stresses encountered along the way (Ortiz-Urquiza & Keyhani, 2013). Upon reaching the hemocoel, the fungi quickly multiply by successfully competing for nutrients and avoiding host immunity.

Compared with the usual fungal model systems such as the yeast *Saccharomyces cerevisiae*, *Metarhizium* and *Beauveria* are extraordinarily versatile. *Metarhizium*, for example, contains species that are narrow host range (eg, *Metarhizium album, Metarhizium acridum*) or broad host range (eg, *Metarhizium robertsii, Metarhizium anisopliae*) pathogens of arthropods, as well as being saprophytes, and colonizers of the rhizosphere and plant root. Consistent with their broad lifestyle options, most *Metarhizium* spp. exhibit an extremely flexible metabolism. This metabolism enables them to grow under various environmental conditions, with sparse nutrients (Rangel, Alston, & Roberts, 2008) and in the presence of compounds lethal to other fungi (Ortiz-Urquiza & Keyhani, 2015; Roberts & St. Leger, 2004).

Because of their mode of infection through the cuticle, fungi function as contact insecticides (Thomas & Read, 2007). Biocontrol researchers have therefore made a tremendous effort to find naturally occurring fungal pathogens capable of controlling mosquitoes and other pest insects. This typically involved the selection of strains pathogenic to target insects without considering the mechanisms involved or the role of these fungi in their natural habitats. Only recently has progress been made in determining the factors that influence the distribution, population structure, and econutritional characteristics of any entomopathogenic fungi, even *Metarhizium* and *Beauveria*. These fungi are known to employ a vast array of metabolites to aid in infection and to outcompete other microbes, and recent genetic studies have revealed the mechanisms and importance of a portion of these metabolites for virulence, but much still remains unknown (see Donzelli and Krasnoff, 2016). These deficiencies have hindered realization of the potential of these fungi as classical biocontrol agents that persist in the environment and recycle through pest populations (Hajek, McManus, & Delalibera, 2007; Hajek & Tobin, 2011; Wang & Feng, 2014).

As well as their direct benefit to agriculture and vector control as insect pathogens, *Cordyceps/Ophiocordyceps* spp. are medicinally valued and insect pathogens in general are prolific producers of enzymes and diverse secondary metabolites with activities against insects, fungi, bacteria, viruses, and cancer cells (Hu et al., 2013; Isaka, Kittakoop, Kirtikara, Hywel-Jones, & Thebtaranonth, 2005; Kim et al., 2010; Liu et al., 2015; Zheng et al., 2011). Enzymes from *Metarhizium* and *Beauveria* spp. are frequently exploited as industrial catalysts (Pereira, Noronha, Miller, & Franco, 2007; Silva et al., 2009), even though the responsible genes for these products are rarely identified.

Comparative genomics offers a way forward by disentangling common themes of fungal biology from specific components involved in insect pathology and allowing broad host range pathogens to be studied in the context of narrow host range pathogens (Wang & St. Leger, 2014). It is also extremely valuable for assessing poorly characterized species such as *Ophiocordyceps sinensis*. Comparative genomics has facilitated identifying fungal fitness traits and the selective forces that act upon them to improve our understanding of how and why entomopathogenic fungi interact with insects and other components of their environments. Thus, sequence data can provide crucial information on the poorly understood ways that these organisms reproduce and persist in different environments. Alongside the recent availability of genomic resources, the wide array of experiments that can be performed with entomopathogenic fungi make them ideal models for answering basic questions on the genetic and genomic processes behind adaptive phenotypes (a "Holy Grail" in biology). Key challenges for fungi as models for other eukaryotes include identifying the genes involved in ecologically relevant traits and understanding the nature, timing, and architecture of the genomic changes governing the origin and processes of local adaptation (Gladieux et al., 2014). These outstanding evolutionary questions are particularly important for biocontrol agents and address fundamental, yet poorly understood, issues by asking: what roles do different kinds of mutations play in adaptation? When organisms adapt to new environments, do they do so because of changes in few genes or many? Are the same genes or networks involved in independent cases of adaptation to the same environment? What is the timescale at which evolutionary processes happen?

2. EVOLUTIONARY RELATIONSHIPS OF ENTOMOPATHOGENIC FUNGI

Fungal species of different phyla like Microsporidia, Chytridiomycota, Entomophthoromycota, Basidiomycota, and Ascomycota are known to infect and kill insects (Shang, Feng, & Wang, 2015; Sung et al., 2007). The two best-studied groups are the ascomycete entomopathogens and the Entomophthoromycota. Researchers studying Entomophthoromycota, which are not easy to mass-produce, have focused on the ecology of these organisms and their role as causative agents of mass epizootics. As of February 2016, one published Entomophthoromycota genome (*Conidiobolus coronatus* (Chang et al., 2015)) and nine incomplete Entomophthoromycota genome sequencing projects were listed in the Genomes OnLine Database (GOLD) (these are discussed in De Fine Licht, Hajek, Eilenberg & Jensen, 2016). Ascomycete entomopathogens are usually developed as inundative control agents which are applied *en masse* to a pest population (Wang, Fan, Li, & Butt, 2004). To date, there is much more genomic information on ascomycete insect pathogens, as sequences are available from nine *Metarhizium* strains (Gao et al., 2011; Hu et al., 2014; Pattemore et al., 2014; Staats et al., 2014), *Beauveria bassiana* (Xiao et al., 2012), *Cordyceps militaris* (Zheng et al., 2011), *Ophiocordyceps sinensis* (anamorph, *Hirsutella sinensis*) (Hu et al., 2013), *Ophiocordyceps unilateralis* (de Bekker et al., 2015),

Tolypocladium inflatum (Bushley et al., 2013), and *Hirsutella thompsonii* (Agrawal, Khatri, Subramanian, & Shenoy, 2015). *Ophiocordyceps unilateralis* is a specialized fungus responsible for zombie ant behavior and is discussed in detail by Hughes et al., (2016).

The most well-studied insect ascomycete pathogens fall into three families within the order Hypocreales: Cordycipitaceae, Clavicipitaceae, and Ophiocordycipitaceae. From the reconstructed phylogeny (Fig. 1), it is evident that entomopathogenicity evolved independently in these families and that genera of hypocrealean entomopathogens cluster among closely related phytopathogens, endophytes, and mycoparasites. These ancestral associations are consistent with repeated transitions (host switching) between plant, fungi, and insect hosts, as suggested by Suh, Noda, and Blackwell (2001) in their study on *Cordyceps* spp. A comparative genome analysis of

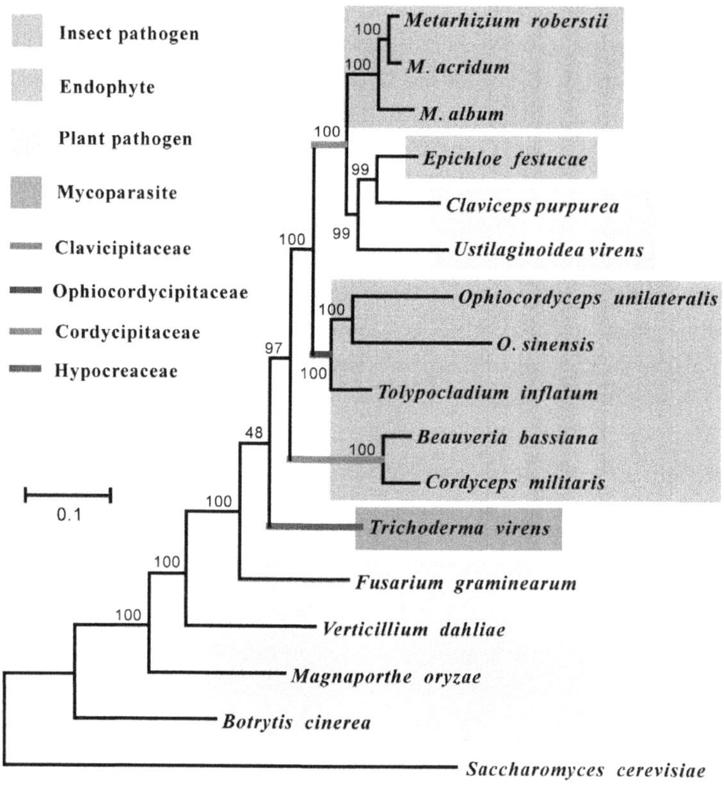

Figure 1 Phylogenomic relationships of insect pathogenic fungi with other fungi. Note: Some *Metarhizium* and *Beauveria* spp. have recently been recognized to be endophytes (reviewed by Moonjely, Barelli & Bidochka, 2016).

seven *Metarhizium* (Clavicipitaceae) genomes confirmed the genus as a monophyletic lineage that diverged from clavicipetacean plant pathogens and endophytes about 231 million years ago (MYA), and placed the hemipteran-specific *M. album* as basal to the *Metarhizium* clade with an estimated divergence time about 117 MYA (Hu et al., 2014). It was suggested that the close physical proximity of the plant-associated ancestor of *M. album* to plant-sap sucking hemipteran bugs may have facilitated this particular host switch to entomopathogenicity (Hu et al., 2014).

Several *Metarhizium* spp. maintain complicated relationships with plants and colonize plant roots and the rhizosphere (the layer of soil influenced by root metabolism) (Hu & St. Leger, 2002; Pava-Ripoll et al., 2011). Different species/strains of *Metarhizium* show differing abilities to form associations with different plant species (Bidochka, Kamp, Lavender, Dekoning, & Amritha De Croos, 2001; Steinwender et al., 2015), but the genetic underpinning of this specificity is unknown. However, *M. robertsii* has mechanisms for rapidly adapting to new soil habitats and plants which involves changes in expression of cell wall and stress response genes (Wang, O'Brien, Pava-Ripoll, & St. Leger, 2011). As shown by their antagonism to plant pathogenic fungi (Kang, Goo, Gyu, & Heon, 1996), ability to survive exposure to lead and other heavy metals (Rhee, Hillier, & Gadd, 2012), and pathogenicity to soil amoebae (Bidochka, Clark, Lewis, & Keyhani, 2010), at least some *Metarhizium* isolates have additional unpredicted flexibility in their trophic capabilities. *Beauveria bassiana* (Cordycipitaceae) also forms intimate endophytic relationships with a broad range of plant species, although it usually colonizes aerial parts of the plant (Biswas, Dey, Satpathy, & Satya, 2012; Brownbridge, Reay, Nelson, & Glare, 2012; Ownley, Gwinn, & Vega, 2010; Vega et al., 2008; Wagner & Lewis, 2000), suggesting that flexibility of lifestyle is a theme in hypocrealean insect pathogens.

3. EVOLUTION OF SEX IN ENTOMOPATHOGENIC FUNGI

As robust and diverse genetic models, fungi provide invaluable insights into the evolution and mechanism of eukaryote sexuality (Dyer & O'Gorman, 2012; Ni, Feretzaki, Sun, Wang, & Heitman, 2011). Fungi exhibit diverse reproductive modes that often determine the rates and patterns of genome evolution (Ni et al., 2011; Whittle, Nygren, & Johannesson, 2011; Zheng, Xia, Zhang, & Wang, 2013), and, as exemplified by entomopathogens, are linked as cause or effect with pathogenic strategies.

There are three modes of sexual reproduction in ascomycetous fungi, ie, heterothallic, homothallic, and pseudohomothallic, which are governed by the mating-type (*MAT*) locus, a miniature fungal version of a sex chromosome coding for transcription factors (TFs) that induce the production of pheromones and pheromone receptors (Fraser et al., 2004; Idnurm, 2011; Kronstad & Staben, 1997; Whittle et al., 2011). Two opposite loci are named as *MAT 1-1* and *MAT 1—2*, respectively (or *MATA* and *MATa*, in different fungi) (Alby, Schaefer, & Bennett, 2009; Whittle et al., 2011; Zheng et al., 2013). The haploid genome of heterothallic species carries only one of the *MAT* loci, thus they are self-sterile, requiring a haploid partner with a compatible *MAT* locus to complete the sexual cycle. Homothallic species (self-fertile) have both loci in their haploid genomes, while the pseudohomothallic (also called secondary homothallic) species are similarly self-fertile, but they contain two compatible haploid nuclei within their sexual spores (Ni et al., 2011; Zheng et al., 2013). Respective genomes of entomopathogenic fungi show single mating type loci in *T. inflatum* (Ophiocordycipitaceae) (Bushley et al., 2013), *Metarhizium* spp. (Gao et al., 2011), *C. militaris* (Zheng et al., 2011), and *B. bassiana* (Xiao et al., 2012), suggesting that most members of the three hypocrealean entomopathogen families (Clavicipitaceae, Cordycipitaceae, and Ophiocordycipitaceae) are heterothallic and are potentially therefore outcrossing fungi. Syntenic analysis of these fungi showed that, except for the idiomorphic regions, the genes flanking the mating-type locus are highly conserved, especially between *B. bassiana* and *C. militaris* (Xiao et al., 2012). However, unlike *B. bassiana* and many *Metarhizium* species (Kepler et al., 2012; Li, Li, Huang, & Fan, 2001; Liu, Liang, Whalley, Yao, & Liu, 2001), *C. militaris* commonly performs sexual reproduction (Zheng et al., 2011, 2013).

There are reproductive mode switches in the evolution of fungal sexuality, eg, analysis of 43 species of *Neurospora* identified at least six independent switches from heterothallism to homothallism (Nygren et al., 2011), while within the genus *Aspergillus*, *Aspergillus nidulans* is homothallic while its relations *Aspergillus fumigatus* and *Aspergillus oryzae* are heterothallic (Galagan et al., 2005; Peterson, 2008). It also seems that most asexual *Aspergillus* lineages arose from sexual lineages (Geiser, Timberlake, & Arnold, 1996). Signs of sex in supposedly asexual species include footprints of repeat-induced point (RIP) mutations, an irreversible fungal genome defense mechanism specific to fungi, occurring only during the sexual stages on repeated sequences (Galagan et al., 2003). The consequences of RIP are that repeated DNA segments, such as would result from the transposition of a retrotransposon, or the duplication of a gene, are inactivated by mutations. Calculations of RIP indices indicated that RIP occurs in narrow host range *M. album* and *M. acridum*, but not in the broad host range species such as *M. robertsii* (Hu et al., 2014). RIP functions only during meiosis, which suggests retention of sexuality in specialists, although their sexual stages have not been verified (Hu et al., 2014). Intriguingly, therefore, asexuality is associated with broad host ranges and sexuality with narrow host ranges. Consistent with this, broad host range generalists have many more heterokaryon incompatibility proteins (HETs) than specialists; this is indicative of tighter controls for achieving reproductive isolation (Hu et al., 2014). HET gene function may have been dispensable in specialists, as their life histories expose them to fewer alternative genotypes than generalists, but, conversely, this suggests selection for retention of reproductive isolation in nonspecialists. Reproductive isolation prevents genetic homogenization and is crucial for speciation: in which case HET genes may have played a key role in the evolution and diversification of *Metarhizium* spp.

Unlike *C. militaris*, which is specific to lepidopteran pupae, *B. bassiana* has a wide host range. Comparative genomics confirmed that asexual *B. bassiana* is closely related to sexual *Cordyceps* spp. (Fig. 1), and a teleomorph (a fungus with a sexual reproductive stage) of *B. bassiana* has been identified as *Cordyceps bassiana* (Li et al., 2001), but it is rarely observed in the field. *Beauveria bassiana*, like generalist *Metarhizium* spp., lacks the RIP mechanism, consistent with the sexual cycle being rare (Zheng et al., 2013). Instead, *B. bassiana* reproduces clonally in the environment like broad host range *Metarhizium* spp. (Meyling, Lübeck, Buckley, Eilenberg, & Rehner, 2009; S.B. Wang, Fang, Wang, & St. Leger, 2011). Individual isolates of *B. bassiana* carry either the *MAT1-1* or *MAT1-2* mating-type locus (Yokoyama, Arakawa, Yamagishi, & Hara, 2006; Zheng et al., 2013), indicating they retain the potential at least for heterothallic sexual activity. Furthermore, many genes functioning in mating processes, karyogamy, meiosis, and fruiting-body development in other fungi are present in *B. bassiana* (Xiao et al., 2012). However, *B. bassiana* lacks a meiosis-specific topoisomerase, SPO11 which may contribute to its infrequent sexual cycle, as SPO11 is crucial for initiating meiotic recombination by generating DNA double-strand breaks (Xiao et al., 2012).

Despite their close phylogenetic relationship, there is an unexpectedly high degree of genome structure divergence between *B. bassiana* and *C. militaris*. Transposable elements (TEs) are a major force driving genetic variation and genome evolution (Cordaux & Batzer, 2009; Daboussi & Capy, 2003). *Beauveria bassiana* has many more TEs than *C. militaris* (88 and 4, respectively) (Xiao et al., 2012). RIP is incompatible with gene duplication events, so its absence is consistent with expanded gene families and more TEs in the *B. bassiana* and *M. robertsii* genomes, relative to *C. militaris* and *M. acridum*. However, the genomes of *Metarhizium* species are highly syntenic in spite of large differences in the number of TEs (148 TEs in *M. robertsii* vs 20 TEs in *M. acridum*) (Gao et al., 2011). *Cordyceps militaris* often reproduces sexually (Zheng et al., 2011), whereas the sexual cycle in *M. acridum* is presumably less common, as it has not been observed in nature despite detection of RIP signatures. Sexuality facilitates genome structure reorganization due to frequent genetic and/or chromosomal recombination. These differences in life cycle may have led to the considerable genome structure disparities between *B. bassiana* and *C. militaris* (Xiao et al., 2012). In fact, syntenic relationships between *B. bassiana* and *Metarhizium* spp. are closer than to *C. militaris* (Xiao et al., 2012). This implies that genome reorganization, and presumably therefore sexuality, in *C. militaris* has accelerated since it shared a common ancestor with *B. bassiana*.

An extreme case of specialization is provided by the caterpillar fungus *O. sinensis*, which mummifies ghost moth larvae (*Thitarodes* spp.) exclusively in Tibetan Plateau alpine ecosystems. Touted as "Himalayan Viagra", the fungus' sexual fruiting-body is highly prized due to its medical benefits and dwindling supply (Hu et al., 2013). Attempts to culture the fruiting-body have failed, and the huge market demand has led to severe devastation of local ecosystems and has driven the fungus toward extinction. The route of infection is unknown but probably occurs at the first instar larval stage. The host caterpillars live underground for 4–5 years and have seven to nine instars. During most of this time, the fungus is believed to remain dormant and is only observed in the insect in later instars, just preceding the host's death (Cannon et al., 2009). The fungus then fully colonizes the cadaver and produces a sexual structure. Before sequencing the *O. sinensis* genome, the molecular basis for this lifestyle was entirely unknown, as was the sex mode of *O. sinensis*.

In contrast to the other sequenced insect pathogens, which are all heterothallic, the *O. sinensis* genome contains two compatible *MAT* loci and is sexually self-fertile, ie, homothallic (Hu et al., 2013). It is likely that inbreeding is an adaptation by *O. sinensis* to its small population size resulting from a very specialized lifestyle and the extreme environmental conditions in its small geographical range. Consistent with this, *O. sinensis* has fewer HETs

(5 genes) than other insect pathogens (≥15 genes) suggesting that, like specialist *Metarhizium* spp., it encounters fewer genetically distinct individuals than its more opportunistic relatives and, therefore, does not need barriers to vegetative fusions. It was also determined that *O. sinensis* resembles biotrophic plant pathogens (obligate pathogens that colonize living plant tissue and obtain nutrients from living host cells) in having a genome shaped by retrotransposon-driven expansions. This contrasts sharply with the genome of the related *T. inflatum* (Ophiocordycipitaceae), with a lower proportion (1.24%) of repeat sequences compared to other ascomycetes (Bushley et al., 2013). Due to its repeat-driven expansion, the *O. sinensis* genome size is approximately three times larger (~120 Mb) than the median of other ascomycete insect pathogens but contains only 6972 protein coding genes as compared to more than 9500 genes in other insect pathogens (Hu et al., 2013) (9998 in *T. inflatum* (Bushley et al., 2013)). The RIP mechanism is dysfunctional in *O. sinensis*, which has probably contributed to the massive proliferation of retrotransposable elements, and thus genome size inflation. Related TEs were clustered together in gene-poor or gene-free regions of the *O. sinensis* genome indicative of repeated rounds of retrotransposition, and the large number of retrotransposed and fragmented pseudogenes in the genome implicates retrotransposition in most of the gene losses in *O. sinensis*. The categories of pseudogenized genes are consistent with a loss of capacity to adapt to heterogenous environments. For example, the single *O. sinensis* nitrate reductase gene was pseudogenized, and the fungus also lacks nitric oxide reductase, suggesting it cannot assimilate nitrate. An inability to assimilate nitrate is also a feature of obligate plant pathogens (Hu et al., 2013).

The data reported by Hu et al. (2013) are consistent with *O. sinensis* having a biphasic pathogenic mechanism beginning with stealth pathogenesis in early host instars and a lethal stage in late instars. It has been proposed that plant pathogenic fungal lineages with large and flexible genomes are likely to adapt faster during coevolution with hosts (Raffaele & Kamoun, 2012). It is reasonable to assume that for the inbreeding *O. sinensis*, the massive proliferation of TEs provides a trade off between advantages of increased genetic variation independent of sexual recombination and deletion of genes dispensable for its specialized pathogenic lifestyle. As *O. sinensis* has lost many genes for opportunism, future transitions away from its current lifestyle seem unlikely, indicating that while retrotransposition may facilitate rapid adaptation, it may also contribute to stabile host interactions.

4. EVOLUTION OF FUNGAL HOST SPECIFICITY

Comparative genomics utilizing the *O. sinensis* genome has provided an unparalleled opportunity to develop a deeper understanding of how this unique pathogen interacts with insects within its ecosystem. It is clear that host–pathogen interactions are a major driving force for diversification, but the genomic basis for speciation and host shifting remains unclear. A major reason much remains unknown regarding speciation is because there are few applicable model systems. The genus *Metarhizium* has been subdivided into 12 different species according to the sequences of several genes (Bischoff, Rehner, & Humber, 2009). Some of these species have a wide host range, whereas others show specificity for certain insect families and can be used to test hypotheses regarding speciation and host specificity. Comparative genomic analyses of seven species revealed a directional speciation continuum from specialists with narrow host ranges (ie, *M. album* and *M. acridum* specific to hemipterans and acridids, respectively), to transitional species with intermediate host ranges (*Metarhizium majus* and *Metarhizium guizhouense* both have host ranges limited to two insect orders), and then to generalists (ie, *M. anisopliae*, *M. robertsii*, and *Metarhizium brunneum*) (Hu et al., 2014). Besides host range, generalist and specialist *Metarhizium* species differ in the way they colonize hosts (Kershaw, Moorhouse, Bateman, Reynolds, & Charnley, 1999). Generalists, like *M. robertsii*, typically kill hosts quickly via toxins and grow saprophytically in the cadaver. In contrast, the specialist *M. acridum* causes a systemic infection of host tissues before the host dies. This may reflect greater adaptation by the specialists to subverting or evading the immune systems of their particular hosts so they do not need to kill quickly. Generalists also have mechanisms for evading host immunity, but perhaps by being "jack of all trades" they are less able to subvert immune responses specific to certain insects. Lack of specific adaptations could have selected for rapid killing of hosts before the host can mount an enfeebling immune response. As described later in this chapter, the gain and loss of the insecticidal cyclopeptide destruxin gene cluster is correlated with host specificity in *Metarhizium* spp. (Hu et al., 2014; Wang, Kang, Lu, Bai, & Wang, 2012).

Specialization in *Metarhizium* is associated with retention of sexuality and rapid evolution of existing protein sequences, whereas generalization is associated with protein family expansion, loss of genome-defense mechanisms, genome restructuring, horizontal gene transfer, and loss of sexuality

(Hu et al., 2014). As mentioned previously, the close physical proximity of the plant-associated ancestor of *M. album* to plant-sap sucking hemipteran bugs likely facilitated the switch from plant endophyte to entomopathogenicity (Hu et al., 2014). The acridid-specific *M. acridum* split from the other lineages 48 MYA (Fig. 2) within the mid-Eocene when recently evolved grasses were growing along riverbanks and grass-feeding acridids first appeared (Stidham & Stidham, 2000). Cocladogenesis such as this implies host-driven divergence (Jackson, 2004; Schulze-Lefert & Panstruga, 2011). The transitional *M. majus* and *M. guizhouense* split 15 MYA (Fig. 2) followed by the generalists *M. robertsii* and *M. anisopliae* that diverged from each other only 7 MYA (Fig. 2), implying accelerated speciation associated with increased phenotypic plasticity (Hu et al., 2014). Their radiations coincided with climate change that was critical for massive diversification of flowering plants, trees, and associated insects (Hay, Soeding, DeConto, & Wold, 2002). *Metarhizium majus* and *M. guizhouense* genome sizes and encoding capacities are similar to those of generalist species, and larger than the specialists.

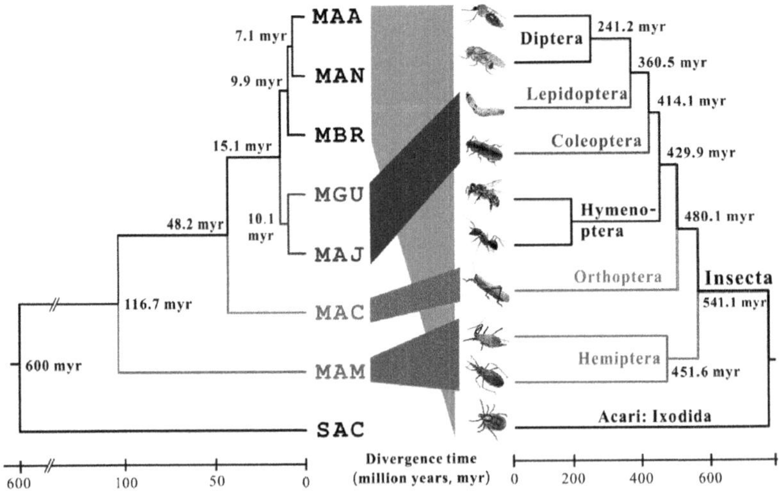

Figure 2 Reconstructed phylogeny of *Metarhizium* showing their insect host ranges and divergence time. *MAA, Metarhizium robertsii; MAN, Metarhizium anisopliae; MBR, Metarhizium brunneum; MGU, Metarhizium guizhouense; MAJ, Metarhizium majus; MAC, Metarhizium acridum; MAM, Metarhizium album; SAC, Saccharomyces cerevisiae. Rebuilt from Hu, X., Xiao, G., Zheng, P., Shang, Y., Su, Y., Zhang, X., … Wang, C. S. (2014). Trajectory and genomic determinants of fungal-pathogen speciation and host adaptation.* Proceedings of the National Academy of Sciences of the United States of America, 111, *16796–16801.* (See color plate)

Overall then, *Metarhizium* speciation provides a model for the evolution of host preference, specificity, and virulence in other pathogens. An analysis of three *Fusarium* species also suggested a transition from specialist to generalist coupled with genome and protein family expansions (Cuomo et al., 2007; Ma et al., 2010). However, a few oomycete plant pathogenic *Phytophthora* spp. appear to have transitioned from generalist to specialist (Blair, Coffey, Park, Geiser, & Kang, 2008), so the general applicability of the data reported by Hu et al. (2014) to other pathogens remains to be determined.

Clearly, there are factors missing from specialists that limit their ability to cause disease in multiple insects, as demonstrated by an increase in host range following transfer of genes from generalist strains to the specialist *M. acridum* (S.B. Wang et al., 2011). Nevertheless, *M. acridum* has a large number of rapidly evolving genes (those with a high abundance of nonsynonymous mutations), compared to generalists showing that it has not remained functionally static; rather, specialization has involved rapid evolution of existing protein sequences rather than the extensive gene duplication present in generalists (Hu et al., 2014). A likely explanation is that these are genes involved in specific locust to *M. acridum* pathogen interactions and by evolving under pairwise coevolution they are often subject to strong balancing or directional selection. By contrast, generalist *Metarhizium* spp. interact with a wide range of hosts in multiple environmental conditions and should therefore be considered as evolving under "diffuse" interactions (Juenger & Bergelson, 1998). Diffuse coevolution to many insect hosts potentially explains why signatures of positive selection were observed less frequently in the genome of *M. robertsii* and other generalists.

Classical theory predicts specialists will be more likely than generalists to lose sexuality in order to prevent shuffling of well-adapted gene combinations (Sun & Heitman, 2011), but lack of RIP in generalist *Metarhizium* species suggests that specialists are more likely to have a recent sexual history. It is possible that an increase in gene family size and protein-coding composition following loss of RIP in generalists facilitated their opportunistic lifestyles for different insect hosts. Similarly, the broad host range plant pathogen *Fusarium oxysporum* lacks RIP whereas the cereal-specialists *Fusarium graminearum* and *Fusarium verticillioides* have strong RIP effects (Ma et al., 2010), suggesting this pattern may be common in related fungal species with a wide range of host preferences. The vast majority (c. 95%) of cataloged strains of *Metarhizium* species in the ARSEF culture collection belong to generalist species (http://www.ars.usda.gov/News/docs.htm?docid=12125), suggesting that this is where most biodiversity within the *Metarhizium* genus resides and that a broad host range is linked to ecological fitness.

5. PROTEIN FAMILY EXPANSIONS AND CONTRACTIONS

Comparative genomics is crucial to understanding how fungi have evolved to occupy diverse ecological niches. The secretome, ie, the entirety of all proteins secreted by an organism, is of particular importance, as by these proteins fungi acquire nutrients and communicate with their surroundings. Secreted proteins are therefore on the front line of host–fungal interactions. With the exception of *O. sinensis*, insect pathogenic ascomycetes have a two- to threefold higher proportion of their genome (~17%) devoted to secreted products than other ascomycetes, including plant pathogens and the mycoparasitic *Trichoderma* spp. (Xiao et al., 2012). The large secretomes of insect pathogens probably reflect the many habitats they must adapt to *in insecta*, including the cuticle and the hemolymph, as well as additional environmental habitats in the soil and with plants. These complex lifestyles are reflected in the transcriptomes of insect pathogens; many different genes are induced during adaptation to host cuticle, hemolymph, or root exudate (Fang, Pava-Ripoll, Wang, & St. Leger, 2009; Feng, Shang, Cen, & Wang, 2015; Freimoser, Hu, & St. Leger, 2005; C. Wang, Butt, & St. Leger, 2005; C.S. Wang, Butt, & St. Leger, 2005; Wang & St. Leger, 2005; Xiao et al., 2012). Some of these genes encode regulators such as the protein kinase A that controls expression of many secreted virulence factors (Fang et al., 2009), an osmosensor that signals to penetrant hyphae that they have reached the hemocoel (Wang, Duan, & St. Leger, 2008), and perilipin, glycerol-3-phosphate acyltransferase, and cell autophagy-related proteins that regulate lipolysis, turgor pressure, and formation of infection structures (Duan et al., 2013; Gao, Shang, Huang, & Wang, 2013; C. Wang & St. Leger, 2007). The interactions between multiple receptors and their signal transduction pathways finely tune gene transcription to adapt to different environmental niches and sites in the host. The insect epicuticle or waxy layer comprises a heterogeneous mixture of long-chain alkanes, wax esters, and fatty acids and represents the first barrier against fungal attack. *Metarhizium robertsii* and *B. bassiana* highly transcribe cytochrome P450 (CYP) subfamily CYP52 enzymes (for metabolism of insect epicuticular lipids) and lipase genes as they germinate on the cuticle surface (Lin et al., 2011; Zhang et al., 2012). Chitin constitutes up to 40% of the procuticle but is absent from the epicuticular layer. Neither *Metarhizium* spp. nor *B. bassiana* significantly upregulate chitinase genes on the epicuticle (Gao et al., 2011; Xiao et al., 2012). Some genes are highly adapted to the specific needs of *Metarhizium*, eg, *Mcl1* (involved in immune evasion) with its collagen domain is so far unique to *Metarhizium* and is only expressed in the hemolymph (Wang & St. Leger, 2006). *Metarhizium robertsii* upregulates a specific plant adhesin in the presence of plants (*Mad2*) and a specific insect adhesin (*Mad1*) in the presence of insect cuticle, demonstrating that it has specialist genes for a bifunctional lifestyle (C.S. Wang & St. Leger, 2007). Other specifically regulated genes required to colonize the rhizosphere and root include a novel oligosaccharide transporter for root-derived nutrients (Fang & St. Leger, 2010a), an RNA-binding protein that has important roles in both saprotrophy and pathogenicity (Fang & St. Leger, 2010b), and an invertase that aids in the regulation of hydrolytic enzymes and provides a plant-derived signal restricting fungal growth (Liao, Fang, Lin, Lu, & St. Leger, 2013). Transcription of these genes is in part controlled by bZIP- or C2H2-type TFs (Huang, Shang, Chen, Cen, & Wang, 2015; Huang, Shang, Chen, Gao, & Wang, 2015).

Within the secretome there are different classes of proteins. Pathogenic microbes evolved multiple enzymes to access nutrients from hosts (Schaible & Kaufmann, 2005). Expansion and/or contraction of protein family size have occurred in different fungal species in association with evolutionary adaptations for different hosts and lifestyles (Table 1). A theme emerges of niche-specific traits, ie, traits shared by fungi that occupy the same niche irrespective of their phylogenetic position (Hu & St. Leger, 2004). For instance, plant pathogens have expanded families of glycoside hydrolases (GHs), carbohydrate esterases, cutinases, and pectin lyases in order to degrade plant materials (Xu, Peng, Dickman, & Sharon, 2006), while mammalian pathogens are enriched for aspartyl proteases and phospholipases (van Asbeck, Clemons, & Stevens, 2009), and mycoparasitic fungi have expanded numbers of chitinases to degrade fungal cell walls (Kubicek et al., 2011). Typically, insect-pathogenic fungi infect insects by breaching the cuticle using a combination of mechanical pressure, exerted by infection structures (appressoria and penetration pegs), and cuticle-degrading enzymes. The cuticle consists mainly of chitin microfibrils embedded in a matrix of proteins and covered in lipids. *Beauveria bassiana*, *Metarhizium* spp., and *C. militaris*, all have greatly expanded families of proteases, chitinases, lipases, fatty acid hydroxylases, and acyl-CoA dehydrogenases (for β-oxidation of fatty acids), illustrating an enhanced capacity to degrade

Table 1 Comparison of selected protein families among different insect pathogens

Protein family	*Ophiocordyceps sinensis*	*Cordyceps militaris*	*Beauveria bassiana*	*Metarhizium roberstii*	*Metarhizium acridum*	*Metarhizium album*
Protein kinases	116	167	159	161	192	127
Dehydrogenases	103	271	252	313	279	191
Cytochrome P450	57	57	83	123	100	74
G protein—coupled receptors	22	29	32	63	37	32
Heterokaryon incompatibility proteins	5	15	21	34	22	10
Horizontally transferred genes	25	45	42	79	56	27
Trypsins	2	12	23	32	17	12
Subtilisins	17	35	43	55	43	28
Aspartic proteases	12	21	21	33	25	22
Chitinases	9	20	20	21	14	12
Concanavalin A-like lectins	6	25	33	20	18	23
Glycoside hydrolases	66	135	145	159	130	108
Secondary metabolite biosynthesis backbone enzymes	28	30	42	61	40	34
Nonribosomal peptide synthetase (NRPS)	6	7	14	11	10	10
Polyketide synthase (PKS)	7	7	10	17	10	9
NPRS-PKS hybrid	1	6	6	8	5	3
Terpene synthase	10	4	6	9	6	5

Major facilitator superfamily	55	132	139	134	107	81
ATP-binding cassette superfamily	35	64	75	55	55	42
Sugar/inositol transporters	10	3	6	5	5	13
Appressorium differentiation protein (MAS-like protein)	2	4	4	6	6	4
Adhesins	0	2	2	2	2	2
Chloroperoxidases	7	0	1	4	2	2
Small secreted cysteine-rich protein	138	255	351	322	240	209

Compiled from Gao, Q., Jin, K., Ying, S. H., Zhang, Y., Xiao, G., Shang, Y., … Wang, C. (2011). Genome sequencing and comparative transcriptomics of the model entomopathogenic fungi Metarhizium anisopliae and M. acridum. *PLoS Genetics, 7*, e1001264; Zheng, P., Xia, Y., Xiao, G., Xiong, C., Hu, X., Zhang, S., … Wang, C.S. (2011). Genome sequence of the insect pathogenic fungus Cordyceps militaris, a valued traditional Chinese medicine. *Genome Biology, 12*, R116; Xiao, G. H., Ying, S-H., Zheng, Z., Wang, Z. L., Zhang, S., Xie, X. Q. … Feng, M.G. (2012). Genomic perspectives on the evolution of fungal entomopathogenicity in Beauveria bassiana. *Scientific Reports, 2*, 483; Hu, X., Zhang, Y. J., Xiao, G. H., Zheng, P., Xia, Y. L., Zhang, X. Y. … Wang, C.S. (2013). Genome survey uncovers the secrets of sex and lifestyle in caterpillar fungus. *Chinese Science Bulletin, 58*, 2846–2854 and Hu, X., Xiao, G., Zheng, P., Shang, Y., Su, Y., Zhang, X., … Wang, C.S. (2014). Trajectory and genomic determinants of fungal-pathogen speciation and host adaptation. *Proceedings of the National Academy of Sciences of the United States of America, 111*, 16796–16801.

cuticles and other potential targets in insect hosts (Gao et al., 2011; Xiao et al., 2012; Zheng et al., 2011). The *B. bassiana*/*C. militaris* lineage evolved into insect pathogens independently of the *Metarhizium* lineage, so similar expansion of these gene families is likely associated with functions necessary for insect pathogenesis and reflects convergent evolution.

Analysis of entomopathogen protein families shows a dynamic loss and gain of genes, but overall the generalists have more of these enzymes than do the narrow host range species (Table 1). Thus, compared to *M. album* and *M. acridum*, *M. robertsii* has evolved many expanded gene families of proteases, chitinases, cytochrome P450s, polyketide synthases (PKSs), and nonribosomal peptide synthetases (NRPSs) for cuticle-degradation, detoxification, and toxin biosynthesis, which may facilitate its ability to adapt to heterogenous environments (Gao et al., 2011). The basal *M. album* genome in particular highlights the early expansion of genes involved in cuticle degradation, as it has greater than threefold more trypsin genes than related plant endophytes and phytopathogens (Hu et al., 2014). However, compared with *M. album* (87 proteases) and *M. acridum* (116 proteases), there has been additional expansion of proteolytic capacity in other *Metarhizium* species (average 165 proteases) (Table 1). The dramatic expansion of proteases in *B. bassiana* resembles that of *M. robertsii*, suggesting that this is also an adaptation to a broad host range. Thus, *B. bassiana* has more subtilisins (43 vs 35) and trypsins (23 vs 12) than *C. militaris* and has four proteolytic subfamilies that *C. militaris* lacks (Table 1, Hu et al., 2013).

Most of the secreted effector-type proteins of plant pathogens are small (<300 amino acids) and contain four or more cysteine residues (Martin et al., 2008). Plant pathogens usually have more of these small, secreted cysteine-rich proteins (SSCPs) than saprophytes (Ohm et al., 2012). Therefore, the identification and analysis of SSCPs has been highlighted in genomic studies assessing many plant pathogenic and symbiotic fungi (Cheng et al., 2014). A survey of SSCPs in 11 fungal species including entomopathogens revealed 396 clusters of which 12 contained SSCPs shared by insect pathogens and the plant endophyte *Epichloë festucae*, and 26 are found in insect and plant pathogens (Xiao et al., 2012). Ninety-one of the 396 SSCP clusters are shared exclusively by the insect pathogens, suggesting many shared strategies for interacting with plants and insects are currently unknown. Of these 91 clusters, 52 contain genes from *B. bassiana*. Relative to other insect pathogens (average 307), the *B. bassiana* genome encodes more SSCPs (373), and many of them are species specific (154 vs an average of 95). As with other classes of secreted proteins, generalist insect pathogens have more

SSCPs than specialists (Table 1), indicating that SSCPs play roles in adaptation to diverse lifestyle options.

However, aside chitinases and hydrophobins (cell wall proteins that facilitate adhesion and morphological changes (Bidochka, St. Leger, & Roberts, 1995)), most of the entomopathogen SSCPs are of unknown function. Of 373 *B. bassiana* SSCPs, only 130 contained conserved domains. Of these, six *B. bassiana* SSCPs were identified as concanavalin A-like lectins, which could potentially function in interactions with both insects and plants. *Beauveria bassiana* also has four genes encoding proteins containing eight cysteine-containing extracellular membrane (CFEM) domains resembling pathogenicity determinants in plant pathogens (Xiao et al., 2012). As for the rest, aside unknown roles in entomopathogenicity, the functions identified for SSCPs in other fungi suggest they could play a mutualistic role in interactions with plants but also may be involved in antagonization of other fungi, bacteria, and higher soil organisms.

Ophiocordyceps sinensis, which probably infects insects through their spiracles (breathing holes) or orally, provides an illuminating exception to insect pathogens having large secretomes: gene families encoding epicuticular degrading CYP52 enzymes, cuticle-degrading proteases, and chitinases were greatly reduced in size (Table 1). In addition, protein families involved in adhesion to cuticles and formation of appressoria were absent or reduced in *O. sinensis* (Hu et al., 2013, Table 1). These gene losses are consistent with the inability of *O. sinensis* to breach intact cuticle. Hydrolytic enzymes, particularly proteases, can elicit host immune defenses (St. Leger, Joshi, Bidochka, & Roberts, 1996). In which case, the reduced number of cuticle-degrading enzymes in *O. sinensis* might also be an adaptation to avoid detection by the host during its extended latent phase. Copy number reduction was also evident for genes encoding known pathogen-associated molecular patterns such as lectins (6 in *O. sinensis* vs average 24 in other insect pathogens), further suggesting that selection for "stealth" (avoidance of host defenses) was a major force driving *O. sinensis* evolution. Evidently, as with plant pathogens (Ma et al., 2010; Stukenbrock et al., 2011), the differences found among the insect pathogens in protein family size are related to their modus operandi and host range.

5.1 Signal Transduction

To recognize and adapt to invertebrate environments, such as the insect cuticle, hemolymph, and cadaver, insect pathogens need to rapidly respond to changes in nutrient availability, osmolarity, and the host immune system

(Luo et al., 2012; Wang & St. Leger, 2006; Wang et al., 2008; Zhang et al., 2009). BLASTing insect pathogen genomes against the pathogen–host interaction gene database (a collection of experimentally verified pathogenicity, virulence, and effector genes from fungi, oomycetes, and bacteria) reveal large numbers of G protein–coupled receptors (GPCRs), protein kinases (PKs), and TFs that share similar sequences in the entomopathogen genomes. Fungal GPCRs sense extracellular cues and transmit the signals to distinct trimeric G protein subunits (Xue, Hsueh, & Heitman, 2008). Most of the insect pathogen GPCRs resemble Pth11-like proteins of the rice-blast fungus *Magnaporthe oryzae* (Xiao et al., 2012). Compared with *M. album* and *M. acridum*, there was a major expansion of GPCR-related proteins in the generalist *Metarhizium* spp., consistent with being able to recognize and respond to many more environmental triggers. In particular, the generalists showed a more than twofold expansion of Pth11-like receptors (average 51 vs 23 in specialists), with particular expansion in the subfamilies that are developmentally upregulated during early *Metarhizium* infection processes (germination and formation of infection structures on host cuticles) (Gao et al., 2011; Xiao et al., 2012). Consistent with early host recognition events being key to establishing specificity, *M. acridum*, but not *M. robertsii*, transcribed different *Pth11*-like GPCR genes on locust and cockroach cuticles indicating a role in recognizing specific hosts, or at least that these genes have a function that varies between strains with different host ranges. Comparatively analyzing the host-invading transcriptomes of *M. acridum* and *M. robertsii* showed recognition by regulatory controls that exclusively limit expression of genes for pathogenicity-related developmental processes to individual hosts (Gao et al., 2011).

Relative to *Metarhizium* spp., *B. bassiana* and *C. militaris* have fewer Pth11-like receptors (average 21). In particular, they lack a GPR1-like GPCR, which in yeast is activated during nitrogen starvation (Xue, Batlle, & Hirsch, 1998) and is upregulated by *M. robertsii* on the cuticle surface in response to the nitrogen poor conditions that are an important trigger for inducing production of infection structures (Gao et al., 2011; St. Leger, Bidochka, & Roberts, 1994; St. Leger et al., 1992). Thus, *B. bassiana* and *Metarhizium* have evolved different mechanisms for nutrient sensing.

Functional kinome analysis of the plant pathogen *F. graminearum* indicated that many PKs are involved in fungal growth, conidiation, pathogenesis, stress responses, toxin production, and/or sexual reproduction (C. Wang et al., 2011). Excluding *O. sinensis* (116 kinases), the insect pathogens average 170 PKs as compared to an average of 145 in the plant pathogens (Xiao et al., 2012). In contrast to GPCRs, the specialist *M. acridum* has more PKs (192) than the generalist *M. robertsii* (161) (Table 1). Based on just the *M. robertsii* and *M. acridum* genomes, Gao et al. (2011) suggested that the additional kinases in *M. acridum* may allow a more stringent discrimination between potential insect hosts and subsequent control of cell differentiation; consistent with this, *B. bassiana* has fewer PKs (159) than *C. militaris* (167) (Xiao et al., 2012). However, the additional data from the *M. album* (127 kinases) and *O. sinensis* genomes rule out a simple relationship between evolution of narrow host range and/or sexuality and proliferation of PKs.

Like other pathogenic fungi, insect pathogens have large numbers of TFs. However, *B. bassiana* has more (10) GATA-type TFs than *C. militaris* (5) and *Metarhizium* species (4–5) (Gao et al., 2011; Xiao et al., 2012). Fungal GATA-type TFs are involved in multiple functions, including nitrogen metabolism, light induction, siderophore biosynthesis, mating-type switching, and chromatin rearrangement (Scazzocchio, 2000). *Metarhizium robertsii* contains an array of 24 bZIP domain–containing TFs; gene deletion of one of these (*MBZ1*) revealed that it contributes to negative regulation of subtilisin proteases, but positive control of adhesin *MAD1* (Huang, Shang, Chen, Cen, et al., 2015), consistent with transcriptomic data showing little subtilisin production while spores are in the process of adhering to cuticle (Gao et al., 2011). Functional study of an *M. robertsii* C2H2-type TF (*MrPacC*) indicated that it was highly activated in alkaline conditions and positively controlled chitinase genes (Huang, Shang, Chen, Gao, et al., 2015). Cultures of *M. robertsii* always show a rapid increase in pH during production of chitinases (St. Leger, Cooper, & Charnley, 1986a), and *M. robertsii* alkalinizes a proteinaceous microenvironment, such as the insect cuticle, by producing ammonia (St. Leger, Nelson, & Screen, 1999). Digestion of cuticle proteins exposes the underlying chitin to enzymolysis (St. Leger, Cooper, & Charnley, 1986b), so linking alkalinization with chitinase production is adaptive for the fungus. It is likely that continued dissection of the roles of individual TFs in the manner of *MBZ1* and *MrPacC* will untangle the complexity and illuminate the interactions between what seem currently disconnected strands of biochemical and molecular data.

5.2 Carbohydrate-Active Enzymes

Chitin is the second most abundant polymer in insect cuticle. The necessity to degrade it is reflected in an abundance of GH18 family chitinases in *B. bassiana, C. militaris,* and *Metarhizium* spp. (on average 19) compared to plant pathogens (on average 11). Fungal chitinases are subdivided into three subgroups (Seidl, 2008). Xiao et al. (2012) found that eight of the 20 *B. bassiana* chitinases belong to subgroup A (without a chitin-binding domain, CBM), four belong to subgroup B (one CBM at the C-terminal), and eight belong to subgroup C chitinases (possessing LysM chitin—binding modules). Insect and plant pathogens have similar numbers of subgroup A chitinases, but the entomopathogens have many more chitinases with CBMs (average 11 vs 2). Phylogenetic analyses of subgroup B and C chitinases revealed that most of the gene duplication events have occurred since *B. bassiania/ C. militaris*, *Metarhizium* spp., and *Trichoderma* spp. diverged from a common ancestor, suggesting their abundance in each clade is due to convergent evolution (Xiao et al., 2012).

Many plant pathogens need GHs, pectate lyases, and cutinases to degrade the plant cuticle (waxy layer) and cell wall (Xu et al., 2006). Whole genome analyses (Fig. 1) indicate that ascomycete insect pathogens evolved from fungi adapted to grow on plants even though they now infect insects. This inference is supported by the consistent existence of genes for plant-degrading enzymes within their genomes (Gao et al., 2011; Hu et al., 2014). The number of GHs possessed by *M. robertsii* (159), *B. bassiana* (145), and *M. acridum* (130) is more than the endophytic close relation of *Metarhizium*, *E. festucae* (98) but is fewer than plant pathogens (average 199) (Xiao et al., 2012). This is because overall, insect pathogens have fewer genes associated with plant utilization than plant pathogens. This includes fewer cellulases (average of 10 in insect pathogens vs 25 in plant pathogens), oxidative lignin enzymes (29 vs 40), carbohydrate esterases (9 vs 33), cutinases (4 vs 12), and pectin lyases (8 vs 20) (Xiao et al., 2012). Two well-known virulence factors for plant pathogens are cutinases and pectin lyases (Hugouvieux-Cotte-Pattat, Condemine, Nasser, & Reverchon, 1996). *Metarhizium* spp., particularly *M. robertsii*, showed overall profiles of carbohydrate-active enzymes (CAZymes) more similar to plant pathogens than to other insect pathogens (Bushley et al., 2013). In part, this is because, compared to *Metarhizium* spp., *B. bassiana* and *C. militaris* have fewer enzymes for degrading xylan; although unlike *C. militaris* (Zheng et al., 2011), *B. bassiana* has a phosphoketolase required for xylose metabolism and full virulence in *Metarhizium* (Duan, Shang, Gao, Zheng, & Wang, 2009). Thus, in contrast to *C. militaris*, *B. bassiana* grows on xylose, albeit very weakly when compared to *M. robertsii* (Xiao et al., 2012). Interestingly, *E. festucae* lacks the same xylanases as *B. bassiana* (the GH11 family) (Xiao et al., 2012). As an alternate endophyte, *B. bassiana* presumably possesses mechanisms to avoid stimulating plant defenses. Fungal endoxylanases are known to trigger plant immune responses and induce necrosis of infected plant tissue (Noda, Brito, & González, 2010). Therefore, lack of GH11 in *B. bassiana* and *E. festucae* could be an adaptation limiting induction of necrosis by these endophytes, and thereby facilitating immune evasion.

To put the plant-degrading genes of entomopathogens in context, fungal pathogens of humans that are seldom recovered from soil, such as *Coccidioides*, exhibit few of these enzymes or none (Sharpton et al., 2009). Even necrophytes such as *Trichoderma reesei* lack many families of plant cell wall–degrading enzymes (Martinez et al., 2008), and the existence of such families in insect pathogens that are associated with plants implies that these species are able to utilize living plant tissues. Potentially, these enzymes could also facilitate colonization of plant surfaces, but this must remain speculative because the genetic basis for rhizosphere competence is largely obscure in fungi in general (St. Leger, 2008). Identification of the full repertoire of *Metarhizium* and *Beauveria* genes will help to identify those responsible for life on plant roots and shoots.

As in many other aspects of its biology, *O. sinensis* represents a special case with respect to CAZymes, with many fewer GHs (66 vs average 139 in other insect fungi) (Hu et al., 2013, Table 1). Most of the missing enzymes are devoted to degradation of plant materials indicating that, unlike related insect pathogens, *O. sinensis* may be exclusively insect parasitic.

5.3 Secondary Metabolites and Host Interaction

Most hypocrealean insect pathogenic fungi produce large numbers of secondary metabolites that seem to have no role in normal fungal metabolism but are highly active in insect tissues (Xu, Luo, Gao, Shang, & Wang, 2015). These are assumed to be part of an ongoing evolutionary arms race between fungi and insects. As pointed out by Donzelli and Krasnoff (2016), there are a lot of *Metarhizium* secondary metabolites that cannot currently be connected to genes and lots of genes in entomopathogen genomes that cannot be connected to known chemistry. The key fungal secondary metabolite-producing enzymes are NRPSs, PKSs, and terpene cyclases (TCs). Hypocrealean fungi are known to synthesize a number of products with activity against insects via both NRPSs and PKSs, such as destruxins (*M. robertsii*) (Pal, St. Leger, & Wu, 2007; Wang et al., 2012), efrapeptins (*Tolypocladium* spp.; Krasnoff, Gupta, St. Leger, Renwick, & Roberts, 1991), and ergot alkaloids (clavicipitalean endophytes of grasses including *Claviceps* and *Epichloë* spp.; Torres, Singh, Vorsa, & White, 2008). Many of these compounds also have pharmaceutical applications and/or roles in antibiosis, pathogenesis, and competitive interactions between organisms (Pal et al., 2007).

In terms of PKS genes and putative polyketides, generalist *Metarhizium* spp., such as *M. robertsii*, appear to possess a greater potential for the production of secondary metabolites than specialist strains and other sequenced ascomycetes, even *Fusarium* species (Chen, Feng, Shang, Xu, & Wang, 2015; Gao et al., 2011; Hu et al., 2014). Donzelli and Krasnoff (2016) report that the lowest number of secondary metabolite genes is found in *M. album* (48 in total) and the highest is in *M. guizhouense* (93 in total). Acting collectively with the secretome, the number and diversity of these effectors may contribute to the ability of generalists to infect and kill a much wider variety of insects than a specialist, such as *M. acridum*. This also relates neatly to pathogenic strategies because, as described previously, *M. robertsii* kills hosts quickly via toxins and grows saprophytically in the cadaver. In contrast, like many specialists, *M. acridum* causes a systemic infection of host tissues before the host dies which suggests a much smaller role for toxins (Kulkarni, Thon, Pan, & Dean, 2005). The best-studied *Metarhizium* toxins are the destruxins, which suppress insect cellular and humoral immune responses (Pal et al., 2007; Wang et al., 2012). The gene cluster responsible, *dtxS1-dtxS4* (Wang et al., 2012), is present in all surveyed *Metarhizium* species except for *M. acridum* and *M. album*. Intriguingly, minor differences in virulence were found between applications (Donzelli, Krasnoff, Sun-Moon, Churchill, & Gibson, 2012; Wang et al., 2012) of *M. robertsii* DTX mutants and wild-type strains. When the removal of a redundant toxin cluster fails to markedly reduce virulence, this points to a complexity of host-range determination far beyond individual toxic metabolites (Donzelli and Krasnoff, 2016).

As reviewed by Donzelli and Krasnoff (2016), *B. bassiana* and *C. militaris* (Cordycipitaceae) share very few secondary metabolite pathways with *Metarhizium* (Clavicipitaceae); more often, sequenced species from Ophiocordicipitaceae have more pathways in common with *Metarhizium*. *Beauveria* is well known for producing a large array of biologically active secondary metabolites (eg, oosporein, bassianin, tenellin, beauvericin, bassianolides, and beauveriolides) and secreted metabolites involved in pathogenesis and virulence (eg, oxalic acid) that have potential or realized industrial, pharmaceutical, and agricultural uses (Molnár, Gibson, & Krasnoff, 2010). As reviewed by Ortiz-Urquiza and Keyhani (2016), the pathways for a number of these *B. bassiana* secondary metabolites have been examined, including oosporein, bassianolide, and beauvericin. In most cases, targeted gene inactivation of key genes in these pathways resulted in a significant reduction in virulence that, in the case of oosporein, may be accounted for by an impaired suppression of insect host immunity (Feng et al., 2015). Silkworm larvae infected by *B. bassiana* (batryticated silkworms, also called *Bombycis corpus* or white-stiff silkworm) have for centuries been a traditional Chinese medicine, and their potential as a medicine has been validated by modern technologies, eg, water extract of batryticated silkworms protect against α-amyloid-induced neurotoxicity (Koo et al., 2003). *Cordyceps militaris* produces cordycepin and cordycepic acid, which have been linked to a variety of benefits from antiaging to sleep-regulating effects (Paterson, 2008). Cordycepin has been investigated for use against cancers (Liao et al., 2015). Likewise, the pyridine alkaloid fumosorinone identified from *Isaria fumosorosea* has potential to treat type II diabetes and other associated metabolic syndromes (Liu et al., 2015).

In this chapter, we have seen that there are genomic similarities between *O. sinensis* and obligate plant pathogens. However, *O. sinensis* differs from the plant pathogens, which have lost secondary metabolites, in having multiple PKSs, modular NRPSs, and TCs for production of an array of secondary metabolites (Table 1). These most likely play roles after the latent period when the fungus is colonizing and killing the host, and they are also likely candidates for production of the pharmacologically active compounds that this fungus is famous for. *Ophiocordyceps sinensis* encodes four terpenoid synthases and one terpenoid cyclase which are absent in other fungi, indicating that this fungus, likened to "natural Viagra," produces novel terpenoids. However, many putative secondary metabolism clusters were conserved between *O. sinensis* and other insect pathogens, providing singular exceptions to the restructuring of the *O. sinensis* genome by repeat elements and suggesting that the physical linkage of secondary metabolite biosynthetic genes has strong adaptive significance for entomopathogenicity. Other members of the family Ophiocordycipitaceae host a variety of secondary metabolites involved in pathogenesis with potential usefulness to humans. *Hirsutella thompsonii* produces Hirsutellin A, which has been tested against honey bee parasitic *Varroa* mites (https://www.google.com/patents/US6277371). *Tolypocladium inflatum* produces a range of insecticidal compounds: most notably cyclosporine, an immunosuppressant in humans, as well as in insects, which is exploited in that role to prevent transplant rejection. Phylogenomic analyses revealed complex patterns of homology between the NRPS that encodes for cyclosporin synthetase and those of other secondary metabolites with activities against insects (eg, beauvericin, destruxins, etc.) and demonstrated the roles of module duplication and gene fusion of distantly related NRPS modules in diversification of NRPSs (Bushley et al., 2013). Famously, *O. unilateralis s.l.* alters infected ant behavior (reviewed by Hughes et al., 2016), instigating them to climb and bite down on elevated vegetation to assist the fungus in spore transmission (Shang et al., 2015). Recent genomic and transcriptomic data revealed that *O. unilateralis* secretes a plethora of putative compounds that have an effect on behavior, including ergot, enterotoxins, alkaloids, polyketides, and nonribosomal peptides (de Bekker et al., 2015). Ergot famously causes serotonergic overstimulation of the central nervous system in humans and livestock leading to ergotism (Eadie, 2003), and polyketides, nonribosomal peptides, and alkaloids have been suggested to be neuromodulatory agents (Molnár et al., 2010).

5.4 Protein Families Involved in Detoxification and Stress Responses

As a plant symbiont and an entomopathogen, *Metarhizium* encounters many different selective pressures. During these two lifestyles, the fungus is confronted with a wide range of abiotic (eg, UV, pH, temperature, etc.) and biotic stresses (eg, insect host defenses), which the fungus has adapted to overcome. The specific genes involved in these processes are discussed in Lovett and St. Leger (2015).

Relative to plant pathogens, amidohydrolases, glyoxalases, and monooxygenases are all expanded in insect pathogens implying that consistent with their broad lifestyle options, the latter are better able to detoxify corresponding compounds. Dehydrogenases and CYPs are also abundant in insect pathogens and are reflective of the lifestyle of the organism. They are involved in multiple essential physiological processes, including degradation/detoxification of xenobiotics, biotransformations, and the biosynthesis of bioactive metabolites (Crešnar & Petrič, 2011). The profile of dehydrogenases can be used to highlight metabolic pathways and metabolic flexibility, ie, the ability to switch between energy sources. By this measure, *M. robertsii* has the greatest flexibility (313 dehydrogenases), and *O. sinensis* the least (103 dehydrogenases; Table 1), which is consistent with their known lifestyles. *Beauveria bassiana* has a comparatively modest 252 dehydrogenases, suggesting fewer lifestyle options than *M. robertsii*. Entomopathogen dehydrogenases with known functions are widely distributed. Thus, mannitol-1-phosphate dehydrogenase and mannitol dehydrogenase contribute to *B. bassiana*'s tolerance to H_2O_2, ultraviolet radiation, and heat stresses by regulating mannitol accumulation (Zhang, Xia, & Keyhani, 2011). Homologs of these genes are present in *Metarhizium* spp. genomes. In fact, *M. robertsii* is particularly enriched in dehydrogenases (17 in *M. roberstii*; 7 in *M. acridum*) required for the biosynthesis of mannitol (Gao et al., 2011).

The genome of *M. robertsii* encodes 123 highly divergent CYP genes versus an average of 118 in ascomycete plant pathogens (Xiao et al., 2012), 100 in *M. acridum*, 83 in *B. bassiana*, and 57 in both *C. militaris* and *O. sinensis* (Table 1). CYP genes are involved in oxygenation steps during alkane assimilation and the biosynthesis of secondary metabolites, as well as with detoxification. Twenty-four CYP families present in *Metarhizium* spp. are absent in *B. bassiana* and *C. militaris*, including nitric oxide reductases (CYP family 55) used for anaerobic denitrification (Xiao et al., 2012). Thus, unlike *Metarhizium* spp., *B. bassiana* and *C. militaris* may not be able to respond to hypoxic conditions in a host by fermenting nitrate or ammonia. *Ophiocordyceps sinensis* also lacks CYP55 (Hu et al., 2013), unlike the closely related *H. thompsonii* (Agrawal et al., 2015), confirming that this enzyme has been lost repeatedly in different lineages of insect pathogens.

Outer membrane transporters serve as the means by which nutrients are absorbed by the cell and as a mechanism for the export of toxic molecules. Most transporters belong to the major facilitator superfamily (MFS) and the ATP-binding cassette (ABC) superfamily. In fungi, many ABC transporters have been implicated in drug resistance and could, therefore, serve to defend the fungus from host-produced secondary metabolites, whereas MFS proteins are typically involved in the transport of a wide range of substrates and may function as nutrient sensors (Morschhauser, 2010). Interestingly, entomopathogen species in general have many more amino acid and peptide transporters than other fungi, consistent with their ability to access a range of protein degradation products from insect sources (Gao et al., 2011; Xiao et al., 2012). Otherwise, as with other categories of genes involved in stress responses and detoxification, generalist genomes are particularly enriched in transporters (Table 1). *Metarhizium robertsii* and *B. bassiana* have a similar repertoire of MFS transporters, but *B. bassiana* has a clear advantage in ABC transporters compared to *M. robertsii* (75 versus 55, respectively). Ortiz-Urquizza and Keyhani review the impacts on virulence and stress resistance imparted to *B. bassiana* by ABC transporters in this volume.

Among insect pathogens, *B bassiana* has the highest numbers of epoxide hydrolases (7 vs 4–5 in others), nitrilases (11 vs 7–10 in others), and monooxygenases (36 vs 16–30 in others) (Xiao et al., 2012). Along with its enhanced profile of ABC transporters and comparative paucity of CYP genes, this indicates that *B. bassiana* may have evolved unique solutions to overcoming environmental challenges in plants and insects. The monooxygenases in particular might be involved in rapid elimination of insect polyphenolics by *ortho*-hydroxylation of phenols to catechols. *Beauveria bassiana* is a well-known whole cell catalyst in industrial applications, and some of these enzymes have the potential to be involved in bioconversions. Epoxide hydrolases are highly versatile biocatalysts for the hydrolysis of epoxide, and nitrilases are valuable alternatives to chemical catalysts for biotransformation of various organic nitriles (Xiao et al., 2012).

6. HORIZONTAL GENE TRANSFER

Horizontal gene transfer (HGT) is the transfer of genetic material directly from the genome of one organism to that of another, rather than between parent and offspring. HGT between distantly related bacteria is a commonplace process that allows them to quickly share adaptive genes, ie, an antibiotic-resistance set of genes to adapt to an antibiotic. It also contributes significantly to the emergence of new bacterial pathogens, but HGT is usually thought to play a minor role in eukaryotes. However, data from multiple genomic sequences suggest that HGT has also occurred from prokaryotes to eukaryotes and between eukaryotes (Crisp, Boschetti, Perry, Tunnacliffe, & Micklem, 2015). Among the eukaryotes, the chitin cell wall of fungi is considered a structural barrier, likely reducing the frequency of HGT, but even so, there are numerous instances of HGT in fungi. The amount of transferred DNA ranges from single genes, to secondary metabolite gene clusters and even to whole chromosomes (Fitzpatrick, 2012). Richards, Leonard, Soanes, and Talbot (2011) analyzed the functions of 323 genes probably originating from HGT into fungi and concluded that HGT played a role in expanding the nutrient acquisition and environmental colonization capacities of many fungi.

Detection of horizontally transferred genes traditionally depends on incongruences in phylogenies: the phylogeny of a given gene does not match a known species phylogeny. However, discrepancies in phylogeny can have other explanations besides HGT, including gene loss, eg, a gene that existed in a distant ancestor could have simply been lost in many relatives. A phylogenomic survey of fungal gene family evolution suggested that individual protease genes have been lost many times independently in different lineages, and that flux of genes is an ongoing process (Hu & St. Leger, 2004).

Issues distinguishing between HGT and gene loss can be potentially circumvented using fully sequenced genomes and filters based on known phylogenetic relationships to look for genes that have been horizontally transferred since speciation.

To evaluate the extent of gene transfer into insect pathogenic fungi, Hu et al. (2014) analyzed seven *Metarhizium* genomes to quantitate how closely a gene aligned to a nonfungal sequence compared to a bacterial or metazoan sequence. Based on this score, they determined that the broad host range species have obtained more bacterial genes than the specialists (on average 63 vs 27), indicating considerable variability in the rates of gain and/or loss of bacterial genes within the estimated 117 million year divergence of the *Metarhizium* species. Orthology analysis revealed that only eight HGT genes are present in all *Metarhizium* species, whereas the remaining genes were probably acquired clade- or species specifically because they are absent from multiple species. In a few cases the importance of the transferred genes has been functionally demonstrated with gene knockouts. For example, experimentally verified, functionally important, bacterial-like proteins in *M. robertsii* encode a chymotrypsin (Screen & St. Leger, 2000), a pentose metabolizing phosphoketolase (Duan et al., 2009), a cold shock protein (Fang & St. Leger, 2010b), and a putative chitinase that are present in all seven *Metarhizium* species, indicating that these genes were acquired before the *Metarhizium* lineages radiated. These genes are all most similar to sequences in soil- and insect-dwelling bacteria, providing an ecological niche overlap between the bacteria and the fungus. Not surprisingly, therefore, hosts are also a source of adaptive traits. A virulence factor (Mr-NPC2a) of *M. robertsii* was horizontally acquired from an insect and allows *Metarhizium* to compete with insect hosts for the sterols necessary to maintain cell membrane integrity (Zhao et al., 2014). This sequence was also present in all seven *Metarhizium* species, suggesting it was acquired by a common ancestor and that HGT from hosts, as well as associated bacteria, has played a role in shaping functional diversity of the gene repertoire in entomopathogens. More recent acquisitions may include a putative insect-like trypsin that is absent only in *M. album* and a putative TcdB-like insecticidal toxin protein that is unique to *M. robertsii* (Hu et al., 2014). Likewise, around half of the gene clusters encoding secondary metabolites in *M. album* (11/19) and *M. acridum* (11/25) are species-specific (vs an average of 7 in the other species), although most are also present in other fungal genera. This implies that they were laterally acquired from other fungi during *Metarhizium* speciation (Hu et al., 2014). For example, Hu et al. (2014) suggest that the destruxin gene cluster may have been acquired 15 MYA by *Metarhizium* lineages that were broadening their host range.

These results suggest that *Metarhizium* genomes have all been subject to low levels of HGT; some of the adaptive HGT events are ancient (>100 MYA), but generalist species have been more susceptible than specialists to HGT events in the last 30 MY, consistent with their being exposed to a greater variety of potential donor DNA. Thus, throughout their evolution, *Metarhizium* spp. may have acquired new virulence mechanisms by HGT and presumably increases in pathogen virulence through this process evolved in a stepwise manner.

7. CONCLUSIONS AND FUTURE PERSPECTIVES

Sequencing related entomopathogen species that have evolved specialist or generalist lifestyles has increased their utility as models and provided insights into the evolution of pathogenicity. Thus, cross-species comparative analysis has identified novel and specialized virulence mechanisms and, compared to experimental methods, has allowed for more rapid identification of genes encoding biologically active molecules and genes responsible for interactions between fungi, plants, and insects. These analyses also show how much we still need to do to comprehensively understand these interactions. For example, most of the abundant SSCPs produced by entomopathogens are of unknown function, highly species-specific, and lack similarity to known proteins. Undoubtedly, the information from comparative genomics will benefit future functional studies of insect—fungus interactions. Entomopathogenic fungi present a vast reservoir of biopolymer-degrading enzymes adapted to a wide range of temperatures and environments. The new information on these abundant enzymes will also facilitate more extensive work to determine mechanisms of the biotransformation reactions that make these fungi such useful industrial catalysts, as well as reveal potentially useful compounds for medicinal use. Overall, entomopathogen genome sequences will help realize the still undeveloped potential possessed by these fungi both as insect pathogens and as microbial biocatalysts, as well as illuminate their, as yet, poorly understood role as endophytes and plant symbionts (reviewed by Moonjely et al., 2016).

Given that specialization has occurred many times in *Metarhizium*, it provides an opportunity to study a genus with species containing a large number of independently evolved models of adaptation and response. Multispecies exploration of genome evolution of *Metarhizium* spp. has already led to better understanding of mechanisms by which novel pathogens emerge with either wide or narrow host ranges (Hu et al., 2014). However, the thousands of different entomopathogen species display extraordinary diversity, especially given the number of different products that can be studied (Isaka et al., 2005). In this context, we can look forward to the completion of the current Entomophthoromycota sequencing projects. These are very virulent pathogens that are highly adapted to their hosts and often devastate pest populations in spectacular epizootics. In order to increase the power and accuracy of comparative analyses, we need to sequence a much more extensive sampling of entomopathogenic fungi.

ACKNOWLEDGMENTS

This work was supported by the US National Science Foundation Grant IOS-1257685 and the NIAID of the National Institutes of Health under award number RO1 AI106998 to Raymond St. Leger and by the National Nature Science Foundation of China Grants 31530001 and 31225023 to Chengshu Wang.

REFERENCES

Agrawal, Y., Khatri, I., Subramanian, S., & Shenoy, B. D. (2015). Genome sequence, comparative analysis, and evolutionary insights into chitinases of entomopathogenic fungus *Hirsutella thompsonii*. *Genome Biology and Evolution, 7*, 916–930.

Alby, K., Schaefer, D., & Bennett, R. J. (2009). Homothallic and heterothallic mating in the opportunistic pathogen *Candida albicans*. *Nature, 480*, 890–893.

van Asbeck, E. C., Clemons, K. V., & Stevens, D. A. (2009). *Candida* parapsilosis: a review of its epidemiology, pathogenesis, clinical aspects, typing and antimicrobial susceptibility. *Critical Reviews in Microbiology, 35*, 283–309.

de Bekker, C., Ohm, R. A., Loreto, R. G., Sebastian, A., Albert, I., Merrow, M.… Hughes, D. P. (2015). Gene expression during zombie ant biting behavior reflects the complexity underlying fungal parasitic behavioral manipulation. *BMC Genomics, 16*, 620.

Bidochka, M. J., Clark, D. C., Lewis, M. W., & Keyhani, N. O. (2010). Could insect phagocytic avoidance by entomogenous fungi have evolved via selection against soil amoeboid predators? *Microbiology, 156*, 2164–2171.

Bidochka, M. J., Kamp, A. M., Lavender, T. M., Dekoning, J., & Amritha De Croos, J. N. (2001). Habitat association in two genetic groups of the insect-pathogenic fungus *Metarhizium anisopliae*: uncovering cryptic species? *Applied and Environmental Microbiology, 67*, 1335–1342.

Bidochka, M. J., St. Leger, R. J., & Roberts, D. W. (1995). The rodlet layer from aerial and submerged conidia of the entomopathogenic fungus *Beauveria bassiana* contains hydrophobin. *Mycological Research, 99*, 403–406.

Bischoff, J. F., Rehner, S. A., & Humber, R. A. (2009). A multilocus phylogeny of the *Metarhizium anisopliae* lineage. *Mycologia, 101*, 512–530.

Biswas, C., Dey, P., Satpathy, S., & Satya, P. (2012). Establishment of the fungal entomopathogen *Beauveria bassiana* as a season long endophyte in jute (*Corchorus olitorius*) and its rapid detection using SCAR marker. *BioControl, 57*, 565–571.

Blair, J. E., Coffey, M. D., Park, S. Y., Geiser, D. M., & Kang, S. (2008). A multi-locus phylogeny for *Phytophthora* utilizing markers derived from complete genome sequences. *Fungal Genetics and Biology, 45*, 266–277.
Brownbridge, M., Reay, S. D., Nelson, T. L., & Glare, T. R. (2012). Persistence of *Beauveria bassiana* (Ascomycota: Hypocreales) as an endophyte following inoculation of radiata pine seed and seedlings. *Biological Control, 61*, 194–200.
Bushley, K. E., Raja, R., Jaiswal, P., Cumbie, J. S., Nonogaki, M., Boyd, A. E.… Spatafora, J. W. (2013). The genome of *Tolypocladium inflatum*: evolution, organization, and expression of the cyclosporin biosynthetic gene cluster. *PLoS Genetics, 9*, e1003496.
Cannon, P. F., Hywel-Jones, N. L., Maczey, N., Norbu, L., Samdup, T. T., & Lhendup, P. (2009). Steps towards sustainable harvest of *Ophiocordyceps sinensis* in Bhutan. *Biodiversity and Conservation, 18*, 2263–2281.
Chang, Y., Wang, S., Sekimoto, S., Aerts, A. L., Choi, C., Clum, A.… Berbee, M. L. (2015). Phylogenomic analyses indicate that early fungi evolved digesting cell walls of algal ancestors of land plants. *Genome Biology and Evolution, 7*, 1590–1601.
Chen, Y. X., Feng, P., Shang, Y. F., Xu, Y.-J., & Wang, C. S. (2015). Biosynthesis of non-melanin pigment by a divergent polyketide synthase in *Metarhizium robertsii*. *Fungal Genetics and Biology, 81*, 142–149.
Cheng, Q., Wang, H., Xu, B., Zhu, S., Hu, L., & Huang, M. (2014). Discovery of a novel small secreted protein family with conserved N-terminal IGY motif in Dikarya fungi. *BMC Genomics, 15*, 1151.
Cordaux, R., & Batzer, M. A. (2009). The impact of retrotransposons on human genome evolution. *Nature Reviews Genetics, 10*, 691–703.
Crešnar, B., & Petrič, S. (2011). Cytochrome P450 enzymes in the fungal kingdom. *Biochimica et Biophysica Acta, 1814*, 29–35.
Crisp, A., Boschetti, C., Perry, M., Tunnacliffe, A., & Micklem, G. (2015). Expression of multiple horizontally acquired genes is a hallmark of both vertebrate and invertebrate genomes. *Genome Biology, 16*, 50.
Cuomo, C. A., Güldener, U., Xu, J. R., Trail, F., Turgeon, B. G., Di Pietro, A.… Kistler, H. C. (2007). The *Fusarium graminearum* genome reveals a link between localized polymorphism and pathogen specialization. *Science, 317*, 1400–1402.
Daboussi, M. J., & Capy, P. (2003). Transposable elements in filamentous fungi. *Annual Review of Microbiology, 57*, 275–299.
De Fine Licht, H. H., Hajek, A. E., Eilenberg, J., & Jensen, A. B. (2016). Utilizing Genomics to Study Entomopathogenicity in the Fungal Phylum Entomophthoromycota: A Review of Current Genetic Resources. *Advances in Genetics, 94*, 41–65.
Donzelli, B. G. G., Krasnoff, S. B., Sun-Moon, Y., Churchill, A. C. L., & Gibson, D. M. (2012). Genetic basis of destruxin production in the entomopathogen *Metarhizium robertsii*. *Current Genetics, 58*, 105–116.
Donzelli, B. G. G., & Krasnoff, S. B. (2016). Molecular Genetics of Secondary Chemistry in Metarhizium Fungi. *Advances in Genetics, 94*, 365–436.
Duan, Z. B., Chen, Y. X., Huang, W., Shang, Y. F., Chen, P. L., & Wang, C. S. (2013). Linkage of autophagy to fungal development, lipid storage and virulence in *Metarhizium robertsii*. *Autophagy, 9*, 538–549.
Duan, Z. B., Shang, Y. F., Gao, Q., Zheng, P., & Wang, C. S. (2009). A phosphoketolase Mpk1 of bacterial origin is adaptively required for full virulence in the insect-pathogenic fungus *Metarhizium anisopliae*. *Environmental Microbiology, 11*, 2351–2360.
Dyer, P. S., & O'Gorman, C. M. (2012). Sexual development and cryptic sexuality in fungi: insights from *Aspergillus* species. *FEMS Microbiology Review, 36*, 165–192.
Eadie, M. J. (2003). Convulsive ergotism: epidemics of the serotonin syndrome? *Lancet Neruology, 2*, 429–434.

Fang, W. G., Pava-Ripoll, M., Wang, S. B., & St. Leger, R. J. (2009). Protein kinase A regulates production of virulence determinants by the entomopathogenic fungus, *Metarhizium anisopliae*. *Fungal Genetic and Biology, 46*, 277–285.

Fang, W. G., & St. Leger, R. J. (2010a). Mrt, a gene unique to fungi, encodes an oligosaccharide transporter and facilitates rhizosphere competency in *Metarhizium robertsii*. *Plant Physiology, 154*, 1549–1557.

Fang, W. G., & St. Leger, R. J. (2010b). RNA binding proteins mediate the ability of a fungus to adapt to the cold. *Environmental Microbiology, 12*, 810–820.

Feng, P., Shang, Y. F., Cen, K., & Wang, C. S. (2015). Fungal biosynthesis of the bibenzoquinone oosporein to evade insect immunity. *Proceedings of the National Academy of Sciences of the United States of America, 112*, 11365–11370.

Fitzpatrick, D. A. (2012). Horizontal gene transfer in fungi. *FEMS Microbiology Letters, 329*, 1–8.

Fraser, J. A., Diezmann, S., Subaran, R. L., Allen, A., Lengeler, K. B., Dietrich, F. S. ... Heitman, J. (2004). Convergent evolution of chromosomal sex-determining regions in the animal and fungal kingdoms. *PLoS Biology, 2*, e384.

Freimoser, F. M., Hu, G., & St. Leger, R. J. (2005). Variation in gene expression patterns as the insect pathogen *Metarhizium anisopliae* adapts to different host cuticles or nutrient deprivation in vitro. *Microbiology, 151*, 361–371.

Galagan, J. E., Calvo, S. E., Borkovich, K. A., Selker, E. U., Read, N. D., Jaffe, D. ... Birren, B. (2003). The genome sequence of the filamentous fungus *Neurospora crassa*. *Nature, 422*, 859–868.

Galagan, J. E., Calvo, S. E., Cuomo, C., Ma, L. J., Wortman, J. R., Batzoglou, S. ... Birren, B. W. (2005). Sequencing of *Aspergillus nidulans* and comparative analysis with *A. fumigatus* and *A. oryzae*. *Nature, 438*, 1105–1115.

Gao, Q., Jin, K., Ying, S. H., Zhang, Y., Xiao, G., Shang, Y. ... Wang, C. (2011). Genome sequencing and comparative transcriptomics of the model entomopathogenic fungi *Metarhizium anisopliae* and *M. acridum*. *PLoS Genetics*, 7, e1001264.

Gao, Q., Shang, Y. F., Huang, W., & Wang, C. S. (2013). Glycerol-3-phosphate acyltransferase contributes to triacylglycerol biosynthesis, lipid droplet formation and host invasion in *Metarhizium robertsii*. *Applied and Environmental Microbiology, 79*, 7646–7653.

Geiser, D. M., Timberlake, W. E., & Arnold, M. L. (1996). Loss of meiosis in *Aspergillus*. *Molecular Biology and Evolution, 13*, 809–817.

Gladieux, P., Ropars, J., Badouin, H., Branca, A., Aguileta, G., de Vienne, D. M. ... Giraud, T. (2014). Fungal evolutionary genomics provides insight into the mechanisms of adaptive divergence in eukaryotes. *Molecular Ecology, 23*, 753–773.

Hajek, A. E., McManus, M. L., & Delalibera, I., Jr. (2007). A review of introductions of pathogens and nematodes for classical biological control of insects and mites. *Biological Control, 41*, 1–13.

Hajek, A. E., & Tobin, P. C. (2011). Introduced pathogens follow the invasion front of a spreading alien host. *Journal of Animal Ecology, 80*, 1217–1226.

Hay, W. W., Soeding, E., DeConto, R. M., & Wold, C. N. (2002). The late cenozoic uplift-climate change paradox. *International Journal of Earth Science, 91*, 746–774.

Hu, G., & St. Leger, R. J. (2002). Field studies using a recombinant mycoinsecticide (*Metarhizium anisopliae*) reveal that it is rhizosphere competent. *Applied and Environmental Microbiology, 68*, 6383–6387.

Hu, G., & St. Leger, R. J. (2004). A phylogenomic approach to reconstructing the diversification of serine proteases in fungi. *Journal of Evolutionary Biology, 17*, 1204–1214.

Hu, X., Xiao, G., Zheng, P., Shang, Y., Su, Y., Zhang, X. ... Wang, C. S. (2014). Trajectory and genomic determinants of fungal-pathogen speciation and host adaptation. *Proceedings of the National Academy of Sciences of the United States of America, 111*, 16796–16801.

Hu, X., Zhang, Y. J., Xiao, G. H., Zheng, P., Xia, Y. L., Zhang, X. Y. … Wang, C. S. (2013). Genome survey uncovers the secrets of sex and lifestyle in caterpillar fungus. *Chinese Science Bulletin, 58*, 2846–2854.

Huang, W., Shang, Y. F., Chen, P. L., Cen, K., & Wang, C. S. (2015). Basic leucine zipper (bZIP) domain transcription factor MBZ1 regulates cell wall integrity, spore adherence, and virulence in *Metarhizium robertsii*. *Journal of Biological Chemistry, 290*, 8218–8231.

Huang, W., Shang, Y. F., Chen, P. L., Gao, Q., & Wang, C. S. (2015). MrpacC regulates sporulation, insect cuticle penetration and immune evasion in *Metarhizium robertsii*. *Environmental Microbiology, 17*, 994–1008.

Hughes, D. P., Araujo, J. P. M., Loreto, R. G., Quevillon, L., de Bekker, C., & Evans, H. C. (2016). From So Simple a Beginning: The Evolution of Behavioral Manipulation by Fungi. *Advances in Genetics, 94*, 437–469.

Hugouvieux-Cotte-Pattat, N., Condemine, G., Nasser, W., & Reverchon, S. (1996). Regulation of pectinolysis in *Erwinia chrysanthemi*. *Annunal Review and Microbiology, 50*, 213–257.

Idnurm, A. (2011). Sex and speciation: the paradox that non-recombining DNA promotes recombination. *Fungal Biology Reviews, 25*, 121–127.

Isaka, M., Kittakoop, P., Kirtikara, K., Hywel-Jones, N. L., & Thebtaranonth, Y. (2005). Bioactive substances from insect pathogenic fungi. *Accounts of Chemical Research, 38*, 813–823.

Jackson, A. P. (2004). A reconciliation analysis of host switching in plant-fungal symbioses. *Evolution, 58*, 1909–1923.

Juenger, T., & Bergelson, J. (1998). Pairwise versus diffuse natural selection and the multiple herbivores of scarlet gilia, *Ipomopsis aggregata*. *Evolution, 52*, 1583–1592.

Kang, C. S., Goo, B. Y., Gyu, L. D., & Heon, Y. (1996). Antifungal activities of *Metarhizium anisopliae* against *Fusarium oxysporum, Botrytis cinerea* and *Alternaria solani*. *Korean Journal of Mycology, 24*, 49–55.

Kepler, R. M., Sung, G. H., Ban, S., Nakagiri, A., Chen, M. J., Huang, B. … Spatafora, J. W. (2012). New teleomorph combinations in the entomopathogenic genus *Metacordyceps*. *Mycologia, 104*, 182–197.

Kershaw, M. J., Moorhouse, E. R., Bateman, R., Reynolds, S. E., & Charnley, A. K. (1999). The role of destruxins in the pathogenicity of *Metarhizium anisopliae* for three species of insects. *Journal of Invertebrate Pathology, 74*, 213–223.

Kim, H. G., Song, H., Yoon, D. H., Song, B. W., Park, S. M., Sung, G. H. … Kim, T. W. (2010). *Cordyceps pruinosa* extracts induce apoptosis of HeLa cells by a caspase dependent pathway. *Journal of Ethnopharmacology, 128*, 342–351.

Koo, B. S., An, H. G., Moon, S. K., Lee, Y. C., Kim, H. M., Ko, J. H., & Kim, C. H. (2003). *Bombycis corpus* extract (BCE) protects hippocampal neurons against excitatory amino acid-induced neurotoxicity. *Immunopharmacology and Immunotoxicology, 25*, 191–201.

Krasnoff, S. B., Gupta, S., St. Leger, R. J., Renwick, J. A. A., & Roberts, D. W. (1991). Antifungal and insecticidal properties of the efrapeptins: metabolites of the fungus *Tolypocladium niveum*. *Journal of Invertebrate Pathology, 58*, 180–188.

Kronstad, J. W., & Staben, C. (1997). Mating type in filamentous fungi. *Annual Review of Genetics, 31*, 245–276.

Kubicek, C. P., Herrera-Estrella, A., Seidl-Seiboth, V., Martinez, D. A., Druzhinina, I. S., Thon, M. … Grigoriev, I. V. (2011). Comparative genome sequence analysis underscores mycoparasitism as the ancestral life style of *Trichoderma*. *Genome Biology, 12*, R40.

Kulkarni, R. D., Thon, M. R., Pan, H., & Dean, R. A. (2005). Novel G-protein-coupled receptor-like proteins in the plant pathogenic fungus *Magnaporthe grisea*. *Genome Biology, 6*, R24.

Li, Z., Li, C., Huang, B., & Fan, M. (2001). Discovery and demonstration of the teleomorph of *Beauveria bassiana* (Bals.) Vuill., an important entomogenous fungus. *Chinese Science Bulletin, 46*, 751–753.

Liao, X., Fang, W., Lin, L., Lu, H. L., & St. Leger, R. J. (2013). *Metarhizium robertsii* produces an extracellular invertase (MrINV) that plays a pivotal role in rhizospheric interactions and root colonization. *PLoS One, 8*, e78118.

Liao, Y., Ling, J., Zhang, G., Liu, F., Tao, S., Han, Z. … Le, H. (2015). Cordycepin induces cell cycle arrest and apoptosis by inducing DNA damage and up-regulation of p53 in Leukemia cells. *Cell Cycle, 14*, 761–771.

Lin, L. C., Fang, W. G., Liao, X. G., Wang, F. Q., Wei, D. Z., & St. Leger, R. J. (2011). The MrCYP52 cytochrome P450 monoxygenase gene of *Metarhizium robertsii* is important for utilizing insect epicuticular hydrocarbons. *PLoS One, 6*, e28984.

Liu, L. S., Zhang, J., Chen, C., Teng, J. T., Wang, C. S., & Luo, D. Q. (2015). Structure and biosynthesis of fumosorinone, a new protein phosphatase 1B inhibitor firstly isolated from the entomogenous fungus *Isaria fumosorosea*. *Fungal Genetics and Biology, 81*, 191–200.

Liu, Z. Y., Liang, Z. Q., Whalley, A. J., Yao, Y. J., & Liu, A. Y. (2001). *Cordyceps brittlebankisoides*, a new pathogen of grubs and its anamorph, *Metarhizium anisopliae* var. majus. *Journal of Invertebrate Pathology, 78*, 178–182.

Lovett, B., & St. Leger, R. J. (2015). Stress is the rule rather than the exception for *Metarhizium*. *Current Genetics, 61*, 253–261.

Luo, X., Keyhani, N. O., Yu, X., He, Z., Luo, Z., Pei, T., & Zhang, Y. (2012). The MAP kinase Bbslt2 controls growth, conidiation, cell wall integrity, and virulence in the insect pathogenic fungus *Beauveria bassiana*. *Fungal Genetics and Biology, 49*, 544–545.

Ma, L. J., van der Does, H. C., Borkovich, K. A., Coleman, J. J., Daboussi, M.-J., & Pietro, A. D. (2010). Comparative genomics reveals mobile pathogenicity chromosomes in *Fusarium*. *Nature, 464*, 367–373.

Martin, F., Aerts, A., Ahren, D., Brun, A., Danchin, E. G. J., Duchaussoy, F. … Grigoriev, I. V. (2008). The genome of *Laccaria bicolor* provides insights into mycorrhizal symbiosis. *Nature, 452*, 88–92.

Martinez, D., Berka, R. M., Henrissat, B., Saloheimo, M., Arvas, M., Baker, S. E. … Brettin, T. S. (2008). Genome sequencing and analysis of the biomass-degrading fungus *Trichoderma reesei* (syn. *Hypocrea jecorina*). *Nature Biotechnology, 26*, 553–560.

Meyling, N. V., Lübeck, M., Buckley, E. P., Eilenberg, J., & Rehner, S. A. (2009). Community composition, host range and genetic structure of the fungal entomopathogen *Beauveria* in adjoining agricultural and seminatural habitats. *Molecular Ecology, 18*, 1282–1293.

Moonjely, S., Barrelli, R., & Bidochka, M. J. (2016). Insect Pathogenic Fungi as Endophytes. *Advances in Genetics, 94*, 107–136.

Molnár, I., Gibson, D. M., & Krasnoff, S. B. (2010). Secondary metabolites from entomopathogenic Hypocrealean fungi. *Natural Product Reports, 27*, 1241–1275.

Morschhauser, J. (2010). Regulation of multidrug resistance in pathogenic fungi. *Fungal Genetics and Biology, 47*, 94–106.

Ni, M., Feretzaki, M., Sun, S., Wang, X., & Heitman, J. (2011). Sex in fungi. *Annual Review of Genetics, 45*, 405–430.

Noda, J., Brito, N., & González, C. (2010). The *Botrytis cinerea* xylanase Xyn11A contributes to virulence with its necrotizing activity, not with its catalytic activity. *BMC Plant Biology, 10*, 38.

Nygren, K., Strandberg, R., Wallberg, A., Nabholz, B., Gustafsson, T., García, D. … Johannesson, H. (2011). A comprehensive phylogeny of *Neurospora* reveals a link between reproductive mode and molecular evolution in fungi. *Molecular Phylogenetics and Evolution, 59*, 649–663.

Ohm, R., Feau, N., Henrissat, B., Schoch, C. L., Horwitz, B. A., Barry, K. W. … Grigoriev, I. (2012). Diverse lifestyles and strategies of plant pathogenesis encoded in the genomes of eighteen Dothideomycetes fungi. *PLoS Pathogens, 8*, e1003037.

Ortiz-Urquiza, A., & Keyhani, N. O. (2013). Action on the surface: entomopathogenic fungi versus the insect cuticle. *Insects, 4*, 357–374.
Ortiz-Urquiza, A., & Keyhani, N. O. (2015). Stress response signaling and virulence: insights from entomopathogenic fungi. *Current Genetics, 61*, 239–249.
Ortiz-Urquiza, A., & Keyhani, N. O. (2016). Molecular Genetics of Beauveria bassiana Infection of Insects. *Advances in Genetics, 94*, 165–250.
Ownley, B. H., Gwinn, K. D., & Vega, F. E. (2010). Endophytic fungal entomopathogens with activity against plant pathogens: ecology and evolution. *BioControl, 55*, 113–128.
Pal, S., St. Leger, R. J., & Wu, L. P. (2007). Fungal peptide Destruxin A plays a specific role in suppressing the innate immune response in *Drosophila Melanogaster*. *Journal of Biological Chemistry, 292*, 8969–8977.
Paterson, R. R. (2008). *Cordyceps*: a traditional Chinese medicine and another fungal therapeutic biofactory? *Phytochemistry, 69*, 1469–1495.
Pattemore, J. A., Hane, J. K., Williams, A. H., Wilson, B. A., Stodart, B. J., & Ash, G. J. (2014). The genome sequence of the biocontrol fungus *Metarhizium anisopliae* and comparative genomics of *Metarhizium* species. *BMC Genomics, 15*, 660.
Pava-Ripoll, M., Angelini, C., Fang, W. G., Wang, S., Posada, F. J., & St. Leger, R. J. (2011). The rhizosphere-competent entomopathogen *Metarhizium anisopliae* expresses a specific subset of genes in plant root exudate. *Microbiology, 157*, 47–55.
Pereira, J. L., Noronha, E. F., Miller, R. N., & Franco, O. L. (2007). Novel insights in the use of hydrolytic enzymes secreted by fungi with biotechnological potential. *Letters in Applied Microbiology, 44*, 573–581.
Peterson, S. W. (2008). Phylogenetic analysis of *Aspergillus* species using DNA sequences from four loci. *Mycologia, 100*, 205–226.
Raffaele, S., & Kamoun, S. (2012). Genome evolution in filamentous plant pathogens: why bigger can be better. *Nature Reviews Microbiology, 10*, 417–430.
Rangel, D. E., Alston, D. G., & Roberts, D. W. (2008). Effects of physical and nutritional stress conditions during mycelial growth on conidial germination speed, adhesion to host cuticle, and virulence of *Metarhizium anisopliae*, an entomopathogenic fungus. *Mycological Research, 112*, 1355–1361.
Rhee, Y. J., Hillier, S., & Gadd, G. M. (2012). Lead transformation to pyromorphite by fungi. *Current Biology, 22*, 237–241.
Richards, T. A., Leonard, G., Soanes, M., & Talbot, N. J. (2011). Gene transfer into the fungi. *Fungal Biology Reviews, 25*, 98–110.
Roberts, D. W., & St. Leger, R. J. (2004). *Metarhizium* spp., cosmopolitan insect-pathogenic fungi: mycological aspects. *Advances in Applied Microbiology, 54*, 1–70.
Scazzocchio, C. (2000). The fungal GATA factors. *Current Opinion in Microbiology, 3*, 126–131.
Schaible, U. E., & Kaufmann, S. H. E. (2005). A nutritive view on the host-pathogen interplay. *Trends in Microbiology, 13*, 373–380.
Schulze-Lefert, P., & Panstruga, R. (2011). A molecular evolutionary concept connecting nonhost resistance, pathogen host range, and pathogen speciation. *Trends in Plant Science, 16*, 117–125.
Screen, S. E., & St. Leger, R. J. (2000). Cloning, expression and substrate specificity of a fungal chymotrypsin. *Journal of Biological Chemistry, 275*, 6689–6694.
Seidl, V. (2008). Chitinases of filamentous fungi: a large group of diverse proteins with multiple physiological functions. *Fungal Biology Reviews, 22*, 36–42.
Shang, Y., Feng, P., & Wang, C. (2015). Fungi that infect insects: altering host behavior and beyond. *PLoS Pathogens, 11*, e1005037.
Sharpton, T. J., Stajich, J. E., Rounsley, S. D., Gardner, M. J., Wortman, J. R., Jordar, V. S.… Taylor, J. W. (2009). Comparative genomic analyses of the human fungal pathogens *Coccidioides* and their relatives. *Genome Research, 19*, 1722–1731.

Silva, W. O. B., Santi, L., Berger, M., Pinto, A. F. M., Guimaraes, J. A., Schrank, A.... Vainstein, M. H. (2009). Characterization of a spore surface lipase from the biocontrol agent *Metarhizium anisopliae*. *Processing Biochemistry, 44*, 829–834.

St. Leger, R. J. (2008). Studies on adaptations of *Metarhizium anisopliae* to life in the soil. *Journal of Invertebrate Pathology, 98*, 271–276.

St. Leger, R. J., Bidochka, M. J., & Roberts, D. W. (1994). Germination triggers of *Metarhizium anisopliae* conidia are related to host species. *Microbiology, 140*, 1651–1660.

St. Leger, R. J., Cooper, R. M., & Charnley, A. K. (1986a). Cuticle-degrading enzymes of entomopathogenic fungi: regulation of production of chitinolytic enzymes. *Journal of General Microbiology, 132*, 1509–1517.

St. Leger, R. J., Cooper, R. M., & Charnley, A. K. (1986b). Cuticle-degrading enzymes of entomopathogenic fungi: cuticle degradation in vitro by enzymes from entomopathogens. *Journal of Invertebrate Pathology, 47*, 167–177.

St. Leger, R. J., Joshi, L., Bidochka, M. J., & Roberts, D. W. (1996). Construction of an improved mycoinsecticide overexpressing a toxic protease. *Proceedings of the National Academy of Sciences of the United States of America, 93*, 6349–6354.

St. Leger, R. J., May, B., Allee, L. L., Frank, D. C., Staples, R. C., & Roberts, D. W. (1992). Genetic differences in allozymes and in formation of infection structures among isolates of the entomopathogenic fungus *Metarhizium anisopliae*. *Journal of Invertebrate Pathology, 60*, 89–101.

St. Leger, R. J., Nelson, J. O., & Screen, S. E. (1999). The entomopathogenic fungus *Metarhizium anisopliae* alters ambient pH, allowing extracellular protease production and activity. *Microbiology, 145*, 2691–2699.

St. Leger, R. J., Wang, C. S., & Fang, W. G. (2011). New perspectives on insect pathogens. *Fungal Biology Reviews, 25*, 84–88.

Staats, C. C., Junges, A., Guedes, R. L., Thompson, C. E., de Morais, G. L., Boldo, J. T., & Schrank, A. (2014). Comparative genome analysis of entomopathogenic fungi reveals a complex set of secreted proteins. *BMC Genomics, 15*, 822.

Steinwender, B. M., Enkerli, J., Widmer, F., Eilenberg, J., Kristensen, H. L., Bidochka, M. J., & Meyling, N. V. (2015). Root isolations of *Metarhizium* spp. from crops reflect diversity in the soil and indicate no plant specificity. *Journal of Invertebrate Pathology, 132*, 142–148.

Stidham, T. A., & Stidham, J. A. (2000). A new Miocene band-winged grasshopper (Orthoptera: Acrididae) from Nevada. *Annals of the Entomological Society of America, 93*, 405–407.

Stukenbrock, E. H., Bataillon, T., Dutheil, J. Y., Hansen, T. T., Li, R., & Zala, M. (2011). The making of a new pathogen: insights from comparative population genomics of the domesticated wheat pathogen *Mycosphaerella graminicola* and its wild sister species. *Genome Research, 21*, 2157–2166.

Suh, S. O., Noda, H., & Blackwell, M. (2001). Insect symbiosis: derivation of yeast-like endosymbionts within an entomopathogenic filamentous lineage. *Molecular Biology and Evolution, 18*, 995–1000.

Sun, S., & Heitman, J. (2011). Is sex necessary? *BMC Biology, 9*, 56–59.

Sung, G. H., Hywel-Jones, N. L., Sung, J. M., Jennifer Luangsa-ard, J., Shrestha, N., & Spatafora, J. W. (2007). Phylogenetic classification of *Cordyceps* and the clavicipitaceous fungi. *Studies in Mycology, 57*, 5–59.

Thomas, M. B., & Read, A. F. (2007). Can fungal biopesticides control malaria? *Nature Reviews Microbiology, 5*, 377–383.

Torres, M. S., Singh, A. P., Vorsa, N., & White, J. F., Jr. (2008). An analysis of ergot alkaloids in the Clavicipitaceae (Hypocreales, Ascomycota) and ecological implications. *Symbiosis, 46*, 11–19.

Vega, F. E., Posada, F., Aime, M. C., Pava-Ripoll, M., Infante, F., & Rehner, S. A. (2008). Entomopathogenic fungal endophytes. *Biological Control, 46*, 72–82.

Wagner, B. L., & Lewis, L. C. (2000). Colonization of corn, Zea mays, by the entomopathogenic fungus *Beauveria bassiana*. *Applied and Environmental Microbiology, 66*, 3468–3473.

Wang, B., Kang, Q. J., Lu, Y. Z., Bai, L., & Wang, C. (2012). Unveiling the biosynthetic puzzle of destruxins in *Metarhizium* species. *Proceedings of the National Academy of Sciences of the United States of America, 109*, 1287–1292.

Wang, C., Butt, T. M., & St. Leger, R. J. (2005). Colony sectorization of *Metarhizium anisopliae* is a sign of ageing. *Microbiology, 151*, 3223–3236.

Wang, C., Fan, M. Z., Li, Z. Z., & Butt, T. M. (2004). Molecular monitoring and evaluation of the application of the insect-pathogenic fungus *Beauveria bassiana* in southeast China. *Journal of Applied Microbiology, 96*, 861–870.

Wang, C., & Feng, M.-G. (2014). Advances in fundamental and applied studies in China of fungal biocontrol agents for use against arthropod pests. *Biological Control, 68*, 129–135.

Wang, C., & St. Leger, R. J. (2006). A collagenous protective coat enables *Metarhizium anisopliae* to evade insect immune responses. *Proceedings of the National Academy of Sciences of the United States of America, 103*, 6647–6652.

Wang, C., & St. Leger, R. J. (2007). The *Metarhizium anisopliae* perilipin homolog MPL1 regulates lipid metabolism, appressorial turgor pressure, and virulence. *Journal of Biological Chemistry, 282*, 21110–21115.

Wang, C., Zhang, S., Hou, R., Zheng, Q., Xu, Q., Zheng, D. … Xu, J. (2011). Functional analysis of the kinome of the wheat scab fungus *Fusarium graminearum*. *PLoS Pathogens, 12*, e1002460.

Wang, C. S., Duan, Z. B., & St. Leger, R. J. (2008). MOS1 osmosensor of *Metarhizium anisopliae* is required for adaptation to insect host hemolymph. *Eukaryotic Cell, 7*, 302–309.

Wang, C. S., Hu, G., & St. Leger, R. J. (2005). Differential gene expression by *Metarhizium anisopliae* growing in root exudate and host (*Manduca sexta*) cuticle or hemolymph reveals mechanisms of physiological adaptation. *Fungal Genetics and Biology, 42*, 704–718.

Wang, C. S., & St. Leger, R. J. (2005). Developmental and transcriptional responses to host and non-host cuticles by the specific locust pathogen *Metarhizium anisopliae* sf. *acridum*. *Eukaryotic Cell, 4*, 937–947.

Wang, C. S., & St. Leger, R. J. (2007). The MAD1 adhesin of *Metarhizium anisopliae* links adhesion with blastospore production and virulence to insects, and the MAD2 adhesin enables attachment to plants. *Eukaryotic Cell, 6*, 808–816.

Wang, C. S., & St. Leger, R. J. (2014). Genomics of entomopathogenic fungi. In F. Martin (Ed.), *The ecological genomics of fungi* (pp. 243–260). John Wiley & Sons, Inc.

Wang, S. B., Fang, W. G., Wang, C. S., & St. Leger, R. J. (2011). Insertion of an esterase gene into a specific locust pathogen (*Metarhizium acridum*) enables it to infect caterpillars. *PLoS Pathogens, 7*, e1002097.

Wang, S., O'Brien, T. R., Pava-Ripoll, M., & St. Leger, R. J. (2011). Local adaptation of an introduced transgenic insect fungal pathogen due to new beneficial mutations. *Proceedings of the National Academy of Sciences, 108*, 20449–20454.

Whittle, C. A., Nygren, K., & Johannesson, H. (2011). Consequences of reproductive mode on genome evolution in fungi. *Fungal Genetics and Biology, 48*, 661–667.

Xiao, G. H., Ying, S.-H., Zheng, Z., Wang, Z. L., Zhang, S., Xie, X. Q. … Feng, M. G. (2012). Genomic perspectives on the evolution of fungal entomopathogenicity in *Beauveria bassiana*. *Scientific Reports, 2*, 483.

Xu, J. R., Peng, Y. L., Dickman, M. B., & Sharon, A. (2006). The dawn of fungal pathogen genomics. *Annual Reviews of Phytopathology, 44*, 337–366.

Xu, Y. J., Luo, F. F., Gao, Q., Shang, Y. F., & Wang, C. S. (2015). Metabolomics reveals insect metabolic responses associated with fungal infection. *Analytical and Bioanalytical Chemistry, 407*, 4815–4821.

Xue, C., Hsueh, Y. P., & Heitman, J. (2008). Magnificent seven: roles of G protein-coupled receptors in extracellular sensing in fungi. *FEMS Microbiology Review, 32*, 1010–1032.

Xue, Y., Batlle, M., & Hirsch, J. P. (1998). GPR1 encodes a putative G protein-coupled receptor that associates with the Gpa2p G-alpha subunit and functions in a Ras-independent pathway. *EMBO Journal, 17*, 1996–2007.

Yokoyama, E., Arakawa, M., Yamagishi, K., & Hara, A. (2006). Phylogenetic and structural analyses of the mating-type loci in *Clavicipitaceae*. *FEMS Microbiology Letter, 264*, 182–191.

Zhang, S., Widemann, E., Bernard, G., Lesot, A., Pinot, F., Pedrini, N., & Keyhani, N. O. (2012). CYP52X1, representing new cytochrome P450 subfamily, displays fatty acid hydroxylase activity and contributes to virulence and growth on insect cuticular substrates in entomopathogenic fungus *Beauveria bassiana*. *Journal of Biological Chemistry, 287*, 13477–13486.

Zhang, S. Z., Xia, Y. X., & Keyhani, N. O. (2011). Contribution of the gas1 gene of the entomopathogenic fungus *Beauveria bassiana*, encoding a putative glycosylphosphatidylinositol-anchored beta-1,3-glucanosyltransferase, to conidial thermotolerance and virulence. *Applied and Environmental Microbiology, 77*, 2676–2684.

Zhang, Y., Zhao, J., Fang, W., Zhang, J., Luo, Z., Zhang, M. … Pei, Y. (2009). Mitogen-activated protein kinase hog1 in the entomopathogenic fungus *Beauveria bassiana* regulates environmental stress responses and virulence to insects. *Applied and Environmental Microbiology, 75*, 3787–3795.

Zhao, H., Xu, C., Lu, H. L., Chen, X., St. Leger, R. J., & Fang, W. (2014). Host-to-pathogen gene transfer facilitated infection of insects by a pathogenic fungus. *PLoS Pathogens, 10*, e10040.

Zheng, P., Xia, Y., Xiao, G., Xiong, C., Hu, X., Zhang, S. … Wang, C. S. (2011). Genome sequence of the insect pathogenic fungus *Cordyceps militaris*, a valued traditional Chinese medicine. *Genome Biology, 12*, R116.

Zheng, P., Xia, Y., Zhang, S., & Wang, C. (2013). Genetics of *Cordyceps* and related fungi. *Applied Microbiology and Biotechnology, 97*, 2797–2804.

CHAPTER FOUR

Insect Pathogenic Fungi as Endophytes

S. Moonjely, L. Barelli and M.J. Bidochka[1]
Brock University, St. Catharines, ON, Canada
[1]Corresponding author: E-mail: mbidochka@brocku.ca

Contents

Abstract

In this chapter, we explore some of the evolutionary, ecological, molecular genetics, and applied aspects of a subset of insect pathogenic fungi that also have a lifestyle as endophytes and we term endophytic insect pathogenic fungi (EIPF). We focus particularly on *Metarhizium* spp. and *Beauveria bassiana* as EIPF.

The discussion of the evolution of EIPF challenges a view that these fungi were first and foremost insect pathogens that eventually evolved to colonize plants. Phylogenetic evidence shows that the lineages of EIPF are most closely related to grass endophytes that diverged c. 100 MYA. We discuss the relationship between genes involved in "insect pathogenesis" and those involved in "endophytism" and provide examples of genes with potential importance in lifestyle transitions toward insect pathogenicity. That is, some genes for insect pathogenesis may have been coopted from genes

Advances in Genetics, Volume 94
ISSN 0065-2660
http://dx.doi.org/10.1016/bs.adgen.2015.12.004

involved in endophytic colonization. Other genes may be multifunctional and serve in both lifestyle capacities.

The interactions of EIPF with their host plants are discussed in some detail. The genetic basis for rhizospheric competence, plant communication, and nutrient exchange is examined and we highlight, with examples, the benefits of EIPF to plants, and the potential reservoir of secondary metabolites hidden within these beneficial symbioses.

1. INTRODUCTION

Insect pathogenic fungi (IPF) encompass over 1000 fungal species found in most major fungal taxonomic groups from Chytridiomycetes to Basidiomycetes. However, phylogenetic data provide little evidence for a single common origin for insect pathogenesis even within a major taxonomic group (Humber, 2008; Wang, St. Leger, & Wang, 2016). There is also a wide range of host specificity in IPF. For example, some fungal species within the Entomophthoromycota (eg, *Entomophthora*, *Strongwellsea*, and *Entomophaga*) are obligate pathogens of a single or very few taxonomically related insect species. Even within a single genus there is a wide range of host specificity. For example, within the hypocrealean fungal genus, *Metarhizium*, there are facultative pathogens with a wide range of host insects (eg, *Metarhizium robertsii* and *Metarhizium brunneum*), as well as species with a narrow host range (eg, *Metarhizium acridum* and *Metarhizium flavoviride*) (Faria & Wraight, 2007).

IPF generally infect their host through transcuticular penetration and need not be ingested. Once they enter the insect hemocoel, they ramify within the insect. After that, fungal hyphae emerge from the insect. The outgrowth of fungal structures from infected, mummified insects is the stuff of science fiction. These fungi serve as models for the mechanics, molecular biology, and evolution of pathogenesis (Wang & St. Leger, 2014; Wang et al., 2016). Furthermore, the potential of agricultural applications of *Metarhizium* and *Beauveria* for insect biocontrol has been known since the early 1900s (Pilat, 1938; Pospelov, 1938), and there are many commercially available formulations (Castrillo, Griggs, Ranger, Reding, & Vandenberg, 2011; Faria & Wraight, 2007). Several IPF are also being developed as pathogens of human disease vectors. For example, natural and transgenic strains of *Metarhizium* are being explored as pathogens of mosquitoes in order to combat malaria and other diseases (Blanford et al., 2005; Fang & St. Leger, 2012; Scholte, Knols, & Takken, 2006; Zhao, Lovett, & Fang, 2016).

Adding further ecological and evolutionary complexity and intrigue to a subset of hypocrealean IPF is a multifunctional lifestyle that includes their role as endophytic symbionts of plants. Two model IPF that have also demonstrated endophytic capability are *Metarhizium* and *Beauveria* (Sasan & Bidochka, 2012; Wagner & Lewis, 2000; Wang & St. Leger, 2014). The genus *Metarhizium* is particularly interesting as an evolutionary model since, as previously mentioned, not only does this genus represent species with narrow and broad insect host ranges, but they also vary in their endophytic capabilities. In this chapter, we explore some of the evolutionary, ecological, and applied aspects of these IPF in light of their endophytic capabilities.

2. EVOLUTION OF ENDOPHYTIC INSECT PATHOGENIC FUNGI

Phylogenetic studies have shown that species of IPF, such as *Metarhizium*, *Beauveria*, *Lecanicillium*, and *Paecilomyces*, are most closely related to the fungal grass endosymbionts *Claviceps* and *Epichloë* (Behie, Padilla-Guerrero, & Bidochka, 2013; Spatafora, Sung, Sung, Hywel-Jones, & White, 2007). In addition, comparative genomic analyses proposed that the *Metarhizium* lineage diverged from the lineage of the plant endophyte *Epichloë festucae* approximately 88–114 MYA (Gao et al., 2011). *Metarhizium* and *Beauveria* are more closely related to endophytes and plant pathogens than to animal pathogens. The genome of *Metarhizium* shows a large number of genes for plant-degrading enzymes (Gao et al., 2011). This evidence supports the idea that the *Metarhizium* lineage evolved from plant-associated fungi, and insect pathogenicity is a more recently acquired adaptation. Hypothetical mechanisms by which genes involved in insect pathogenesis may have arisen have been theorized as being coopted, evolved, or acquired by horizontal gene transfer from other fungi or host insects (Screen & St. Leger, 2000).

The evolution of IPF must have involved adaptations that allowed degradation of insect cuticle and host body components, as indicated by the large number of proteases, lipases, and chitinases present within genomes of endophytic IPF (EIPF). In this chapter, we also explore the mechanisms of insect pathogenicity, and how genes involved in insect pathogenicity are related to those involved in endophytism. Furthermore, refinement of mutualistic plant colonization and the loss of genes involved in plant pathogenesis would have also been driving factors in the evolution of *Metarhizium* as an endophyte (Wyrebek & Bidochka, 2013).

Interkingdom host jumping by EIPF from plants back to arthropods and then back to plants has been proposed for the evolution of insect pathogenicity (Humber, 2008). In contrast, we suggest that many of these fungi evolved to infect insects while maintaining their mutualistic endosymbiosis with plants. We cannot reenact the ecological conditions that allowed for niche diversification by the ancestors of these endophytic fungi, as they evolved insect pathogenic capabilities c. 100 MYA. However, we hypothesize that the driving force behind this evolution was the host plant demanding reciprocal nutrient exchange from the fungus in exchange for access to plant carbohydrates in the rhizosphere. IPF would be able to provide the plant with a source of nitrogen, or other growth-limiting nutrients, derived from insect parasitism.

3. MULTIFUNCTIONAL LIFESTYLES

3.1 Insect Pathogenicity

IPF are ubiquitous in nature, and they range from narrow host range to broad host range. The underlying molecular mechanisms of the infection process are well studied and documented particularly for *Metarhizium* spp. and *Beauveria bassiana* (Bidochka, St. Leger, & Roberts, 1997; Ortiz-Urquiza & Keyhani, 2013; St. Leger, Wang, & Fang, 2011). These fungi have been established as model organisms for general infection processes. In the following, we relate mechanisms of insect pathogenesis in light of the evolutionary history of these fungi as plant associates. That is, are there similarities in genes involved in insect pathogenesis and those involved in plant associations?

3.1.1 Adhesion

The conidia of *Metarhizium* and *Beauveria* are generally hydrophobic, and adhesion of these fungi to the host insect cuticle is mediated by means of specific adhesion genes or nonspecific (principally hydrophobic) interactions. The surface structure and composition of the insect exoskeleton influence the adherence of fungal conidia to the cuticle. The outermost layer of the insect integument is a lipid layer, which is hydrophobic in nature, facilitating attachment of fungal propagules (Ortiz-Urquiza, Luo, & Keyhani, 2015; Pedrini, Crespo, & Juárez, 2007). In *Metarhizium* spp., conidial adherence to the insect cuticle is facilitated by both specific and nonspecific interactions. The key genes involved are the adhesin-like protein, *Mad1*, and *ssgA/HYD1*, a hydrophobin (Boucias, Pendland, & Latge, 1988; St. Leger,

Staples, Roberts, 1992; Wang & St. Leger, 2007a). Two other hydrophobin genes, *HYD3* and *HYD2*, which code for class I and class II hydrophobins respectively, have been reported in *M. brunneum* and aid in conidial adhesion to insect epicuticle and affect virulence (Sevim et al., 2012). In *B. bassiana*, the initial adhesion to insect cuticle is mediated via nonspecific hydrophobic interactions. Studies have shown that in *Beauveria* two hydrophobin genes, *hyd1* and *hyd2* encode for class I hydrophobins and play a significant role in adhesion and virulence (Zhang, Xia, Kim, & Keyhani, 2011).

It is interesting, from an endophytic perspective, that the mechanisms of insect infection and plant infection show remarkable similarities (Gao et al., 2011). Comparative analysis of the *Metarhizium* hydrophobin genes, using the pathogen–host interaction (PHI) gene database, has shown marked similarity with sequences from plant pathogenic fungi (Gao et al., 2011). For example, hydrophobins have also been implicated in plant infection processes in the rice blast fungus, *Magnaporthe grisea* (Whiteford & Spanu, 2002). In particular, the *M. grisea* class I hydrophobin gene, *Mpg1* acts as an adhesion factor initiating the infection process in plants (Whiteford & Spanu, 2002), similar to the role of ssgA (hydrophobin; *Metarhizium*) enabling adherence to insect cuticle (Wang & St. Leger, 2007a).

3.1.2 Penetration

The insect integument is formed of several layers—epicuticle, procuticle, and epidermis, which are composed mainly of lipids, chitins, and proteins. The conidial surface proteins act synergistically to aid in germination through recognition of insect-specific components and subsequent cuticle degradation (Ortiz-Urquiza & Keyhani, 2013; Pedrini et al., 2007; Santi et al., 2010). The major carbon sources utilized for conidial germination are endogenous or cuticular lipids (Moraes, Schrank, & Vainstein, 2003; Ortiz-Urquiza & Keyhani, 2013; Pedrini et al., 2007; Wang, Fang, Wang, & St. Leger, 2011).

Once the fungal conidia successfully adhere to the insect cuticle, they germinate to form hyphae on the insect cuticle and express hydrolytic enzymes, such as proteases, esterases, N-acetylglucosaminidases, chitinases, and lipases (Schrank & Vainstein, 2010; St. Leger, Charnley, & Cooper, 1987, St. Leger, Joshi, Bidochka, Rizzo, & Roberts, 1996). In addition to enzymatic degradation, mechanical pressure through formation of specialized hyphal structures (appressoria) has also been implicated for successful cuticular penetration. For example, the expression of *Mpl1* (perilipin) plays a significant role in the transport and breakdown of endogenous lipids that increases turgor pressure and thereby aids the formation of appressoria (Wang & St. Leger, 2007b).

The sequential expression of different degradative enzymes in combination with mechanical pressure accelerates cuticular penetration by fungal hyphae toward the insect hemocoel (Bidochka et al., 1997; Moraes et al., 2003; Ortiz-Urquiza & Keyhani, 2013; Pedrini et al., 2007; St. Leger, 1995; Wang et al., 2011). Transcriptomic analysis revealed that *Metarhizium* spp. express a diverse array of secreted proteases, including subtilisin-like proteases, trypsins, carboxypeptidase, aspartic protease, threonine protease, cysteine protease, and metalloproteases as penetration commences (Gao et al., 2011). However, generalist and specialist strains of *Metarhizium* (*M. robertsii* and *M. acridum*, respectively) express different proteases on the cuticles of their hosts, implying a role for proteases in virulence, host recognition, and in the different stages of pathogenesis.

The expression of high amounts of proteases presumably enables *Metarhizium* to adapt to different habitats (Gao et al., 2011). Studies have shown that different proteases expressed by *Metarhizium* are important pathogenicity determinants (St. Leger, 1995). Gao et al. (2011) also reported that *M. robertsii* expresses many more trypsins than pathogens of plants, animals, or fungi. Screening of trypsins from *Metarhizium* using the PHI database of phytopathogen virulence determinants revealed a similarity with glucanase-inhibitor proteins (GIP), a trypsin from the soybean pathogen *Phytophthora sojae* (Gao et al., 2011). *Phytophthora sojae* secretes GIP when infecting plants and thereby reduces the induction of plant-derived defense responses (Rose, Ham, Darvill, & Albersheim, 2002). These studies provide important insights into mechanisms of insect pathogenesis, but genetic similarities to plant pathogens and endophytes tempt further investigation into the multifunctional roles of these genes, particularly in an endophytic capacity.

3.1.3 Proliferation, Immune Avoidance, and Insect Death

Through the combination of mechanical pressure and enzymatic processes, fungal hyphae penetrate the insect cuticle and eventually reach the insect hemocoel, where they differentiate to form yeast-like bodies called blastospores. Insect hemolymph is rich in nutrients with the most abundant carbohydrate in hemolymph being trehalose (Bidochka et al., 1997).

Upon reaching the insect hemolymph, the fungal hyphae switch phenotypes to blastospores and short hyphal lengths called hyphal bodies (Ortiz-Urquiza & Keyhani, 2015; Small & Bidochka, 2005). *Metarhizium* expresses a collagen-like protein (*mcl1*) which functions as a defensive coat that prevents hyphal bodies from being phagocytosed or encapsulated by host immune cells (Wang & St. Leger, 2006), however, gene-sequencing studies have shown that a homolog of the collagen-like protein is absent in *Beauveria* (Xiao et al., 2012).

Metarhizium and *Beauveria* can also survive phagocytosis by amoeboid predators in the soil (Bidochka, Clark, Lewis, & Keyhani, 2010); a survival strategy also exhibited by several mammalian fungal and bacterial pathogens. This suggests that the ability of fungal insect and mammalian pathogens to survive host phagocytic cells may be a consequence of adaptations that originally evolved in order to avoid predation by soil amoeba. This potentially speaks to strategies of survival of these EIPF fungi as they persist in the rhizosphere as endophytes (Hu & St. Leger, 2002).

Beauveria and *Metarhizium* also produce insecticidal metabolites, including beauvericin and destruxins, respectively, that allow blastospores to proliferate inside the hemolymph (Bidochka et al., 1997). Besides cuticle degradation, proteases are involved in nutrient acquisition, degradation of antifungal proteins to bypass the host immune response, and in the regulation of microenvironmental pH (St. Leger, Nelson, & Screen, 1999).

3.1.4 Conidiation on the Surface of the Insect Cadaver

Blastospores proliferating within the hemolymph kill the insect host within 3–7 days by absorbing hemocoelic nutrients and through toxic metabolite production (Schrank & Vainstein, 2010; Small & Bidochka, 2005). After fungal hyphae ramify throughout the dead infected host, they reemerge from the insect and conidiate on the insect cadaver. *Metarhizium* produces green conidia on the surface of the insect cadaver, from which the designation of "green muscardine disease" arose; *Beauveria* infections result in white conidia. Suppression subtractive hybridization has shown significant upregulation of the gene, *cag7 (Pr1 protease)* in *Metarhizium* at the onset of conidiation (Small & Bidochka, 2005).

The complex process of insect pathogenesis requires these fungi to undergo morphological switching from conidia to hyphae, to appressoria, to single-celled blastospores, and finally conidiogenous cells, all of which utilize overlapping subsets of genes. Furthermore, these fungi are constantly sampling the environment and relaying those messages through signaling mechanisms for gene transcription. Microarray analysis has shown that in *Metarhizium* and *Beauveria*, overlapping subsets of genes are differentially expressed when they are grown in media containing insect cuticle, insect hemolymph, or plant root exudate. The abundance of proteases, and the differential expression of genes by *Metarhizium* in particular reveals the phenotypic plasticity that enables this fungus to survive both as an insect pathogen and as a plant endophyte (Gao et al., 2011; Wang, Hu, & St. Leger, 2005).

3.1.5 Proteins and Signaling Mechanisms Involved in Insect Pathogenesis

A number of membrane proteins, transcription factors (TFs), and biochemical pathways implicated in insect pathogenesis have been characterized in the model IPF *Metarhizium* and *Beauveria* (Ortiz-Urquiza & Keyhani, 2015 and Wang et al., 2016). Tetraspanin, a membrane protein involved in membrane signaling, is required to breach the insect cuticle, as *M. acridum* mutants impaired for tetraspanin (*pls1*) showed reduced virulence (Luo, He, Cao, & Xia, 2013). Tetraspanin appeared to be involved in the initial infection stages including conidial germination, formation of appressoria, and production of cuticle-degrading enzymes (Luo et al., 2013).

Msn2, a well-characterized stress response TF in yeast, has also been identified in *Beauveria* and *Metarhizium* and is crucial for virulence, conidiation, and the stress response; a deletion mutant showed a marked repression of diverse virulence-associated genes (Liu, Ying, Li, Tian, & Feng, 2013). Functional characterization found that a basic leucine-zipper domain of the *M. robertsii* TF MBZ1 is essential for proper functionality, as gene deletion resulted in decreased virulence toward wax moth larvae (Liu et al., 2013). Additionally, MBZ1 positively regulates the adhesin gene *Mad1*, as the expression of *Mad1* was found to be downregulated fourfold in *Metarhizium* MBZ1 deletion mutants (Huang, Shang, Chen, Cen, & Wang, 2015). Characterization of a homolog of the PacC TF (pH-responsive TF) in *Metarhizium* showed that PacC contributes to fungal virulence by impacting cuticle penetration, evasion of the host immune response, and mycosis (Huang, Shang, Chen, Gao, & Wang, 2015). Mutant strains lacking a functional *PacC* gene were found to be unable to penetrate the insect cuticle and proliferate in the hemocoel (Huang et al., 2015). In terms of virulence, two other TFs characterized in *B. bassiana* are CreA (Luo, Qin, Pei, & Keyhani, 2014) and a multiprotein binding factor (MBF1) (Ying, Ji, Wang, Feng, & Keyhani, 2014).

Several of the signaling pathways that contribute to the virulence of IPF have been characterized in both *Metarhizium* and *Beauveria*. These include

G-protein-coupled receptor (GPCR) signaling, mitogen-activated protein kinase (MAP kinase), and cAMP-PKA pathways. Fungal GPCRs are involved in niche recognition and nutrient sensing. Insect bioassays using *Beauveria* mutants defective in the GPCR3 gene revealed reduced virulence on topical application, indicating the role of GPCRs in initial infection stages (Ying, Feng, & Keyhani, 2013). Targeted gene disruption of the *Metarhizium anisopliae* regulators of G-protein signaling (RGS) gene, *cag8* (conidiation-associated gene) showed that *cag8* plays a significant role in virulence and hydrophobin synthesis (Fang, Pei, & Bidochka, 2007). The role MAP kinase pathways play in insect virulence has been studied in *M. robertsii* and *B. bassiana* through genetic analysis. Targeted deletion of the MAPK1 gene in *B. bassiana* demonstrated that MAPK1 protein is involved in fungal adhesion and penetration of insect cuticle (Zhang et al., 2010). Characterization of another MAPK in *B. bassiana* encoding Bb *hog1* showed the key role of functional HOG1 MAPK in pathogenicity toward insects (Zhang, Zhao, et al., 2009). Similarly, *M. acridum* mutants defective for hog1 MAPK signaling showed greatly reduced virulence (Jin, Ming, & Xia, 2012). *Beauveria bassiana slt2* encodes an slt2 family MAPK and is likewise involved in insect virulence. Mutant strains that lack slt2 show a marked decrease in virulence in insect bioassays using either topical application or hemocoel injection of spores (Luo et al., 2012). Targeted gene expression studies with *Beauveria* defective for adenylate cyclase, the key enzyme that catalyzes the conversion of adenosine triphosphate (ATP) to cyclic adenosine 3′5′monophosphate (cAMP), showed that cAMP-dependent processes were crucial for virulence (Wang, Zhou, Ying, & Feng, 2014). Fang, Pava-Ripoll, Wang, and Leger (2009) investigated the role of the cAMP-dependent protein kinase A (PKA) subunit in insect pathogenesis and reported that *M. robertsii* deficient in the pka1 subunit were greatly reduced in virulence (>90%) against wax moth larvae. Moreover, microarray analysis on mutant strains that lack the pka1 subunit showed downregulation of 244 genes involved in cuticular infection process (Fang, Pava-Ripoll, et al., 2009).

Despite infection processes being similar overall for *Metarhizium* and *Beauveria*, there are differences in key molecular aspects that arbitrate virulence toward insect hosts (Ortiz-Urquiza & Keyhani, 2015). MAP kinases and other signaling mechanisms, homologous to those in IPF, are well known in plant pathogenic fungi (Mehrabi, Zhao, Kim, & Xu, 2009), and our knowledge of the molecular interplay between plant and microbial symbionts has developed over the last decade (Kawaguchi & Minamisawa, 2010). The signaling mechanisms that have been studied for insect pathogenesis in IPF could have the same or parallel pathways in their endophytic capacities.

4. RELATIONSHIP BETWEEN INSECT PATHOGEN GENES AND ENDOPHYTISM

The endophytic abilities of these fungi add a level of complexity to their role as insect pathogens. In light of this, the genomes of these fungi are extremely interesting (Wang et al., 2016). There needs to be a change in the approach to the analysis of these organisms, their genetic makeup, and their ecological interactions as insect pathogens and plant associates. For example, the mutants in signaling genes discussed in the previous chapter should also be analyzed for their endophytic capacities.

The multifunctional lifestyles of these fungi require genotypic plasticity when exposed to diverse environments for survival. For example, to establish adherence to insect cuticle or plant roots, *Metarhizium* differentially expresses the adhesin proteins MAD1 and MAD2, respectively (Wang & St. Leger, 2007a). Tracing genes pertinent to all lifestyles, those that overlap in functionality, would help to elucidate the evolutionary history of these fungi. Another example is the subtilisin-like protease, Pr1A, from *Metarhizium* that is highly expressed in media of both insect and plant origin (Wang et al., 2005). It has been suggested that adaptation to various hosts is the result of gene duplication events or horizontal gene transfer (Bagga, Hu, Screen, & St. Leger, 2004; Screen & St. Leger, 2000; Xiao et al., 2012). Further investigations of the molecular mechanisms for the symbiosis between plants and EIPF will help to uncover the genetic events that lead to insect pathogenicity. For example, the involvement of the subtilisin-like family of *Metarhizium* proteases in endophytism may have preadapted the ancestor of *Metarhizium* to insect pathogenesis.

If we accept the hypothesis that some IPF evolved from plant symbionts, from where did the genes involved in insect pathogenicity evolve? As an example, the protease, Atl1, from the grass endophyte *Acremonium typhinum*, facilitates fungal colonization by assisting in the degradation of the plant cell wall (Reddy, Lam, & Belanger, 1996). Homologous proteases have been identified in species of the EIPF fungi *Metarhizium* and *Beauveria*, the mycoparasite *Trichoderma harzianum*, and the nematode-trapping fungus *Arthrobotrys oligospora*, all of which are also endophytes (Geremia et al., 1993; Reddy et al., 1996; Tunlid, Rosen, & Rask, 1994). It is possible that the IPF proteases had initially been utilized for endophytic colonization and were subsequently coopted for their utility in insect pathogenesis. This may be particularly true for the subtilisin-like proteases found in *Metarhizium* that may have arisen by gene duplication events (Hu & St. Leger, 2004; Li et al., 2010) from an ancestral "endophytic protease." Since some of these IPF are also endophytes, delving into the evolutionary relationships between "insect pathogenic" genes and "endophytic" genes in a single organism could be extremely informative. Examples previously mentioned are the subtilisin-like protease genes and the adhesin genes, but that is probably the tip of the iceberg and many more genes could be identified. What, if any, are the relationships between these genes? Did they arise by duplication events and subsequent specialization? Or were they gained through horizontal gene transfer? If so, what was the potential source?

4.1 Plant Root Colonization by Insect Pathogenic Fungi

Traditionally IPF were considered solely as insect pathogens, but are now also being investigated as endophytes (Ownley, Gwinn, & Vega, 2010; Sasan & Bidochka, 2012; Vega, 2008) and rhizosphere colonizers (Behie, Jones, & Bidochka, 2015; Ownley et al., 2008; Pava-Ripoll et al., 2011). This dual lifestyle provides promising opportunities for EIPF, not only as biocontrol agents, but also as biofertilizers and for general plant protection.

Why adopt a dual role as insect pathogens and endophytes? What drove evolution from a plant colonizer to an insect pathogen? Previously, we showed that plants were able to reacquire nitrogen from insects through a partnership with the endophytic, insect pathogenic fungus *M. robertsii* (Behie & Bidochka, 2014; Behie et al., 2013). That is, the endophytic capability and insect pathogenicity of *M. robertsii* and *B. bassiana* are coupled so that these fungi act as conduits to provide insect-derived nitrogen to plant hosts. We suggest that EIPF initially evolved as plant colonizers and that this relationship was encouraged by the ability of these fungi to provide nitrogen, or other nutrients, first as nutrient scavengers, then through adaptations as insect pathogens. The consequence of this is that plants forming relationships with IPF would have had an advantage, particularly in nutrient poor soils.

The role of *Metarhizium* and *Beauveria* as insect pathogens and endophytes results in healthier plants (Jha, Jha, & Chourasia, 2015; Khan et al., 2012; Liao, O'Brien, Fang, & St. Leger, 2014; Sasan & Bidochka, 2012; Vega, 2008). The discovery of this lifestyle as plant colonizers has led to a reevaluation of the ecology of other IPF. For example, *Ophiocordyceps sinensis* is a pathogen specific to larvae of the ghost moth *Thitarodes* (Zhang, Xu, et al., 2009). It is believed that without specific nutrients from the larval tissue, *O. sinensis* cannot complete the teleomorph stage of its life cycle (Zhang, Xu, et al., 2009), and the genome of *O. sinensis* has lost many of the genes that allow *Metarhizium* spp. to interact with plants (Hu et al., 2013, 2014). However, Zhong et al. (2014) detected *O. sinensis* on the roots of as many as 23 different plants. The finding that such a host-specific IP fungus can also colonize plants suggests that numerous IPF play larger roles in their ecosystems than previously thought, and that plant host range may influence evolution of these IPF (Behie et al., 2013). We have previously suggested that host-specific IPF may not be capable of forming associations with plants since there would be selective pressure toward host insect specificity (Behie et al., 2013). However, the occurrence of *O. sinensis* on the roots of numerous plants suggests that even pathogenic fungi with a very narrow host range to insects can form plant associations.

The type of association with the host plant(s) of *O. sinensis* is currently unknown (eg, asymptomatic, beneficial, pathogenic, or, quite possibly, coincidental) but may be potentially beneficial as in several other plant-associating IPF (eg, *Metarhizium*, *Beauveria*) (Behie & Bidochka, 2014; Behie, Zelisko, & Bidochka, 2012).

Although the complex interactions and ecology of the soil is still largely a "black box," isolation of fungal propagules from rhizospheric soil is potentially a strong indication of rhizospheric competence (Hu & St. Leger, 2002; St. Leger et al., 2011). The rhizosphere (the region of soil influenced by root chemistry) is a competitive habitat for nutrients as many nutrients are limited by being bound in unusable forms, requiring enzymatic digestion for absorption (Pickles & Pither, 2014). Survivability within the rhizosphere is also dictated by the ability of an organism to overcome abiotic stressors, such as temperature, pH, osmotic stress, and salinity (Alam, 1999). Identifying and modifying genes responsible for tolerance to such stresses could allow increased survivability of IPF within soil. Conidia of a genetically engineered strain of *M. robertsii* overexpressing heat shock protein 25 (HSP25) had increased tolerance to extreme temperature and osmotic stress (Liao et al., 2014). However, the survivability of this engineered strain of *Metarhizium* was dependent on the presence of plant roots as the survivability within bulk soil was unaffected, but was significantly increased in rhizospheric soils (Liao et al., 2014). This highlights the fact that natural environmental conditions must be taken into account when determining the effectiveness of genetically engineered strains.

The ability of a fungus to persist in the rhizosphere is directly influenced by the plant root exudate and microbial diversity within the rhizosphere (Jaeger, Lindow, Miller, Clark, & Firestone, 1999). The types of compounds found in root exudate, although categorically similar (ie, they all contain carbohydrates, amino acids, vitamins, etc.), differ in their proportions and specific makeup, and will alter the pH, organic matter composition, and thus the microbial community (Rovira, 1969). Several IPF have been shown to endophytically associate with plant roots and/or aboveground tissues (eg, *Metarhizium*, *Beauveria*, *Lecanicillium*) (Behie & Bidochka, 2014; Ownley et al., 2010) and benefit the plant in many ways. Soil sampling in a region of Ontario, Canada, revealed three species of *Metarhizium*, *M. robertsii*, *M. brunneum*, and *Metarhizium guizhouense*, that associate with grasses, shrubs, and trees, respectively (Wyrebek, Huber, Sasan, & Bidochka, 2011), and five species of *Metarhizium* were able to transfer insect nitrogen to monocots as well as dicots (Behie & Bidochka, 2014). The ability of an EIPF to colonize a plant host would be dictated by its compatibility with the rhizospheric conditions created by root exudate and its ability to compete within this niche. Thus, any conclusions made that an IP fungus is unable to persist as a plant endophyte must come from exhaustive examination of a range of potential plant hosts, as some IP fungal strains may be more restricted than others in plant host range. For instance, *T. harzianum*, a notable mycoparasite that is also able to infect insects, was deemed rhizosphere-incompetent when evaluated for colonization of cucumber and radish roots (Ahmad & Baker, 1987). However, *T. harzianum* is fully capable of colonizing and promoting the health of maize plants and is routinely used as a biopesticide against plant pathogenic fungi (Harman, 2006; Shakeri & Foster, 2007).

Rhizospheric competence is a desirable attribute for a biopesticide, as the ability of it to persist in the environment long after application would help to maintain crop protection throughout the year and minimize application costs. In order to manipulate naturally occurring endophytes, IPF genomic analyses are being conducted to elucidate genes specific to initiation and maintenance of plant–fungal symbioses (Pava-Ripoll et al., 2011), and certain fungal genes have already been identified as being essential for plant root symbiosis.

The *Mad2* gene of *M. robertsii* encodes a plant adhesin that is essential for attachment to plant material and is upregulated in bean root exudate (Wang & St. Leger, 2007a). The gene *Mrt*, encoding an oligosaccharide (raffinose) transporter, was found to be required for rhizospheric competence of *M. robertsii*, as deletion-mutant germlings failed to develop branching hyphae, and deletion mutants produced 11-fold fewer colony-forming units in rhizospheric soil than the wild type (Fang & St. Leger, 2010). The role of this sugar transporter is clearly significant to the ability of *Metarhizium* to maintain rhizospheric competence. Gene knockout strains of *M. robertsii* for invertase (*MrINV*) resulted in severe reduction of growth in root exudates, but improved colonization on roots. The increased colonization could be due to lower availability of sugar in invertase mutants reducing carbon catabolite repression of enzymes that allow the fungus to colonize roots (Liao, Lu, Fang, & St Leger, 2013). Carbohydrates are the most abundant component of root exudates (Jaeger et al., 1999), and could therefore have the biggest impact on the metabolic requirements and limitations of an EIPF.

Although *Metarhizium* appears to be restricted to colonization of plant roots, other EIPF are able to colonize aboveground tissues of plants as well. In an experiment with haricot bean, *M. robertsii* was found to localize to the root system below the hypocotyl in comparison to *B. bassiana* that was isolated from the roots, hypocotyl, stem, and leaves, after 60 days of growth (Behie et al., 2015). The finding that *Beauveria* can colonize all tissues of a plant has been seen with tomato, cotton, corn, and snap bean as well (Ownley et al., 2008; Wagner & Lewis, 2000), and the mechanism of entry is similar to that utilized for insect infection (Pekrul & Grula, 1979; Wagner & Lewis, 2000).

The mechanism of plant colonization for many EIPF species is unknown, as the discovery of this ability is still relatively recent. How is a plant able to distinguish between a potentially beneficial partner and a pathogenic one? Early recognition events must take place as timing of the initiating defense mechanism is believed to be the difference between infection and resistance to a potential pathogen (Kuc & Strobel, 1992), and the initial phases of infection and colonization of pathogens, mutualists, and commensals are identical for many fungi (Rodriguez, Redman, & Henson, 2005). A recent discovery with the arbuscular mycorrhizal fungus, *Glomus intraradices*, revealed a diffusible factor that stimulated root hair development in the legume, *Medicago truncatula* (Maillet et al., 2011). The structure of this communication molecule was determined to be the same as Nod factors released from rhizobia, a lipochitooligosaccharide (LCO). Molecules released from mycorrhizal fungi that prime the plant for colonization, termed myc factors, often induce activation of the SYM (symbiotic) signaling pathway to cause morphological changes (eg, root hair growth) that increase contact between roots and fungus (Oldroyd, Harrison, &

Paszkowski, 2009; Kosuta et al., 2003). In cultures of switchgrass inoculated with conidia of *Metarhizium*, there was extensive root hair development that was not observed in fungal-free cultures, indicating that *Metarhizium* is able to communicate to the plant prior to colonization, possibly with a myc-like factor (Sasan & Bidochka, 2012). As these LCOs appear to be utilized by both rhizobia and mycorrhizal fungi, it is a fair assumption that EIPF may utilize a similar molecule for communication to their host plants to distinguish themselves as beneficial symbionts.

4.2 Tripartite Interactions of Endophytic Insect Pathogenic Fungi

EIPF potentially colonize plants in order to exploit a carbon source. In mycorrhizal symbioses, plants exchange photosynthetically derived carbohydrates for nutrients that would otherwise be unavailable for uptake (ie, nitrogen and phosphorus) (Govindarajulu et al., 2005; Guether et al., 2009; Jakobsen, Abbott, & Robson, 1992). The EIPF *M. robertsii* was recently shown to provide its plant hosts with insect-derived nitrogen (Behie & Bidochka, 2014), and in return *Metarhizium* receives plant-derived carbohydrates (Fang & St. Leger, 2010). This represents a previously unknown method of nitrogen acquisition for plants, as well as defining the association between the EIPF *M. robertsii* and its plant host as mutualistic. A question that remains is whether or not a continued symbiotic relationship, between plant and EIPF, is specifically reliant on nutrient exchange. That is, if the fungus were to suddenly stop transferring nitrogen, would the plant perceive the fungus as pathogenic and eject the fungal partner? In mycorrhizal symbioses between *Medicago* and various species of *Glomus*, the exchange of phosphorus and carbon is bidirectionally controlled (Kiers et al., 2011). Kiers et al. (2011) tested the cooperation of both the plant and fungal partner in maintaining a successful symbiosis when each was supplied differing concentrations of nutrients for exchange. The results showed that both *Medicago* and *Glomus* were able to detect the nutrients provided by the partner and allocate the reciprocal resource based on the most rewarding root/hyphae (Kiers et al., 2011). To this end, the ability of a plant to shift resource allocation to reciprocating hyphae and thus away from less-cooperative hyphae would limit the growth of the fungal symbiont and its survival in the absence of providing nutrients. This same mechanism was seen with regard to both nitrogen and carbon exchange (Fellbaum et al., 2012). The dynamics that sustain the association between facultative EIPF and plants, as compared to obligate mycorrhizal symbionts, may reflect similar control mechanisms, but this remains to be determined. If nutrient transfer were as tightly controlled in a bidirectional fashion with EIPF, it would lend support to the hypothesis that fungal evolution toward insect parasitism occurred to increase the competitive advantage by permitting access to a larger, insect-derived supply of nitrogen available in exchange for plant carbohydrates.

5. APPLICATION OF ENDOPHYTIC INSECT PATHOGENIC FUNGI

5.1 Insect Pathogenic Endophytes as Biocontrol Agents

Insect pest control in an agricultural setting has historically relied heavily on chemical pesticides; however, pest control practices have been shifting from chemical insecticides to the use of biological control agents (Vega et al., 2009). Among these, IPF of genera including *Metarhizium*, *Beauveria*, *Lecanicillium*, *Isaria*, *Sporothrix*, *Hirsutella*, *Aschersonia*, *Paecilomyces*, *Tolypocladium*, and *Nomuraea* have been traditionally known and extensively studied as insect pest control agents in agriculture (Faria & Wraight, 2007; Vega et al., 2009). Of these *Metarhizium*, *Beauveria*, *Lecanicillium*, and *Isaria* are among the most commercially available insect pathogens (Vega et al., 2009).

Recent work has specifically focused on genetic engineering to increase overall fungal virulence. For example, Wang and St. Leger (2007b) genetically modified *M. robertsii* to express an insect-specific neurotoxin, a gene harvested from the scorpion, *Androctonus australis*. Compared to the wild-type fungus, the neurotoxin-expressing strain reduced the survival time of the tobacco hornworm by 28% (Wang & St Leger, 2007c). More recently, Fang, Lu, King, and St. Leger (2014) found that engineering *M. acridum* to express a combination of four insect-specific toxins produced a synergistic effect and overall reduced the LT50 for acridids by 48% in comparison to the wild-type strain. When the engineered strains of *M. acridum* and the wild type were challenged with non-locust hosts (eg, cockroach, mosquito), all were unable to cause infection and death (Fang et al., 2014). This example of host specificity is of the utmost importance when introducing a biopesticide into the environment. The careful evaluation of genetically engineered mycoinsecticides is required to prevent the accidental death of nontarget organisms, thus *M. acridum* was a prime candidate for engineered virulence, as its host range is more specific when compared to some other *Metarhizium* species (eg, *M. robertsii*) (Wang & St. Leger, 2005). Host specificity has been an extremely important factor in mycoinsecticide research, and species of IPF with narrow host specificity are excellent candidates to overcome time-lapse problems associated with biological control (Fan, Borovsky, Hawkings, Ortiz-Urquiza, & Keyhani, 2012).

Studies have also reported the involvement of chitin deacetylase in cuticle softening during pathogenesis and suggest its potential use in mycopesticide formulation to accelerate the eradication of insect pests from agricultural fields. Chitin deacetylase converts cuticular chitin to the easily degradable glucosamine polymer chitosan and thus facilitates easier penetration of the cuticle during insect pathogenesis (Nahar, Ghormade, & Deshpande, 2004). Genetic manipulations of *Beauveria* and *Metarhizium* have shown improved virulence against numerous insect hosts. Improved cuticle penetration was observed in *B. bassiana* strains that express engineered proteases and chitinases (Fan et al., 2010; Fang, Feng, et al., 2009). For example, chitin degradation was increased by a hybrid chitinase constructed by fusing a chitinase from *B. bassiana* with a chitin-binding domain from the silkworm (Fan et al., 2007).

Despite their positive influence on the environment, biocontrol agents are often slow acting in comparison to chemical applications (Ujjan & Shahzad, 2014). The commercially available biopesticide, Green Muscle employs *M. acridum* to control locust and grasshopper populations; however, this treatment can take between one and three weeks to kill adult pests (Jackson, Dunlap, & Jaronski, 2010), and during this time, crop loss can be substantial. It is for this reason that the majority of myco-control research has been focused on improving the virulence of fungal pathogens (Fang et al., 2014; Liao et al., 2014; Peng & Xia, 2014; Wang & St. Leger, 2007a). Furthermore, environmental factors, such as UV radiation, temperature, and humidity, affect the population of fungi in natural habitats, which in turn limits their utility as biocontrol agents (Ruijter et al., 2003).

Research could also focus on the identification of secondary metabolites involved in different stages of the fungal life cycle that can be utilized for strain improvement with increased stress tolerance (Carollo et al., 2010). LC-MS analysis of *Metarhizium* conidial extraction showed higher amounts of mannitol and a novel secondary metabolite, tyrosine betaine (Carollo et al., 2010). Mannitol is the main storage carbohydrate present in most fungi and is involved in improving oxidative and temperature stress tolerance in *Aspergillus nidulans* (Ruijter et al., 2003). Tyrosine betaine is conserved in *Metarhizium* spp. and hence is speculated to be important for their biology; however, more analysis is needed to confirm this (Carollo et al., 2010). Ortiz-Urquiza and Keyhani (2015) discussed the major signaling pathways involved in stress responses in both *Metarhizium* and *Beauveria*, thus providing insights into future directions that can be pursued for strain improvement. More research is needed toward effective pest management tools by developing persistent fungal strains with improved virulence and host range for field applications.

The endophytic capabilities of these EIPF have great potential in terms of biocontrol of soil-dwelling pests. For example, the redheaded cockchafer larvae, *Adoryphorus couloni*, is a common pasture pest in South Eastern Australia with subterranean feeding habits. This feeding habit of the redheaded cockchafer larvae renders aboveground-applied insecticides useless. Application of *Metarhizium* in the soil reduced the infestation of the larvae and increased pasture productivity (Berg et al., 2014; Srivastava, Maurya, Sharma, & Mohan, 2009). EIPF can live endophytically without causing any negative effects on plants, and plant colonization has been established by applying fungal propagules on either seeds or roots, thereby, providing opportunities to control soil-borne insects that cannot be easily controlled by chemical insecticides.

5.2 Plant Protection and Improvement

A fungal endophyte is defined as the occurrence of a fungal species living asymptomatically within the tissues of a plant (Aly, Debbab, & Proksch, 2011). The recent identification and investigation of IPF living as endophytes in various plants has revealed that in many cases this association is not just asymptomatic, but beneficial. A summary of the impacts and benefits conferred to plants via EIPF is shown in Fig. 1. The benefits conferred to a plant host depends on the fungal species involved and can be a single benefit or a combination of several benefits, including increase in plant biomass and productivity, alleviation of abiotic stresses (eg, drought, salinity, temperature fluctuations) and improved resistance to biotic stress (eg, herbivory and fungal disease). Vega (2008) and Quesada-Moraga, Herrero, and Zabalgogeazcoa (2014) provide comprehensive summaries of EIPF and plant hosts found to occur naturally, as well as successful symbioses obtained through artificial inoculation. *Metarhizium*, *Beauveria*, and *Isaria* are the most commonly isolated EIPF from temperate soils. In many instances, increased growth of the host plant occurs directly as a result of acquisition of growth-limiting nutrients (ie, nitrogen and phosphorus) in exchange for plant-derived carbohydrates (Bonfante & Genre, 2010) or indirectly through alleviation of stress that would otherwise limit growth potential.

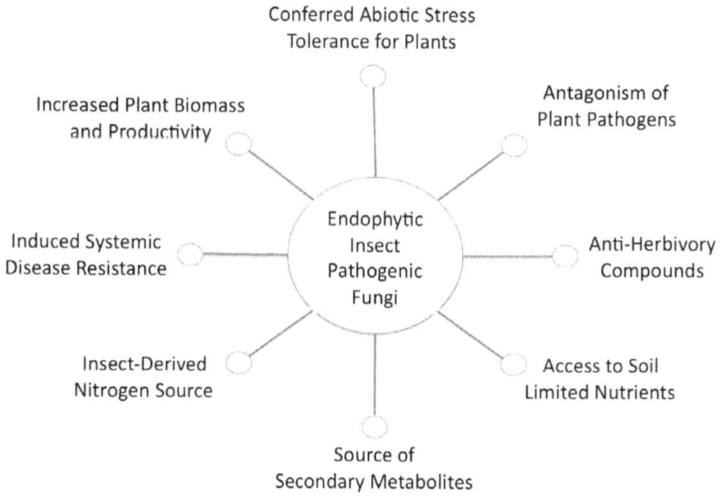

Figure 1 The applications, benefits, and impacts of endophytic insect pathogenic fungi colonization of plants.

Recent studies with *Beauveria* and *Metarhizium* have demonstrated plant health—promoting properties (Sasan & Bidochka, 2012) and the important roles of these fungi in the ecological cycling of insect-derived nitrogen to plant communities (Behie & Bidochka, 2014; Behie et al., 2012). *Metarhizium* has been shown to increase plant biomass and mitigate salt stress in soybean in comparison to plants lacking fungal colonization (Khan et al., 2012). Tomato, haricot bean, switchgrass, and soybean plants had increased root lengths and shoot/root dry weights (Behie & Bidochka, 2014; Elena, Beatriz, Alejandro, & Lecuona, 2011; Sasan & Bidochka, 2012) when colonized by various species of *Metarhizium*. The grass endophyte *Neotyphodium* has been shown to prevent defoliation and increase seed dispersal and vegetative yield in *Bromus auleticus* (Iannone & Cabral, 2006); it also protected *Bromus setifolius* from leaf-cutting ants (White et al., 2001). EIPF are capable of successfully protecting plants from microbial pathogens by suppressing disease-causing agents or increasing plant defense responses. The fungal disease powdery mildew (*Sphaerotheca*) was suppressed by species of *Lecanicillium* (formerly *Verticillium*) and *Isaria* in cucumber and strawberry plants (Kim et al., 2008; Miller, Gubler, Laemmlen, Geng, & Rizzo, 2004), and *Metarhizium* was shown to be antagonistic toward the root-rot fungus, *Fusarium solani* (Sasan & Bidochka, 2013). *Beauveria* and *Lecanicillium* have been shown to induce the expression and production of plant defense compounds in the date palm, *Phoenix dactylifera*

(Gómez-Vidal, Salinas, Tena, Lopez-Llorca, 2009) and may contribute to plant "priming" against plant pathogens and increased plant growth (Ownley et al., 2010), as seen with the endophytic mycoparasite *Trichoderma* (Harman, Howell, Viterbo, Chet, & Lorito, 2004). *Lecanicillium* is also epiphytic and may prevent fungal disease by competitively growing on the surface of leaves, thus instilling spatial restrictions, producing antimicrobial compounds, limiting available nutrients, and being mycoparasitic, in addition to inducing plant responses while colonizing plant roots (Ownley et al., 2010; reviewed by Quesada-Moraga et al., 2014).

With the knowledge that many EIPF have broad host ranges, in respect to both insect and plant hosts (eg, *Metarhizium*, *Beauveria*, *Lecanicillium*), research should focus on establishing a method by which to measure the strength of symbiosis between certain species in order to ascertain the best candidate for pest control in a particular plant population.

6. SECONDARY METABOLITES

Aside from their role as biocontrol agents and as plant growth promoters, insect pathogens are rich sources of natural products/secondary metabolites, as well. Many gene clusters involved in secondary metabolite production have been predicted in IPF (see Donzelli and Krasnoff, 2016). Analysis of secondary metabolite gene clusters in *Metarhizium* strains indicated that there are 85 and 57 core genes present in *M. robertsii* and *M. acridum*, respectively (Gibson, Donzelli, Krasnoff, & Keyhani, 2014). A variety of pharmaceutically relevant and insecticidal metabolites have been reported to date in *Metarhizium* including destruxins, fusarin-like compounds (NG39x), helvolic acid, cytochalasin, swainsonine, serinocyclins, viridoxins, and valicin (Fuji et al., 2000; Gupta et al., 1993; Krasnoff et al., 2006, 2007; Kuboki et al., 1999; Lee, Kinoshita, Ihara, Igarashi, & Nihira, 2008; Liu & Tzeng, 2012; Patrick, Adlard, & Keshavarz, 1993). Genome analysis predicts that the *B. bassiana* genome contains 45 core genes putatively involved in secondary metabolite synthesis (Gibson et al., 2014) with the best-studied secondary metabolites in *Beauveria* being beauvericin (Wang & Xu, 2012) and bassianolide (Scharf, Heinekamp, & Brakhage, 2014). Beauvericin has shown to have antimicrobial, insecticidal, and cytotoxic activities (Wang & Xu, 2012). Bassianolide, also produced by other fungal species including *Lecanicillium*, has insecticidal and pharmacological properties (Gibson et al., 2014).

In addition to the potential for secondary metabolite biosynthesis, these insect pathogens, specifically *Beauveria*, can be utilized as biocatalytic units for the transformation of chemical substrates (Grogan & Holland, 2000). In biotransformation, the enzymatic repertoire of the fungus is exploited for the transformation or modification of chemical compounds. *Beauveria* spp. are the most frequently used insect pathogens as biocatalysts for the transformation of chemicals. The primary reactions catalyzed by *Beauveria* are demethylation, hydroxylation, and glycosidations (Gibson et al., 2014). Although a number of secondary gene clusters have been predicted in these insect pathogens, the natural products emanating from these gene clusters remain to be identified.

Fungal endophytes are capable of synthesizing natural products/secondary metabolites when associated with plants, which have huge pharmaceutical potential. A well-studied example is the anticancer drug, Taxol produced by endophytic fungi when associated with yew trees belonging to the *Taxus* family (Garyali & Reddy, 2013). *Metarhizium anisopliae* produces one of the highest yields of Taxol so far reported from an endophyte, 0.85 mg/L of fermentation broth (H-27 Accession #FJ375161, see Donzelli and Krasnoff, 2016). A more thorough understanding of the natural products produced during endophytic association by these fungi will contribute to the broader use of EIPF beyond their agricultural applications.

REFERENCES

Ahmad, J. S., & Baker, R. (1987). Rhizosphere competence of *Trichoderma harzianum*. *Phytopathology, 77*, 182–189.

Alam, S. M. (1999). Nutrient uptake by plants under stress conditions. In M. Pessakakli (Ed.), *Handbook of plant and crop stress* (pp. 285–314). New York: Marcel Dekker.

Aly, A. H., Debbab, A., & Proksch, P. (2011). Fungal endophytes: unique plant inhabitants with great promises. *Applied Microbiology and Biotechnology, 90*, 1829–1845.

Bagga, S., Hu, G., Screen, S. E., & St. Leger, R. J. (2004). Reconstructing the diversification of subtilisins in the pathogenic fungus *Metarhizium anisopliae*. *Gene, 324*, 159–169.

Behie, S. W., & Bidochka, M. J. (2014). Ubiquity of insect-derived nitrogen transfer to plants by endophytic insect-pathogenic fungi: an additional branch of the soil nitrogen cycle. *Science, 80*, 1553–1560.

Behie, S. W., Jones, S. J., & Bidochka, M. J. (2015). Plant tissue localization of the endophytic insect pathogenic fungi *Metarhizium* and *Beauveria*. *Fungal Ecology, 13*, 112–119.

Behie, S. W., Padilla-Guerrero, I. E., & Bidochka, M. J. (2013). Nutrient transfer to plants by phylogenetically diverse fungi suggests convergent evolutionary strategies in rhizospheric symbionts. *Communicative and Integrative Biology, 6*, e22321.

Behie, S. W., Zelisko, P. M., & Bidochka, M. J. (2012). Endophytic insect-parasitic fungi translocate nitrogen directly from insects to plants. *Science, 336*, 1576–1577.

Berg, G., Faithfull, I. G., Powell, K. S., Bruce, R. J., Williams, D. G., & Yen, A. L. (2014). Biology and management of the redheaded pasture cockchafer *Adoryphorus couloni*

(Burmeister)(Scarabaeidae: Dynastinae) in Australia: a review of current knowledge. *Austral Entomology, 53*, 144–158.

Bidochka, M. J., Clark, D. C., Lewis, M. W., & Keyhani, N. O. (2010). Could insect phagocytic avoidance by entomogenous fungi have evolved via selection against soil amoeboid predators? *Microbiology, 156*, 2164–2171.

Bidochka, M. J., St. Leger, R. J., Roberts, D. W. (1997). Mechanisms of deuteromycete fungal infections in grasshoppers and locusts: an overview. In M. S. Goettel & D. L. Johnson (Eds.), Microbial control of grasshoppers and locusts. *Memoirs of the Entomological Society of Canada, 171*, 213–224.

Blanford, S., Chan, B. H. K., Jenkins, N., Sim, D., Turner, R. J., Read, A. F., & Thomas, M. B. (2005). Fungal pathogen reduces potential for malaria transmission. *Science, 308*, 1638–1641.

Bonfante, P., & Genre, A. (2010). Mechanisms underlying beneficial plant–fungus interactions in mycorrhizal symbiosis. *Nature Communications, 1*, 48.

Boucias, D. G., Pendland, J. C., & Latge, J. P. (1988). Nonspecific factors involved in attachment of entomopathogenic deuteromycetes to host insect cuticle. *Applied and Environmental Microbiology, 54*, 1795–1805.

Carollo, C. A., Calil, A. L. A., Schiave, L. A., Guaratini, T., Roberts, D. W., Lopes, N. P., & Braga, G. U. (2010). Fungal tyrosine betaine, a novel secondary metabolite from conidia of insect pathogenic *Metarhizium* spp. fungi. *Fungal Biology, 114*, 473–480.

Castrillo, L. A., Griggs, M. H., Ranger, C. M., Reding, M. E., & Vandenberg, J. D. (2011). Virulence of commercial strains of *Beauveria bassiana* and *Metarhizium brunneum* (Ascomycota: Hypocreales) against adult *Xylosandrus germanus* (Coleoptera: Curculionidae) and impact on brood. *Biological Control, 58*, 121–126.

Donzelli, B. G. G., & Krasnoff, S. B. (2016). Molecular Genetics of Secondary Chemistry in *Metarhizium* Fungi. *Advances in Genetics, 94*, 365–436.

Elena, G. J., Beatriz, P. J., Alejandro, P., & Lecuona, R. E. (2011). *Metarhizium anisopliae* (Metschnikoff) Sorokin promotes growth and has endophytic activity in tomato plants. *Advanced Biomedical Research, 5*, 22–27.

Fan, Y., Borovsky, D., Hawkings, C., Ortiz-Urquiza, A., & Keyhani, N. O. (2012). Exploiting host molecules to augment mycoinsecticide virulence. *Nature Biotechnology, 30*, 35–37.

Fan, Y. H., Fang, W. G., Guo, S. J., Pei, X. Q., Zhang, Y. J., Xiao, Y. H. ... Pei, Y. (2007). Increased insect virulence in *Beauveria bassiana* strains over expressing an engineered chitinase. *Applied and Environmental Microbiology, 73*, 295–302.

Fan, Y. H., Pei, X. Q., Guo, S. J., Zhang, Y. J., Luo, Z. B., Liao, X. G., & Pei, Y. (2010). Increased virulence using engineered protease-chitin binding domain hybrid expressed in the insect pathogenic fungus *Beauveria bassiana*. *Microbial Pathogenesis, 49*, 376–380.

Fang, W. G., Feng, J., Fan, Y. H., Zhang, Y. J., Bidochka, M. J., Leger, R. J. S., & Pei, Y. (2009). Expressing a fusion protein with protease and chitinase activities increases the virulence of the insect pathogen *Beauveria bassiana*. *Journal of Invertebrate Pathology, 102*, 155–159.

Fang, W., Lu, H. L., King, G. F., & St. Leger, R. J. (2014). Construction of a hypervirulent and specific mycoinsecticide for locust control. *Scientific Reports, 4*.

Fang, W., Pava-Ripoll, M., Wang, S., & Leger, R. S. (2009). Protein kinase A regulates production of virulence determinants by the entomopathogenic fungus, *Metarhizium anisopliae*. *Fungal Genetics and Biology, 46*, 277–285.

Fang, W., Pei, Y., & Bidochka, M. J. (2007). A regulator of a G protein signalling (RGS) gene, *cag8*, from the insect-pathogenic fungus *Metarhizium anisopliae* is involved in conidiation, virulence and hydrophobin synthesis. *Microbiology, 153*, 1017–1025.

Fang, W., & St Leger, R. J. (2010). *Mrt*, a gene unique to fungi, encodes an oligosaccharide transporter and facilitates rhizosphere competency in *Metarhizium robertsii*. *Plant Physiology, 154*, 1549–1557.

Fang, W., & St. Leger, R. J. (2012). Enhanced UV resistance and improved killing of malaria mosquitoes by photolyase transgenic entomopathogenic fungi. *PLoS One, 7*(8), e43069. http://dx.doi.org/10.1371/journal.pone.0043069.
Faria, M. R., & Wraight, S. P. (2007). Mycoinsecticides and mycoacaricides: a comprehensive list with worldwide coverage and international classification of formulation types. *Biological Control, 43*, 237–256.
Fellbaum, C. R., Gachomo, E. W., Beesetty, Y., Choudhari, S., Strahan, G. D., Pfeffer, P. E.... Bücking, H. (2012). Carbon availability triggers fungal nitrogen uptake and transport in arbuscular mycorrhizal symbiosis. *Proceedings of the National Academy of Sciences of the United States of America, 109*, 2666–2671.
Fujii, Y., Tani, H., Ichinoe, M., & Nakajima, H. (2000). Zygosporin D and two new cytochalasins produced by the fungus *Metarhizium anisopliae*. *Journal of Natural Products, 63*, 132–135.
Gao, Q., Jin, K., Ying, S. H., Zhang, Y., Xiao, G., Shang, Y.... Wang, C. (2011). Genome sequencing and comparative transcriptomics of the model insect pathogenic fungi *Metarhizium anisopliae* and *M. acridum*. *PLoS Genetics*. http://dx.doi.org/10.1371/journal.pgen.1001264.
Garyali, S., & Reddy, M. S. (2013). Taxol production by an endophytic fungus, *Fusarium redolens*, isolated from Himalayan yew. *Journal of Microbiology and Biotechnology, 23*, 1372–1380.
Geremia, R. A., Goldman, G. H., Jacobs, D., Ardiles, W., Vila, S. B., Van Montagu, M., & Herrera-Estrella, A. (1993). Molecular characterization of the proteinase-encoding gene, *prbl*, related to mycoparasitism by *Trichoderma harzianum*. *Molecular Microbiology, 8*, 603–613.
Gibson, D. M., Donzelli, B. G., Krasnoff, S. B., & Keyhani, N. O. (2014). Discovering the secondary metabolite potential encoded within insect pathogenic fungi. *Natural Product Reports, 31*, 1287–1305.
Gómez-Vidal, S., Salinas, J., Tena, M., & Lopez-Llorca, L. V. (2009). Proteomic analysis of date palm (*Phoenix dactylifera* L.) responses to endophytic colonization by entomopathogenic fungi. *Electrophoresis, 30*, 2996–3005.
Govindarajulu, M., Pfeffer, P. E., Jin, H., Abubaker, J., Douds, D. D., Allen, J. W.... Shachar-Hill, Y. (2005). Nitrogen transfer in the arbuscular mycorrhizal symbiosis. *Nature, 435*, 819–823.
Grogan, G. J., & Holland, H. L. (2000). The biocatalytic reactions of *Beauveria* spp. *Journal of Molecular Catalysis B: Enzymatic, 9*, 1–32.
Guether, M., Neuhäuser, B., Balestrini, R., Dynowski, M., Ludewig, U., & Bonfante, P. (2009). A mycorrhizal-specific ammonium transporter from *Lotus japonicus* acquires nitrogen released by arbuscular mycorrhizal fungi. *Plant Physiology, 150*, 73–83.
Gupta, S., Krasnoff, S. B., Renwick, J. A. A., Roberts, D. W., Steiner, J. R., & Clardy, J. (1993). Viridoxins A and B: novel toxins from the fungus *Metarhizium flavoviride*. *Journal of Organic Chemistry, 58*, 1062–1067.
Harman, G. E. (2006). Overview of mechanisms and uses of *Trichoderma* spp. *Phytopathology, 96*, 190–194.
Harman, G. E., Howell, C. R., Viterbo, A., Chet, I., & Lorito, M. (2004). *Trichoderma* species—opportunistic, avirulent plant symbionts. *Nature Reviews Microbiology, 2*, 43–56.
Hu, G., & St. Leger, R. J. (2002). Field studies using a recombinant mycoinsecticide (*Metarhizium anisopliae*) reveal that it is rhizosphere competent. *Applied and Environmental Microbiology, 68*, 6383–6387.
Hu, G., & St. Leger, R. J. (2004). A phylogenomic approach to reconstructing the diversification of serine proteases in fungi. *Journal of Evolutionary Biology, 17*, 1204–1214.
Hu, X., Xiao, G., Zheng, P., Shang, Y., Su, Y., Zhang, X.... Wang, C. S. (2014). Trajectory and genomic determinants of fungal-pathogen speciation and host adaptation. *Proceedings of the National Academy of Sciences of the United States of America, 111*, 16796–16801.

Hu, X., Zhang, Y., Xiao G, Zheng P., XIA Y., & Zhang X. (2013). Genome survey uncovers the secrets of sex and lifestyle in caterpillar fungus. *Chinese Science Bulletin, 58*, 2846–2854.
Huang, W., Shang, Y., Chen, P., Cen, K., & Wang, C. (2015). Regulation of bZIP transcription factor MBZ1 on cell wall integrity, spore adherence, and virulence in *Metarhizium robertsii*. *Journal of Biological Chemistry*. http://dx.doi.org/10.1074/jbc.M114.630939.
Huang, W., Shang, Y., Chen, P., Gao, Q., & Wang, C. (2015). MrpacC regulates sporulation, insect cuticle penetration and immune evasion in *Metarhizium robertsii*. *Environmental Microbiology, 17*, 994–1008.
Humber, R. A. (2008). Evolution of insect pathogenicity in fungi. *Journal of Invertebrate Pathology, 98*, 262–266.
Iannone, L. J., & Cabral, D. (2006). Effects of the *Neotyphodium* endophyte status on plant performance of *Bromus auleticus*, a wild native grass from South America. *Symbiosis, 41*, 61–69.
Jackson, M. A., Dunlap, C. A., & Jaronski, S. T. (2010). Ecological considerations in producing and formulating fungal insect pathogens for use in insect biocontrol. In *The ecology of fungal insect pathogens* (pp. 129–145). Netherlands: Springer.
Jaeger, C. H., Lindow, S. E., Miller, W., Clark, E., & Firestone, M. K. (1999). Mapping of sugar and amino acid availability in soil around roots with bacterial sensors of sucrose and tryptophan. *Applied and Environmental Microbiology, 65*, 2685–2690.
Jakobsen, I., Abbott, L. K., & Robson, A. D. (1992). External hyphae of vesicular-arbuscular mycorrhizal fungi associated with *Trifolium subterraneum* L. 1. Spread of hyphae and phosphorus inflow into roots. *New Phytologist, 120*, 371–379.
Jha, M. N., Jha, S., & Chourasia, S. K. (2015). Agroecology and agromicrobes. In N. Benkeblia (Ed.), *Agroecology, ecosystems, and sustainability* (pp. 81–102). Boca Raton, FL: CRC Press.
Jin, K., Ming, Y., & Xia, Y. X. (2012). MaHog1, a Hog1-type mitogen-activated protein kinase gene, contributes to stress tolerance and virulence of the insect pathogenic fungus *Metarhizium acridum*. *Microbiology Sgm, 158*, 2987–2996.
Kawaguchi, M., & Minamisawa. (2010). Plant-microbe communications for symbiosis. *Plant and Cell Physiology, 51*, 1377–1380.
Khan, A. L., Hamayun, M., Khan, S. A., Kang, S. M., Shinwari, Z. K., Kamran, M.... Lee, I. J. (2012). Pure culture of *Metarhizium anisopliae* LHL07 reprograms soybean to higher growth and mitigates salt stress. *World Journal of Microbiology and Biotechnology, 28*, 1483–1494.
Kiers, E. T., Duhamel, M., Beesetty, Y., Mensah, J. A., Franken, A., Verbruggen, E.... Bücking, H. (2011). Reciprocal rewards stabilize cooperation in the mycorrhizal symbiosis. *Science, 333*, 880–882.
Kim, J. S., Roh, J. Y., Choi, J. Y., Shin, S. C., Jeon, M. J., & Je, Y. H. (2008). Insecticidal activity of *Paecilomyces fumosoroseus* SFP-198 as a multi-targeting biological control agent against the greenhouse whitefly and the two-spotted spider mite. *International Journal of Industrial Entomology, 17*, 181–187.
Kosuta, S., Chabaud, M., Lougnon, G., Gough, C., Dénarié, J., Barker, D. G., & Bécard, G. (2003). A diffusible factor from arbuscular mycorrhizal fungi induces symbiosis-specific *MtENOD11* expression in roots of *Medicago truncatula*. *Plant Physiology, 131*, 952–962.
Krasnoff, S. B., Keresztes, I., Gillilan, R. E., Szebenyi, D. M., Donzelli, B. G., Churchill, A. C., & Gibson, D. M. (2007). Serinocyclins A and B, cyclic heptapeptides from *Metarhizium anisopliae*. *Journal of Natural Products, 70*, 1919–1924.
Krasnoff, S. B., Sommers, C. H., Moon, Y. S., Donzelli, B. G., Vandenberg, J. D., Churchill, A. C., & Gibson, D. M. (2006). Production of mutagenic metabolites by *Metarhizium anisopliae*. *Journal of Agricultural and Food Chemistry, 54*, 7083–7088.

Kuboki, H., Tsuchida, T., Wakazono, K., Isshiki, K., Kumagai, H., & Yoshioka, T. (1999). Mer-f3, 12-hydroxy-ovalicin, produced by *Metarhizium* sp. f3. *Journal of Antibiotics, 52*, 590–593.

Kuc, J., & Strobel, N. E. (1992). Induced resistance using pathogens and nonpathogens. In *Biological control of plant diseases* (pp. 295–303). US: Springer.

Lee, S. Y., Kinoshita, H., Ihara, F., Igarashi, Y., & Nihira, T. (2008). Identification of novel derivative of helvolic acid from *Metarhizium anisopliae* grown in medium with insect component. *Journal of Bioscience and Bioengineering, 105*, 476–480.

Liao, X., Lu, H. L., Fang, W., & St Leger, R. J. (2013). Overexpression of a *Metarhizium robertsii HSP25* gene increases thermotolerance and survival in soil. *Applied Microbiology and Biotechnology, 98*, 777–783.

Liao, X., O'Brien, T. R., Fang, W., & St. Leger, R. J. (2014). The plant beneficial effects of *Metarhizium* species correlate with their association with roots. *Applied Microbiology and Biotechnology, 98*, 7089–7096.

Liu, B. L., & Tzeng, Y. M. (2012). Development and applications of destruxins: a review. *Biotechnology Advances, 30*, 1242–1254.

Liu, Q., Ying, S. H., Li, J. G., Tian, C. G., & Feng, M. G. (2013). Insight into the transcriptional regulation of Msn2 required for conidiation, multi-stress responses and virulence of two insect pathogenic fungi. *Fungal Genetics and Biology, 54*, 42–51.

Li, J., Yu, L., Yang, J., Dong, L., Tian, B., Yu, Z. … Zhang, K. (2010). New insights into the evolution of subtilisin-like serine protease genes in *Pezizomycotina. BMC Evolutionary Biology, 10*, 68.

Luo, S., He, M., Cao, Y., & Xia, Y. (2013). The tetraspanin gene *MaPls1* contributes to virulence by affecting germination, appressorial function and enzymes for cuticle degradation in the insect pathogenic fungus, *Metarhizium acridum. Environmental Microbiology, 15*, 2966–2979.

Luo, X. D., Keyhani, N. O., Yu, X. D., He, Z. J., Luo, Z. B., Pei, Y., & Zhang, Y. J. (2012). The MAP kinase Bbslt2 controls growth, conidiation, cell wall integrity, and virulence in the insect pathogenic fungus *Beauveria bassiana. Fungal Genetics and Biology, 49*, 544–555.

Luo, Z. B., Qin, Y. Q., Pei, Y., & Keyhani, N. O. (2014). Ablation of the creA regulator results in amino acid toxicity, temperature sensitivity, pleiotropic effects on cellular development and loss of virulence in the filamentous fungus *Beauveria bassiana. Environmental Microbiology, 16*, 1122–1136.

Maillet, F., Poinsot, V., André, O., Puech-Pagès, V., Haouy, A., Gueunier, M. … Dénarié, J. (2011). Fungal lipochitooligosaccharide symbiotic signals in arbuscular mycorrhiza. *Nature, 469*, 58–63.

Mehrabi, R., Zhao, X., Kim, Y., & Xu, J.-R. (2009). The cAMP signaling and MAP kinase pathways in plant pathogenic fungi. In H. Deising (Ed.) (2nd ed., *The Micota: vol V. Plant relationships* (pp. 157–172). Springer-Verlag.

Miller, T. C., Gubler, W. D., Laemmlen, F. F., Geng, S., & Rizzo, D. M. (2004). Potential for using *Lecanicillium lecanii* for suppression of strawberry powdery mildew. *Biocontrol Science and Technology, 14*, 215–220.

Moraes, C. K., Schrank, A., & Vainstein, M. H. (2003). Regulation of extracellular chitinases and proteases in the insect pathogen and acaricide *Metarhizium anisopliae. Current Microbiology, 46*, 0205–0210.

Nahar, P., Ghormade, V., & Deshpande, M. V. (2004). The extracellular constitutive production of chitin deacetylase in *Metarhizium anisopliae*: possible edge to insect pathogenic fungi in the biological control of insect pests. *Journal of Invertebrate Pathology, 85*, 80–88.

Oldroyd, G. E. D., Harrison, M. J., & Paszkowski, U. (2009). Reprogramming plant cells for endosymbiosis. *Science, 324*, 753–754.

Ortiz-Urquiza, A., & Keyhani, N. O. (2013). Action on the surface: insect pathogenic fungi versus the insect cuticle. *Insects, 4*, 357–374.

Ortiz-Urquiza, A., & Keyhani, N. O. (2015). Stress response signaling and virulence: insights from insect pathogenic fungi. *Current Genetics, 61*, 239–249.

Ortiz-Urquiza, A., Luo, Z., & Keyhani, N. O. (2015). Improving mycoinsecticides for insect biological control. *Applied Microbiology and Biotechnology, 99*, 1057–1068.

Ownley, B. H., Griffin, M. R., Klingeman, W. E., Gwinn, K. D., Moulton, J. K., & Pereira, R. M. (2008). *Beauveria bassiana*: endophytic colonization and plant disease control. *Journal of Invertebrate Pathology, 98*, 267–270.

Ownley, B. H., Gwinn, K. D., & Vega, F. E. (2010). Endophytic fungal insect pathogens with activity against plant pathogens: ecology and evolution. *BioControl, 55*, 113–128.

Patrick, M., Adlard, M. W., & Keshavarz, T. (1993). Production of an indolizidine alkaloid, swainsonine by the filamentous fungus, *Metarhizium anisopliae*. *Biotechnology Letters, 15*, 997–1000.

Pava-Ripoll, M., Angelini, C., Fang, W., Wang, S., Posada, F. J., & St. Leger, R. J. (2011). The rhizosphere-competent insect pathogen *Metarhizium anisopliae* expresses a specific subset of genes in plant root exudate. *Microbiology, 157*, 47–55.

Pedrini, N., Crespo, R., & Juárez, M. P. (2007). Biochemistry of insect epicuticle degradation by entomopathogenic fungi. *Comparative Biochemistry and Physiology Part C: Toxicology and Pharmacology, 146*, 124–137.

Pekrul, S., & Grula, E. A. (1979). Mode of infection of the corn earworm (*Heliothis zea*) by *Beauveria bassiana* as revealed by scanning electron microscopy. *Journal of Invertebrate Pathology, 34*, 238–247.

Peng, G. X., & Xia, Y. X. (2014). Expression of scorpion toxin LqhIT2 increases the virulence of *Metarhizium acridum* towards *Locusta migratoria manilensis*. *Journal of Industrial Microbiology and Biotechnology, 41*(11), 1659–1666.

Pickles, B. J., & Pither, J. (2014). Still scratching the surface: how much of the 'black box' of soil ectomycorrhizal communities remains in the dark? *New Phytologist, 201*, 1101–1105.

Pilat, M. V. (1938). Permeability of the chitin of insects to entomogenous fungi. *Lenin Academy of Agricultural Sciences (Medicine) Russia*, 73–75.

Pospelov, V. P. (1938). Methods of infecting insects with entomogenous fungi. *Lenin Academy of Agricultural Sciences Russia*, 64–67.

Quesada-Moraga, E., Herrero, N., & Zabalgogeazcoa, I. (2014). Insect pathogenic and nematophagous fungal endophytes. In V. C. Verma, & A. C. Gange (Eds.), *Advances in endophytic research* (pp. 85–99). New Delhi, India: Springer.

Reddy, P. V., Lam, C. K., & Belanger, F. C. (1996). Mutualistic fungal endophytes express a proteinase that is homologous to proteases suspected to be important in fungal pathogenicity. *Plant Physiology, 111*, 1209–1218.

Rodriguez, R. J., Redman, R. S., & Henson, J. M. (2005). Symbiotic lifestyle expression by fungal endophytes and the adaptation of plants to stress: unraveling the complexities of intimacy. *Mycology Series, 23*, 683.

Rose, J. K., Ham, K. S., Darvill, A. G., & Albersheim, P. (2002). Molecular cloning and characterization of glucanase inhibitor proteins: coevolution of a counter defense mechanism by plant pathogens. *Plant Cell, 14*, 1329–1345.

Rovira, A. D. (1969). Plant root exudates. *Botanical Review, 35*, 35–57.

Ruijter, G. J. G., Bax, M., Patel, H., Flitter, S. J., van de Vondervoort, P. J. I., de Vries, R. P. … Visser, J. (2003). Mannitol is required for stress tolerance in *Aspergillus niger* conidiospores. *Eukaryotic Cell, 2*, 690–698.

Santi, L., da Silva, W. O. B., Berger, M., Guimarães, J. A., Schrank, A., & Vainstein, M. H. (2010). Conidial surface proteins of *Metarhizium anisopliae*: source of activities related with toxic effects, host penetration and pathogenesis. *Toxicon, 55*, 874–880.

Sasan, R. K., & Bidochka, M. J. (2012). The insect-pathogenic fungus *Metarhizium robertsii* (Clavicipitaceae) is also an endophyte that stimulates plant root development. *American Journal of Botany, 99*, 101–107.

Sasan, R. K., & Bidochka, M. J. (2013). Antagonism of the endophytic insect pathogenic fungus *Metarhizium robertsii* against the bean plant pathogen *Fusarium solani* f. sp. *phaseoli*. *Canadian Journal of Plant Pathology, 35*, 288–293.
Scharf, D. H., Heinekamp, T., & Brakhage, A. A. (2014). Human and plant fungal pathogens: the role of secondary metabolites. *PLoS Pathogens*. http://dx.doi.org/10.1371/journal.ppat.1003859.
Scholte, E. J., Knols, B. G., & Takken, W. (2006). Infection of the malaria mosquito *Anopheles gambiae* with the entomopathogenic fungus *Metarhizium anisopliae* reduces blood feeding and fecundity. *Journal of Invertebrate Pathology, 91*, 43–49.
Schrank, A., & Vainstein, M. H. (2010). *Metarhizium anisopliae* enzymes and toxins. *Toxicon, 56*, 1267–1274.
Screen, S. E., & St. Leger, R. J. (2000). Cloning, expression, and substrate specificity of a fungal chymotrypsin - evidence for lateral gene transfer from an actinomycete bacterium. *Journal of Biological Chemistry, 275*, 6689–6694.
Sevim, A., Donzelli, B. G., Wu, D., Demirbag, Z., Gibson, D. M., & Turgeon, B. G. (2012). Hydrophobin genes of the entomopathogenic fungus, *Metarhizium brunneum*, are differentially expressed and corresponding mutants are decreased in virulence. *Current Genetics, 58*, 79–92.
Shakeri, J., & Foster, H. A. (2007). Proteolytic activity and antibiotic production by *Trichoderma harzianum* in relation to pathogenicity to insects. *Enzyme and Microbial Technology, 40*, 961–968.
Small, C. L. N., & Bidochka, M. J. (2005). Up-regulation of *Pr1*, a subtilisin-like protease, during conidiation in the insect pathogen *Metarhizium anisopliae*. *Mycological Research, 109*, 307–313.
Spatafora, J. W., Sung, G. H., Sung, H. M., Hywel-Jones, L., & White, J. F. (2007). Phylogenetic evidence for an animal pathogen origin of ergot and the grass endophytes. *Molecular Ecology, 16*, 1701–1711.
Srivastava, C. N., Maurya, P., Sharma, P., & Mohan, L. (2009). Prospective role of insecticides of fungal origin: Review. *Entomological Research, 39*, 341–355.
St Leger, R. J. (1995). The role of cuticle-degrading proteases in fungal pathogenesis of insects. *Canadian Journal of Botany, 73*, S1119–S1125.
St. Leger, R. J., Charnley, A. K., & Cooper, R. M. (1987). Characterization of cuticle-degrading proteases produced by the entomopathogen *Metarhizium anisopliae*. *Archives of Biochemistry and Biophysics, 253*, 221–232.
St. Leger, R. J., Joshi, L., Bidochka, M. J., Rizzo, N. W., & Roberts, D. W. (1996). Characterization and ultrastructural localization of chitinases from *Metarhizium anisopliae*, *M. flavoviride*, and *Beauveria bassiana* during fungal invasion of host (*Manduca sexta*) cuticle. *Applied and Environmental Microbiology, 62*, 907–912.
St. Leger, R. J., Nelson, J. O., & Screen, S. E. (1999). The insect pathogenic fungus *Metarhizium anisopliae* alters ambient pH, allowing extracellular protease production and activity. *Microbiology, 145*, 2691–2699.
St. Leger, R. J., Staples, R. C., & Roberts, D. W. (1992). Cloning and regulatory analysis of starvation-stress gene, *ssgA*, encoding a hydrophobin-like protein from the insect pathogenic fungus, *Metarhizium anisopliae*. *Gene, 120*, 119–124.
St. Leger, R. J., Wang, C., & Fang, W. (2011). New perspectives on insect pathogens. *Fungal Biology Reviews, 25*, 84–88.
Tunlid, A., Rosen, S. E. B., & Rask, L. (1994). Purification and characterization of an extracellular serine protease from the nematode trapping fungus *Arthrobotrys oligospora*. *Microbiology, 140*, 1687–1695.
Ujjan, A. A., & Shahzad, S. (2014). Insecticidal potential of *Beauveria bassiana* strain PDRL1187 and imidacloprid to mustard aphid (*Lipaphis erysimi*) under field conditions. *Pakistan Journal of Zoology, 46*, 1277–1281.

Vega, F. E. (2008). Insect pathology and fungal endophytes. *Journal of Invertebrate Pathology, 98*, 277–279.

Vega, F. E., Goettel, M. S., Blackwell, M., Chandler, D., Jackson, M. A., Keller, S. … Roy, H. E. (2009). Fungal entomopathogens: new insights on their ecology. *Fungal Ecology, 2*, 149–159.

Wagner, B. L., & Lewis, L. C. (2000). Colonization of corn, *Zea mays*, by the insect pathogenic fungus *Beauveria bassiana*. *Applied and Environmental Microbiology, 66*, 3468–3473.

Wang, J. B., St. Leger, R. J., & Wang, C. (2016). Advances in Genomics of Insect Pathogenic Fungi. *Advances in Genetics, 94*, 67–105.

Wang, S., Fang, W., Wang, C., & St. Leger, R. J. (2011). Insertion of an esterase gene into specific locust pathogen (*Metarhizium acridum*) enables it to infect caterpillars. *PLoS Pathogens, 7*(6), e1002097.

Wang, C., Hu, G., & St. Leger, R. J. (2005). Differential gene expression by *Metarhizium anisopliae* growing in root exudate and host (*Manduca sexta*) cuticle or hemolymph reveals mechanisms of physiological adaptation. *Fungal Genetics and Biology, 42*, 704–718.

Wang, C., & St. Leger, R. J. (2005). Developmental and transcriptional responses to host and nonhost cuticles by the specific locust pathogen *Metarhizium anisopliae* var. *acridum*. *Eukaryotic Cell, 4*, 937–947.

Wang, C., & St. Leger, R. J. (2006). A collagenous protective coat enables *Metarhizium anisopliae* to evade insect immune responses. *Proceedings of the National Academy of Sciences of the United States of America, 103*, 6647–6652.

Wang, C., & St. Leger, R. J. (2007a). The MAD1 adhesin of *Metarhizium anisopliae* links adhesion with blastospore production and virulence to insects, and the MAD2 adhesin enables attachment to plants. *Eukaryotic Cell, 6*, 808–816.

Wang, C., & St. Leger, R. J. (2007b). The *Metarhizium anisopliae* perilipin homolog MPL1 regulates lipid metabolism, appressorial turgor pressure, and virulence. *Journal of Biological Chemistry, 282*, 21110–21115.

Wang, C., & St Leger, R. J. (2007c). A scorpion neurotoxin increases the potency of a fungal insecticide. *Nature Biotechnology, 25*, 1455–1456.

Wang, C., & St. Leger, R. J. (2014). Genomics of entomopathogenic fungi. In F. Martin (Ed.), *The ecological genomics of fungi* (pp. 243–260). John Wiley & Sons, Inc.

Wang, Q., & Xu, L. (2012). Beauvericin, a bioactive compound produced by fungi: a short review. *Molecules, 17*, 2367–2377.

Wang, J., Zhou, G., Ying, S. H., & Feng, M. G. (2014). Adenylate cyclase orthologues in two filamentous insect pathogens contribute differentially to growth, conidiation, pathogenicity, and multistress responses. *Fungal Biology, 118*, 422–431.

Whiteford, J. R., & Spanu, P. D. (2002). Hydrophobins and the interactions between fungi and plants. *Molecular Plant Pathology, 3*, 391–400.

White, J. F., Sullivan, R. F., Balady, G. A., Gianfagna, T., Yue, Q., & Meyer, W. (2001). A fungal endosymbiont of the grass *Bromus setifolius*: distribution in some Andean populations, identification and examination of beneficial properties. *Symbiosis, 31*, 241–257.

Wyrebek, M., & Bidochka, M. J. (2013). Variability in the insect and plant adhesins, Mad1 and Mad2, within the fungal genus *Metarhizium* suggest plant adaptation as an evolutionary force. *PLoS One*. http://dx.doi.org/10.1371/journal.pone.0059357.

Wyrebek, M., Huber, C., Sasan, R. K., & Bidochka, M. J. (2011). Three sympatrically occurring species of *Metarhizium* show plant rhizosphere specificity. *Microbiology, 157*, 2904–2911.

Xiao, G., Ying, S. H., Zheng, P., Wang, Z. L., Zhang, S., Xie, Q., & … … Feng, M. G. (2012). Genomic perspectives on the evolution of fungal insect pathogenicity in *Beauveria bassiana*. *Scientific Reports, 2*, 1–10.

Ying, S. H., Feng, M. G., & Keyhani, N. O. (2013). A carbon responsive G-protein coupled receptor modulates broad developmental and genetic networks in the insect pathogenic fungus, *Beauveria bassiana*. *Environmental Microbiology, 15*, 2902–2921.

Ying, S. H., Ji, X. P., Wang, X. X., Feng, M. G., & Keyhani, N. O. (2014). The transcriptional co-activator multiprotein bridging factor 1 from the fungal insect pathogen, *Beauveria bassiana*, mediates regulation of hyphal morphogenesis, stress tolerance and virulence. *Environmental Microbiology, 16*, 1879–1897.

Zhang, S., Xia, Y. X., Kim, B., & Keyhani, N. O. (2011). Two hydrophobins are involved in fungal spore coat rodlet layer assembly and each play distinct roles in surface interactions, development and pathogenesis in the insect pathogenic fungus, *Beauveria bassiana*. *Molecular Microbiology, 80*, 811–826.

Zhang, Y., Xu, L., Zhang, S., Liu, X., An, Z., Wang, M., & Guo, Y. (2009). Genetic diversity of *Ophiocordyceps sinensis*, a medicinal fungus endemic to the Tibetan Plateau: implications for its evolution and conservation. *BMC Evolutionary Biology, 9*, 290.

Zhang, Y. J., Zhang, J. Q., Jiang, X. D., Wang, G. J., Luo, Z. B., Fan, Y. H.... Pei, Y. (2010). Requirement of a mitogen-activated protein kinase for appressorium formation and penetration of insect cuticle by the insect pathogenic fungus *Beauveria bassiana*. *Applied and Environmental Microbiology, 76*, 2262–2270.

Zhang, Y. J., Zhao, J. H., Fang, W. G., Zhang, J. Q., Luo, Z. B., Zhang, M.... Pei, Y. (2009). Mitogen-activated protein kinase hog1 in the insect pathogenic fungus *Beauveria bassiana* regulates environmental stress responses and virulence to insects. *Applied and Environmental Microbiology, 75*, 3787–3795.

Zhao, H., Lovett, B., & Fang, W. (2016). Genetically Engineering Entomopathogenic Fungi. *Advances in Genetics, 94*, 137–163.

Zhong, X., Peng, Q. Y., Li, S. S., Chen, H., Sun, H. X., Zhang, G. R., & Liu, X. (2014). Detection of *Ophiocordyceps sinensis* in the roots of plants in alpine meadows by nested-touchdown polymerase chain reaction. *Fungal Biology, 118*, 359–363.

CHAPTER FIVE

Genetically Engineering Entomopathogenic Fungi

H. Zhao*, B. Lovett[§] and W. Fang*[,1]
*Zhejiang University, Hangzhou, Zhejiang, China
[§]University of Maryland, College Park, MD, United States
[1]Corresponding author: E-mail: wfang1@zju.edu.cn

Contents

Abstract

Entomopathogenic fungi have been developed as environmentally friendly alternatives to chemical insecticides in biocontrol programs for agricultural pests and vectors of disease. However, mycoinsecticides currently have a small market share due to low virulence and inconsistencies in their performance. Genetic engineering has made it possible to significantly improve the virulence of fungi and their tolerance to adverse conditions. Virulence enhancement has been achieved by engineering fungi to express insect proteins and insecticidal proteins/peptides from insect predators and other

Advances in Genetics, Volume 94
ISSN 0065-2660
http://dx.doi.org/10.1016/bs.adgen.2015.11.001

insect pathogens, or by overexpressing the pathogen's own genes. Importantly, protein engineering can be used to mix and match functional domains from diverse genes sourced from entomopathogenic fungi and other organisms, producing insecticidal proteins with novel characteristics. Fungal tolerance to abiotic stresses, especially UV radiation, has been greatly improved by introducing into entomopathogens a photoreactivation system from an archaean and pigment synthesis pathways from nonentomopathogenic fungi. Conversely, gene knockout strategies have produced strains with reduced ecological fitness as recipients for genetic engineering to improve virulence; the resulting strains are hypervirulent, but will not persist in the environment. Coupled with their natural insect specificity, safety concerns can also be mitigated by using safe effector proteins with selection marker genes removed after transformation. With the increasing public concern over the continued use of synthetic chemical insecticides and growing public acceptance of genetically modified organisms, new types of biological insecticides produced by genetic engineering offer a range of environmentally friendly options for cost-effective control of insect pests.

1. INTRODUCTION

Fungi are the commonest pathogens in insects, with approximately 1000 species known to cause disease in arthropods (Roberts & St. Leger, 2004). Unlike microsporidia, bacteria and viruses, that infect insects through the gut, most entomopathogenic fungi infect insects by direct penetration through the cuticle. Infection is via conidia that adhere to the insect cuticle and produce germ tubes that meander across the cuticle until they find a suitable site for penetration. They then cease polar growth and penetrate the insect cuticle; before penetration occurs, the hyphal tips of many fungi, such as *Metarhizium* spp., differentiate into swollen "holdfasts" called appressoria that facilitate the penetration of the cuticle. In the hemocoel, the fungus changes morphology from filamentous to yeast-like hyphal bodies, and the insect is killed by proliferation of the yeast-like cells and the production of toxins. Hyphae subsequently reemerge from the cadaver to produce conidia. Unlike bacteria and viruses entomopathogenic fungi do not require ingestion by the host, so they can target sucking insects, such as mosquitoes and aphids. Several entomopathogenic fungi, such as *Metarhizium* spp. and *Beauveria* spp., have been developed as environmentally friendly alternatives to chemical insecticides in biocontrol programs for agricultural pests and vectors of disease (Federici, Bonning, & St. Leger, 2008; Thomas & Read, 2007). Industrial production of *Metarhizium* spp. is highly automated; the price of commercialized *Metarhizium acridum* for locust control in Africa,

Australia, and China is about 20 USD/ha when applied at 50 g/ha, which is comparable to the price of conventional chemical insecticides (Langewald & Kooyman, 2007). The technology for mass production of the other widely applied entomopathogenic fungus *Beauveria bassiana* is also available. Before the 1990s, hundreds of small farm-owned or county-owned workshops were set up throughout China to produce mainly *B. bassiana*. Approximately 0.8–1.3 million hectares of forests and cornfields in China were treated annually with *B. bassiana* for the control of Masson's pine caterpillar *Dendrolimus punctatus* and the Asian corn borer *Ostrinia furnacalis* (Feng, Poprawski, & Khachatourians, 1994; Wang & Feng, 2014).

However, fungal pathogens currently have a small market share due to low virulence (slow kill and high inoculum load) compared to the chemical insecticides with which they compete, and due to inconsistencies in their performance (Fang, Azimzadeh, & St. Leger, 2012). Low virulence may be inbuilt as an evolutionary balance may have developed between microorganisms and their hosts so that quick kill, even at high doses, is not adaptive for the pathogen, in which case cost-effective biocontrol will require genetic modification of the fungi (Gressel, 2007). The inconsistencies in performance in the field mainly result from the sensitivity of entomopathogenic fungi to environmental stresses (Lovett & St. Leger, 2015). Genetic engineering has been proved to be an efficient tool to improve the efficacy of mycoinsecticides by improving their tolerance to environmental stresses and their virulence. In addition, some methods appear to mitigate safety concerns regarding genetically engineered entomopathogenic fungi. In this chapter, we will extensively review the progress on and rationale behind the genetic engineering of entomopathogenic fungi, and we discuss methods that have potential to ensure the success and safety of genetically modified strains.

2. IMPROVING VIRULENCE

Genetic engineering to improve virulence has focused on reducing both lethal conidial dosage and time to kill. Reducing conidial dosage improves infection rates; allowing control to be achieved with less product. It also increases effective persistence of the biopesticides because as conidia decay there is a greater probability that an insect will come into contact with enough propagules of the genetically engineered fungus to exceed the inoculum threshold. Based on the sources of genes for genetic engineering (Fig. 1 and Table 1), four major strategies are currently being exploited to improve virulence of entomopathogenic fungi.

2.1 Strategy I: Using the Pathogen's Own Genes to Improve Virulence

Our current understanding of molecular mechanisms for fungal pathogenesis (primarily *Metarhizium* spp. *and B. bassiana*), has allowed many pathogenicity-related genes to be characterized, and these genes can be used as a resource for genetically enhancing entomopathogenic fungi. The insect cuticle is mostly composed of chitin fibrils embedded in a matrix of proteins. To penetrate this barrier, entomopathogenic fungi produce proteases and chitinases. Under native regulation, expression of most of these genes is under tight control (Fang, Pava-Ripoll, Wang, & St. Leger, 2009). Constitutively overexpressing the gene encoding the subtilisin-like protease Pr1A

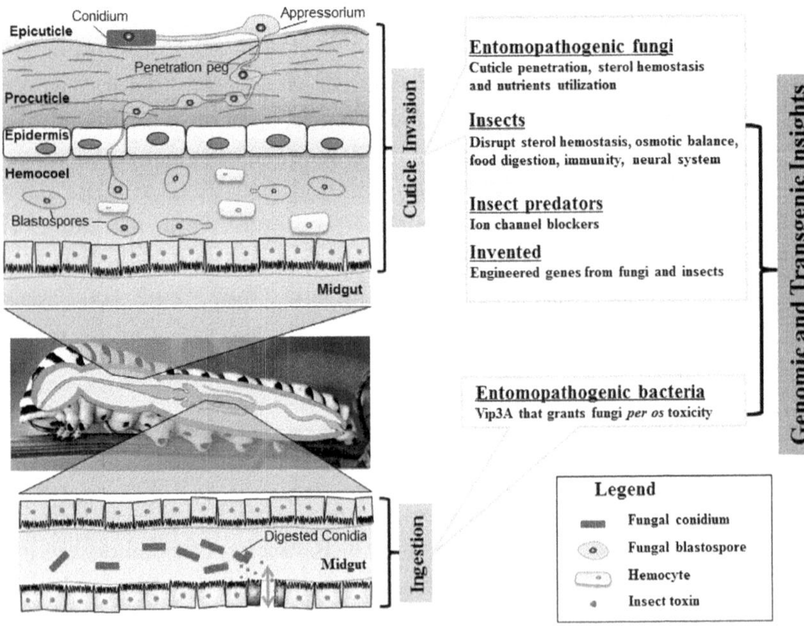

Figure 1 Strategies for genetically engineering to improve the virulence of entomopathogenic fungi. Left panel: infection pathways of entomopathogenic fungi through cuticle contact (upper zoom) and per os toxicity to insects of ingested transgenic strains expressing Bt toxins such as Vip3A (lower zoom). Right panel: currently exploited sources of genes for improving virulence of entomopathogenic fungi. (See color plate)

Table 1 Genes and metabolic pathways that have been used to improve fungal virulence and tolerance to abiotic stresses

Aim	Type	Source
Improve virulence		
Genes from entomopathogenic fungi		
Pr1A	Subtilisin-like protease	*Metarhizium robertsii*
CDEP1	Subtilisin-like protease	*Beauveria bassiana*
Bbchit1	Chitinase	*Beauveria bassiana*
Mr-Npc2a	Sterol carrier	*Metarhizium robertsii*
ATM1	Trehalase	*Metarhizium acridum*
Mr-Ste1	Esterase	*Metarhizium robertsii*
BbBqrA	Benzoquinone oxidoreductase	*Beauveria bassiana*
Genes from insects		
MSDH	Diuretic hormone	*Manduca sexta*
Serpin	Serine proteinease inhibitor	*Drosophila melanogaster*
TMOF	Trypsin modulating oostatic factor	*Anopheles aegypti, Sarcophaga bullata*
β-NP	Pyrokinin β-neuropeptide	*Solenopsis invicta*
Genes from insect predators		
AaIT1	Sodium channel blocker	*Androctonus australis*
BmKit	Sodium channel blocker	*Buthus martensi*
LqhIT2	Sodium channel blocker	*Leiurus quinquestriatus hebraeus*
BjαIT	Sodium channel blocker	*Buthotus judaicus*
ω-HXTX-Hv1a	Calcium channel blocker	*Atrax robustus*
κ-HXTX-Hv1c	Ca^{2+}-activated K^+ channel blocker	*Hadronyche versuta*
Hybrid-toxin	CaV and KCa channel blocker	*Hadronyche versuta*
Invented genes		
CDEP1:Bbchit1	Protease and chitinase activity	Engineered
CDEP1:CBD	Chitin-binding domain fused to a protease	Engineered
Bbchit1:CBD	Chitin-binding domain fused to a chitinase	Engineered

(*Continued*)

Table 1 Genes and metabolic pathways that have been used to improve fungal virulence and tolerance to abiotic stresses—cont'd

Aim	Type	Source
Genes from other insect pathogens		
Vip3A	Vegetative insecticidal protein	*Bacillus thuringiensis*
Improve tolerance to abiotic stresses		
Try	Tryosinase	*Aspergillus fumigatus*
BbSOD1	Superoxide dismutase	*Beauveria bassiana*
DHN-melanin synthesis pathway	Three genes	*Alternaria alternate*
MrPhr1	CPD photolyase	*Metarhizium robertsii*
HsPHR2	CPD photolyase	*Halobacterium salinarum*
trxA	Thioredoxin	*Escherichia coli*
HSP25	Heat shock protein 25	*Metarhizium robertsii*

increased the virulence of *Metarhizium anisopliae* to *Manduca sexta*, and the recombinant strain showed a 25% mean reduction in survival time (ST_{50}) toward the insect as compared to the parent wild-type (WT) strain (St. Leger, Joshi, Bidochka, & Roberts, 1996a). Importantly, a Pr1A overexpressing strain of *M. anisopliae* was used in the first EPA approved field trial of a transgenic fungal pathogen, providing a precedent paving the way for future trials (Hu & St. Leger, 2002). Similarly, constitutive overproduction of *B. bassiana*'s chitinase CHIT1 improved virulence by 23% (Fang et al., 2005). Expression of Pr1A from *M. anisopliae* also increases the killing speed of *B. bassiana* (Gongora, 2004), showing that pathogenicity-related genes from one entomopathogenic fungus can be used to improve virulence of other entomopathogenic fungi.

For yeast-like cells to proliferate in the hemolymph, entomopathogenic fungi need to increase synthesis of enzymes that allow them to utilize carbohydrates. Trehalose is the main sugar in the hemolymph of insects and is a key nutrient source for entomopathogenic fungi (Thompson & Borchardt, 2003). *Metarhizium acridum* secretes an acid trehalase (ATM1) to utilize trehalose in the locust hemolymph (Jin, Peng, Liu, & Xia, 2015). Overexpression of the acid trehalase ATM1 accelerated the growth of *M. acridum* in the hemocoel of locusts, reducing the number of conidia causing 50% mortality (LC_{50}) by 8.3-fold compared with the WT strain (Peng, Jin, Liu, & Xia, 2015). Modifying the way entomopathogenic fungi exploit their hosts for nutrition is therefore a feasible way to improve virulence with many potential mechanisms that can be tried.

The range of entomopathogenic fungi endogenous genes suitable for genetic engineering is likely to be enormous as adhesins, species-specific toxin-encoding genes and systems for evading host immunity have evolved independently in many insect pathogens (Fang & St. Leger, 2010a; Gao et al., 2011; Lin et al., 2011; Wang & St. Leger, 2006, 2007a, 2007b). Recently, significant insights into the molecular mechanisms controlling host selectivity by entomopathogenic fungi have been achieved, and a couple of host-range related genes were characterized (Pedrini et al., 2015; Wang, Fang, Wang, & St. Leger, 2011). In the future, combining the available genomes from *B. bassiana* and several *Metarhizium* species with robust genetic manipulation technologies (Fang, Pei, & Bidochka, 2006; Fang et al., 2004; Gao et al., 2011; Hu et al., 2014; Xiao et al., 2012; Xu et al., 2014), will allow characterization of the full range of pathogenicity and host-specificity-related genes. With these genes, entomopathogenic fungi with novel combinations of insect specificity and virulence will be created. The potential of this approach has already been shown. Transfer of an esterase gene (Mest1) from the generalist *Metarhizium robertsii* to the locust specialist *M. acridum* enabled the latter strain to expand its range to infect caterpillars (Wang, Fang, et al., 2011). An enduring theme in pathogenic microbiology is poor understanding of the mechanisms of host specificity. These technologies will help remedy that deficiency and increase confidence that it is possible to produce transgenic strains with known, clearly defined host ranges that will not cause collateral damage to populations of nontarget insects.

2.2 Strategy II: Insect's Proteins for Genetically Engineering Entomopathogenic Fungi

Genome-wide analysis of horizontal gene transfer (HGT) events revealed that, as in other fungi (Wisecaver, Slot, & Rokas, 2014), *Metarhizium* species acquired diverse genes from bacteria and archaea and even arthropods, plants, and vertebrates (Hu et al., 2014). An example is a sterol transporter (Mr-NPC2a) that several strands of evidence suggest was acquired from an insect by HGT and which allows the fungus to compete with the host for growth limiting sterols in the hemolymph. To our knowledge, this is first example of HGT from host to a eukaryotic pathogen, and the host gene ultimately improved the infectivity of the pathogen (Zhao et al., 2014). Hence, HGT provided entomopathogenic fungi with host genetic material that impaired the host's normal physiological processes. This evolutionary event was reproduced by transferring the sterol carrier gene into

B. bassiana, which lacks an endogenous Mr-NPC2a homolog, improving its pathogenicity (Zhao et al., 2014).

Before publication of the discovery of the HGT acquired Mr-NPC2a, insect molecules had been used for genetically engineering entomopathogenic fungi. Driven by a hypothesis that disruption of the expression pattern of key regulators of insects could impair normal physiological functions and increase susceptibility to microbial pathogens, Dr. Keyhani's group (University of Florida) has used transgenic expression of several insect molecules to improve the virulence of *B. bassiana* (Ortiz-Urquiza, Luo, & Keyhani, 2015). Diuretic hormones regulate insect water–salt balance; expression of the *M. sexta* diuretic hormone (MSDH) significantly increased the virulence of *B. bassiana* against various Lepidopteran targets (eg, *M. sexta* and *Galleria mellonella*) as well as mosquitoes (*Anopheles aegypti*) (Fan, Borovsky, Hawkings, Ortiz-Urquiza, & Keyhani, 2012). Expression of an inhibitory regulator of a key immune-related signal pathway (the Toll signaling pathway) also increased *B. bassiana*'s virulence against two taxonomically different insect species; *G. mellonella* (wax moth) and adult *Myzus persicae* (green peach aphid) (Yang et al., 2014). Therefore, a conserved molecule among taxonomically distant insect species can increase the virulence of entomopathogenic fungi against various insect species. On the other hand, by using molecules only found in target insects, it is feasible to increase the virulence of entomopathogenic fungi against specific insects. Trypsin modulating oostatic factors (TMOFs) are deca-and hexa-peptides that circulate in the hemolymph and bind to gut receptors on the hemolymph side of the gut. This binding inhibits synthesis of a trypsin that digests food in the mosquito gut after a blood meal. Expression of TMOF from *A. aegypti* significantly increases the virulence of *B. bassiana* against adults and larvae of the mosquito (Fan, Borovsky, et al., 2012). This TMOF expressing strain was also potent against the malaria mosquito *Anopheles gambiae* (Kamareddine, Fan, Osta, & Keyhani, 2013), but not the gray flesh fly, *Sarcophaga bullata*. Likewise, expression of the TMOF from *S. bullata* had no impact on the virulence of *B. bassiana* against mosquitoes (Fan, Borovsky, et al., 2012). The narrow host range targeted by the engineered *B. bassiana* could be due to divergence of TMOF sequences from different insect families (Ortiz-Urquiza et al., 2015). Similarly, expression of the species-specific pyrokinin β-neuropeptide from fire ants (*Solenopsis invicta*) increased the virulence of *B. bassiana* against fire ants, but not against Lepidopteran hosts *G. mellonella* and *M. sexta* (Fan, Pereira, Kilic, Casella, & Keyhani, 2012).

To date, the insect molecules that have been used to increase virulence of entomopathogenic fungi are involved in five types of biological processes: sterol hemostasis, osmotic balance, food digestion, immunity, and the neural system. Theoretically, any biological process in insets could be potential targets for disruption; it is thus possible that insect biology can be exploited by mixing and matching different insect molecules to create more virulent fungal strains with high specificity against target insect pests.

2.3 Strategy III: Genes from Insect Predators and Other Insect Pathogens for Genetically Engineering Entomopathogenic Fungi

Insect predatory arthropods inject victims with venom, which kills or paralyzes the prey. The venom is usually a cocktail of potent compounds (neurotoxins, enzyme inhibitors, etc.), each causing a different effect; over one million such peptide toxins have been isolated from arachnids and scorpions, independently evolved for the express purpose of killing arthropod prey. With their natural ability to penetrate into insects, entomopathogenic fungi can be used as the equivalent of stings to deliver antiinsect compounds from the predators into insects, improving the virulence of the fungi. The scorpion sodium channel blocker AaIT1 is well studied and very potent, so it was the first arthropod toxin tested in the broad host-range *M. anisopliae* strain ARSEF 549 (Wang & St. Leger, 2007a, 2007b). The modified fungus achieved the same mortality rates in tobacco hornworm (*M. sexta*) at 22-fold lower conidial doses than the WT fungus, and survival times at some doses were reduced by 40%. Similar results were obtained with mosquitoes (LC_{50} reduced 9-fold) and the coffee berry borer beetle (LC_{50} reduced 16-fold) (Pava-Ripoll, Posada, Momen, Wang, & St. Leger, 2008; Wang & St. Leger, 2007a, 2007b). The scorpion toxin BmKit from *Buthus martensi* was expressed in another broad host-range entomopathogenic fungus *Lecanicillium lecanii*, and the recombinant strain achieved the same mortality rates in cotton aphids (*Aphis gossypii*) at 7.1-fold lower conidial doses than the WT strain, and the median survival time (ST_{50}) for a transformant (BmKit-12) was reduced by 26.5% compared with WT (Pava-Ripoll et al., 2008; Xie et al., 2015). Mycelial growth, conidiation, and conidial germination of transgenic strains expressing BmKit were not significantly different from WT, so expression of BmKit will not impact future industrial production of the strain; however, conidial yields of transformants were reduced on insect cadavers, possibly because of the reduced mycelial growth in the host hemocoel before death (Xie et al., 2015). Therefore, the accelerated killing speed conferred by the arthropod toxin can have trade-off effects on reproduction and infection by *L. lecanii*. Nevertheless, conidiation of *M. anisopliae* expressing AaIT1 on insect cadavers was the same as WT on some, but not all, insect hosts

(Pava-Ripoll et al., 2008; Wang & St. Leger, 2007a, 2007b), suggesting that the impact of expression of toxins on fungal reproduction could be dependent on fungal and insect species or toxins.

The generalists, such as *M. anisopliae* and *L. lecanii*, provided suitable vehicles to test the effects of toxins on several pests, but it would be safer to employ natural strains with narrower host ranges for insect pest control to avoid nontarget effects on the ecosystem. Specialists, such as the locust pathogen *M. acridum*, could benefit from this technology as they kill more slowly and produce fewer toxins than generalist fungi. Along with AaIT1, several spider toxins have been tested: κ-HXTX-Hv1c (an inhibitor of Ca^{2+}-activated K^{+} channels) and hybrid-toxin (a self-synergizing peptide toxin that blocks both CaV and KCa channels) from *Hadronyche versuta*, and ω-HXTX-Hv1a (a blocker of insect voltage-gated calcium channels) from *Atrax robustus* (Fang, Lu, King, & St. Leger, 2014). Expression of the four insect-specific neurotoxins improved the efficacy of *M. acridum* against acridids by reducing lethal dose, time to kill and food consumption. Co-inoculating recombinant strains expressing AaIT1 and the hybrid-toxin, produced synergistic effects, including an 11.5-fold reduction in LC_{50}, 43% reduction in LT_{50}, and a 78% reduction in food consumption, which was significantly greater than expression of either one of the two toxins alone. However, host specificity was retained, as the recombinant strains did not cause disease in nonacridids, probably because specificity is determined by events at the cuticle surface (St. Leger & Screen, 2001). In addition, the toxins are all insect-selective, and the most potent hybrid-toxin, also known as Versitude, has been recently approved by the US EPA for control of lepidopteran pests. This specifically indicates they are safe to human beings and other animals. Expression of the toxins was limited to the hemocoel of the target insects by using the insect-hemolymph-inducible promoter *PMcl1*, so this precludes casual release of the toxins by *M. acridum* surviving saprophytically or symbiotically with plants. Therefore, the genetically engineered *M. acridum* strains inherently contain many well-understood safety nets preventing harm of nontarget organisms, though this would require ecophysiological testing to confirm both increased efficacy and safety in field conditions.

Beyond AaIT1 and the three spider toxins, the virulence of *M. acridum* was also improved using the insect-specific neurotoxin LqhIT2 from the Israeli yellow scorpion *Leiurus quinquestriatus hebraeus* and the neurotoxin BjαIT from the Judean black scorpion *Buthotus judaicus* (Peng & Xia, 2014, 2015). These data confirm that the arsenal for improving the virulence of entomopathogenic fungi using different insect predators is enormous.

In addition, bacterial and viral pathogens of insects could provide additional toxins with novel modes of action. Insect-toxins from *Bacillus thuringiensis* (Bt) can be processed in the midgut and the resulting proteins are able to kill insects. Aside from the Bt crystal proteins that are widely used for genetically engineering crops, Bt vegetative insecticidal proteins (Vips) have insecticidal activities only in insect intestines. Vip3A has been proven to kill a broad spectrum of lepidopteran insects by lysing the midgut epithelium. Transgenic plants expressing Vip3A are more resistant to the cotton bollworm *Helicoverpa armigera* (Schnepf et al., 1998; Warren, 1997). Expression of Vip3A did not increase the infectivity of *B. bassiana* conidia applied to host cuticles, but in contrast to the WT strain, the transformants were lethal following ingestion and hosts showed typical symptoms of Vip3A action. Therefore, using transgenic conidia insects can be killed by a combination of cuticle infection and per os toxicity (Qin, Ying, Chen, Shen, & Feng, 2010). A full-season field trial showed that the control efficacy of the recombinant strain was similar to a chemical counterpart, and it had no adverse effect on nontarget arthropods and low environmental persistence (Liu, Liu, Ying, Liu, & Feng, 2013). Notwithstanding this, Bt toxins expressed by *B. bassiana* may be more environmentally stable than those expressed by, for example, *Pseudomonas* and *Rhizobium* bacteria (Nambiar, Ma, & Iyer, 1990; Obukowicz, Perlak, Kusano-Kretzmer, Mayer, & Waturd, 1986), and the transgenic *B. bassiana* may be effective at lower concentrations than the WT. When fungal conidia are sprayed onto plants in the field, only some of them come into contact with insect cuticle to start infection via cuticular penetration; however, additional insects may eat some conidia. Since, except for microsporidia, entomopathogenic fungi cannot infect insects through the gut, any WT conidia ingested are essentially wasted because they are either digested as food or are excreted without hurting the insects. The strains expressing Vip3A are still able to attack insects through the gut after ingestion ensuring more conidia are effective at reducing pest populations in the field and, thus, reducing the cost of application of mycoinsecticides.

2.4 Strategy IV: Invented Proteins for Genetically Engineering Entomopathogenic Fungi

Scientists today are not limited to the immense diversity of peptide and protein sequences that already occur in nature. They can further invent, through protein engineering, synthetic, multifunctional genes that are hybrids of different activities found in other genes. The cuticle was, again, the first target for protein engineering efforts. As described above, overexpression of either the Pr1A protease or the Bbchit1 chitinase resulted in an increase in fungal virulence (Fang et al., 2005; St. Leger et al., 1996a). Expression of the fusion protein CDEP1:Bbchit1 that contains the Pr1A-like protease CDEP1and the chitinase Bbchit1 accelerated cuticular penetration by *B. bassiana* when compared to the WT strain or transformants overexpressing each gene singly (Fang, Feng, et al., 2009). Thus, this hybrid gene improved the virulence of the fungus to a greater extent than overexpression of either Pr1A or Bbchit1, presumably as a result of the composite structure of the insect cuticle, which contains chitin embedded in protein. WT strains of entomopathogenic fungi need to digest cuticular proteins first in order to expose the chitin and induce the expression of fungal chitinases (Screen, Hu, & St. Leger, 2001; St. Leger, Joshi, Bidochka, & Roberts, 1996b). The proteolytic activity of the fusion protein digests the cuticular proteins to expose the chitin, which is then immediately available to the attached chitinase. The accelerated collapse of the chitin matrix in turn increases the vulnerability of the cuticle to further proteolysis, as cuticle depleted of chitin is more vulnerable to protease activity (Hassan & Charnley, 1989). Accelerating fungal penetration of the cuticle may potentially improve the utility of fungal pathogens as biocontrol agents by reducing the time of exposure of the fungus to potentially debilitating environmental conditions, such as UV, and to constitutive and inducible insect defenses, such as melanization. The fungus may, therefore, penetrate the cuticle before the host has time to mount a defense of sufficient magnitude to block it.

The fusion of insect components to genes endogenous to entomopathogenic fungi can also produce genes with novel virulence-enhancing characteristics compared to native fungal virulence genes. The chitinase Bbchit1 identified from *B. bassiana* lacks chitin-binding domains. Fan et al. (2007) constructed several *B. bassiana* hybrid chitinases where Bbchit1 was fused to chitin-binding domains derived from plant, bacterial, or insect sources. A hybrid chitinase containing the chitin-binding domain from the silkworm *Bombyx mori* chitinase fused to the *B. bassiana* chitinase showed the greatest ability to bind to chitin and the insect cuticle compared to the WT Bbchit1 and other hybrid chitinases. Expression of this hybrid chitinase gene by *B. bassiana* reduced time to death of insect hosts by 23% compared to the WT fungus, and this reduction was also greater than that achieved by overexpressing the native chitinase. Interestingly, the same group fused the chitin-binding domain to the Pr1A-like protease BbCDEP1 so that the protein bound to chitin. Compared to the native BbCDEP, the hybrid protease

(CDEP1-BmChBD) released greater amounts of peptides/proteins from insect cuticles (Fan et al., 2010). Expression of the hybrid protease in *B. bassiana* also significantly increased fungal virulence compared to WT and strains overexpressing the native protease.

The above research exemplifies how protein engineering can be used to mix and match genes from entomopathogenic fungi and other organisms to produce proteins with novel virulence-enhancing characteristics.

In addition, expression systems in *Metarhizium* and *Beauveria* are as easy to manipulate as commercially available yeasts, but have the added advantage of providing a delivery system into the insects. Insect pathogenic fungi could, thus, provide a tractable model system for screening novel effectors invented with protein engineering. The obtained potent effectors could be used to improve the efficacy of entomopathogenic fungi or to develop other transgenic pest control technologies, such as insecticidal crops.

3. IMPROVING THE EFFICACY OF MYCOINSECTICIDES TO CONTROL VECTOR-BORNE DISEASES

Insects and arthropods vector many human, animal, and plant diseases including malaria, bluetongue, dengue fever, and Pierce's disease, and most of these vectors are susceptible to entomopathogenic fungi (Fang, Azimzadeh, et al., 2012). Entomopathogenic fungi, such as *M. anisopliae* and *B. bassiana*, can naturally kill adult mosquitoes, albeit slowly. However, it takes about 14 days for *Plasmodium* to develop from ingested gametocytes to infectious sporozoites. Mosquitoes can be killed in time to block malaria transmission as long as they are infected with fungi at their first or second blood meal. However, the high coverage required for early infection of most mosquitoes in a population may be hard to achieve in the field because of issues such as user resistance (Read, Lynch, & Thomas, 2009). Fungal strains with accelerated killing speed could block malaria transmission by mosquitoes even with a late-stage *Plasmodium* infection. As described above, the virulence of *M. anisopliae* can be increased to a remarkable extent by expressing a sodium channel blocker (AaIT1). Virulence can be further improved by simultaneously expressing AaIT1 and Hybrid-toxin, including an over 150-fold reduction in LC_{50} and 50% reduction in feeding activity (Fang et al., unpublished data). However, mosquitoes are notoriously adept at out-evolving control strategies, so they may develop resistance to the fungal strains with increased virulence. In contrast, the slow speed of kill with WT entomopathogenic fungi enables mosquitoes to achieve part of their lifetime reproductive output, which could reduce selection pressure on mosquitoes to develop resistance to the biopesticide (Read et al., 2009). Fungal strains engineered to greatly reduce mosquito infectiousness without increasing speed of kill could improve disease control without increasing the spread of resistance. To achieve this effect, recombinant strains were produced expressing effectors that target sporozoites as they travel through the hemolymph to the salivary glands. Three antimalaria effectors were tested: salivary gland and midgut peptide 1 (SM1) that blocks attachment of sporozoites to salivary glands, a single-chain antibody that agglutinates sporozoites, and an antimicrobial toxin called scorpine (Fang et al., 2011). The virulence of the recombinant strains against mosquitoes was found to be the same as the WT strain, so the expression of the antimalaria proteins does not add extra evolutionary pressure on the mosquitoes. Meanwhile, even with a late-stage infection, the recombinant strains significantly reduced the titer of sporozoites in the salivary glands of the mosquitoes, ie, blocking malaria transmission. Eleven days after a *Plasmodium*-infected blood meal, mosquitoes were inoculated with *M. anisopliae* expressing a combination of antimalaria effectors (scorpine/SM1: scorpine and scorpine/PfNPNA-1), which reduced the sporozoite intensity by approximate 98% six days after infection. A potential problem with relying on antimalarial effects is that they might, in the long run, suffer from the evolution of resistant malaria parasites. The single-chain antibody (PfNPNA-1) specifically recognizes the repeat region (Asn-Pro-Asn-Ala) of the *Plasmodium falciparum* surface circumsporozoite protein (Chappel, Rogers, Hoffman, & Kang, 2004), so multiple mutations would be required to achieve resistance in this case. In addition, reducing the probability of emergence of resistance to one mechanism could be achieved in this system by expressing multiple antimalaria effectors with different modes of action.

In south Asia, human vector mosquitoes feed predominantly on domestic animals and only secondarily on human beings, and applying deltamethrin insecticide to cattle reduced human malaria transmission to the same extent as indoor spraying, but at 80% less cost (Rowland et al., 2001). *Metarhizium*-based insecticides have been developed and applied to sheep to control mange vector Psoroptes mites (Abolins et al., 2007; Brooks et al., 2004), so potentially, transgenic fungal strains could be applied to livestock to simultaneously improve their health and control human malaria. Fungi can attack almost all known disease vectors; various transgenic fungi could be constructed to express different effector proteins that each attacks one or several vector-borne diseases of humans, animals, and plants. For example, a fungus expressing the antimicrobial, scorpion-toxin scorpine could control livestock and poultry malaria responsible for significant economic loss.

As a tractable model, *Metarhizium* is suitable for evaluating and delivering proteins to block transmission of malaria. The selected potent proteins could be mixed and matched to improve the efficacy of entomopathogenic fungi to block malaria transmission. These potent proteins could also be used to improve the efficacy of other microorganisms employed to block malaria transmission. As described above, expression of scorpine increases the ability of *M. anisopliae* to block malaria transmission. This work inspired the expression of scorpine in *Pantoea agglomerans*, a common mosquito symbiotic bacterium. These engineered *P. agglomerans* strains inhibited development of the human malaria parasite *P. falciparum* and rodent malaria parasite *Plasmodium berghei* by up to 98% (Wang et al., 2012).

4. IMPROVE TOLERANCE TO ABIOTIC STRESSES

Abiotic stresses such as ultraviolet (UV) radiation, high temperature, and low water activity result in inconsistent performances by mycoinsecticides in the field, limiting their use. Tolerance to these abiotic stresses can be improved by selecting optimal growth substrate and conditions for the production of the conidia (Rangel et al., 2015). Recent studies have shown that fungal tolerance to the abiotic stresses can also be greatly improved by genetic engineering.

4.1 Improve Tolerance to UV Radiation

UV radiation from sunlight is probably the most detrimental environmental factor affecting the viability of entomopathogenic fungi applied to solar-exposed sites (eg, leaves) for pest control. UV radiation primarily causes DNA damage through induction of chemical base modifications: cyclobutane pyrimidine dimers (CPDs) and pyrimidine (6-4) photoproducts [(6-4)PPs] (Sinha & Hader, 2002). Removal of these photolesions from the DNA is performed by the nucleotide excision repair (NER) pathway and by photoreactivation, with the latter playing a major role in *M. robertsii* (Fang & St. Leger, 2012). In *M. robertsii*, the native (6-4) photoproduct photolyase can fix (6-4)PPs produced by daily sunlight exposure. However, its native CPD photolyase is insufficient to fix CPD lesions (the major photoproduct) and prevent the loss of viability caused by 7 h of solar radiation (11 am to 6 pm). We found that overexpressing the native *M. robertsii* photolyase or expressing the photolyase of a highly UV tolerant *Halobacterium* (McCready & Marcello, 2003) imparted increased UV tolerance, but the *Halobacterium* enzyme was much more effective: achieving greater than 30-fold improvement in survivability of *M. robertsii* and *B. bassiana* to sunlight. Unlike WT strains, *M. robertsii* or *B. bassiana* expressing the *Halobacterium* photolyase retained virulence against the malarial vector *A. gambiae* even after several hours of exposure to sunlight. In the field, this improved persistence should translate into much more effective pest control over a longer time frame (Fang & St. Leger, 2012).

UV radiation causes not only DNA damage but also produces reactive oxidative species (ROS) that elevate oxidative stress in cells (Lesser, 1996). Overexpression of a superoxide dismutase (SOD) increased the ability of *B. bassiana* to detoxify ROS, enhancing UV tolerance (Xie, Wang, Huang, Ying, & Feng, 2010). Similarly, expression of thioredoxin (trxA) from the bacterium *Escherichia coli* also increased the tolerance of *B. bassiana* to UV-B irradiation, oxidation, and heat (Ying & Feng, 2011).

Pigments such as melanin are effective absorbers of UV light and can dissipate the absorbed UV radiation. Pigments on conidial cell surfaces usually act as a coat to protect fungal cells from UV damage; however, *B. bassiana* does not produce visible pigments in the conidia. Expression of a tyrosinase from *Aspergillus fumigatus* activated the production of pigments in *B. bassiana* and thus increased the tolerance of this fungus to UV radiation (Shang, Duan, Huang, Gao, & Wang, 2012). *Metarhizium robertsii* produces dark green pigments in conidia, but it does not produce the DHN-melanin that contributes to the tolerance of many other fungi to various abiotic stresses. Tseng, Chung, and Tzean (2011) transferred the DHN-synthesis pathway of *Alternaria alternata* into *M. anisopliae*. Compared to the WT strain, the transformant showed a 2-fold greater tolerance to UV radiation. Interestingly, the transformants also displayed a 1.3-fold greater tolerance to thermal stress (35°C) and a 3-fold greater tolerance to low water activity (aw = 97.1%), indicating that the melanin provides resistance to multiple abiotic stresses in *M. anisopliae*. In a follow-up paper, this group reported that production of DHN-melanin also enhanced the virulence of *M. anisopliae* to insects by promoting germination, appressorium formation, and increasing the expression of virulence genes (Tseng, Chung, & Tzean, 2014). The increased virulence could result from the enhanced resistance to stresses encountered by the fungus during its pathogenesis in insects. Another possible reason is that the genes in the pathway of DHN-melanin synthesis participate in the infection of insects. In *M. robertsii*, the laccase gene involved in synthesis of conidial pigment is also related to cell wall rigidity of appressorium and, thus, is involved in pathogenicity (Fang, Fernandes, Roberts, Bidochka, & St. Leger, 2010).

4.2 Improve Tolerance to Heat Stress

Temperature extremes are an important adverse factor limiting the effectiveness of microbial pest control agents by reducing virulence and persistence in field conditions. Tolerance of entomopathogenic fungi to heat stress can be increased through experimental evolution via continuous culture of fungal cells under heat stress. In these conditions, thermotolerant variants of *M. robertsii* were obtained, which displayed robust growth at 37°C and WT level of pathogenicity (de Crecy, Jaronski, Lyons, Lyons, & Keyhani, 2009). Tolerance to heat stress by entomopathogenic fungi can also be improved by transferring single genes. Similar to UV radiation, heat stress produces ROS. Expression of ROS scavengers, such as SOD and bacterial thioredoxin, increases the heat tolerance in entomopathogenic fungi (Xie et al., 2010; Ying & Feng, 2011). Small heat shock proteins have been shown to confer thermotolerance in many organisms, and HSP25 expression was found to be upregulated when *M. robertsii* was grown at extreme temperatures or in the presence of oxidative or osmotic agents. Overexpressing HSP25 in *M. robertsii* increased fungal growth under heat stress either in nutrient-rich medium or on locust wings and enhanced the tolerance of heat shock-treated conidia to osmotic stress (Liao, LU, Fang, & St. Leger, 2014).

Entomopathogenic fungi, such as *M. robertsii* and *B. bassiana*, and mammalian fungal pathogens have many parallels with respect to pathogenesis, including breaching proteinaceous integuments and evading host innate immune systems (Scully & Bidochka, 2006). An important caveat to creating transgenic strains with hypertolerance to high temperature, especially above mammalian body temperature, is that they could gain the potential to infect humans and other mammals. The high mammalian body temperature acts as a thermal barrier to infection by many fungi. Risk assessment should be particularly rigorous before such transgenic fungal strains are applied in the field for pest control.

5. PROMOTERS USED FOR GENETIC ENGINEERING OF ENTOMOPATHOGENIC FUNGI

A promoter is a DNA fragment that regulates gene transcription by allowing RNA polymerase and transcription factors to bind to it. In genetic engineering, the promoter is, thus, an important element in expression vectors: driving the expression of target genes. Strong promoters are usually employed in the construction of genetically engineered strains. In the early stages of research on the genetic engineering of entomopathogenic fungi, the constitutive promoter of the glyceraldehyde-3-phosphate dehydrogenase gene (*gpdA*) from *Aspergillus nidulans* was usually used, and reasonable expression levels of target genes were achieved (Fan et al., 2007; Fang, Feng, et al., 2009; Fang et al., 2005; St. Leger et al., 1996a). Native promoters of *gpd* genes from *B. bassiana* and *M. acridum* were later identified, and they display significantly stronger activity in their respective species than the *A. nidulans gpd* promoter (Cao, Jiao, & Xia, 2012; Liao et al., 2008). Besides the promoters of *gpd* genes, other native promoters with stronger activity were also identified. A gene encoding *B. bassiana* class I hydrophobin (*Hyd1*) was found to be expressed in almost all developmental stages of *B. bassiana* (Cho, Kirkland, Holder, & Keyhani, 2007). The ability of its full-length promoter to drive the reporter gene GFP was significantly greater than *PgpdA* from *A. nidulans*; a truncated 1290-bp fragment of this promoter had even stronger activity. The expression level of the insect midgut-specific toxin Vip3A driven by this truncated promoter was significantly higher than the *PgpdA* promoter from *A. nidulans* and the full-length promoter of *Hyd1*, thus, increasing per os virulence of *B. bassiana* to *Spodoptera litura* larvae to a greater extent than the earlier two promoters (Wang, Ying, & Feng, 2013).

Development-stage-specific promoters facilitate directed gene expression at certain stages of fungal development, while leaving other stages unmodified. Using this kind of promoter, the expression of target genes in genetically engineered fungal strains can be tightly regulated. The expression of *Metarhizium* collagen-like protein MCL1 is only induced by insect hemolymph, and in insect hemolymph it is expressed at a high level. Using the promoter of *Mcl1* for genetically engineering *Metarhizium*, the expression of the target genes can be limited to the hemocoel of the target insects, ensuring the specificity and targeted expression of the genetically engineered fungal strains (Wang & St. Leger., 2006). The *Mcl1* promoter is to date the only development-stage-specific promoter used in the genetic engineering of entomopathogenic fungi. With genome-wide characterization of fungal development and pathogenesis in insects, more genes regulated by development-stage-specific promoters are being identified in entomopathogenic fungi. For example, the aforementioned sterol carrier gene Mr-npc2a was identified from analysis of a random T-DNA insertion library. The impact of this gene would not have been detected without screening random mutants against insects. Of particular interest from a bioengineering perspective, using a Mr-NPC2a promoter-GFP reporter gene fusion, it was demonstrated that Mr-NPC2a is expressed exclusively in living insects (Zhao et al., 2014). While the MCl1 promoter also expresses in isolated hemolymph in a test tube, this is not so for the Mr-NPC2a promoter, which would thus give maximum selectivity of transgene expression.

6. METHODS TO MITIGATE THE SAFETY CONCERNS OF GENETICALLY MODIFIED ENTOMOPATHOGENIC FUNGI

As with other genetically modified organisms, such as transgenic plants, safety concerns exist regarding genetically engineered entomopathogenic fungi. As touched on previously, several methods are available to mitigate these concerns.

To date, it has been axiomatic that a recombinant pathogen should not persist and reproduce in the field. Reduction in ecological fitness of a fungal strain will make the strain lose competitiveness with native strains in the field and possibly go extinct in the environment (Committee on the Biological confinement of genetically engineered organisms, National Research Council, 2004). We found that removal of the *M. robertsii* photoreactivation system resulted in a strain that was highly sensitive to UV radiation (Fang & St. Leger, 2012). UV hypersensitive strains with hypervirulence from expression of antiinsect or antimalarial proteins will be especially useful in controlling mosquitoes and malaria transmission because the genetically engineered strains cannot spread into the environment from the application site (eg, mosquito traps or houses). With a better understanding of the mechanisms entomopathogenic fungi use to adapt to the environment, more targets could be manipulated to reduce their ecological fitness in different environments. For example, the ability of *Metarhizium* to survive over winter in a temperate climate could be greatly reduced by disrupting the RNA binding proteins that allow it to adapt to the cold (Fang & St. Leger, 2010a). *Metarhizium robertsii* is a potent insect pathogen that is also ubiquitous in the soil community, where it establishes mutualistic interactions with plants as a rhizospheric fungus (Behie, Zelisko, & Bidochka, 2012; Hu & St. Leger, 2002). The distribution of genetic groups of *M. robertsii* depends on their local adaptations to specific soils and plant types, rather than their pathogenicity to insects (Bidochka, Kamp, Lavender, Dekoning, & De Croos, 2001). We found that an oligosaccharide transporter (MRT) is important to the rhizosphere competency of *M. robertsii* (Fang & St. Leger, 2010b). Disrupting *Mrt* significantly reduced rhizosphere competence, but not its pathogenicity to insects, demonstrating that *Mrt* is exclusively involved in *M. robertsii*'s interactions with plants. The reduction in rhizosphere competence could significantly reduce persistence of the fungus in the environment (Wang, O'Brien, Pava-Ripoll, & St. Leger, 2011). By using the *Mrt* mutant strain as a starting strain for genetically engineering improved virulence, the resultant strains will be hypervirulent but scarcely able to survive in the environment. There could be many other genes involved in the interaction between *M. robertsii* and plants, and they could also be potential targets for manipulation to reduce rhizosphere competency of transgenic strains in the field (Pava-Ripoll et al., 2011).

Selectable marker genes (SMGs) and selection agents are useful tools in the production of genetically engineered fungi by selecting transformed cells from a matrix consisting of mostly untransformed cells. Most SMGs express protein products that confer antibiotic or herbicide resistance traits, and the presence of these genes in transgenic fungal strains is of concern; they are subject to special government regulation in many countries. As such, it is worthwhile to use alternatives to SMGs for transformation or to remove SMGs in the final genetically engineered strains (Luo et al., 2007). The obvious alternatives to SMGs are auxotrophic mutants. A *B. bassiana* mutant with its orotidine 5′-phosphate decarboxylase (*ura3*) gene mutated was obtained using a positive screening protocol with 5′-fluoro-orotate. The native *ura3* gene can complement the mutant and thus is a suitable marker for genetic engineering of *B. bassiana*. Theoretically, many genes can be transformed in *B. bassiana* by repeating such a positive screening protocol (Ying, Feng, & Keyhani, 2013). However, the positive screening protocol randomly generates mutations in the genome including many undesired, harmful mutations. Even the reported *ura3* mutant had a lower conidial yield than WT, which would increase the cost of production. Therefore, it would be impractical to rapidly acquire a library of suitable auxotrophic mutants for genetic engineering, especially when repeated transformations are needed to transfer several genes in the same strain.

An SMG-removal method has been applied to transgenic plants for increased precision of genome modification (Luo et al., 2007). Similar strategies, such as Cre-loxP recombination system, can be applied to genetically engineered entomopathogenic fungi. In *M. robertsii*, expression of the Cre recombinase results in excision of an integrated SMG that is flanked by loxP sites via anastomosis. The resulting strains are free of SMG and have only a short DNA fragment of the LoxP recognition sites. Moreover, Cre-loxP recombination system could allow unlimited recycling of loxP-flanked SMGs and foster the generation of transgenic strains with the expression of multiple target genes (Zhang, Lu, Liao, St. Leger, & Nuss, 2013).

7. CONCLUSION

The application of mycoinsecticides has been limited by their low tolerance to abiotic stresses and slow killing speed. Tolerance to abiotic stresses has been enhanced using the entomopathoge's own genes, as well as exogenous genes from other fungi, bacteria, and archaea. Likewise, virulence of entomopathogenic fungi has also been significantly improved by genetic engineering using naturally existing genes from entomopathogenic fungi and bacteria, insects and insect predators. Virulence was also improved using synthetic genes generated by mixing and matching functional domains of fungal pathogenicity-related genes and host components. With ever more virulence genes and insect components being identified, protein engineering could generate an inexhaustible arsenal of antiinsect proteins, with which new transgenic strains can be developed that stay one step ahead of evolving resistance in insects. The potent antiinsect effectors vetted in entomopathogenic fungi could also be used for genetic engineering of other organisms (eg, microbes or plants).

Fungal strains with reduced ecological fitness are already available for use as recipients for genetic engineering. Coupled with their natural insect specificity, safety concerns on genetically modified strains could be mitigated by using safe effector proteins with their selection marker genes removed after transformation. Although the initial hypervirulent fungal products will have features providing biological containment, it should also be possible to develop second generation recombinant microbes that show narrow specificity for target pests, eg, mosquitoes, and that persist in the environment, providing sustainable cheap control for much longer periods than existing chemicals. Producing such an organism could not have been envisioned as recently as a decade ago. However, the molecular biological knowledge and techniques that will make creation of these microbes highly feasible are being assembled. Furthermore, the precision and malleability of molecular techniques could allow design of multiple pathogens with different strategies to be used for different ecosystems and avoid resistance. Given the increasing public acceptance of genetically modified organisms, particularly crops expressing *B. thuringiensis* toxins, field application of genetically modified insecticidal microbes should have a bright future if our understanding of their unique biology continues to grow and care is taken to ensure social acceptance through rigorous and transparent risk—benefit analyses.

ACKNOWLEDGMENTS

This work was funded by the National Natural Science Foundation of China (31471818 and 31272097) and Zhejiang Provincial Natural Science Foundation of China (LR13C010001) and "1000 Young Talents Program of China" to WF. This work was also supported by the Biotechnology Risk Assessment Program competitive grant number 2015-33522-24107—from the USDA—National Institute of Food and Agriculture, and by NIAID of the National Institutes of Health under award number RO1 AI106998 to Raymond St. Leger.

REFERENCES

Abolins, S., Thind, B., Jackson, V., Luke, B., Moore, D., Wall, R., et al. (2007). Control of the sheep scab mite *Psoroptes ovis* in vivo and in vitro using fungal pathogens. *Veterinary Parasitology, 148*, 310—317.

Behie, S. W., Zelisko, P. M., & Bidochka, M. J. (2012). Endophytic insect-parasitic fungi translocate nitrogen directly from insects to plants. *Science, 336*, 1576—1577.

Bidochka, M. J., Kamp, A. M., Lavender, T. M., Dekoning, J., & De Croos, J. N. (2001). Habitat association in two genetic groups of the insect-pathogenic fungus *Metarhizium anisopliae*: uncovering cryptic species? *Applied and Environmental Microbiology, 67*, 1335—1342.

Brooks, A. J., de Muro, M. A., Burree, E., Moore, D., Taylor, M. A., & Wall, R. (2004). Growth and pathogenicity of isolates of the fungus *Metarhizium anisopliae* against the parasitic mite, *Psoroptes ovis*: effects of temperature and formulation. *Pest Management Science, 60*, 1043—1049.

Cao, Y., Jiao, R., & Xia, Y. (2012). A strong promoter, PMagpd, provides a tool for high gene expression in entomopathogenic fungus, *Metarhizium acridum*. *Biotechnology Letters, 34*, 557—562.

Chappel, J. A., Rogers, W. O., Hoffman, S. L., & Kang, A. S. (2004). Molecular dissection of the human antibody response to the structural repeat epitope of *Plasmodium falciparum* sporozoite from a protected donor. *Malaria Journal, 3*, 28—39.

Cho, E. M., Kirkland, B. H., Holder, D. J., & Keyhani, N. O. (2007). Phage display cDNA cloning and expression analysis of hydrophobins from the entomopathogenic fungus *Beauveria (Cordyceps) bassiana*. *Microbiology, 153*, 3438—3447.

Committee on the Biological confinement of genetically engineered organisms, National Research Council. (2004). *Biological confinement of genetically engineered organisms*, ISBN 0-309-52778-3.

de Crecy, E., Jaronski, S., Lyons, B., Lyons, T. J., & Keyhani, N. O. (2009). Directed evolution of a filamentous fungus for thermotolerance. *BMC Biotechnology, 9*, 74.

Fan, Y., Borovsky, D., Hawkings, C., Ortiz-Urquiza, A., & Keyhani, N. O. (2012). Exploiting host molecules to augment mycoinsecticide virulence. *Nature Biotechnology, 30*, 35–37.

Fan, Y., Fang, W., Guo, S., Pei, X., Zhang, Y., Xiao, Y., et al. (2007). Increased insect virulence in *Beauveria bassiana* strains overexpressing an engineered chitinase. *Applied and Environmental Microbiology, 73*, 295–302.

Fan, Y., Pei, X., Guo, S., Zhang, Y., Luo, Z., Liao, X., et al. (2010). Increased virulence using engineered protease-chitin binding domain hybrid expressed in the entomopathogenic fungus *Beauveria bassiana*. *Microbial Pathogenesis, 49*, 376–380.

Fan, Y., Pereira, R. M., Kilic, E., Casella, G., & Keyhani, N. O. (2012). Pyrokinin β-neuropeptide affects necrophoretic behavior in fire ants (*S. invicta*), and expression of β-NP in a mycoinsecticide increases its virulence. *PLoS One, 7*, e26924.

Fang, W., Azimzadeh, P., & St. Leger, R. J. (2012). Strain improvement of fungal insecticides for controlling insect pests and vector-borne diseases. *Current Opinion in Microbiology, 15*, 232–238.

Fang, W., Feng, J., Fan, Y., Zhang, Y., Bidochka, M. J., St. Leger, R. J., et al. (2009). Expressing a fusion protein with protease and chitinase activities increases the virulence of the insect pathogen *Beauveria bassiana*. *Journal of Invertebrate Pathology, 102*, 155–159.

Fang, W., Fernandes, E. K., Roberts, D. W., Bidochka, M. J., & St. Leger, R. J. (2010). A laccase exclusively expressed by *Metarhizium anisopliae* during isotropic growth is involved in pigmentation, tolerance to abiotic stresses and virulence. *Fungal Genetics and Biology, 47*, 602–607.

Fang, W., Leng, B., Xiao, Y., Jin, K., Ma, J., Fan, Y., et al. (2005). Cloning of *Beauveria bassiana* chitinase gene Bbchit1 and its application to improve fungal strain virulence. *Applied and Environmental Microbiology, 71*, 363–370.

Fang, W., Lu, H. L., King, G. F., & St. Leger, R. J. (2014). Construction of a hypervirulent and specific mycoinsecticide for locust control. *Scientific Reports, 4*, 7345.

Fang, W., Pava-Ripoll, M., Wang, S., & St. Leger, R. J. (2009). Protein kinase A regulates production of virulence determinants by the entomopathogenic fungus, *Metarhizium anisopliae*. *Fungal Genetics and Biology, 46*, 277–285.

Fang, W., Pei, Y., & Bidochka, M. J. (2006). Transformation of *Metarhizium anisopliae* mediated by *Agrobacterium tumefaciens*. *Canadian Journal of Microbiology, 52*, 623–626.

Fang, W., & St. Leger, R. J. (2010a). RNA binding proteins mediate the ability of a fungus to adapt to the cold. *Environmental Microbiology, 12*, 810–820.

Fang, W., & St. Leger, R. J. (2010b). Mrt, a gene unique to fungi, encodes an oligosaccharide transporter and facilitates rhizosphere competency in *Metarhizium robertsii*. *Plant Physiology, 154*, 1549–1557.

Fang, W., & St. Leger, R. J. (2012). Enhanced UV resistance and improved killing of malaria mosquitoes by photolyase transgenic entomopathogenic fungi. *PLoS One, 7*, e43069.

Fang, W., Vega-Rodríguez, J., Ghosh, A. K., Jacobs-Lorena, M., Kang, A., & St. Leger, R. J. (2011). Development of transgenic fungi that kill human malaria parasites in mosquitoes. *Science, 331*, 1074–1077.

Fang, W., Zhang, Y., Yang, X., Zheng, X., Duan, H., Li, Y., et al. (2004). *Agrobacterium tumefaciens*-mediated transformation of *Beauveria bassiana* using an herbicide resistance gene as a selection marker. *Journal of Invertebrate Pathology, 85*, 18–24.

Federici, B. A., Bonning, B. C., & St.Leger, R. J. (2008). Improvement of insect pathogens as insecticides through genetic engineering. In C. Hill, & R. Sleator (Eds.), *PathoBiotechnology*. Landes Bioscience.

Feng, M. G., Poprawski, T. J., & Khachatourians, G. G. (1994). Production, formulation and application of the entomopathogenic fungus *Beauveria bassiana* for insect control: current status. *Biocontrol Science and Technology, 4*, 3–34.

Gao, Q., Jin, K., Ying, S. H., Zhang, Y., Xiao, G., Shang, Y., et al. (2011). Genome sequencing and comparative transcriptomics of the model entomopathogenic fungi *Metarhizium anisopliae* and *M. acridum*. *PLoS Genetics, 7*, e1001264.

Gongora, C. E. (2004). Transformacion de *Beauveria bassiana* cepa Bb9112 con les genes de la proteina verde fluorescente y la protease pr1A de M. anisopliae. *Revista Colombiana de Entomologia, 30*, 1–5.

Gressel, J. (2007). Failsafe mechanisms for preventing gene flow and organism dispersal of enhanced microbial biocontrol agents. In M. Vurro, & J. Gressel (Eds.), *Novel biotechnologies for biocontrol agent enhancement and management*. Springer.

Hassan, A., & Charnley, A. K. (1989). Ultrastructural study of the penetration by *Metarhizium anisopliae* through dimilin-affected cuticle of *Manduca sexta*. *Journal of Invertebrate Pathology, 54*, 117–124.

Hu, G., & St. Leger, R. J. (2002). Field studies using a recombinant mycoinsecticide (*Metarhizium anisopliae*) reveal that it is rhizosphere competent. *Applied and Environmental Microbiology, 68*, 6383–6387.

Hu, X., Xiao, G., Zheng, P., Shang, Y., Su, Y., Zhang, X., et al. (2014). Trajectory and genomic determinants of fungal-pathogen speciation and host adaptation. *Proceedings of the National Academy of Sciences of the United States of America, 111*, 16796–16801.

Jin, K., Peng, G., Liu, Y., & Xia, Y. (2015). The acid trehalase, ATM1, contributes to the in vivo growth and virulence of the entomopathogenic fungus, *Metarhizium acridum*. *Fungal Genetics and Biology, 77*, 61–67.

Kamareddine, L., Fan, Y., Osta, M. A., & Keyhani, N. O. (2013). Expression of trypsin modulating oostatic factor (TMOF) in an entomopathogenic fungus increases its virulence towards *Anopheles gambiae* and reduces fecundity in the target mosquito. *Parasites & Vectors, 6*, 22.

Langewald, J., & Kooyman, C. (2007). Green muscle, a fungal biopesticide for control of grasshoppers and locusts in Africa. In C. Vincent, M. S. Goettel, & G. Lazarovitis (Eds.), *Biological control a global perspective*. CABI.

Lesser, M. P. (1996). Elevated temperatures and ultraviolet radiation cause oxidative stress and inhibit photosynthesis in symbiotic dinoflagellates. *Limnology and Oceanography, 41*, 271–283.

Liao, X., LU, H. L., Fang, W., & St. Leger, R. J. (2014). Overexpression of a *Metarhizium robertsii* HSP25 gene increases thermotolerance and survival in soil. *Applied Microbiology and Biotechnology, 98*, 777–783.

Liao, X. G., Fang, W. G., Zhang, Y. J., Fan, Y. H., Wu, X. W., Zhou, Q., et al. (2008). Characterization of a highly active promoter, PBbgpd, in *Beauveria bassiana*. *Current Microbiology, 57*, 121–126.

Lin, L., Fang, W., Liao, X., Wang, F., Wei, D., & St. Leger, R. J. (2011). The MrCYP52 cytochrome P450 monoxygenase gene of *Metarhizium robertsiii* important for utilizing insect epicuticular hydrocarbons. *PLoS One, 16*, e28984.

Liu, Y. J., Liu, J., Ying, S. H., Liu, S. S., & Feng, M. G. (2013). A fungal insecticide engineered for fast *per os* killing of caterpillars has high field efficacy and safety in full-season control of cabbage insect pests. *Applied and Environmental Microbiology, 79*, 6452–6458.

Lovett, B., & St. Leger, R. J. (2015). Stress is the rule rather than the exception for Metarhizium. *Current Genetics, 61*, 253–261.

Luo, K., Duan, H., Zhao, D., Zheng, X., Deng, W., Chen, Y., et al. (2007). 'GM-gene-deletor': fused loxP-FRT recognition sequences dramatically improve the efficiency of FLP or CRE recombinase on transgene excision from pollen and seed of tobacco plants. *Plant Biotechnology Journal, 5*, 263–274.

McCready, S., & Marcello, L. (2003). Repair of UV damage in *Halobacterium salinarum*. *Biochemical Society Transactions, 31*, 694–698.

Nambiar, P. T. C., Ma, S. W., & Iyer, V. N. (1990). Limiting an insect infestation of nitrogen-fixing root nodules of the pigeon pea (*Cajanus cajan*) by engineering the expression of an entomocidal gene in its root nodules. *Applied and Environmental Microbiology, 56*, 2866–2869.

Obukowicz, M. G., Perlak, F. J., Kusano-Kretzmer, K., Mayer, E. J., & Waturd, L. S. (1986). Integration of the delta-endotoxin gene of *Bacillus thuringiensis* into the chromosome of root-colonizing strains of pseudomonads using Tn5. *Gene, 45*, 327–331.

Ortiz-Urquiza, A., Luo, Z., & Keyhani, N. O. (2015). Improving mycoinsecticides for insect biological control. *Applied Microbiology and Biotechnology, 99*, 1057–1068.

Pava-Ripoll, M., Angelini, C., Fang, W., Wang, S., Posada, F. J., & St. Leger, R. J. (2011). The rhizosphere-competent entomopathogen *Metarhizium anisopliae* expresses a specific subset of genes in plant root exudate. *Microbiology, 157*, 47–55.

Pava-Ripoll, M., Posada, F. J., Momen, B., Wang, C., & St. Leger, R. J. (2008). Increased pathogenicity against coffee berry borer, *Hypothenemus hampei* (Coleoptera: Curculionidae) by *Metarhizium anisopliae* expressing the scorpion toxin (AaIT) gene. *Journal of Invertebrate Pathology, 99*, 220–226.

Pedrini, N., Ortiz-Urquiza, A., Huarte-Bonnet, C., Fan, Y., Juárez, M. P., & Keyhani, N. O. (2015). Tenebrionid secretions and a fungal benzoquinone oxidoreductase form competing components of an arms race between a host and pathogen. *Proceedings of the National Academy of Sciences of the United States of America, 112*, E3651–E3660.

Peng, G., Jin, K., Liu, Y., & Xia, Y. (2015). Enhancing the utilization of host trehalose by fungal trehalase improves the virulence of fungal insecticide. *Applied Microbiology and Biotechnology, 99*, 8611–8618.

Peng, G., & Xia, Y. (2014). Expression of scorpion toxin LqhIT2 increases the virulence of *Metarhizium acridum* towards *Locusta migratoria manilensis*. *Journal of Industrial Microbiology & Biotechnology, 41*, 1659–1666.

Peng, G., & Xia, Y. (2015). Integration of an insecticidal scorpion toxin (BjαIT) gene into *Metarhizium acridum* enhances fungal virulence towards *Locusta migratoria manilensis*. *Pest Management Science, 71*, 58–64.

Qin, Y., Ying, S. H., Chen, Y., Shen, Z. C., & Feng, M. G. (2010). Integration of insecticidal protein Vip3Aa1 into *Beauveria bassiana* enhances fungal virulence to *Spodoptera litura* larvae by cuticle and *per Os* infection. *Applied and Environmental Microbiology, 76*, 4611–4618.

Rangel, D. E., Braga, G. U., Fernandes, É. K., Keyser, C. A., Hallsworth, J. E., & Roberts, D. W. (2015). Stress tolerance and virulence of insect-pathogenic fungi are determined by environmental conditions during conidial formation. *Current Genetics, 61*, 383–404.

Read, A. F., Lynch, P. A., & Thomas, M. B. (2009). How to make evolution-proof insecticides for malaria control. *PLoS Biology, 7*, e1000058.

Roberts, D. W., & St. Leger, R. J. (2004). *Metarhizium* spp., cosmopolitan insect-pathogenic fungi: mycological aspects. *Advances in Applied Microbiology, 54*, 1–70.

Rowland, M., Durrani, N., Kenward, M., Mohammed, N., Urahman, H., & Hewitt, S. (2001). Control of malaria in Pakistan by applying deltamethrin insecticide to cattle: a community-randomised trial. *The Lancet, 357*, 1837–1841.

Schnepf, H. E., Crickmore, N., van Rie, J., Lereclus, D., Baum, J., Feilelson, J., et al. (1998). *Bacillus thuringiensis* and its pesticidal proteins. *Microbiology and Molecular Biology Reviews, 62*, 775–806.

Screen, S. E., Hu, G., & St. Leger, R. J. (2001). Transformants of *Metarhizium anisopliae* sf. *anisopliae* overexpressing chitinase from *Metarhizium anisopliae* sf. *acridum* show early induction of native chitinase but are not altered in pathogenicity to *Manduca sexta*. *Journal of Invertebrate Pathology, 78*, 260–266.

Scully, L. R., & Bidochka, M. J. (2006). Developing insect models for the study of current and emerging human pathogens. *FEMS Microbiology Letters, 263*, 1–9.
Shang, Y., Duan, Z., Huang, W., Gao, Q., & Wang, C. (2012). Improving UV resistance and virulence of *Beauveria bassiana* by genetic engineering with an exogenous tyrosinase gene. *Journal of Invertebrate Pathology, 109*, 105–109.
Sinha, R. P., & Hader, D. P. (2002). UV-induced DNA damage and repair: a review. *Photochemical & Photobiological Sciences, 1*, 225–236.
St. Leger, R. J., Joshi, L., Bidochka, M. J., & Roberts, D. W. (1996a). Construction of an improved mycoinsecticide over-expressing a toxic protease. *Proceedings of the National Academy of Sciences of the United States of America, 93*, 6349–6354.
St. Leger, R. J., Joshi, L., Bidochka, M. J., & Roberts, D. W. (1996b). Characterization of chitinases from *Metarhizium anisopliae, M. flavoviridae* and *Beauveria bassiana*, and ultrastructural localization of chitinase production during invasion of insect cuticle. *Applied and Environmental Microbiology, 62*, 907–912.
St. Leger, R. J., & Screen, S. (2001). Prospects for strain improvement of fungal pathogens of insects and weeds. In T. M. Butt, C. Jackson, & N. Morgan (Eds.), *Fungal biocontrol agents: Progress, problems and potential* (pp. 219–238). CAB International.
Thomas, M. B., & Read, A. F. (2007). Can fungal biopesticides control malaria? *Nature Reviews Microbiology, 5*, 377–383.
Thompson, S. N., & Borchardt, D. B. (2003). Glucogenic blood sugar formation in an insect *Manduca sexta* L.: asymmetric synthesis of trehalose from 13C enriched pyruvate. *Comparative Biochemistry and Physiology — Part B: Biochemistry & Molecular Biology, 135*, 461–471.
Tseng, M. N., Chung, P. C., & Tzean, S. S. (2011). Enhancing the stress tolerance and virulence of an entomopathogen by metabolic engineering of dihydroxynaphthalene melanin biosynthesis genes. *Applied and Environmental Microbiology, 77*, 4508–4519.
Tseng, M. N., Chung, C. L., & Tzean, S. S. (2014). Mechanisms relevant to the enhanced virulence of a dihydroxynaphthalene-melanin metabolically engineered entomopathogen. *PLoS One, 9*, e90473.
Wang, C., & Feng, M. G. (2014). Advances in fundamental and applied studies in China of fungal biocontrol agents for use against arthropod pests. *Biological Control, 68*, 129–135.
Wang, C., & St. Leger, R. J. (2006). A collagenous protective coat enables *Metarhizium anisopliae* to evade insect immune responses. *Proceedings of the National Academy of Sciences of the United States of America, 103*, 6647–6652.
Wang, C., & St. Leger, R. J. (2007a). A scorpion neurotoxin increases the potency of a fungal insecticide. *Nature Biotechnology, 25*, 1455–1456.
Wang, C., & St. Leger, R. J. (2007b). The MAD1 adhesin of *Metarhizium anisopliae* links adhesion with blastospore production and virulence to insects, and the MAD2 adhesin enables attachment to plants. *Eukaryotic Cell, 6*, 808–816.
Wang, S., Fang, W., Wang, C., & St. Leger, R. J. (2011). Insertion of an esterase gene into a specific locust pathogen (*Metarhizium acridum*) enables it to infect caterpillars. *PLoS Pathogens*, 7(6), e1002097.
Wang, S., Ghosh, A. K., Bongio, N., Stebbings, K. A., Lampe, D. J., & Jacobs-Lorena, M. (2012). Fighting malaria with engineered symbiotic bacteria from vector mosquitoes. *Proceedings of the National Academy of Sciences of the United States of America, 109*, 12734–12739.
Wang, S., O'Brien, T., Pava-Ripoll, M., & St. Leger, R. J. (2011). Local adaptation of an introduced transgenic insect fungal pathogen due to new beneficial mutations. *Proceedings of the National Academy of Sciences of the United States of America, 108*, 20449–20454.
Wang, Z. L., Ying, S. H., & Feng, M. G. (2013). Recognition of a core fragment of *Beauveria bassiana* hydrophobin gene promoter (P hyd1) and its special use in improving fungal biocontrol potential. *Microbial Biotechnology, 6*, 27–35.

Warren, G. W. (1997). Vegetative insecticidal proteins: novel proteins for control of corn pests. In N. B. Carozzi, & M. Koziel (Eds.), *Advances in insect control: The role of transgenic plants*. London, United Kingdom: Taylors & Francis Ltd.

Wisecaver, J. H., Slot, J. C., & Rokas, A. (2014). The evolution of fungal metabolic pathways. *PLoS Genetics, 10*, e1004816.

Xiao, G., Ying, S. H., Zheng, P., Wang, Z. L., Zhang, S., Xie, X. Q., et al. (2012). Genomic perspectives on the evolution of fungal entomopathogenicity in *Beauveria bassiana. Scientific Reports, 2*, 483.

Xie, M., Zhang, Y. J., Zhai, X. M., Zhao, J. J., Peng, D. L., & Wu, G. (2015). Expression of a scorpion toxin gene BmKit enhances the virulence of *Lecanicillium lecanii* against aphids. *Journal of Pest Science, 88*, 637–644.

Xie, X. Q., Wang, J., Huang, B. F., Ying, S. H., & Feng, M. G. (2010). A new manganese superoxide dismutase identified from *Beauveria bassiana* enhances virulence and stress tolerance when overexpressed in the fungal pathogen. *Applied Microbiology and Biotechnology, 86*, 1543–1553.

Xu, C., Zhang, X., Qian, Y., Chen, X., Liu, R., Zeng, G., et al. (2014). A high-throughput gene disruption methodology for the entomopathogenic fungus *Metarhizium robertsii. PLoS One, 9*, e107657.

Yang, L., Keyhani, N. O., Tang, G., Tian, C., Lu, R., Wang, X., et al. (2014). Expression of a toll signaling regulator serpin in a mycoinsecticide for increased virulence. *Applied and Environmental Microbiology, 80*, 4531–4539.

Ying, S. H., & Feng, M. G. (2011). Integration of *Escherichia coli* thioredoxin (trxA) into *Beauveria bassiana* enhances the fungal tolerance to the stresses of oxidation, heat and UV-B irradiation. *Biological Control, 59*, 255–260.

Ying, S. H., Feng, M. G., & Keyhani, N. O. (2013). Use of uridine auxotrophy (ura3) for markerless transformation of the mycoinsecticide *Beauveria bassiana. Applied Microbiology and Biotechnology, 97*, 3017–3025.

Zhang, D. X., Lu, H. L., Liao, X., St. Leger, R. J., & Nuss, D. L. (2013). Simple and efficient recycling of fungal selectable marker genes with the Cre-loxP recombination system via anastomosis. *Fungal Genetics and Biology, 61*, 1–8.

Zhao, H., Xu, C., Lu, H. L., Chen, X., St. Leger, R. J., & Fang, W. (2014). Host-to-pathogen gene transfer facilitated infection of insects by a pathogenic fungus. *PLoS Pathogens, 10*, e1004009.

CHAPTER SIX

Molecular Genetics of *Beauveria bassiana* Infection of Insects

A. Ortiz-Urquiza and N.O. Keyhani[1]
University of Florida, Gainesville, FL, United States
[1]Corresponding author: E-mail: keyhani@ufl.edu

Contents

Abstract

Research on the insect pathogenic filamentous fungus, *Beauveria bassiana* has witnessed significant growth in recent years from mainly physiological studies related to its insect biological control potential, to addressing fundamental questions regarding the underlying molecular mechanisms of fungal development and virulence. This has been in part due to a confluence of robust genetic tools and genomic resources for the fungus, and recognition of expanded ecological interactions with which the fungus engages. *Beauveria bassiana* is a broad host range insect pathogen that has the ability to form intimate symbiotic relationships with plants. Indeed, there is an increasing

Advances in Genetics, Volume 94
ISSN 0065-2660
http://dx.doi.org/10.1016/bs.adgen.2015.11.003

realization that the latter may be the predominant environmental interaction in which the fungus participates, and that insect parasitism may be an opportunist lifestyle evolved due to the carbon- and nitrogen-rich resources present in insect bodies. Here, we will review progress on the molecular genetics of *B. bassiana*, which has largely been directed toward identifying genetic pathways involved in stress response and virulence assumed to have practical applications in improving the insect control potential of the fungus. Important strides have also been made in understanding aspects of *B. bassiana* development. Finally, although increasingly apparent in a number of studies, there is a need for progressing beyond phenotypic mutant characterization to sufficiently investigate the molecular mechanisms underlying *B. bassiana's* unique and diverse lifestyles as saprophyte, insect pathogen, and plant mutualist.

1. INTRODUCTION

From the initial modern investigations of insect pathology conducted by the 19th-century Italian scientist Agostino Bassi, after whom the fungus received its eponym, to the recent elucidation of the genomic sequence of *Beauveria bassiana*, there has been significant interest in organisms that can parasitize and kill insects as means for pest control (Glare et al., 2012; Goettel, Hajek, Siegel, & Evans, 2001; Hajek & Delalibera, 2010; Inglis, Goettel, Butt, & Strasser, 2001; Jackson, Dunlap, & Jaronski, 2010; Jaronski, 2010; Leathers, Gupta, & Alexander, 1993). These fungi are also tractable systems with which to explore evolutionary and environmental processes impacting the life history traits of both the fungal pathogen and their target insect hosts (Boomsma, Jensen, Meyling, & Eilenberg, 2014; de Crecy, Jaronski, Lyons, Lyons, & Keyhani, 2009; Meyling & Hajek, 2010; Roy, Steinkraus, Eilenberg, Hajek, & Pell, 2006; Vega et al., 2009). *Beauveria* species are readily cultivatable, facultative insect pathogens that can grow as saprophytes and can easily be isolated from almost all environmental ecosystems. Bait protocols using various insect larvae and environmental sampling from field-harvested insect cadavers have led to the isolation of thousands of *Beauveria* isolates from all major insect orders, as well as across Arthropoda including ticks and mites (see strain collection (Humber, 2015) and isolates used in the phylogenetic characterization of *Beauveria* (Rehner & Buckley, 2005; Rehner et al., 2011), also (Kirkland, Cho, & Keyhani, 2004; Kirkland, Westwood, & Keyhani, 2004; Zimmermann, 1986)). It has long been recognized that most strains, even if isolated from a particular insect host, display a broad host range across insect (and even arthropod) orders, although strain variation in virulence is often reported (Wraight, Ramos, Avery, Jaronski, & Vandenberg, 2010). The extent to which variation in virulence is an intrinsic property of individual strains is unclear and can be confounded by issues of passage (Brownbridge, Costa, & Jaronski, 2001; Samsinakova & Kalalova, 1983). It is known, for example, that strains displaying low virulence against a particular insect, if (re)-isolated from the cadaver of the target insect, will then display increased virulence against that particular host, ie, passage through an insect can (but not always) lead to increased virulence (Song & Feng, 2011). Conversely, repeated culturing on synthetic media can lead to decreased (but reversible) virulence (Safavi, 2011). Despite this degree of plasticity in host targeting, a picture of some of the critical fungal processes involved in insect virulence and how this knowledge may be applied toward improving the insect biological control properties of the fungus has emerged (Ortiz-Urquiza & Keyhani, 2013, 2014; Ortiz-Urquiza, Luo, & Keyhani, 2015). With the development and application of molecular techniques, numerous publications have examined the roles of a wide range of genes and their protein products, not only in relation to insect pathogenesis, but also as they may contribute to fungal development and stress response. It is these recent molecular characterizations that will be the focus of this chapter. Although outside the scope of this chapter, *B. bassiana*, from a phylogenetic and evolutionary perspective, is closely related to plant-associated fungi and appears to be able to form stable and/or mutualistic interactions with certain plants, either generally in the soil, within the plant phylloplane, and/or potentially as a plant endophyte. Here, we will provide a description of the infection process, and then present a systematic survey of the state-of-the-art molecular and physiological studies of *B. bassiana*, focusing on work regarding the molecular dissection of genes implicated in fungal development and virulence. A number of milestones in the genomic examination of *B. bassiana* include isolating genomic and mitochondrial DNA (Pfeifer & Khachatourians, 1989), DNA karyotyping and telomeric fingerprinting (Viaud, Couteaudier, Levis, & Riba, 1996), the elucidation of the mitochondrial DNA sequence (Jin-Zhu, Bo, Chang-Sheng, & Zeng-Zhi, 2009; Pfeifer, Hegedus, & Khachatourians, 1993) (including a recent report of the mitochondrial sequence of *Beauveria pseudobassiana* (Oh, Kong, & Sung, 2015)), identification of transposable elements (Maurer, Rejasse, Capy, Langin, & Riba, 1997), and has culminated with the publication of the genomic sequence (Xiao et al., 2012), with much of the genetic dissection in *B. bassiana* having occurred in the last 5 years. A summary of the genes examined thus far in *B. bassiana*, their known and/or putative functions, and phenotype of mutants (typically via targeted gene knockout) is given in Table 1.

Table 1 Molecular dissection: gene knockouts in *Beauveria bassiana*[a]

Gene	Function	Mutant phenotype	References
Cuticle targeting and other enzymes			
Chit1	Chitinase	No knockout mutant available.	Fang et al. (2005)
Chit2	Chitinase	No publication yet.	
Cdep1	Protease	No knockout mutant available.	Zhang et al. (2008)
Cyp52X1	Cytochrome P450 (fatty acid hydroxylase)	*ΔBbcyp52X1*: (1) reduced virulence if applied topically. (2) Impaired germination on insect epicuticle. (3) No growth phenotype on rich media, minimal media, or minimal media supplemented with alkanes or fatty acids.	Zhang et al. (2012)
Cyp p450s	Putative alkane oxidation	*ΔBbcyp655C1*, *ΔBbcyp5337A1*, *ΔBbcyp52G11*, *ΔBbcyp616N1*, and *ΔBbcyp53A26* displayed no obvious phenotype. No knockout mutants available for *BbCyp539B5* and *BbCyp884Q1*.	Pedrini et al. (2013)
Insect defense			
BqrA	Benzoquinone reductase	*ΔBbBqrA* showed impaired ability to (1) germinate and grow in the presence of 1,4-benzoquinone. (2) Infect beetles producing benzoquinone-containing cuticular secretions. Overexpression of *BbBqrA* led to increased virulence versus the red flour beetle, *Tribolium castaneum*, but not to any of the non-quinone-producing insects tested.	Pedrini et al. (2015)

Stress response			
CatA	Spore-specific catalase	*ΔBbCatA*: (1) Colony tolerance[a] to oxidative stress (H_2O_2 and menadione). (2) Conidial sensitivity to wet heat (at 45°C), UV-B, and H_2O_2. (3) Decrease in virulence if applied topically.	Wang, Zhang, et al. (2013)
CatB	Secreted catalase	*ΔBbCatB766*: (1) Colony tolerance to H_2O_2 and menadione. (2) Conidial tolerance to wet heat and reduced conidial tolerance to UV-B and H_2O_2. (3) No decrease in virulence in topical applications.	
CatC	Cytoplasmic catalase	*ΔBbCatC*: (1) Reduced colony and conidial tolerance to H_2O_2. (2) Conidial tolerance to heat and UV-B irradiation, and colony tolerance to menadione. (3) No decrease in virulence in topical applications.	
CatD	Secreted/peroxisomal catalase	*ΔBbCatD*: (1) Reduced colony tolerance to H_2O_2 and complete tolerance to menadione. (2) Reduced conidial tolerance to H_2O_2, heat, and UV-B irradiation. (3) Decrease in virulence when applied topically.	
CatP	Peroxisomal catalase	*ΔBbCatP*: (1) Very reduced colony and conidial tolerance to H_2O_2. (2) Conidial tolerance to UV-B irradiation and colony tolerance to menadione oxidation. (3) Conidial tolerance to wet heat (at 45°C). (4) Decrease in virulence in topical applications.	

(Continued)

Table 1 Molecular dissection: gene knockouts in *Beauveria bassiana*[a]—cont'd

Gene	Function	Mutant phenotype	References
Sod1	Cytosolic Cu/Zn superoxide dismutase	*ΔBbSod1*: (1) Delay in sporulation and increased conidial viability (shorter mean germination time $GT_{50})^a$. (2) No effect on colony growth. (3) Increased colony sensitivity to H_2O_2 and menadione and tolerance to NaCl and Congo red. (4) Conidial sensitivity to UV-A, UV-B, H_2O_2, and menadione. (5) Slightly decrease in virulence after topical application, greater decrease in virulence when conidia were injected.	Li et al. (2015a)
Sod2	(Cytosolic) Mn^{2+}-superoxide dismutase	*ΔBbSod2*: (1) Delayed colony growth and sporulation. (2) Increased blastospores and colony sensitivity to menadione and H_2O_2. (3) Tolerance to NaCl and heat (45°C). (4) Reduced conidial viability in the presence of UV-A, UV-B, menadione, and H_2O_2. (5) Decreased virulence when applied topically.	Xie et al. (2012) and Xie et al. (2010)
Sod3	(Mitochondrial) Mn^{2+}-superoxide dismutase	*ΔBbSod3* exhibited the same phenotype as *ΔBbSod2*.	
Sod2-Sod3	Cytosolic and mitochondrial Mn^{2+}-superoxide dismutase (double RNAi mutant)	*BbSod2-Sod3-DKD* (RNAi) showed an additive effect on the same phenotypes exhibited by *ΔBbSod2* and *ΔBbSod2*.	
Sod4	Mitochondrial Fe-superoxide dismutase	*Sod4-KD* (RNAi mutant): (1) No effect on colony growth, sporulation or conidial viability (GT_{50}). (2) Increased colony sensitivity to H_2O_2 and menadione and tolerance to NaCl and Congo red. (3) Reduced conidial tolerance to UV-A, UV-B, H_2O_2, and menadione. (4) No decrease in virulence in topical applications.	Li et al. (2015a)

Sod5	Cell wall-anchored Cu/Zn superoxide dismutase	*ΔBbSod5* displayed the same phenotype as *BbSod1*.	
Trx1	Thioredoxin (cytoplasmic)	*ΔBbTrx1*: (1) No differences compared to the wild type, on sporulation, conidial viability (GT_{50}), and conidial tolerance to heat and UV-B. (2) Decreased virulence when conidia were injected and topically applied.	Zhang et al. (2015)
Trx2	Thioredoxin (cytoplasmic)	*ΔBbTrx1*: (1) Reduced sporulation and conidial viability (increased GT_{50}). (2) Conidia appeared to be tolerant to heat but not to UV-B. (3) Decrease in virulence when conidia were injected and topically applied.	
Trx3	Thioredoxin (cytoplasmic)	*ΔBbTrx3*: (1) No differences compared to the wild type on sporulation, conidial viability (GT_{50}), and conidial tolerance to UV-B. (2) Decreased conidial viability with heat. (3) Decrease in virulence when conidia were injected and topically applied.	
Trx4	Thioredoxin (cytoplasmic)	*ΔBbTrx4*: (1) Increased conidial yield and increased conidial viability (shorter GT_{50}). (2) No differences in conidial viability compare to the wild type, when conidia were exposed to heat or UV-B. (3) Decrease in virulence when conidia were both injected and topically applied.	
Trx5	Thioredoxin (nuclear membrane)	*ΔBbTrx5*: (1) No differences compared to the wild type on sporulation, conidial viability (GT_{50}), or conidial tolerance to UV-B. (2) Reduced conidial tolerance to heat. (3) Reduced virulence when conidia were both injected and topically applied.	

(Continued)

Table 1 Molecular dissection: gene knockouts in *Beauveria bassiana*[a]—cont'd

Gene	Function	Mutant phenotype	References
Trx6	Thioredoxin (mitochondria)	*ΔBbTrx6*: (1) No differences on sporulation and conidial viability (GT_{50}) when compare to the wild type. (2) increased conidial sensitivity to heat and UV-B. (3) Reduced virulence when conidia were injected or topically applied.	
Ohmm	Mitochondrial membrane protein (oxidative stress)	*ΔBbOhmm*: (1) Greater tolerance to osmotic stress (sorbitol and NaCl) and oxidative stress (H_2O_2 and menadione). (2) Abnormal conidiogenesis, such as reduced sporulation, accumulation of ROS, thicker walls with increased contents of sugars. (3) Increased conidial viability (shorter GT_{50}) with germination tubes growing faster. (4) Increase in virulence when conidia where injected or topically applied.	He et al. (2014)
Cp15	Conidial protein	*Cp15-KD* (RNAi): (1) Reduced tolerance to heat (48°C) and oxidative stress (menadione). (2) Tolerance to UV-B and osmotic stress (NaCl).	Ying and Feng (2011)
Signal transduction			
Rgs1	Regulator of G-protein signaling	*ΔBbRGS1*: (1) Colony collapse after 15 days growing in M100 plates. (2) No effect on cell wall integrity when stained with calcofluor white. (3) Reduced mycelia growth and sporulation on artificial media and insect cadavers. However, conidia germinated faster on YPD broth or insect cuticle media. Ability to germinate on minimal media, while the WT could not. (4) Longer hyphae. (5) Reduced conidial tolerance to heat and reduced conidial viability over time (reduced shelf life).	Fang et al. (2008)

Csa1	(Neuronal) calcium sensor-1	*ΔBbCsa1*: (1) Delayed acidification of the media when growing only in CZB. No major changes in the production of organic acids (ie, citrate, oxalate, formate, and lactate). Downregulation of the plasma membrane proton pump only when grown in CZB. (2) Reduced growth in PDA and CZA supplemented with $CaCl_2$ and sorbitol. No significant phenotype observed with calcofluor white, Congo red, KCl, SDS, or EDTA. (3) Impairment of normal vacuole morphology and development. Germinating conidia showed a greater number of vacuoles and 40% more enlarged vacuoles than the WT. Vacuoles in hyphal bodies varied more in size.	Fan et al. (2012)
Mpk1	MAPK, YERK1 family	*ΔBbMpk1*: (1) Decreased ability to adhere to the insect cuticle and form appressoria on it. (2) Reduced sporulation, but no effect on germination or colony growth. (3) Slight impairment of the hyphae to breach the liquid interface. (4) Complete loss of virulence when conidia were applied topically, but not when conidia were injected into the hemocoel. *ΔBbmpk1* was unable to penetrate the cuticle outward and sporulate on the cadavers. (5) No alteration of the expression profiles of cuticle-degrading enzymes.	Zhang, Zhang, et al. (2010)

(Continued)

Table 1 Molecular dissection: gene knockouts in *Beauveria bassiana*[a]—cont'd

Gene	Function	Mutant phenotype	References
Hog1	MAPK	*ΔBbHog1*: (1) Lower conidial yield and conidial viability (longer GT_{50}). Impaired ability of the conidia to adhere the insect cuticle and to form appressoria. (2) Reduced tolerance to temperature (32°C) and oxidative stress (H_2O_2) and remarkable tolerance to the fungicide fludioxonil. (3) Sensitivity to osmotic stress (altered hyphae, swollen and vacuolated). Exposition to osmotic stress did not affect intracellular accumulation of trehalose, mannitol, and glycerol. However, osmotic stress did reduced intracellular accumulation of erythritol and arabitol. (4) Decreased in virulence when conidia were applied topically. (5) Downregulation of *hyd1* and *hyd2*. No alteration of the expression profiles of the cuticle-degrading enzymes BbChit1, Pr1, or CDEP1.	Zhang et al. (2009)
Slt2	MAPK (cell wall integrity)	*ΔBbSlt2*: (1) Reduced growth, sporulation, and conidial viability (longer GT_{50}) when grown in minimal media. In rich media, *ΔBbSlt2* grew and sporulated faster. Shorter shelf life at 4°C. (2) Conidia showed altered cell wall and composition. Conidia appeared swollen with distorted cell wall and alterations in cell surface carbohydrate epitopes (ie, α-glucose and α-N-acetylglucosamine). (3) Hypersensitivity to Congo red, osmotic stress (NaCl and sorbitol), UV-B, menadione, H_2O_2, and the fungicides carbendazim, tricyclazole, and	Chen et al. (2014b) and Luo et al. (2012)

		ethirimol. Tolerance to heat. Greater susceptibility to cell wall lytic enzymes. Driselase caused abundant release of protoplast from mycelia. No difference on germ tubes or hyphae when stained with calcofluor white. (4) No observed effect on intracellular accumulation of mannitol, erythritol, arabitol, and glycerol with temperature or sorbitol. In general, lower content of trehalose compared to the WT. However, sorbitol and temperature increased the levels of trehalose in *ΔBbSlt2*. This effect is due to changes in the expression of trehalase Ntl1. (5) Decreased virulence when conidia were applied topically or injected.	
Mkk1	MAPKK (CWI)	*ΔBbMkk1*: (1) Reduced growth, sporulation, and conidial viability when grown in minimal media. In rich media, *ΔBbMkk1* grew and sporulated faster. Although, conidia exhibited shorter shelf life at 4°C. (2) Hypersensitivity to Congo red, SDS and caffeine, osmotic stress (NaCl and sorbitol), UV-B, H_2O_2, menadione, heat, and the fungicides carbendazim, tricyclazole, and ethirimol. Greater susceptibility to cell wall lytic enzymes than the WT parent. (3) Distorted cell wall in hyphae and conidia. However, only minimal changes were observed on conidial surface carbohydrate epitopes. (4) Reduced intracellular levels of mannitol and trehalose. (5) Decreased virulence when conidia were applied topically and injected.	Chen et al. (2014b)

(Continued)

Table 1 Molecular dissection: gene knockouts in *Beauveria bassiana*[a]—cont'd

Gene	Function	Mutant phenotype	References
Bck1	MAPKKK (CWI)	*ΔBbBck1*: (1) No effect on growth when grown in minimal media. In rich media, *ΔBbBck1* grew and sporulated faster. Although, conidia exhibited shorter shelf life at 4°C. (2) Hypersensitivity to Congo red, SDS and caffeine, osmotic stress (NaCl and sorbitol), UV-B, H_2O_2, menadione, heat, and the fungicides carbendazim and tricyclazole. Greater susceptibility to cell wall lytic enzymes than the WT parent. (3) Distorted cell wall in hyphae and conidia with changes on conidial surface carbohydrate epitopes. (4) Reduced intracellular levels of mannitol and trehalose. (5) Decreased virulence when conidia were applied topically and injected.	Chen et al. (2014b)
Ssk1	His to asp phosphorelay protein (feeds into Ssk2 → Pbs2 → Hog1)	*ΔBbSsk1*: (1) Reduced conidial yield. (2) Slight sensitivity to UV-B and some metal ions such as Ca^{2+}, Fe^{3+}, and Zn^{2+}. Reduced tolerance to osmotic stress (NaCl and KCl), oxidative stress (H_2O_2 and menadione), and fungicides (carbendazim and dimethachlone). Sensitivity to cell wall stressors (Congo red, calcofluor white, and SDS). (3) Decreased virulence when conidia were applied topically.	Wang, Li, Li, and Feng (2014)
Ac	Adenylate cyclase	*ΔBbAc*: (1) Generalized reduced colony growth on rich, minimal, and media supplemented with cAmp, and different sources of carbon and nitrogen. (2) Increased conidiation, with conidia exhibiting reduced viability (longer GT_{50}). Tolerance of conidia	Wang, Zhou, et al. (2014)

		to wet heat (45°C) and slight intolerance to UV-B. Reduced conidial tolerance to NaCl, H_2O_2, menadione, and Congo red. (3) Low colony tolerance to osmotic stress (NaCl), oxidative stress (H_2O_2 and menadione), wall stressors (Congo red and EDTA), and metal ions (Cu^{2+}, Mn^{2+}, Fe^{3+}, and Zn^{2+}). (4) Decreased in virulence when conidia were applied topically.	
Gpcr	G-protein-coupled receptor	*ΔBbGpcr*: (1) No effect on growth on SDAY. Reduced growth on various media containing different carbon or nitrogen sources. (2) Generalized reduced conidiation on media containing different sugars. (3) Low tolerance to osmotic stress (NaCl, sorbitol), oxidative stress (H_2O_2, menadione), and cell wall stress (Congo red). (4) Downregulation of heat shock and antioxidant factors, compatible solute-forming enzymes, and cell wall biosynthesis/remodeling proteins. (5) Inability to produced blastospores in liquid media supplemented with different sugars (glucose, trehalose, sucrose, fructose, glycerol, and lactate). Addiction of cAmp did not increase blastospores production. This phenomenon seemed to worsen as glucose concentration increased. (6) Reduced virulence in topical assays but not in intrahemocoel injection assays.	Ying et al. (2013a)

(Continued)

Table 1 Molecular dissection: gene knockouts in *Beauveria bassiana*[a]—cont'd

Gene	Function	Mutant phenotype	References
CnA1	Calcineurin catalytic subunit A 1 (calcium/calmodulin-activated serine/threonine protein phosphatase)	*ΔBbCnA1*: (1) Reduced growth on SDAY. (2) Reduced conidiation, but no effect on conidial viability. Alteration on conidial cell surface carbohydrate epitopes. (3) Sensitivity to metal ions (Ca^{2+}, Zn^{2+}, Mn^{2+}, and Mg^{2+}), oxidative stress (H_2O_2, menadione), Cell wall stressors (Congo red, calcofluor white, SDS), and fungicides (carbendazim). Tolerance to osmotic stress (KCl, NaCl). (4) Increased conidial sensitivity to wet heat (45°C) and UV-B. (5) Reduced accumulation of intracellular mannitol and trehalose. (6) Downregulation of superoxide dismutases and catalases. (7) Reduced virulence in topical assays.	Li, Wang, et al. (2015)
CnA2	Calcineurin catalytic subunit A 1	(Calcium/calmodulin-activated serine/threonine protein phosphatase) *ΔBbCnA2* showed the same phenotype as *ΔBbCnA1*.	
CnB	Calcineurin regulatory subunit B	*ΔBbCnB* exhibited essentially the same phenotype as *ΔBbCnA1* and *ΔBbCnA2* with two differences: (1) No alterations on conidia cell surface carbohydrate epitopes. (2) Abnormal branching of germinating conidia.	
Crz1	Downstream calcineurin-responsive zinc finger transcription factor	*ΔBbCrz1* displayed the same phenotype as *ΔBbCnA1* and *ΔBbCnA2* with two differences: (1) No effect on growth on SDAY. (2) Slightly reduced conidial viability (longer GT_{50}).	

Ras1	Ras GTPase	*ΔBbRas1* construction failed in multiple attempts. *mRas1*: (WT expressing *Ras1*). No apparent phenotype observed. *mRas1*^{G19V}: (WT expressing a dominant active form of *Ras1* (*Ras1*^{G19V})). (1) No growth phenotype on SDAY. (2) Reduced conidial viability (increased GT_{50}). (3) Germ tubes slimmer and curlier at the tips. (4) Partial tolerance to oxidative stress (sensitive to H_2O_2 but not to menadione). Tolerant to carbendazim. Reduced tolerance to cell wall stressors (Congo red and SDS) and to osmotic stress (NaCl). (5) Conidial tolerance to carbendazim and wet heat (45°C). Reduced conidial tolerance to oxidative stress (menadione and H_2O_2), cell wall stressors (Congo red and SDS), osmotic stress (NaCl), and UV-B/UV-A. (6) Dramatic reduce in virulence in topical assays. *mRas1*^{D126A}: (WT expressing a dominant negative form of *Ras1* (*Ras1*^{D126A})). (1) Reduced growth on SDAY. (2) Germ tubes slimmer and curlier at the tips. (3) Greater tolerance to oxidative stress (menadione and H_2O_2), cell wall stressors (Congo red and SDS), and carbendazim. Tolerance to osmotic stress (NaCl). (4) Greater conidial tolerance to oxidative stress and cell wall stressors. Conidial tolerance to carbendazim, NaCl, UV (A and B), heat. (5) No effect on virulence observed in topical assays.	Xie et al. (2013)

(Continued)

Table 1 Molecular dissection: gene knockouts in *Beauveria bassiana*[a]—cont'd

Gene	Function	Mutant phenotype	References
Ras2	Ras GTPase	*ΔRas2*: (1) No growth phenotype on SDAY. (2) Decreased conidial viability (increased GT_{50}). (3) Two or more germ tubes budding from each conidium, abnormal branching, and growth of hyphae. (4) Reduced tolerance to oxidative stress, cell wall stressors, and osmotic stress. Tolerance to carbendazim. (5) Increased conidial sensitivity to oxidative stress, cell wall stressors, osmotic stress, and UV-B/UV-A. No apparent sensitivity of conidia to carbendazim and heat. (6) Decreased virulence in topical assays. *ΔRas2/Ras1*: (1) No growth phenotype on SDAY. (2) Decreased conidial viability. (3) Two or more germ tubes budding from each conidium, abnormal branching, and hyphal growth. (4) Low tolerance to oxidative stress and osmotic stress. No apparent sensitivity to cell wall stressors and carbendazim. (5) Conidial intolerance to menadione and H_2O_2, cell wall stressors osmotic stress, and UV (A and B) irradiation. Complete tolerance to carbendazim and heat. (6) Decreased virulence in topical assays. *ΔRas2/Ras1*^{G19V}: (1) No growth phenotype on SDAY. (2) Decreased conidial viability. (3) Two or more germ tubes budding from each conidium, abnormal branching, and hyphal growth. (4) Reduced tolerance to oxidative stress, cell wall	Xie et al. (2013)

		stressors, osmotic stress, and carbendazim. (5) Conidial sensitivity to oxidative stress, cell wall stressors, osmotic stress, and UV-B/UV-A irradiation. No apparent sensitivity to heat or carbendazim. (6) Dramatic decrease in virulence in topical assays. Δ*Ras2/Ras1*^{D126A}: (1) No growth phenotype on SDAY. (2) Two or more germ tubes budding out from each conidium, abnormal branching, and hyphal growth. (3) Tolerance to oxidative stress, cell wall stressors, and osmotic stress. Greater tolerance than the WT to carbendazim. (4) Conidial sensitivity to menadione but not to H_2O_2. Conidial sensitivity to osmotic stress. Conidial tolerance to cell wall stressors, carbendazim, and UV-B/UV-A irradiation. (5) Decreased virulence in topical application.	
Ras3	Ras GTPase	Δ*Ras3*: (1) Greater growth on SDAY than the WT. Slightly reduced growth on minimal media. (2) Reduced conidial yield and conidial viability. Conidial sensitivity to heat and UV-B irradiation. (3) Reduced colony tolerance to oxidative stress, osmotic stress, cell wall stressors, carbendazim, and heat (32°C). (4) Downregulation of superoxide dismutases, catalases, and Hog1-related genes. (5) Decreased virulence in topical assays and intrahemocoel injections.	Guan et al. (2015)

(Continued)

Table 1 Molecular dissection: gene knockouts in *Beauveria bassiana*[a]—cont'd

Gene	Function	Mutant phenotype	References
Snf1	SNF1/AMPK protein kinase	*ΔBbSnf1*: (1) No growth effect on SDAY. (2) Significant reduction in number of conidiophores and conidial yield. Conidial yield was significantly increased with exogenous addition of cAmp, but not in a concentration-dependent manner. (3) Reduction in the intracellular content of seven amino acids (arginine, alanine, valine, phenylalanine, lysine, leucine, and glutamic acid). Addition of exogenous amino acids partially recovered conidiation.	He, Wang, Chu, Feng, and Ying (2015) and Wang, He, Feng, and Ying (2014)
Phy	Phytochrome (group VIII histidine kinase N-terminal photosensory core)	*ΔBbPhy*: (1) Reduced growth on rich media and increased growth on several minimal media. Light/dark cycle did not affect growth in rich or minimal media. (2) Reduced conidial yield under full-day dark, full-day red, and full-day far-red lighting conditions. Full-day blue and full-day white light did not affect conidial yield. In summary, exposure to dark light significant reduced conidial yield, while increasing exposure to white light restored conidiation to the WT levels. (3) Low conidial tolerance to wet heat and UV-B. (4) Reduced tolerance to osmotic stress (NaCl and sorbitol), oxidative stress (H_2O_2 and menadione). Little sensitivity to cell wall stressors (Congo red and SDS). Tolerance to fungicides (carbendazim and iprodione). (5) Upregulation, in general, of	Qiu et al. (2014)

		superoxide dismutases and catalases in the presence of H_2O_2. Downregulation of *Ypd1*, *Ssk1*,*hog1*, and *Msn2* in the presence of NaCl. (6) No decrease in virulence when conidia were either applied topically or injected.	
Protein turnover			
Atg5	Autophagy, autophagosome formation	*ΔBbAtg5*: (1) Normal growth in SDAY or OMA media and reduced growth in minimal or starvation media. (2) Reduced conidial yield on rich media. Conidia viability was affected by nutrient limitation. Enlarged and distorted conidia. (3) Reduced ability to produce blastospores in liquid media. (4) Decreased virulence in topical assays.	Zhang et al. (2013)
Lipid storage			
Cal1	Caleosin, lipid body formation/maintenance	*ΔBbCal1*: (1) No effect on (a) growth or colony morphology on SDAY, PDA, and CZA, (b) stress response (osmotic, cell wall), or (c) germination. (2) More compact assemblage of conidia. Reduced and delayed dispersal of conidia. (3) No changes observed in lipid droplets with Nile red. Changes in vacuole and endoplasmic reticulum/multilamellar vesicle-like structures. (4) Altered cellular lipid profiles in conidia grown on PDA and PDA supplemented with olive oil, glyceryl trioleate, oleic acid, and C_{16}. (5) Decreased virulence in topical assays.	Fan et al. (2015)

(*Continued*)

Table 1 Molecular dissection: gene knockouts in *Beauveria bassiana*[a]—cont'd

Gene	Function	Mutant phenotype	References
Glycosyltransferases			
Gas1	GPI-anchored glucanosyltransferase	*ΔBbGas1*: (1) No apparent growth phenotype on different culture media. (2) Tolerance to some wall stressors (calcofluor white and SDS). Sensitiveness to Congo red. Tolerance to sorbitol and reduced tolerance to KCl (colony collapse phenotype noted at high concentrations of KCl). (3) Alteration of conidial surface carbohydrate epitopes. Conidia with loosened or frayed cell wall. Reduced conidial viability at temperatures >40°C. (4) Decreased virulence in topical and intrahemocoel injection assay.	Zhang, Xia, and Keyhani (2011)
Ktr1/Mnt1	α-1,2-Mannosyltransferase	*ΔBbKtr1*: (1) Reduced growth on SDAY, sucrose-free CZA, and $NaNO_3$-free CZA. (2) Reduced conidial yield. Conidia were smaller and more sensitive to cell wall stressors (Congo red, SDS, calcofluor white). Conidia and hyphae appeared to have thinner cell walls. (3) No alteration on surface epitopes observed with GNL and WGA carbohydrate-binding lectins. Decreased conidial hydrophobicity. Reduced conidial viability under osmotic and oxidative stress (NaCl, sorbitol, and menadione) H_2O_2 did not impair germination of *ΔBbKtr1* conidia. Increased conidial sensitivity to wet heat (45°C) and tolerance	Wang, Qiu, Cai et al. (2014)

		to UV-B. (4) Slight colony sensitivity to Congo red and SDS. Low colony tolerance to calcofluor white. Colony sensitivity to osmotic (NaCl and sorbitol) and oxidative (menadione) stresses (*ΔBbKtr1* exhibited tolerance to H_2O_2). (5) Increased susceptibility to lytic enzymes. (6) No effect on virulence when conidia were applied topically or injected.	
Ktr4/Mnt1	α-1,2-Mannosyltransferase	*ΔBbKtr4*: (1) Reduced growth on rich and several minimal media. (2) Delayed onset of conidiation and very reduced conidial yield. Conidia were smaller and more sensitive to cell wall stressors (Congo red, SDS, calcofluor white). Conidia and hyphae appeared to have thinner cell walls. (3) Alteration on conidial surface carbohydrate epitopes observed with ConA, GNL, and WGA. Decreased conidial hydrophobicity. Reduced conidial viability under osmotic and oxidative stress (NaCl, sorbitol, menadione, and H_2O_2). Reduced conidial tolerance to wet heat (45°C), and UV-B. (4) Slight colony sensitivity to Congo red and SDS. Increased colony sensitivity to calcofluor white. Colony sensitivity to osmotic (NaCl and sorbitol) and oxidative (menadione and H_2O_2) stress. (5) Increased susceptibility to lytic enzymes. (6) Decreased virulence with topical and injection assays.	Wang, Qiu, Cai et al. (2014)

(Continued)

Table 1 Molecular dissection: gene knockouts in *Beauveria bassiana*[a]—cont'd

Gene	Function	Mutant phenotype	References
Kre2/Mnt1	α-1,2-Mannosyltransferase	*ΔBbKre2* exhibited the same phenotype as *ΔBbKtr4*.	Wang, Qiu, Cai et al. (2014))
Pmt1	Protein O-mannosyltransferase	*ΔBbPmt1*: (1) Reduced growth on SDAY, several minimal media with different sources of carbon and nitrogen, and starvation media. (2) Reduced sporulation, decreased conidial hydrophobicity, and reduced conidial viability on normal conditions. Conidial viability was reduced by oxidative (menadione and H_2O_2), osmotic (NaCl and sorbitol), and cell wall (Congo red and SDS) stress. Reduced conidial tolerance to wet heat (45°C) and UV-B. Conidia presented thinner walls and alterations on the wall surface carbohydrate epitopes when stained with ConA, GNL, and WGA. (3) Reduced colony tolerance to oxidative, osmotic, and cell wall stress. (4) Decreased virulence only with topical applications.	Wang, Qiu, Chu et al. (2014)
Pmt2	Protein O-mannosyltransferase	*ΔBbPmt2* failed in numerous attempts. *Pmt2-KD* (RNAi mutant) showed essentially the same phenotype as *ΔBbPmt1*, with one difference. *ΔBbPmt2* exhibited tolerance to H_2O_2.	Wang, Qiu, Chu et al. (2014)
Pmt4	Protein O-mannosyltransferase	*ΔBbPmt4* displayed essentially the same phenotype as *ΔBbPmt1*, with one difference. *ΔBbPmt4* exhibited tolerance to NaCl.	Wang, Qiu, Chu et al. (2014)

Metabolic pathways			
Mpd	Mannitol-1-phosphate dehydrogenase	*ΔBbMpd*: (1) Reduced growth on minimal media supplemented with fructose, mannitol, and sorbitol. Addition of sucrose and glucose to the media did not affect growth. (2) Reduced conidial yield on SDAY. (3) Reduced conidial viability on minimal media supplemented with glucose, fructose, or mannitol. (4) Decreased accumulation of mannitol and increased accumulation of trehalose in mycelia and conidia, under no-stress and stress conditions (oxidative, osmotic, thermal). Downregulation of genes involved in trehalose hydrolysis and biosynthesis. (5) Reduced colony tolerance to oxidative stress (H_2O_2). (6) Increased conidial sensitivity to oxidative stress (H_2O_2), wet heat (45°C), and UV-B irradiation. (7) Slight decrease in virulence in topical applications.	Wang et al. (2012)
Mtd	Mannitol dehydrogenase	*ΔBbMtd* exhibited essentially the same phenotype as *ΔBbmpd*, with two differences: (1) No effect on conidial yield on SDAY. (2) No effect on conidial viability on minimal media supplemented with glucose, fructose, or mannitol.	
Nth1	Neutral trehalose (promoter characterization)	No knockout mutant available. They looked at regions in the promoter regulating stress-inducible expression of the *Nth1*.	Liu et al. (2011)

(Continued)

Table 1 Molecular dissection: gene knockouts in *Beauveria bassiana*[a]—cont'd

Gene	Function	Mutant phenotype	References
Transporter			
Jen1	Carboxylic acids	*ΔBbJen1*: (1) Increased conidial yield and a decreased accumulation of carboxylates (oxalate, citrate, pyruvate, and lactate). (2) Decreased virulence in topical applications. *OvBbJen1* (overexpression mutant): (1) Reduced conidial yield and increased accumulation of carboxylates (oxalate, citrate, pyruvate, and lactate). (2) *OvBbJen1* virulence was not assayed.	Jin et al. (2010)
Agt1	α-Glucoside transporter	*ΔBbAgt1*: (1) Decreased growth on CZ media with different sugars (glucose, sucrose, trehalose, maltose, fructose, mannose, mannitol, and glycerol). (2) Reduced conidial yield and conidial viability on CZ media modified with different sugars. (3) Downregulation of sugar transporters when *ΔBbAgt1* was grown on CZ-trehalose, and upregulation of sugar transporters if the mutant was grown on CZ with mannitol, glucose, or sucrose. (4) Decreased virulence when conidia were topically applied or injected.	Wang, Ji, et al. (2013)
Mdr1	ABC transporter (B)	*ΔBbMdr1* displayed: (1) Reduced growth in the presence of the fungicide dimethachlone. No effect on growth with carbendazim, azoxystrobin, ethirimol or itraconazole. (2) Increased sensitivity to	Song, Zhao, Ying, and Feng (2013)

		the drugs cycloheximide (inhibitor of protein biosynthesis) and 4-nitroquinoline N-oxide (induces DNA damage). (3) Tolerant to Congo red. (4) Reduced tolerance to oxidative stress (H_2O_2 and menadione). (5) No effect on virulence in topical applications.	
Mrp1	ABC transporter (C)	*ΔBbMrp1*: (1) Reduced growth in the presence of the fungicides dimethachlone, carbendazim, and azoxystrobin. No effect on growth with ethirimol or itraconazole. (2) Increased sensitivity to 4-nitroquinoline N-oxide (induces DNA damage) and no effect with cycloheximide. (3) Susceptibility to the cell wall stressor Congo red. (4) Reduced tolerance to oxidative stress (H_2O_2 and menadione). (5) Decreased virulence in topical applications.	Song et al. (2013)
Pdr1	ABC transporter (G)	*ΔBbPdr1*: (1) Reduced growth with the fungicides dimethachlone, carbendazim, ethirimol, and itraconozale. No effect with azoxystrobin was observed. (2) Increased sensitivity to cycloheximide and 4-nitroquinoline N-oxide. (3) Tolerant to the cell wall stressor Congo red. (4) Reduced tolerance to oxidative stress (H_2O_2 and menadione). (5) No effect on virulence in topical applications.	Song et al. (2013)

(Continued)

Table 1 Molecular dissection: gene knockouts in *Beauveria bassiana*[a]—cont'd

Gene	Function	Mutant phenotype	References
Pdr2	ABC transporter (G)	*ΔBbPdr2*: (1) Reduced growth with the fungicides dimethachlone and ethirimol. No effect with carbendazim, azoxystrobin, or itraconazole was observed. (2) Increased sensitivity to cycloheximide and resistance to 4-nitroquinoline N-oxide. (3) Susceptibility to the cell wall stressor Congo red. (4) Reduced tolerance to oxidative stress (H_2O_2 and menadione). (5) No effect on virulence.	Song et al. (2013)
Pdr5	ABC transporter (G)	*ΔBbPdr5*: (1) Reduced growth with the fungicides dimethachlone, carbendazim, and azoxystrobin. No effect with ethirimol or itraconazole was observed. (2) Tolerance to cycloheximide and increased sensitivity to 4-nitroquinoline N-oxide. (3) Sensitivity to the cell wall stressor Congo red. (4) Reduced tolerance to oxidative stress (H_2O_2 and menadione). (5) Decreased virulence in topical applications.	Song et al. (2013)
Transcription/regulation			
CreA	TF, "catabolite repressor"	*ΔBbCrea*: (1) Reduced growth at 26°C on CZA-sucrose, PDA, SDA, CZA-0.5% peptone, and minimal-salt media supplemented with glucose, sucrose, mannose, glycerol, trehalose, raffinose, sorbitol, or glycerol. Colonies looked shrunken and shriveled without fluffy mycelia. Slight reduced growth on minimal-salt media with melibiose, galactose, N-acetylglucosamine, or maltose.	Luo, Qin, et al. (2014)

	Minimal-salt media with arabinose, xylose, or lactose did not show any effect on growth. (2) Even a more reduced effect on growth when media was complemented with peptone, casamino acids, arginine, asparagine, and proline. This defective growth is due to lysis occurring on the germinating conidia. Threonine, lysine, isoleucine, alanine, valine, leucine, histidine, phenylalanine, tyrosine, tryptophan, glutamine, glutamic acid, serine, or glycine did not cause cell lysis. (3) The defective growth and the cell lysis were even more dramatic at 32°C. However, these effects were not seen when *ΔBbCreA* was grown on CZA. (4) Absence of the yellow pigment produced by the fungus on minimal media when the media was supplemented with glucose. (5) Altered acidification phenotype. (6) Slight sensitivity to cell wall stressors (SDA, Congo red, calcofluor white) and osmotic stress (KCl and sorbitol). (7) Earlier onset of conidiation, although conidial yield was reduced. (8) Slower onset of blastospore production. Blastospores showed morphological defects (irregular shape, and transparent/hollow appearance), although propidium iodide staining revealed no phenotype. (9) Derepression of proteases and lipases. (10) Decreased virulence in topical applications and intrahemocoel injections. Impaired ability to penetrate the cuticle outward and failure to sporulate on the cadavers.

(*Continued*)

Table 1 Molecular dissection: gene knockouts in *Beauveria bassiana*[a]—cont'd

Gene	Function	Mutant phenotype	References
Msn2	TF	*ΔBbMsn2*: (1) Reduced growth on rich media (PDA, SDA, or SDAY), minimal-salt media, and CZA-sucrose. Such growth reduction could be partially rescued with sorbitol. Reduced growth on minimal media supplemented with glucose, glycerol, mannitol, erythritol, or trehalose. pH-dependent growth phenotype; poor growth at lower pH (4–5) and maximum growth at pH 7–8. (2) Lower conidial yield on CZA plates. (3) Increased sensitivity to cell wall stressors (calcofluor white), and oxidative stress (H_2O_2). (4) More rapid acidification of the media when grown on CZA plates. No differences on acidification of the media observed in culture supernatants. (5) Derepression of lipases and proteases. (6) Earlier onset of oosporein production. Mutant was able to produce oosporein within a wider range of pH values than the WT. (7) Decreased virulence in topical applications and intrahemocoel injections. Altered and scarce hyphal bodies observed in the hemolymph. Reduced sporulation on cadavers.	Liu et al. (2013) and Luo, Li, et al. (2014)
Fkh2	Forkhead transcription factor	*ΔBbFkh2*: (1) Propidium iodine stained showed altered cell cycle (longer G_2/M and S, but shorter G_0/G_1 phases). (2) Increased conidial yield. However, germination was impaired in general on CZA, CZA + different C and N sources, and starvation media. Conidia were smaller and less dense. Reduced	Wang et al. (2015)

		conidial hydrophobicity. (3) 20% increased in blastospore production on SDB. Blastospores were smaller and less dense. (4) Shorter and wider hyphae with denser septa. Some of them were aberrantly branched. (5) Downregulation of G_2/M transition genes and upregulation of genes require for septum formation. (6) Reduced superoxide and catalase activity. (7) Decreased in intracellular compatible solutes (trehalose and mannitol). (8) Colony sensitivity to oxidative stress (H_2O_2 and menadione), osmotic stress (only sorbitol), and the fungicide carbendazim and dimethachlone. Tolerance to NaCl, Congo red, and SDS. (9) Reduced conidial tolerance to menadione, H_2O_2, NaCl, sorbitol, carbendazim, dimethachlone, temperature (45°C), and UV-B irradiation. Conidial tolerance to Congo red and SDS. (10) Decreased virulence in topical applications and intrahemocoel injections.	
WetA	Developmental regulator	*ΔBbWetA*: (1) Milder reduced growth on ¼ SDAY, dramatic reduced growth on CZA media supplemented with different sources of C and N. (2) Reduced conidial yield and reduced conidial viability on SDAY. (3) Conidial alterations on surface carbohydrate epitopes (observed with ConA, GNL, and WGA). Conidia were less hydrophobic. (4) Decreased virulence in topical applications and intrahemocoel injections. Dramatic decrease in virulence when conidia were injected.	Li et al. (2015b)

(Continued)

Table 1 Molecular dissection: gene knockouts in *Beauveria bassiana*[a]—cont'd

Gene	Function	Mutant phenotype	References
VosA	Velvet protein	*ΔBbVosA* showed basically the same phenotype as *ΔBbWetA*, with one difference. *ΔBbVosA* exhibited no changes in conidial hydrophobicity.	Li et al. (2015b)
Mbf1	Multiprotein-bridging factor-1	*ΔBbMbf1*: (1) Reduced growth on CZA supplemented with glucose, trehalose, NaAc, methionine, proline, or N-acetyl-glucosamine. No differences on growth observed on SDA, or CZA supplemented with sucrose, glycerol, olive oil, oleic acid, gelation, or chitin. (2) Impairment of hyphal morphogenesis and development. Hyphae appeared swollen and distorted 2 days after germination on CZ, CZ-trehalose, CZ-glucose, CZ-sucrose, and CZ-glycerol. Aberrant hyphal morphogenesis phenotype could be rescued on rich media (SDAY). (3) Reduced production of blastospores on CZ-glucose, CZ-sucrose, CZ-trehalose, and CZ-glycerol. This phenomenon was restored on SDAY. (4) Sensitivity to osmotic (NaCl and sorbitol), oxidative (H_2O_2 and menadione), thermal (40°C), and cell wall (Congo red) stress. (5) Low conidial tolerance to UV irradiation and temperature. (6) Decreased virulence in topical applications and intrahemocoel injections.	Ying et al. (2014)
Cell cycle			
Cdc14	Nuclear phosphatase (cytokinesis regulating)	*ΔBbCdc14*: (1) Reduced growth on SDAY, CZA, and CZA with different C and N sources. (2) Reduced conidial and blastospore yield on SDAY and SDB,	Wang, Liu, et al. (2013)

		respectively. Failure to produce zigzag conidiophore clusters and spore balls. (3) Increased number of nuclei observed on hyphae. (4) Colony sensitivity to osmotic stress (NaCl), oxidative stress (menadione and H_2O_2), cell wall stressors (Congo red and SDS), and fungicides (carbendazim). (5) Increased conidial sensitivity to NaCl, H_2O_2, and menadione, Congo red, SDS, temperature, and UV-B. (6) General downregulation of genes related to hyphal development, conidiogenesis, and signal transduction. Downregulation of superoxide dismutases and catalases. (7) Decreased virulence with topical applications.	
Cdk1	Cycle-dependent kinase 1	No knockout mutant available. Disruption of *Cdk1* failed in multiple attempts. *Tcdk1* (Cdk1 was overexpressed in the WT): (1) Propidium iodine stained showed same cell cycle as the wild type. (2) No apparent phenotype seen on hyphae. (3) Reduced blastospore yield and larger blastospores. (4) No effect on growth (SDAY or CZA), conidial yield, conidial viability, or changes in conidium size and branching. (5) Tolerant to NaCl, sorbitol, H_2O_2, menadione, Congo red, SDS. Sensitive to hydroxyurea and carbendazim. (6) Conidial tolerance to heat and UV-B. (7) Decreased virulence in topical assays and intrahemocoel injections.	Qiu et al. (2015)

(Continued)

Table 1 Molecular dissection: gene knockouts in *Beauveria bassiana*[a]—cont'd

Gene	Function	Mutant phenotype	References
		Tcdk1^{AF} (mutated *Cdk1*; The^{14} and Tyr^{15} were mutated to Ala and Phe): (1) Propidium iodine stained showed altered cell cycle (longer G_2/M and G_0/G_1 phases, but shorter S phase). (2) Denser septa and shorter hyphae. (3) Decreased blastospore yields and smaller blastospores. (4) Reduced growth on SDA and CZA. No effect on: conidial yield, conidial viability, or changes in conidium size and branching. (5) Tolerant to NaCl, sorbitol, H_2O_2, menadione, Congo red, SDS, hydroxyurea, and carbendazim. (6) Conidial tolerance to heat and UV-B. (7) Decreased virulence in topical assays and intrahemocoel injections.	
Cdc25	Nuclear phosphatase (cyclin-dependent kinase regulating)	*ΔBbCdc25*: (1) Propidium iodine stained showed altered cell cycle (longer G_0/G_1 phase and shorter G_2/M and S phases. (2) Denser septa and slender hyphae. (3) Abnormal mitosis. (4) Altered transcription profiles of genes involved in septum formation. (5) Decreased blastospore yield and larger blastospores. (6) Reduced growth on SDA and CZA. (7) No effect on growth (SDAY or CZA). Reduced conidial yield and increased conidial viability. Conidia appeared to be larger. No abnormal branching was observed. (8) Tolerant to NaCl, sorbitol, H_2O_2, menadione, Congo red, SDS, hydroxyurea, and carbendazim. (9) Conidial	Qiu et al. (2015)

		tolerance to heat and UV-B. (10) Decreased virulence in topical assays and intrahemocoel injections.	
Wee1	Nuclear kinase (cyclin-dependent kinase regulating)	*ΔBbWee1*: (1) Propidium iodine stained showed altered cell cycle (longer G_0/G_1 phase and shorter G_2/M and S phases. (2) Denser septa and shorter hyphae. (3) Abnormal mitosis. (4) Altered transcription profiles of genes involved in septum formation. (5) Decreased blastospore yield and smaller blastospores. (6) No effect on growth on SDA or CZA. (7) Earlier onset of conidiation. Reduced conidial yield and increased conidial viability. Conidia appeared to be larger. Abnormal conidial branching was observed. (8) Tolerant to NaCl, sorbitol, menadione, and Congo red. Sensitivity to H_2O_2, SDS, hydroxyurea, and carbendazim. (9) Conidial sensitivity to heat and UV-B. (10) Decreased virulence in topical assays, but not with intrahemocoel injections.	Qiu et al. (2015)
Cdc25 Wee1	Nuclear phosphatase (cyclin-dependent kinase regulating) Nuclear kinase (cyclin-dependent kinase regulating)	*ΔBbWee1ΔBbCdc25* (double knockout mutant): (1) Propidione iodine stain showed altered cell cycle (longer G_0/G_1 phase and shorter G_2/M and S phases. (2) Denser septa and very short hyphae. (3) Abnormal mitosis. (4) Altered transcription profiles of genes involved in septum formation. (5) Increased blastospore yield and smaller blastospores. (6) Reduced growth on SDA or CZA. (7) No effect on conidial yield, slower conidiation, decreased	Qiu et al. (2015)

(Continued)

Table 1 Molecular dissection: gene knockouts in *Beauveria bassiana*[a]—cont'd

Gene	Function	Mutant phenotype	References
		viability. Conidia appeared to be smaller with abnormal branching. (8) Tolerant to sorbitol and Congo red. Sensitivity to NaCl, H_2O_2, menadione, SDS, hydroxyurea, and carbendazim. (9) Conidial sensitivity to heat and UV-B. (10) Decreased virulence in topical assays and in intrahemocoel injections.	
Cell wall			
Hyd1	Hydrophobin 1	*ΔBbHyd1*: (1) No effect on growth. (2) Incomplete wettable phenotype. (3) Bald conidia devoid of fascicles and bundles. (4) Reduced hydrophobicity index. No effect on conidial binding to hydrophobic or hydrophilic surfaces. (5) Alterations on conidial surface carbohydrate epitopes (seen with ConA, GNL, and WGA). (6) Tolerance to osmotic stress (KCl, sorbitol) and cell wall stressors (Congo red, calcofluor white, SDS). (7) Increased conidial tolerance to heat shock. (8) Decreased virulence in topical applications. Increased virulence if conidia were injected.	Zhang, Xia, Kim, et al. (2011)
Hyd2	Hydrophobin 2	*ΔBbHyd2*: (1) No effect on growth. (2) Incomplete wettable phenotype. (3) Conidia showed visible bundles, but they were distorted and less organized. (4) Reduced hydrophobicity index. Reduced conidial binding to hydrophobic but not to	Zhang, Xia, Kim, et al. (2011)

		hydrophilic surfaces. (5) Small variations on conidial surface carbohydrate epitopes (seen with ConA, GNL, and WGA). (6) Tolerance to osmotic stress (KCl, sorbitol) and cell wall stressors (Congo red, calcofluor white, SDS). (7) No effect on conidial tolerance to heat shock. (8) No effect on virulence in topical applications. Increased virulence if conidia were injected.	
Hyd1 Hyd2	Double mutant	*ΔBbHyd1 ΔBbHyd2*: (1) No effect on growth. (2) Complete wettable phenotype. (3) Bald conidia devoid of bundles. (4) Reduced hydrophobicity index. Reduced conidial binding to hydrophobic but not to hydrophilic surfaces. (5) Alterations on conidial surface carbohydrate epitopes (seen with ConA, GNL, and WGA). (6) Tolerance to osmotic stress (KCl, sorbitol) and cell wall stressors (Congo red, calcofluor white, SDS). (7) Increased conidial tolerance to heat shock. (8) Decreased virulence in topical applications. Increased virulence if conidia were injected.	Zhang, Xia, Kim, et al. (2011)
Ecm33	GPI-anchored cell wall protein	*ΔBbEcm33*: (1) Reduced growth on SDAY. (2) Reduced conidial yield on SDAY. (3) Alterations on conidial surface carbohydrate epitopes (seen with ConA and WGA). (4) Decrease in intracellular compatible solutes (mannitol and trehalose). (5) Increased colony sensitivity to cell wall stressors (SDS and Congo red), oxidative stress (menadione	Chen, Zhu, Ying, and Feng (2014a)

(Continued)

Table 1 Molecular dissection: gene knockouts in *Beauveria bassiana*[a]—cont'd

Gene	Function	Mutant phenotype	References
		and H_2O_2), osmotic stress (NaCl and Ca_2Cl), and the fungicides carbendazim and ethirimol. (6) Increased conidial sensitivity to wet heat and UV-B. (7) No effect on virulence with topical applications.	
Secondary metabolites			
Oosporein	Ops1-7 biosynthetic cluster. *Ops1*, *Ops2*, *Ops3*, *Ops4*, *Ops5*, *Ops6*, *Ops7*	*ΔBbOps1*: (1) Inability to produce oosporein and orsellinic acid. The addition of orsellinic acid restored the production of oosporein. (2) Decreased virulence. Failure to scape insect host immunity. *ΔBbOps2*: (1) Increased oosporein production and produces orsellinic acid. *ΔBbOps3*: (1) No oosporein or orsellinic acid production. (2) Ablation of *Ops3* disabled the expression of *Ops1*, *Ops2*, *Ops4*, *Ops6*, and *Ops7*. *ΔBbOps4*: No oosporein production. *ΔBbOps5*: No oosporein production. *ΔBbOps6*: No oosporein production. *ΔBbOps7*: No oosporein production.	Feng et al. (2015)
Bassianolide	*Bsls* (nonribosomal peptide synthetase)	*ΔBb*Bsls: (1) Complete impaired production of bassianolide. It did no affect biosynthesis of beauvericin. (2) Dramatic decrease in virulence if applied topically.	Xu et al. (2009)

Beauvericin	Beas (nonribosomal peptide synthetase)	*ΔBb*Beas: (1) Complete impaired production of beauvericin. It did no affect biosynthesis of bassianolide. (2) Dramatic decrease in virulence if applied topically.	Xu et al. (2008), and Xu et al. (2007)
Bassianolide beauvericin	*Kivr*: Keitoisovalerate reductase.	*ΔBb*Kivr: (1) Complete impaired production of beauvericin and bassianolide. (2) Dramatic decrease of virulence if applied topically.	
Tenellin	Pks: Polyketide synthase/ NRPS	*ΔBbPks*: (1) Complete impaired production of tenellin. (2) No differences in virulence were observed in topical applications.	Eley et al. (2007), Halo, Heneghan, et al. (2008), Halo, Marshall, et al. (2008), and Jirakkakul et al. (2015)
Additional genes examined			
Pmr1	P-type calcium ATPase	*ΔBbpmr1*: (1) Reduced growth on SDAY, CZA, CZA + different C and N sources, and starvation media. (2) Reduced conidial yield and viability. No morphological changes in budding germ tubes or early growing hyphae were observed. (3) Drastic repression of conidiation-related genes. (4) Ca^{2+}, Zn^{2+}, Cu^{2+}, Fe^{3+}, and EDTA impacted growth negatively. Mn^{2+} increased growth. (5) Colony sensitivity to oxidative stress (H_2O_2 and menadione), osmotic stress (NaCl), cell wall stressors (Congo red and SDS), and carbendazim and rapamycin (fungicides). (6) Reduced conidial tolerance to NaCl, H_2O_2, menadione, SDS, Congo red, heat, and UV-B irradiation. (7) Decreased virulence in topical applications.	Wang, Zhou, Ying, and Feng (2013)

(*Continued*)

Table 1 Molecular dissection: gene knockouts in *Beauveria bassiana*[a]—cont'd

Gene	Function	Mutant phenotype	References
Vcx1A, *Vcx1B*, *Vcx1C*, *Vcx1D*, and *Vcx1E*	Vacuolar calcium exchanger (associated with calcineurin)	*ΔVcx1A*, *ΔVcx1A*, *ΔVcx1A*, *ΔVcx1A*, and *ΔVcx1A* shared the following characteristics: (1) No effect on growth on SDA or CZA. (2) No effect on conidial yield on SDAY. (3) No morphological changes in germinating conidia or hyphae. (4) Upregulation of Ca^{2+} ATPases. (5) No effect on growth in the presence of 0.5 M Ca^{2+} at 25 or 30°C. (6) Reduced growth with 0.5 mM DTT at 25°C. *ΔVcx1A*: (1) Reduced growth with 25 ng/mL of CsA at 25 and 30°C. (Cyclosporine A, drug that inhibits calcineurin). (2) Reduced growth with 25 ng/mL of CsA and 0.5 mM Ca^{2+} and Mn^{2+} at 25°C. (3) Stain with Fura-2-AM (assesses calcium flux in the cells) revealed: (a) Increased of calcium flux at 25°C, and at 30°C with 0.5 M Ca^{2+}, (b) decreased of calcium flux at 30°C, (c) no effect on calcium flux at 25°C with 0.5 M Ca^{2+}. (4) Increased colony sensitivity to Congo red and tolerance to SDS, menadione, H_2O_2, and NaCl. (5) Conidial tolerance to heat. (6) Decreased virulence with topical assays. *ΔVcx1B*: (1) Reduced growth with 25 ng/mL of CsA only at 30°C. (2) Reduced growth with 25 ng/mL of CsA and 0.5 mM Ca^{2+} and Mn^{2+} at 25°C. (3) Stain with Fura-2-AM revealed: (a) Same calcium flux at 25°C, and at 25°C with 0.5 M Ca^{2+} as the WT, (b) decreased calcium flux at 30°C, (c) increased on	Hu et al. (2014)

	calcium flux at 30°C with 0.5 M Ca^{2+}. (4) Colony sensitivity to Congo red, SDS, and H_2O_2, and tolerance to menadione and NaCl. (5) Conidial tolerance to heat. (6) Decreased virulence with topical assays.
	ΔVcx1C: (1) No effect on growth with 25 ng/mL of CsA only at 25 or 30°C. (2) No effect on growth with 25 ng/mL of CsA and 0.5 mM Ca^{2+} and Mn^{2+} at 25°C. (3) Stain with Fura-2-AM revealed: (a) Less calcium flux at 25°C and same calcium flux at 25°C with 0.5 M Ca^{2+} as the WT, (b) increased calcium flux at 30°C, (c) decreased on calcium flux at 30°C with 0.5 M Ca^{2+}. (4) Colony sensitivity to Congo red, menadione and H_2O_2, and tolerance to SDS and NaCl. (5) Conidial tolerance to heat. (6) Decreased virulence with topical assays.
	ΔVcx1D: (1) Reduced growth with 25 ng/mL of CsA at 25 and 30°C. (2) Reduced growth with 25 ng/mL of CsA and 0.5 mM Ca^{2+} and Mn^{2+} at 25°C. (3) Stain with Fura-2-AM revealed: (a) Same calcium flux at 25°C as the WT and increased calcium flux at 25°C with 0.5 M Ca^{2+}, (b) Same calcium flux at 30°C as the WT, (c) Increased on calcium flux at 30°C with 0.5 M Ca^{2+}. (4) Colony sensitivity to Congo red, menadione and SDS, and tolerance to H_2O_2 and NaCl. (5) Conidial tolerance to heat. (6) Decreased virulence with topical assays.

(Continued)

Table 1 Molecular dissection: gene knockouts in *Beauveria bassiana*[a]—cont'd

Gene	Function	Mutant phenotype	References
		ΔVcx1E: (1) No effect on growth with 25 ng/mL of CsA at 25 or 30°C. (2) Reduced growth at 25°C with 25 ng/mL of CsA and 0.5 mM Ca^{2+}, but no effect with Mn^{2+}. (3) Stain with Fura-2-AM revealed: (a) Increased calcium flux at 25°C, 30°C, and at 30°C with 0.5 mM Ca^{2+}, (b) Same calcium flux at 30°C as the WT. (4) Colony sensitivity to Congo red, menadione, H_2O_2, and SDS. Tolerance to NaCl. (5) Reduced conidial tolerance to heat. (6) No effect on virulence.	
Bmh1 and Bmh2	14-3-3 phosphopeptide-Binding protein	*ΔBbBmh1ΔBbmh2* (double knockout mutant) failed in multiple attempts. *ΔBbBmh1* (knockout mutant). *ΔBbmh2* (knockout mutant). *sBmh1ΔBbBmh2* (*Bmh1* RNAi mutant expressed in *ΔBbmh2* knockout mutant). *sBmh2ΔBbBmh1* (*Bmh2* RNAi mutant expressed in *ΔBbmh1* knockout mutant). These mutants shared the following phenotype: (1) Reduced growth on SDAY, CZA, and CZA with different sources of C and N. (2) Decreased blastospore yield. Blastospores were smaller and less dense. (3) Reduced conidial yield. Downregulation of conidiation-related genes (4) Hyphae were shorter and slightly wider than the WT. Hyphae presented abnormal branching and septation. Upregulation of	Liu et al. (2015)

		genes essential for septum formation. (5) Alteration of cell cycle. Longer G_0/G_1 and G_2/M cycles and shorter S cycle than the WT. (6) Colony sensitivity to H_2O_2, menadione, Congo red, SDS, NaCl, carbendazim. (7) Low conidial tolerance to NaCl, H_2O_2, menadione, carbendazim, Congo red, SDS, heat, and UV-B irradiation. (8) Reduced hyphal body production. (9) Decreased virulence in topical assays and intrahemocoel injections. (10) Downregulation of *Ras1*, *Ras2*, *Hog1* (signal transduction), *Hyd1*, *Hyd2*, *Mad1*, *Mad2* (adhesion), and *Cht1*, *Cht2* (chitinases).	
		The only phenotypic difference observed was *ΔBbBmh1 and ΔBbmh2* showed increased conidial viability, while *sBmh1ΔBbBmh2* and *sBmh2ΔBbBmh1* exhibited decreased conidial viability.	
MtrA	Methyltransferase (LaeA-like)	*ΔBbMtrA*: (1) Reduced growth on PDA CZA + glucose, CZA + mannitol, and CZA + trehalose. (2) Increased hyphal branching. (3) Decreased germination. Spore viability dropped over time. (4) No differences on intracellular compatible solutes (ie, trehalose, mannitol, glycerol, glucose, lactate, formate, or pyruvate). (5) Reduced production of secreted proteins (proteases, lipases, peptidases, glycoside hydrolases, and hydrophobin 1). Attenuation in virulence, and failure to grow outward and sporulate on the cadaver.	Qin et al. (2014)

(Continued)

Table 1 Molecular dissection: gene knockouts in *Beauveria bassiana*[a]—cont'd

Gene	Function	Mutant phenotype	References
Ura3	Orotidine 5′-phosphate decarboxylase (uracil biosynthesis)	*ΔBbMtrA*: (1) Reduced growth on SDA and CZA. Growth was restored to the WT levels when the media was supplemented with uridine. (2) Decreased conidial yield). (3) *ΔBbMtrA* was completely avirulent.	Ying et al. (2013b)
Gpd promoter	Glyceraldehyde phosphate dehydrogenase regulatory region	For expression	Liao et al. (2008)

[a]All mutant phenotypes given are relative to the wild-type parent. "Tolerance/tolerant" indicates that the mutant did not show any significant difference(s) from the wild type under the conditions examined. "Increased" indicates that the mutant performed better on a particular phenotypic trait with respect to the wild type (ie, grew better and/or was more resistant), and "decreased" indicates that the mutant performed poorer. In some instances, conidial viability is based on the mean germination times (GT_{50} values), with shorter GT_{50} as compared to wild type used to indicate increased conidial viability, and longer GT_{50} used to indicate decreased conidial viability.

2. THE INFECTION PROCESS

2.1 Attachment and Penetration of the Insect Cuticle

Beauveria bassiana infects insects via attachment of cells, namely spores or aerial conidia, to the host surface (Boucias & Pendland, 1991). All life stages of the fungus appear to be infectious including hyphae, aerial conidia, single-cell blastospores (produced during saprophytic growth under certain nutrient conditions), and submerged conidia (specialized cells produced in minimal liquid media) (Holder, Kirkland, Lewis, & Keyhani, 2007). The extent to which these latter fungal forms infect insects in nature is unknown; the asexually produced (aerial) conidia are considered to be the main dispersal and infectious structures, capable of resisting, to a greater extent than hyphae and blastospores, various abiotic stresses. *Beauveria bassiana* conidia are hydrophobic, binding to the similarly hydrophobic insect epicuticle or waxy layer, a structure rich in hydrocarbons, fatty acids, and wax esters (Holder & Keyhani, 2005). As a nonmotile organism, targeting of insects by the fungus is considered a passive event, with airborne, water dispersed, and/or presence in substrata over which insects would forage, ie, the leaf surface and soil, mediating initial contact with potential hosts. Thus, infection can be viewed as an opportunistic program initiated by conidia that happen to find themselves on a host cuticle. In this respect, although preferential sites of infection, typically those where sclerotization of the cuticle is lower, ie, mouthparts and anus, have been noted for many insects, the fungus can initiate infection essentially anywhere on the host cuticle. This is contrast to other microbial pathogens that have to be ingested and/or utilize a more specific route of entry into the host. Attachment itself appears to include both nonspecific and specific components (Boucias, Pendland, & Latge, 1988; Zhang, Xia, Kim, & Keyhani, 2011). Hydrophobic interactions predominate with *B. bassiana* conidia containing an outermost or rodlet layer comprised of proteins known as hydrophobins. At least two hydrophobin proteins, Hyd1 and Hyd2, have been characterized in *B. bassiana* (Cho, Kirkland, Holder, & Keyhani, 2007; Kirkland & Keyhani, 2011; Zhang, Xia, Kim, et al., 2011b). Experiments using antibodies against the tagged versions of Hyd1 and Hyd2 revealed that Hyd1 is localized to the surface of aerial and submerged conidia, and at the base of germinating conidia, but was not detected on blastospores. Likewise, Hyd2 is found on the surface of aerial conidia and at the base of the germinating conidia. However, Hyd2 could not be found on submerged conidia and blastospores. In addition, neither Hyd1 nor Hyd2 was observed on hyphae or hyphal bodies (Zhang,

Xia, Kim, et al., 2011b). Although both hydrophobins contributed to the rodlet layer structure, and variously to cell surface hydrophobicity and mediating adhesion (to hydrophobic surfaces), only loss of *Hyd1* (*ΔBbHyd1* mutant) resulted in decreased virulence, whereas *ΔBbHyd2* mutants were unaffected in terms of targeting insect hosts, although the double mutant (*ΔBbHyd1ΔBbHyd2*) showed the most pronounced phenotype in terms of reduced adhesion and virulence (Zhang, Xia, Kim, et al., 2011b). A specific adhesion gene (*Mad1*), mediating attachment to insect cuticles has been characterized in the related entomopathogenic fungus, *Metarhizium robertsii* (Wang & St Leger, 2007), and although a homolog has been identified in the *B. bassiana* genome, its contribution(s) to adhesion to insect surfaces and/or virulence has yet to be characterized. Further experimental evidence is needed, however, our current understanding of adhesion to host cuticles involves both nonspecific and specific fungal components that mediate initial binding. The extent to which this initial binding is then consolidated by additional factors and the cue(s) used by the fungus to initiate its infection program remains unknown. For certain plant pathogenic fungi, production of infectious structures (appressoria) appears to require little more than a hard surface, and appressoria formation in *Metarhizium anisopliae* and in some strains of *B. bassiana* can be induced on artificial surfaces in the absence of an insect cuticle (Kolattukudy, Rogers, Li, Hwang, & Flaishman, 1995; Wang and Leger 2007; Zhang, Zhang, et al., 2010). Most, if not all, *Metarhizium* strains appear to form clearly defined appressoria. Whether *B. bassiana* strains also form appressoria remains debatable and may be strain dependent, eg, an SEM study of *B. bassiana* growth on the red palm weevil revealed potential appressoria, which were, however, far less readily notable than comparable studies using *M. anisopliae* (Guerri-Agulloo, Gomez-Vidal, Asensio, Barranco, & Lopez-Llorca, 2010). In most of the strains of the author's (NOK) lab, clearly defined appressoria, ie, pronounced swelling of the growing germ tube/hyphal tip preceding penetration, have not been noted. In addition, the extent to which appressoria produced under synthetic conditions (ie, on a plastic surface) are identical to those produced during germination and growth on the insect cuticle remains unknown.

2.2 Immune Evasion and Growth Within the Hemocoel

Subsequent to adhesion and initial germination and growth of the fungal cell, penetration of the insect cuticle occurs. As perhaps the major structural feature of the insect cuticle is the melanized protein cross-linked chitin exoskeleton, chitinases and proteases have long been considered critical fungal enzymes required for breakdown and penetration of the insect cuticle (Bidochka & Khachatourians, 1987; Charnley, 2003; Gupta, Leathers, Elsayed, & Ignoffo, 1992, 1994; Kucera, 1971; Leopold & Samsinakova, 1970; St Leger, Cooper, & Charnley, 1987). These enzymes have also been used as potential markers for isolation of strains with higher virulence, although the extent to which such enzyme activities correlate to virulence may be low, particularly as their expression is often inducible. A number of secreted chitinases and proteases have been described from *B. bassiana*, and genomic data reveal that *B. bassiana* has 23 trypsins, 43 subtilisins, and 20 chitinases. Overexpression of chitinase and protease activities has led to the construction of more virulent strains, suggesting that these enzyme activities may be (partially) limiting in the wild type (Fan, Fang, Guo, et al., 2007; Fang et al., 2009; Zhang et al., 2008). Despite the supposed importance of these enzymes in the pathogenic program, genetic dissection via characterization of chitinase and/or protease gene knockout mutants has yet to be reported in *B. bassiana*. As these enzymes appear to exist as gene families, such efforts may require targeting of multiple genes, as some may have redundant functions. Alternatively, however, there may be one gene that is predominately involved in targeting the insect cuticle. Prior to reaching the protein—chitinaceous part of the cuticle, the fungus must first deal with the lipid-rich outermost waxy layer. The activities of a number of lipases/esterases have been described in *B. bassiana*, and as with proteases and chitinases, they exist as gene families in the fungus. Again, however, any phenotypes of targeted gene knockouts of *B. bassiana* lipases have yet to be reported. Some information is available concerning hydrocarbon assimilation by *B. bassiana* (Pedrini, Ortiz-Urquiza, Huarte-Bonnet, Zhang, & Keyhani, 2013). Growth on insect lipid extracts can result in conidia displaying higher virulence (Crespo et al., 2002), presumably due to (pre)-induction of lipid assimilatory pathways and a set of alkane P450 monooxygenases, enzymes involved in oxidation of alkanes and fatty acids, an initial step in their assimilation (Pedrini, Zhang, Juarez, & Keyhani, 2010). Of the eight cytochrome P450 enzymes potentially involved in alkane and/or insect epicuticle degradation, individual targeted gene knockouts of most did not result in any noticeable phenotype, potentially due to redundancy (briefly considered in Section 4 of this chapter) (Pedrini et al., 2013). An exception to this was seen in the characterization of *B. bassiana Cyp52X1* (Zhang et al., 2012). Knockout mutants of this gene were impaired in their ability to infect *Galleria mellonella* hosts in topical bioassays, which represent the natural route of infection, but displayed wild-type levels of mortality when injected directly into the host hemocoel. These data suggest that the gene and its protein product are important in penetration events, consistent with a function of the enzyme in targeting cuticular waxy layer components, but that the gene is dispensable when the cuticle has been breached, ie, during within-host growth of the fungus. A critical outstanding point is that these cytochrome P450 enzymes are likely localized to peroxisomal and/or endosomal-like membranes found in the cytosol, leaving open the issue of how the enzyme accesses the substrate(s), ie, insoluble long-chain hydrocarbons and fatty acids. This would suggest the need for (secreted) lipid carrier and transport proteins that have yet to be identified.

2.3 Growth From the Inside Out

Little information exists concerning the events that follow growth and proliferation of the hyphal bodies in the hemocoel. That these hyphal bodies are biochemically and structurally distinct from their in vitro counterparts is clear, and their ability to evade the insect immune system and freely float and disperse within the insect hemocoel has been well characterized (Lewis, Robalino, & Keyhani, 2009; Pendland, Hung, & Boucias, 1993; Tartar & Boucias, 2004; Wanchoo, Lewis, & Keyhani, 2009). In addition, *B. bassiana* conidia have been shown to be able to germinate and grow within insect cells and even after being phagocytosed by soil amoeba, a process that may parallel intracellular microbial pathogens (Bidochka, Clark, Lewis, & Keyhani, 2010; Kurtti & Keyhani, 2008). Within the hemocoel, the current model speculates that, initiated by as yet some unknown signal, the hyphal bodies begin to attach to the integument and work their way from the inside out via hyphal growth. Such a signal may be depletion of critical nutrients found in the hemolymph or even production of a quorum-like sensing molecule by the fungus, although to date no such molecule has been identified. Hyphal growth outward is presumed to involve at least some of the enzyme machinery, ie, chitinases and proteases, involved in the initial penetration event. It is intriguing to speculate that given the diversity of such enzymes in the *B. bassiana* genome, specific chitinase(s) and/or protease(s) may be specialized for inward penetration and others for outward penetration; however, no experimental evidence exists for such partition of functioning and/or regulation. Death of the host occurs after hyphal proliferation in the hemocoel, but whether it is concomitant with the outward penetration phase is unclear. A final aspect of the infection process for which little information exists is the nature and expression of antimicrobial compounds secreted by the fungus during the infection process. That suppression of competing microbes occurs appears likely, but whether such activities are expressed throughout the infection process or only in the latter stages when sufficient fungal biomass has already accumulated in unknown. The final stage of the infection life cycle is the production of conidia on the host cadaver, resulting in the white fluffy ("white muscardine") growth characteristic of *B. bassiana* resultant morbidity.

3. TECHNIQUES FOR MOLECULAR MANIPULATION OF *BEAUVERIA BASSIANA*

Early molecular characterization in *B. bassiana* has included construction and analyses of expressed sequence tagged libraries from cells grown under a variety of in vitro conditions (Cho, Boucias, & Keyhani, 2006; Cho, Liu, Farmerie, & Keyhani, 2006) and has more recently extended to those derived from growth on insect cuticles and from in vivo produced hyphal bodies (Mantilla et al., 2012; Tartar & Boucias, 2004). Genetic studies on *B. bassiana* have been facilitated by the development of robust techniques for transformation of the fungus. Although *B. bassiana* isolates show high levels of resistance to many antibiotics, two major selection markers have been found to be effective for most strains. These include the *Bar* gene encoding for phosphinothricin acetyltransferase activity conferring resistance to the glutamate analog, phosphinothricin (glufosinate, the herbicide biapholos contains phosphinothricin) (Fan, Zhang, Kruer, & Keyhani, 2011; Fang et al., 2004), and the *Sur* gene, coding for an enzyme variant of acetolactate synthase, conferring resistance to the sulfonylurea herbicide, chlorimuron ethyl (Zhang, Fan, Xia, & Keyhani, 2010). Hygromycin has also been used as a selection agent (with resistance mediated by the hygromycin phosphotransferase, *Hph*, gene), although in the experience of the authors, the use of this marker is problematic as many *B. bassiana* strains naturally display high resistance to this antibiotic. Expression of various markers and other genes has been directed mainly using promoters derived from *Aspergillus* (eg, *Aspergillus nidulans* glyceraldehyde 3-phosphate dehydrogenase, *Gpd* and tryptophan, *TrpC* promoters). In several instances, fungal derived termination sequences (eg, the *A. nidulans TrpC* terminator) have also been included in vector constructs, including the ethanol-inducible *Alc* system in which expression levels can be regulated to a certain degree (Liao et al., 2009). More recently, the *B. bassiana Gpd* and other promoter sequences (eg, *Hyd1*) have been characterized (Liao et al., 2008). The availability of two markers has been essential for more complete genetic studies requiring the construction of gene knockouts and complementation strains. Several methods for transformation of *B. bassiana* have been developed: (1) formation of protoplasts and electroporation (Pfeifer & Khachatourians, 1992), (2) conjugative transfer using *Agrobacterium tumefaciens* (Fang et al., 2004; Wu, Ridgway, Carpenter, & Glare, 2008a), and (3) use of in vitro-generated blastospores and a protocol similar to yeast transformation (LiCl/heat shock) (Zhang, Fan, et al., 2010). These methods have been coupled to the use of both selectable and visible markers, ie, green fluorescent protein (Jin et al., 2008). A number of vectors for the former have been used in which intervening sequences defined by left border and right border motifs recognized by the *A. tumefaciens* DNA transfer machinery are mobilized into the fungus. The efficiency (and ease) of the latter blastospore transformation system was dramatically improved with the somewhat accidental observation of enhanced selection as a function of pH (optimum at pH ~6.2) (Fan et al., 2011). The basic strategy for construction of targeted gene knockouts has been similar to that used in other fungal systems (eg, *Aspergillus*) in which upstream and downstream flanking sequences of a gene, usually incorporating a deletion, are cloned into a vector sandwiching the selection marker. Targeted gene inactivation, occurring due to double recombination, is then screened using gene-specific primers in transformants via PCR. Desired (double recombinant) knockout mutants typically occur at frequencies ranging from 1% to 10%, with the rest representing ectopic integrants. Split marker methods, in which correct gene targeting can reach >90%, have also been employed. In most instances, complementation has been achieved via ectopic integration of the entire gene with 1–2 kb of upstream flanking sequence, presumably containing the appropriate regulatory elements, although "knockout/knockin" strategies have also been used (Ma et al., 2009; Qin, Ortiz-Urquiza, & Keyhani, 2014). RNAi-mediated gene expression suppression has also been successfully used in *B. bassiana*, proving useful for probing genes whose activities may be essential (Xie, Li, Ying, & Feng, 2012). Various advances including construction of auxotrophs for markerless selection (Ying, Feng, & Keyhani, 2013b), use of suppressive subtraction protocols for enrichment of mRNAs involved in specific processes (Chantasingh et al., 2013; Wu, Ridgway, Carpenter, & Glare, 2008b; Zhang, Zhang, et al., 2010), and protein tagging with green fluorescent protein or antibody recognizable peptides (eg, Myc, Flag) (He et al., 2014; Ying, Ji, Wang, Feng, & Keyhani, 2014; Zhang, Xia, Kim, et al., 2011) have resulted in facile genetic manipulation, identification of cohorts of mRNAs that contribute to developmental programs, and methods for protein visualization in *B. bassiana*. Indeed, it appears as if most methods developed for other filamentous fungi are likely applicable to *B. bassiana*.

4. WHAT CONSTITUTES A VIRULENCE FACTOR?

Here we shall consider pathogenicity as "the quality or state of being pathogenic" and virulence as "the disease producing power of an organism" (Shapiro-Ilan, Fuxa, Lacey, Onstad, & Kaya, 2005). Within this definition, pathogenicity is a qualitative all-or-none phenomenon applied to groups or species intended to indicate whether an organism is "pathogenic" to a host or not. Virulence is the quantifiable aspect of pathogenicity and reflects the ability to cause disease, with lethal dose (LD) and lethal time (LT) used as measurements of virulence. This definition in which pathogenicity is absolute for a given host–pathogen pair, whereas virulence is variable, is not universal, however, it appears to be the one most consistently used in invertebrate pathology. An interesting reform was suggested in which virulence was defined as equal to number of dying/number of infected, infectivity = number of infected/number exposed, and pathogenicity = number of dying/number exposed = virulence × infectivity (Thomas & Elkinson, 2004). While such a scheme appeals to the authors' sense of quantifiable variables, several issues have been raised concerning these definitions (Shapiro-Ilan et al., 2005) and they have not been widely adopted.

Additional terms that are sometimes used include parasite host range, aggressiveness, infectivity, and fitness, which encompass environmental, ecological, and evolutionary interactions. However, only a few studies have identified factors (genes) that can impact virulence toward specific hosts, and molecular examinations of issues centered on strain attenuation aimed towards increasing our understanding of infectivity and fitness are currently lacking. The simplest currently applied standard of what constitutes a virulence factor is whether or not loss of the gene affects LT and/or LD values. However, such a broad definition runs the risk of conflating processes that contribute to normal (in vitro) fungal growth and fitness, with those that can be considered uniquely involved in the ability of the fungus to parasitize insect hosts. The strictest definition of a virulence factor would be one in which the gene/protein product under investigation is essential for infection, ie, loss of which would lead to no insect mortality, but would not affect saprophytic growth. Such a narrow definition, however, runs the risk of neglecting important contributors to virulence that may also cross "talk" or participate in other cellular responses. The most notable of these would be factors that contribute to stress resistance, ie, osmotic, oxidative, and/or thermal stress. Such factors may contribute to saprophytic growth in mitigating abiotic derived stress, but may also be vital for resistance against similar stressors derived from the host, especially as part of microbial defense systems. The issue of what constitutes a virulence factor is further confounded by three considerations: (1) the strategy *B. bassiana* employs in infecting hosts, (2) the existence of pathways shared by other (noninsect-related) lifestyles, and (3) the existence of gene families and potential redundancy. Unlike many bacteria and certain plant pathogenic fungi, in which targeting of the host involves a small subset of "virulence" genes, our current understanding of *B. bassiana* insect parasitism suggests that the fungus overwhelms insect defenses by secretion of a plethora of degradative enzymes and toxic molecules, fast assimilation of nutrients, inhibition of defense responses, resistance against insect defense responses (stress), and rapid growth. Many of these processes are inherently nonspecific in nature and also contribute to saprophytic growth. Thus, teasing apart in vitro growth from in vivo (in the host) growth may be difficult. Aside from its saprophytic growth, *B. bassiana* also forms intimate associations with plants, surviving near plant roots, on the phylloplane, and even potentially as a plant endophyte (Biswas, Dey, Satpathy, & Satya, 2012; Brownbridge, Reay, Nelson, & Glare, 2012; Inglis, Goettel, & Johnson, 1993; Meyling & Eilenberg, 2006; Vega et al., 2008). Carbon and nitrogen flow between *B. bassiana* and associated plants has been demonstrated and the environmental distribution and persistence of *B. bassiana* strains has been hypothesized to be more closely linked to plant interactions than to insect distribution (Behie & Bidochka, 2014; Behie, Zelisko, & Bidochka, 2012; Bidochka, Menzies, & Kamp, 2002). It is intriguing to speculate that systems important for plant association may also contribute to insect parasitism, especially on the level of transcriptional control, ie, some stress responses may be similar within both the plant and insect contexts. Our discussion has considered what can be called "false-positive" genes that might affect virulence, but are not virulence factors due to impairment of nonvirulence-related processes that allow for normal growth and development. However, the opposite possibility also exists, namely if a knockout mutation of a gene results in no appreciable phenotype with respect to virulence, it may still be an important "virulence" factor. This can occur particularly for activities that are encoded by gene families in *B. bassiana*, ie, the chitinases, proteases, lipases, etc. The observation that a mutation in a particular chitinase, protease, lipase (or alkane cytochrome P450 assimilating enzyme) may not affect virulence could be due to overlapping and redundant activities by other members of the gene family. In such cases, multiple gene knockouts (or a more global RNAi-mediated suppression of the gene family), within the same strain may be needed to reveal the roles of these genes with respect to insect infection. Indeed, lack of a phenotype of single mutants due to functional redundancy may be an indication of the extreme importance of the activities under question. Adaptive evolution in gene families might explain the fungus maintaining multiple paralogs. These considerations put us in the curious position of potentially challenging results that indicate a gene may be virulence factor due to decreased virulence as well as those that indicate a gene may not be a virulence factor due to no effect on virulence. Reconciling these conflicting perspectives may be difficult, and it may not be possible to have a rigid definition of what constitutes a virulence factor in *B. bassiana*. One could define a pathogenicity factor as a gene that contributes to an all-or-nothing infection phenotype and a virulence factor as one that affects LD and/or LT values with minimal effects on saprophytic growth. Regardless, it is the position of the authors that effects of virulence need to be considered within the context of the function of the gene and the pathways it may participate in, and that those genes displaying minor or moderate effects on virulence, while impairing normal growth and development should probably not be categorized as virulence factors.

5. GENETIC DISSECTION IN *BEAUVERIA BASSIANA*

5.1 Cuticle-Degrading Enzymes

Among the first enzymes characterized in *B. bassiana* were chitinases and proteases (Bidochka & Khachatourians, 1987, 1988, 1990; Kucera, 1971; Leopold & Samsinakova, 1970; Smith & Grula, 1983; St Leger, Joshi, Bidochka, Rizzo, & Roberts, 1996). The gene sequences of these enzymes were also among the first to be reported for *B. bassiana* (Fang et al., 2005; Joshi, St Leger, & Bidochka, 1995; Kim et al., 1999) (mitochondrial transfer RNA-encoding genes were reported in 1991 (Hegedus, Pfeifer, MacPherson, & Khachatourians, 1991)). With the elucidation of the gene sequences of these enzymes, there rapidly came genetic engineering efforts at improving the virulence of *B. bassiana* via overexpression of protease and/or chitinase activities that has included construction of hybrid enzymes and even directed evolution of the *B. bassiana* chitinase for improved activity (Fan, Fang, Guo, et al., 2007; Fan, Fang, Xiao, et al., 2007; Fang et al., 2009; St Leger, Joshi, Bidochka, & Roberts, 1996; Zhang et al., 2008). Adaptation of hydrolases to fulfill specific roles in various ecological niches, and their regulation as a function of insect host cuticle and other factors, has also been extensively studied (Bidochka & Khachatourians, 1990, 1994; Dias, Neves, Furlaneto-Maia, & Furlaneto, 2008; Donatti, Furlaneto-Maia, Fungaro, & Furlaneto, 2008; Qazi & Khachatourians, 2008; Rao, Lu, Liu, & Tzeng, 2006; Shimizu, Tsuchitani, & Matsumoto, 1993; St Leger, Joshi, & Roberts, 1997; Urtz & Rice, 2000; Zibaee & Bandani, 2009). Surprisingly, aside from gene expression and some biochemical characterizations, to date, the contributions of specific proteases, chitinases, and/or lipases to cuticle penetration events have not been examined in *B. bassiana* via construction of targeted gene knockouts. As noted in Section 2.2, these activities, as well as a set of cytochrome P450 enzymes implicated in (waxy layer) hydrocarbon assimilation, exist as gene families, and it is possible that clear phenotypes for single mutants may be difficult to discern. Many of the biochemical details of the cuticle degradative/assimilatory process still remain to be elucidated, especially as the substrates are insoluble polymers with significant variability. Open questions include the extent to which enzymes may have substrate specificity for different aspects of the insect exoskeletal structure, ie, do the different *B. bassiana* proteases target different protein cross-linked moieties and how do these proteases reach their intended substrates? Given the observation that the *B. bassiana* chitinases appear to lack chitin-binding domains, how are they targeted (if at all) and how do they function in concert with the proteases and/or lipases? What, if any, role(s) do these enzymes have during internal (within the hemocoel) growth of the fungus? Is expression shut down and then reactivated as the fungus begins to work its way outward? Are the same enzymes responsible for inward and outward penetration? It is interesting to note that the fungal cuticle degradative system appears significantly different from that employed by bacteria, eg, marine bacteria that target copepods, in which a defined cascade of highly active chitinolytic enzymes results in the production of either the monosaccharide N-acetylglucosamine (GlcNAc) or the disaccharide $(GlcNAc)_2$, which are rapidly transported into bacterial cells via efficient transport systems (Keyhani, Li, & Roseman, 2000; Keyhani & Roseman, 1996a, 1996b, 1999; Keyhani, Wang, Lee, & Roseman, 1996). Transport systems for the products of the *B. bassiana* enzymes, ie, GlcNAc mono- and/or oligosaccharides and amino acids have yet to be characterized.

5.2 Stress Response

It can be argued that most organisms are almost continuously exposed to some form of stress. For entomopathogenic fungi the major abiotic stresses include (1) osmotic, from low availability of water/high salt, desiccation, to hypoosmotic stress during periods of rainfall and other high water concentration conditions; (2) temperature, again from low $<10°C$ to high $>32°C$; and (3) exposure to UV irradiation. Many of these stressors result in the production of reactive oxygen species (oxidative stress), providing links and cross talk in the physiological consequences and hence, presumably, in the responses to various abiotic stress. Many abiotic stress conditions have counterparts that occur during infection of insect targets; the insect cuticle represents a nutrient and water poor environment, some insects regulate temperature via physiological but mostly behavioral mechanisms, the latter is a phenomenon known as behavioral fever in which an infected insect will bask in the sun in attempting to eliminate and/or reduce the ability of an infecting microbe to spread. In addition, generation of free radicals (oxidative burst) and other innate and/or adaptive immune responses target foreign cells. It is intriguing to speculate that the associations that *B. bassiana* makes with plants, increasingly being recognized as a vital part of the fungus' ecological and environmental strategy for survival, may be driven by attempts to escape various abiotic stresses. However, it is likely that any fungal association with plants likely also involves adaptive responses to (plant-derived) stress conditions.

The roles of a suite of superoxide dismutases (SODs), catalases (CATs), and thioredoxin (TRXs) proteins have been investigated in *B. bassiana* (Fig. 1, Table 1). SODs catalyze the conversion of reactive oxygen species to peroxide. Five different *B. bassiana* SOD genes have been characterized: *Sod1*, coding for a cytoplasmic Cu^{2+}/Zn^{2+} enzyme; *Sod2*, cytoplasmic Mn^{2+}; *Sod3*, mitochondrial Mn^{2+}; *Sod4*, mitochondrial Fe^{2+}; and *Sod5*, cell wall Cu^{2+}/Zn^{2+} enzyme (Li, Shi, Ying, & Feng, 2015a; Xie et al., 2012; Xie, Wang, Huang, Ying, & Feng, 2010). Somewhat surprisingly, inactivation of any of the *Sod* genes (RNAi knockdown in the case of *Sod4*) resulted in decreased tolerance to peroxide and menadione, and reduced conidial tolerance to oxidative stress and UVA/B exposure suggesting little, if any, occurrence of compensatory effects with respect to the

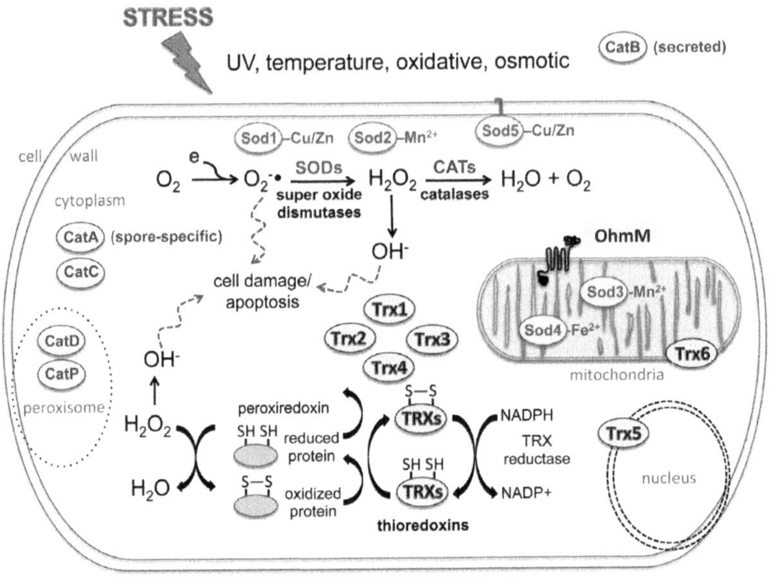

Figure 1 Overview of select *Beauveria bassiana* stress response genes. *SOD*, superoxide dismutase; *CAT*, catalase; *TRX*, thioredoxin; *OhmM*, oxidative homeostasis membrane protein mitochondria.

conditions tested. Targeted gene knockouts of either *Sod1* or *Sod5* resulted in similar phenotypes that included only minor effects on vegetative growth, but delayed sporulation. Only a small decrease in virulence was noted for the mutants when applied topically in insect bioassays, with a greater effect seen when conidia were injected. These data suggest that SOD-mediated detoxification of immune-related (oxidative) defenses occurs mainly during the internal growth phase of the fungus. *Sod2* and *Sod3* showed similar gene knockout phenotypes that differed from *Sod1* or *Sod5* mutants in producing generalized delayed colony growth and sporulation, and decreased virulence in topical application, suggesting that these enzymes may be important in relation to fungal derived free radical generating systems that may act as intracellular signals affecting fungal development. Aside from the phenotypes describe above, RNAi knockdown of *Sod4* had no effect on colony growth, sporulation, germination, or virulence.

CATs catalyze the conversion of peroxide to oxygen and water. The contributions of five different *B. bassiana* CAT genes include *CatA*, spore-specific enzyme; *CatB*, secreted; *CatC*, cytoplasmic; *CatD*, either secreted or peroxisomal; and *CatP*, peroxisomal, to stress response and virulence have been investigated (Wang, Zhang, Ying, & Feng, 2013). Mutants in each gene displayed significantly different phenotypes. Gene knockouts of *CatC*, *CatD* (no effect on menadione tolerance), and *CatP* (more pronounced phenotype), but not in *CatA* or *CatB*, resulted in impaired tolerance to peroxide and menadione (except as noted earlier). Differential effects were also seen with respect to tolerances to wet heat and UV irradiation, ie, reduced for both in *CatA* and *CatD* mutants, only in the latter in Δ*BCatB*, and no effect for either in Δ*BCatC* or Δ*BCatP*. Decreased insect virulence was seen only for the Δ*BCatA*, Δ*BCatD*, and Δ*BCatP* mutants.

TRXs are small molecular weight (~12 kDa) oxidoreductase enzymes that help maintain redox balance, responding to reactive oxygen species to regulate a wide range of signaling and developmental processes. They act as potent antioxidants by catalyzing protein reduction via cysteine thiol—disulfide exchange and are essential in many organisms. Six *B. bassiana* TRX enzymes have been characterized and determined to be localized as follows: $Trx_{(s)}$ 1—4, cytoplasmic; Trx5, nuclear membrane; and Trx6, mitochondrial membrane (Zhang, Tang, Ying, & Feng, 2015). Targeted gene knockouts of any of the TRX genes resulted in varying degrees of reduced virulence in both topical and injection bioassays. However, differential effects were seen with respect to a number of phenotypes including sporulation and germination, conidial tolerance to heat stress and UVB exposure. Notably, loss of either *Trx1*, 3 or 5 did not affect normal growth and sporulation, whereas the *ΔBbTrx4* mutant actually produced more conidia that germinated faster than wild type. Trx3 and 5 appeared to respond to heat stress and Trx2 to UV exposure, suggesting discrimination of certain cellular targets by the different proteins. The observation that Trx4 appears to limit conidial production and germination suggests negative consequences in too rapid proliferation, an observation illustrated even more dramatically in the characterization of the *Ohmm* gene (He et al., 2014). The latter, a unique oxidative homeostasis membrane mitochondrial protein, belongs to a protein family, which although members share significant amino acid and predicted structural homology appear to have diverged into at least two separate functional subfamilies whose members have opposing effects on cellular physiology. The initial member of this family, TmpL, was characterized as a peroxisome-targeted protein whose loss (in either *Aspergillus fumigatus* or *Alternaria brassicicola*) resulted in decreased conidiation, hypersensitivity to oxidative stress and decreased virulence. In sharp contrast to these results, the Ohmm protein in *B. bassiana* was demonstrated to be targeted to the mitochondria and *ΔBbOhmm* mutants displayed increased resistance to oxidative stress and increased virulence. In addition, Ohmm was found to act downstream of the Hog1 mitogen-activated protein kinase (MAPK), responding to oxidative stress but not osmotic stress, proving one of the few discriminatory bifurcations that mediate the various stimuli sensed by the Hog1 MAPK pathway.

5.3 Signal Transduction

All living organisms respond to environmental stimuli via transduction of external signals that can affect cellular targets directly, ie, protein modification/turnover, enzymatic activation, alterations in redox and other processes, or indirectly via activation and/or repression of gene expression. The former is often considered a faster response as the latter requires changes in transcriptional, and where applicable, translational programs and their outputs. The major aspects of these pathways include membrane and soluble receptors, the MAPK cascades, adenylate cycle/protein kinase A, the RAS protein family of GTPases, and their downstream targets that include metabolic adaptor proteins and various transcription factors (TFs) (Fig. 2).

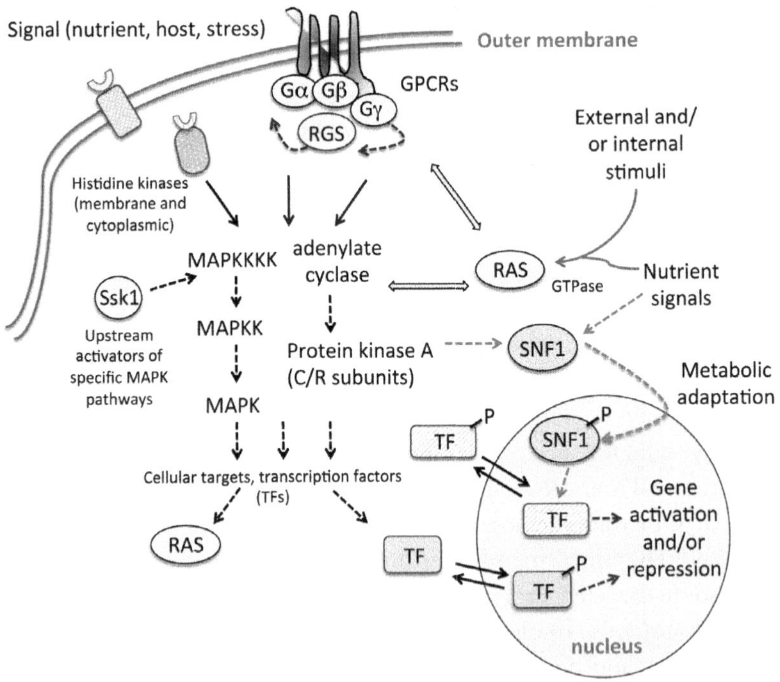

Figure 2 Overview of select *Beauveria bassiana* genes involved in signal transduction. *GPCR*, G-protein-coupled receptor; *MAPKs*, mitogen-activated protein kinase pathways; *RAS*, RAS family of GTPases; *SNF1*, sucrose non-fermenting factor (metabolic adaptor protein); *TF*, transcription factor. (See color plate)

G-protein-coupled receptors (GPCRs) are membrane proteins that respond to external stimuli initiating a cascade of GDP-GTP exchange-mediated signal relay activation that includes the heterotrimeric Gα, Gβ, and Gγ proteins and a regulator of G-protein signaling (RGS) protein. A single GPCR family member, *Gpcr3*, has been shown to link carbohydrate sensing to fungal development (Ying, Feng, & Keyhani, 2013a). In addition, the *Rgs1* gene has been shown to contribute to stress response, conidiation, and virulence (Fang, Scully, Zhang, Pei, & Bidochka, 2008). A family of histidine kinases (HKs) has also been shown to act as initial receptors for a variety of signals. These proteins consist of an N-terminal input or receptor domain, and a C-terminal module comprised of a HK domain coupled to a response regulator domain. These proteins have been classified into 11 different groups, and to date, only a single Group VIII HK (also known as a phytochrome), implicated in mediating fungal responses to light, has been characterized in *B. bassiana* (Qiu, Wang, Chu, Ying, & Feng, 2014). Intriguingly, although many phytochrome HKs are transmembrane proteins, the *B. bassiana* homolog appears to lack a transmembrane domain and is either coupled to the membrane by some other means or functions as a cytoplasmic receptor. The identification of at least eight additional HKs in the *B. bassiana* genome suggests a rich avenue of exploration for those proteins responsible for initial signal perception in this fungus.

G-protein pathway and/or HK-mediated signal perception feeds variously into the MAP pathway and/or protein kinase A signal transduction pathway. MAPK pathways typically involve a cascade of protein phosphorylation events that sequentially travel down a set of three proteins initiated by a MAPK—kinase—kinase (MAPKKK) to a MAPKK, then to a MAPK. Once an MAPK is activated (phosphorylated), the protein, in turn, phosphorylates a number of target proteins. These include various TFs located in the cytoplasm that once phosphorylated, are then targeted to the nucleus, regulating gene expression. Three different MAPK pathways have been examined in *B. bassiana* including members belonging to the Fus3/Kss1 (Ste11 → Ste7 → Mpk1), Slt2 (Bck1 → Mkk1 → Slt2), and Hog1 (Ssk1 → Pbs2 → Hog1) systems. Of these, mutants in the *Mpk1*, *Bck1*, *Mkk1*, *Slt2*, and *Hog1* genes have been examined thus far (Chen, Zhu, Ying, & Feng, 2014b; Luo et al., 2012; Zhang, Zhang, et al., 2010; Zhang et al., 2009). The functions of the GPCR and MAPK pathways, particularly as they pertain to stress response and virulence in *B. bassiana* have been reviewed elsewhere (Ortiz-Urquiza & Keyhani, 2014). An additional upstream signaling protein, Ssk1, that modulates the Ssk2 → Pbs2 → Hog1 pathway has been described in *B. bassiana*. Loss of *Ssk2* resulted in reduced conidial yield, increased sensitivity to osmotic, oxidative, and cell wall stress, and decreased virulence in topical insect bioassays.

The adenylate cyclase gene (*Ac*) has also been examined (Wang, Zhou, Ying, & Feng, 2014). Δ*BbAc* mutant displayed a generalized reduction in growth, increased sensitivity to osmotic, oxidative, and cell wall stress, as well as to the presence of various metal ions. Loss of *Ac* also resulted in decreased virulence in topical insect bioassays. The RAS superfamily of small GTPases feeds into and responds to a large range of cellular pathways. Inputs include signals perceived by membrane tyrosine kinases, phospholipase C, GPCRs, and calcium channels that function to activate RAS proteins that in turn affect phosphoinositide-3 kinase, the RAF serine threonine kinases, phospholipase—protein kinase C, and other targets including downstream MAPK pathways. RAS GTPases play important roles in the regulation of membrane trafficking, growth and differentiation, calcium signaling, and apoptosis. Three *B. bassiana* RAS genes, *Ras1*, *Ras2*, and *Ras3*, have been characterized (Guan, Wang, Ying, & Feng, 2015; Xie, Guan, Ying, & Feng, 2013). An overexpression mutant of Ras1 showed no apparent phenotype, however, expression of a dominant active form, $mRas1^{G19V}$, resulted in abnormal germ tube and hyphal growth, conidia sensitive to multiple stresses, eg, oxidative, osmotic, cell wall, and UV, and dramatically reduced virulence in topical bioassays. In contrast, expression of a dominant negative mutant, $mRas1^{D126A}$, reduced growth on rich media, increased tolerances to a wide variety of stresses, and had no effect on virulence.

Two members of the 14-3-3 family of conserved regulatory proteins were identified in the *B. bassiana* genome (Liu et al., 2015). These proteins are ubiquitously distributed in eukaryotic cells where they bind and modulate the activities of diverse signaling proteins including transmembrane and soluble receptors, kinases, and phosphatases. They operate via common recognition motifs that contain phosphorylated serine or threonine residues, although binding to nonphosphorylated ligands can also occur. Genetic characterization of these proteins involved the construction of four mutant strains: (1) Δ*BbBmh1*, (2) Δ*BbBmh2*, (3) Δ*BbBmh1::sBbBmh2* (RNAi gene knockdown of *BbBmh2* in a Δ*BbBmh1* background), and (4) Δ*BbBmh2::sBbBmh1*. As has been seen in other eukaryotes where attempts to delete the full complement of 14-3-3 proteins was lethal, double Δ*Bbbmh1*Δ*BbBmh2* mutants could not be obtained, suggesting the essentiality of at least one such protein for viability. All three mutant strains displayed a generalized reduction in growth, blastospore development, and conidiation with abnormal hyphal growth and branching evident. Mutants also showed altered cell cycle and decreased stress resistance. Conidial germination was markedly slower in the Δ*BbBmh1::sBbBmh2* and Δ*BbBmh2::sBbBmh1* mutants, but was faster in the Δ*BbBmh1 and* Δ*BbBmh2* strains than in the wild type. All mutants showed decrease virulence in both topical and intrahemocoel injection bioassays and decreased hyphal body proliferation in the insect hosts.

5.4 Transcription and Gene Regulation

It is generally assumed that the major aspects of the transcription machinery in *B. bassiana* are similar to that found in other filamentous fungi and perhaps even yeast. Multiprotein-bridging factor(s) (MBFs) are protein connectors that act as linkers between TATA-binding protein and select TFs. MBFs were first identified in the silkworm *Bombyx mori*. Although some organisms (mainly plants) contain multiple MBF homologs, a unique MBF gene, *Mbf1*, has been characterized in *B. bassiana* (Ying et al., 2014). Loss of *Mbf1* resulted in minor growth defects on certain nutrient sources, however, abnormal hyphal morphogenesis and loss of blastospore development were seen on select media, a phenomenon that was rescued during growth in rich media. Mutants also displayed a generalized increased sensitivity to stress and decreased virulence in both topical and intrahemocoel injection assays.

A number of regulatory TFs have been examined in *B. bassiana*. These include CreA, implicated in catabolite repression; Msn2 involved in stress response; Crz1, target of the calcineurin pathway; Fkh2, a forkhead TF implicated in regulation of cell cycle progression; WetA, developmental regulator; and VosA or velvet protein involved in secondary metabolite regulation (Li, Shi, Ying, & Feng, 2015b; Li, Wang, Zhang, Ying, & Feng, 2015; Liu, Ying, Li, Tian, & Feng, 2013; Luo, Qin, Pei, & Keyhani, 2014; Wang, Qiu, Cai, Ying, & Feng, 2015). *ΔBbCreA* mutants showed a unique amino acid/peptone-dependent growth defect resulting in autolysis that was exacerbated at high (32°C) temperature. Although protease and lipase activities were derepressed, the *ΔBbCreA* mutant showed reduced virulence in both topical and intrahemocoel injection assays, and a decreased ability to sporulate on insect cadavers (Luo, Qin, et al., 2014). These data indicate that impairment of nutrient responsive pathways results in virulence deficiencies that are not overcome by simple derepression of cuticle-degrading enzymes. The Msn2 TF knockout showed a pH-dependent deregulation of oosporein production, poor growth at low pH (4—5), a generalized reduced growth and an increased stress sensitivity phenotype, along with derepression of protease and lipase activities and faster culture acidification when grown in minimal (CZB) media (Liu et al., 2013; Luo, Li, et al., 2014). Similar to the CreA mutant, the *ΔBbMsn2* strain showed decreased virulence in topical and intrahemocoel bioassays, reduced hyphal body formation, and reduced sporulation on host cadavers. Mutants in the forkhead TF (Δ*BbFkh2*) showed a wide range of pleiotropic phenotypes including altered cell cycle progression, increased stress susceptibility, abnormal hyphal growth and branching, increased production of conidia (albeit misshapen), hydrophobicity, germination rate, and overall reduced virulence in both topical and intrahemocoel bioassays (Wang et al., 2015). The WetA developmental regulator is part of a BrlA (C_2H_2 zinc finger TF) → AbaA (leucine zipper TF) → WetA regulatory cascade that controls expression of conidiation genes during conidiophore development and spore maturation. VosA regulator feeds into the BrlA and other pathways, also contributing to normal conidial development. Deleting either *WetA* or *VosA* produced various degrees of impaired growth and alterations in conidial surface features, eg, carbohydrate epitopes and/or hydrophobicity, as well as decreased virulence, particularly in intrahemocoel injection assays, but also when applied topically (Li et al., 2015b).

5.5 Cell Cycle

Protein phosphorylation/dephosphorylation is a ubiquitous modification that results in the modulation of protein activity via a wide range of mechanisms. These include the induction of protein conformational changes, alterations in the ability of proteins to interact with protein partners and other molecules, and changes in the targeting of proteins from one cellular compartment to another. The functions of many nuclear proteins including members of the basal transcription machinery, TFs, and histones are regulated by nuclear protein kinases and phosphatases. Cyclin-dependent kinase 1 (Cdk1) is a highly conserved small (~34 kDa) protein that partners with cyclins (a family of proteins variously expressed through the cell cycle that activate Cdk1) to form complexes that phosphorylate a range of targets critical for cell cycle progression. Cdk1 is part of a biochemical oscillating circuit that includes three interacting positive and double negative feedback loops: (1) —| Cdk1 +→ Cdc25 +→ Cdk1, (2) Cdk1 —| Wee1 —| Cdk1, and (3) Cdk1 —| Myt1 Cdk1 (Pomerening, Ubersax, & Ferrell, 2008). No knockout mutant of *Cdk1* has been obtained, suggesting, as has been observed for other fungi, that the gene is essential (Qiu, Wang, Ying, & Feng, 2015). Overexpression of *Cdk1* resulted in only minor phenotypes, although decreased virulence was noted in both topical and intrahaemocoel bioassays (Qiu et al., 2015). Expression of a phosphorylation site mutant *TCdk1*^{AF}, which short circuits the feedback loops (Pomerening et al., 2008), resulted in altered cell cycle progression, reduced growth, and decreased blastospore production, but has little effect on conidial yield and stress response. Virulence, however, was reduced in both topical and intrahemocoel injection bioassays (Qiu et al., 2015). Two nuclear phosphatases, Cdc14 and Cdc25, implicated in regulation of cytokinesis and other cellular processes, and a single nuclear kinase, Wee1, have been characterized in *B. bassiana* (Qiu et al., 2015; Wang, Liu, Hu, Ying, & Feng, 2013). Deletion of the *Cdc14* gene resulted in a generalized reduced growth phenotype and decreased blastospore and conidial production, which included a failure to form the characteristic rachis (zigzag) structure during the sympodial conidial production mechanism used by *B. bassiana*. There was also a general decrease in stress resistance and decreased virulence in topical bioassays (Wang, Liu, et al., 2013). Cell cycle progression was impaired in the *ΔBbCdc14* strain leading to increased number of nuclei in hyphal segments, and RNAseq analyses indicated reduced expression of genes involved in hyphal development, conidiogenesis, stress response (eg, SODs and CATs), and signal transduction. Altered cell cycle progression was also noted for the *ΔBbCdc25* mutant, where abnormal mitosis, denser septa, and more slender hyphae were produced (Qiu et al., 2015). Both blastospore and conidial yields were decreased, and the cells were larger than their wild-type counterparts. Little to no change in stress response was seen; however, the strain was less virulent in both topical and intrahemocoel injection bioassays. Deletion of the *Wee1* nuclear kinase resulted in phenotypes similar to the *ΔBbCdc25* mutant, although blastospores were smaller than in wild type, and an earlier onset of conidiation was noted. In addition, *ΔBbWee1* cells showed increased sensitivity to some stressors including peroxide, SDS, hydroxyurea, carbendazim, heat, and UVB. Double *ΔBbCdc25ΔBbWee1* mutants showed reduced growth and conidial viability with abnormal branching of germ tubes, but increased production of smaller blastospores with stress and virulence profiles similar to the *ΔBbWee1* mutant (Qiu et al., 2015).

5.6 Glycosyltransferases

Glycosylation and production of glucan chains is an abundant and structurally diverse modification affecting a significant proportion of fungal proteins that are critical for the formation of fungal cell walls. Almost all secreted proteins are glycosylated, and protein mannosylation is an essential modification conserved from yeast to human. Protein mannosylation typically begins in the endoplasmic reticulum (ER) with addition of mannose to the hydroxyl groups on serine and threonine residues (O-mannosylation). These reactions are mediated by O-mannosyltransferases (Dol-P-mannose—protein O-mannosyltrasferases or PMTs) that transfer mannose from dolichol phosphate-activated mannose to proteins in the secretory pathway. Further chain elongation occurs in the Golgi apparatus via a family of α-1,2 mannosyltrasferases (Ktr/Kre/Mnt proteins) that use nucleotide-activated sugars as the donors. Three different O-mannosylatransferases, *Pmt1*, *Pmt2*, and *Pmt4*, have been characterized, as well as three α-mannosylatransferases: *Ktr1*, *Ktr4*, and *Kre2/Mnt1* (Wang, Qiu, Cai, Ying, & Feng, 2014; Wang, Qiu, Chu, Ying, & Feng, 2014). A *ΔBbKrt1* mutant showed severe pleiotropic effects with decreased growth, formation of abnormal conidia and hyphae with thinner cell walls, and increased sensitivity to a wide range of stress conditions. Surprisingly, however, virulence was unaffected in these mutants in either topical or intrahemocoel injection assays. *ΔBbKrt4* mutant also showed growth and conidial defects, with the latter cells displaying decreased hydrophobicity, and reduced viability under osmotic, oxidative, heat, and UV stress. In this case, decreased virulence was seen in both topical and intrahemocoel injection assays. *ΔBbKre2* mutants were generally similar to *ΔBbKrt4* in terms of many phenotypes, including decreased virulence (Wang, Qiu, Cai, et al., 2014). With respect to the protein O-mannosyltransferases, Pmt2 appeared to be essential, as a knockout mutant could not be isolated, so its function was probed via RNAi knockdown (Wang, Qiu, Chu, et al., 2014). *ΔBbPmt1* and *ΔBbPmt4* mutants could be generated, and along with the *Pmt2*—RNAi knockdown mutant, they appeared to broadly share similar phenotypes in disruption of normal development, stress tolerances and responses, and virulence (Wang, Qiu, Chu, et al., 2014). In contrast to the α-mannosyltransferases, Kre4 and Kre2, a decrease in virulence was seen only in topical bioassays, with essentially wild-type virulence seen in intrahemocoel injection assays. As protein O-mannosylation typically precedes the α-mannosyl additions, these data suggest significant functional overlap between the three O-mannosyltransferases with more differentiated targets for the α-mannosyltransferases. Cell wall formation and dynamic changes in cell wall carbohydrate constituents have long been known to affect important fungal development processes including hyphal growth, stress resistance, conidiation, and virulence in the case of pathogens. To date, only a single cell wall-modifying enzyme, Gas1, a putative glycosylphosphatidylinositol-anchored glucanosyltransferase has been characterized (Zhang, Xia, & Keyhani, 2011).

5.7 Transporters

Outer membrane transporters serve as the means by which nutrients are absorbed by the cell and a mechanism for the export of toxic molecules, the latter is of particular importance in terms of (multi-) drug/antibiotic resistance (Fig. 3). Trehalose is an important compatible solute used in many organisms for buffering against osmotic shock. It is also the main carbohydrate constituent of the insect hemolymph, and the ability to utilize trehalose has long been recognized as an important factor in the ability of entomopathogenic fungi to proliferate within the hemocoel. The gene for an α-glucoside transporter, *Agt1*, has been characterized in *B. bassiana* (Wang, Ji, et al., 2013). Deletion of the *Atg1* gene resulted in decreased growth in minimal media supplemented with a variety of carbohydrate substrates, reduced conidial yield, and decreased virulence in both topical and intrahemocoel injection bioassays (Wang, Ji, et al., 2013). In addition, the promoter sequence for a trehalose has been examined, although the consequences of a targeted gene knockout of the enzyme remain to be described (Liu, Ying, & Feng, 2011).

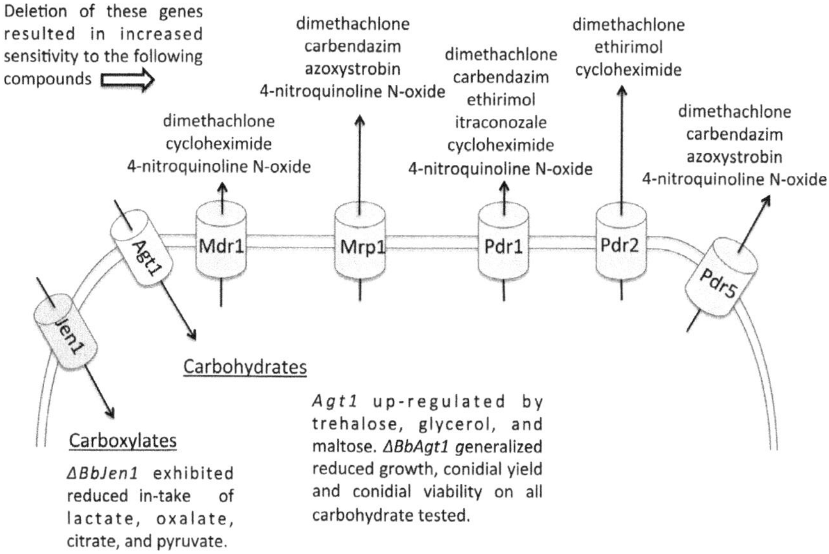

Figure 3 Overview of select *Beauveria bassiana* transporters. Jen1 (light blue), carboxylic acid transporter, Agt1 (light green), carbohydrate transporter, Mdr1, Mrp1, Pdr1, Pdr2, and Pdr5 (gray), ABC-type (multidrug) transporters. (See color plate)

ATP-binding cassette (ABC)-type transporters, in which substrate translocation across membranes is coupled to ATP hydrolysis, constitute an important family of transporters frequently implicated in solute efflux and multidrug resistance. These transporters often function in conjunction with membrane permeability barriers to effectively achieve resistance to broad categories of chemical toxins and antibiotics. Five ABC transporters have been examined in *B. bassiana* including (Liu et al., 2011) one B-type (outer membrane or mitochondria; transports peptides and other molecules, *Mdr1*), one C-type (outer membrane or vacuole, involved in ion transport and/or toxin secretion, *Mrp1*), and three G-type (outer membrane, can transport drug substrates, lipids, steroids, *Pdr1*, *Pdr2*, and *Pdr3*). Single deletion mutants in any of the ABC transporters resulted in increased susceptibility to various toxic chemical compounds and antibiotics. Deletion of any of the transporters examined resulted in increased susceptibility to dimethachlone and, with the exception of the Δ*BbPdr2* mutant, increased sensitivity to 4-nitroquinoline N-oxide, suggesting that the sum activities of multiple transporters contribute to the resistance of the wild-type strain to these two toxins. Carbendazim resistance appeared to be mediated by Mrp1, Pdr1, and Pdr5, with cycloheximide, azoxystrobin, and ethirimol resistances by transporter pairs; Mdr1 and Pdr2, Mrp1 and Pdr5, and Pdr1 and Pdr2, respectively. Resistance to itraconazole appears to be uniquely mediated by Pdr1. Of the five *B. bassiana* ABC transporters examined, only the Δ*BbMrp1* and Δ*BbPdr5* mutants showed decreased virulence in topical bioassays.

5.8 Calcium Transport and Signaling

Calcium is a major intracellular signal molecule, with cytoplasmic levels of the cation carefully regulated and transient fluxes acting to regulate diverse (fungal) cell processes ranging from sensing of external stimuli including nutrients availability, cell cycle progression, stress response, and mating. In filamentous fungi, calcium signaling has also been linked to spore germination, hyphal growth including branching and orientation, and even to circadian rhythms and responses to hosts in the case of pathogens. Cellular components of Ca^{2+} signaling include an array of channels, pumps, and transporters, as calcium-binding proteins that are distributed in the outer membrane, cytoplasm, and the membranes and lumen of most cellular organelles (Fig. 4). Several cytosolic proteins are capable of responding to intracellular calcium fluxes. One of the best characterized is the calmodulin (CaM) pathway in which Ca^{2+} binding to CaM results in association of

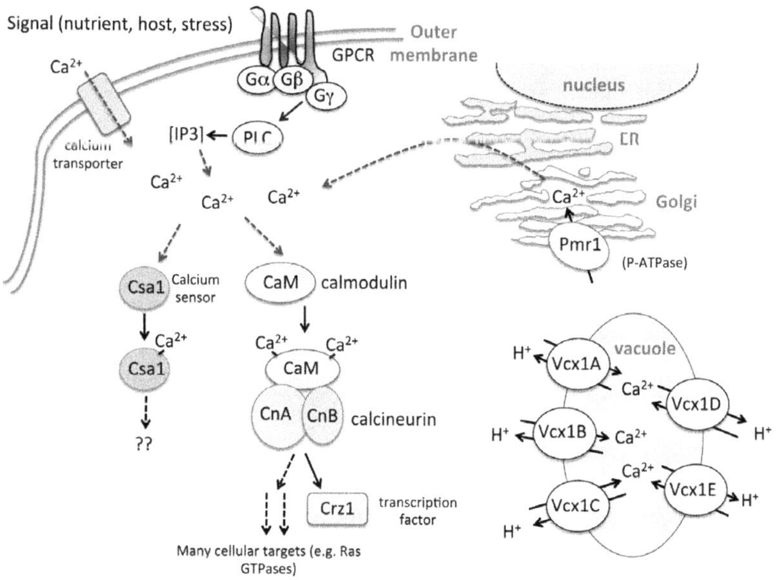

Figure 4 Overview of select *Beauveria bassiana* genes involved in calcium signaling and transport. *GPCR*, G-protein-coupled receptor; *PLC*, phospholipase C; *Csa1*, calcium sensor acidification (Ca^{2+} binding protein); *CaM*, calmodulin; *CnA/B*, calcineurin catalytic and regulatory subunits; *Crz1*, calcineurin responsive transcription factor; *Vcx*, vacuolar Ca^{2+} transporters; *Pmr1*, Golgi Ca^{2+} transporter. (See color plate)

Ca^{2+}-CaM with the calcineurin A and B subunits (CnA-CnB) that activate various downstream targets including the Crz1 TF (Li et al., 2015). CnA is the catalytic subunit that possesses Ca^{2+}-CaM activated serine/threonine protein phosphatase activity, whereas CnB modulates activity of the complex and is known as the "regulatory" subunit. Eukaryotes genomes can contain multiple homologs of the calcineurin genes and, similar to many other filamentous fungi, *B. bassiana* codes for two CnA subunits (CnA1 and CnA2) and one CnB subunit. Deletion of *CnA1* or *CnA2* resulted in mutants that displayed reduced growth and conidial production in rich media and increased sensitivity to stressors (oxidative, cell wall, heat, and UVB stress), as well as to the fungicide carbendazim and divalent metal ions. No effect was seen with respect to osmotic stress, although reduced accumulation of intracellular trehalose and mannitol levels were noted. Knockouts of either gene also resulted in downregulation of SODs and CATs, and the mutant strains showed reduced virulence in topical bioassays. A *ΔBbCnB* mutant was similar to *ΔBbCnA1* and *ΔBbCnA2* mutants with the exception of abnormal branching of germinating conidia and a lack of any alterations in conidial cell surface carbohydrate epitopes. Similarly, the phenotypes of gene knockouts of the Crz1 zinc finger TF were similar to the calcineurin mutant strains, although reduced growth in rich media was not seen and conidial germination time was slightly longer in the *ΔBbCrz1* mutant (Li et al., 2015). In addition, the interplay between calcineurin and the Slt2 MAP-kinase has been linked to conidiation and virulence (Huang et al., 2015). Like CaM, the *B. bassiana* calcium sensor acidification (Csa1, homologous to neuronal calcium sensor proteins), belong to the EF-hand (helix-loop-helix) calcium-binding protein subgroup. The strain harboring the Δ*Bbcsa1* mutation showed changes in the production of organic acids (eg, oxalate, citrate, lactate, and formate) and downregulation of the plasma membrane proton pump that correlated with a delayed acidification phenotype in certain media. Vacuole morphology was also altered in the mutant and decreased virulence was seen in topical bioassays, but not in intrahemocoel injection bioassays (Fan, Ortiz-Urquiza, Kudia, & Keyhani, 2012). As acidification of host tissues, and especially production of oxalic acid, has been implicated in targeting of the insect cuticle, these results suggest that at least some calcium-signaling events participate in regulating organic acid production needed for cuticle penetration, but that this is dispensable once the exoskeleton has been breached. Five different *B. bassiana* vacuolar calcium exchangers (*Vcx1A-E*) implicated as being involved in calcineurin signaling have been examined (Hu, Wang, Ying, & Feng, 2014). These proteins have roles in mediating intracellular calcium ion sequestration via Ca^{2+}/H^{+} exchange across the vacuolar membrane, as well as maintenance of manganese ion homeostasis (vacuolar uptake). In yeast, calcineurin has been shown to act to inhibit Vcx-dependent Ca^{2+}/H^{+} exchange, while promoting Ca^{2+}-dependent ATPases (Cunningham & Fink, 1996). Single gene knockouts of any of the five vacuolar calcium exchangers displayed reduced growth in the presence of DTT (which results in an ER stress response) but had little to no other effects on growth, conidial yield and cellular morphology even in the presence of 0.5 M Ca^{2+}, although all showed upregulation of Ca^{2+}-ATPases. However, most mutants showed increased sensitivity to Congo red with differential sensitivities to other stress conditions. The different mutants, with the exception of the *ΔBbVcx3* strain which showed no effect, also showed various degrees of temperature and/or Ca^{2+}- or Mn^{2+}-dependent growth inhibition in the presence of cyclosporine A (a downstream calcineurin target/inhibitor). In addition, fluorescent staining with the calcium probe Fura-2 revealed various alterations in calcium flux at 25°C and/or 30°C in the presence and absence of exogenous calcium. All of the *ΔBbVcx* mutants, with the exception of *ΔBbVcxE*, showed decreased virulence in topical bioassays (Hu et al., 2014). A putative Golgi localized P-type Ca^{2+} transporter/ATPase, *Pmr1*, has also been characterized in *B. bassiana*. The *ΔBbPmr1* strain showed a generalized reduced growth and conidial yield/viability phenotype, as well as increased sensitivity to oxidative, osmotic, cell wall, and UVB stress. An increased growth rate, however, was seen in media supplemented with various divalent cations, an effect that could be suppressed via addition of EDTA. The *ΔBbPmr1* strain showed decreased virulence in topical insect bioassays.

5.9 Insect Defense Detoxification and Secondary Metabolites

Insect defenses against microbes range from innate and adaptive immune responses to behavioral changes. However, the insect cuticle represents the initial and perhaps major barrier against potentially pathogenic microbes. The nutrient- and water-limiting insect epicuticle, especially when coupled to the secretion of antimicrobial compounds and antiseptic chemical molecules, acts as a potent defense in blocking and/or inhibiting the growth of pathogenic microbes (Ortiz-Urquiza & Keyhani, 2013). Within this context, it has long been recognized that certain insect species are recalcitrant to infection by *B. bassiana*, even though other closely related species are susceptible. The tenebrionid beetle *Tribolium castaneum* (the red flour beetle) is one such resistant insect species. This insect produces copious amounts of benzoquinones that act as defensive compounds against other organisms, including microbes. These benzoquinones can inhibit *B. bassiana* growth, and certain wild-type strains of the fungus result in only 20–25% mortality when tested against *T. castaneum*. The fungus, however, does produce a benzoquinone reductase (BqrA) that can detoxify benzoquinones, and targeted gene knockouts of the *BqrA* gene reduced mortality to ~10%. A strain overexpressing BqrA led to an increase in virulence to ~40–45%. These data were used to suggest a coevolutionary arms race between the beetle and the fungus in the production/detoxification of a cuticular antimicrobial compound; a race in which, at present, the beetle is ahead (Pedrini et al., 2015).

As described throughout this article, *B. bassiana* has evolved a number of mechanisms for breaching the insect cuticle that are thought to include mechanical pressure, production of cuticle-degrading enzymes, and detoxification and/or assimilation of cuticular components. Once inside the insect, rapid proliferation, immune evasion, competition for iron and other compounds, and utilization of host nutrients feed into the ultimately necrotrophic outcome of the interaction. As part of this process, both at the level of the cuticle and during within host fungal growth, the fungus produces a number of secreted fungal toxins and metabolites postulated to contribute to various aspects of the infection (Boucias, Mazet, Pendland, & Hung, 1995; Mazet, Hung, & Boucias, 1994; Ortiz-Urquiza, Keyhani, & Quesada-Moraga, 2013). This may include weakening of the cuticular structure (by organic acids, eg, oxalic acid appears to act as a *B. bassiana* acaricidal factor (Kirkland, Eisa, & Keyhani, 2005)), immune evasion/suppression, and/or inhibition of the growth of competing microbes. However, until recent genetic studies, the specific nature and/or importance of the contributions of secondary (and even primary) metabolites have often been obscure. That *B. bassiana* produces a wide range of secondary metabolites and has a significant genomic capacity for the biosynthesis of many different compounds is clear (Gibson, Donzelli, Krasnoff, & Keyhani, 2014; Molnar, Gibson, & Krasnoff, 2010) and the pathways for a number of these *B. bassiana* secondary metabolites have been examined (Fig. 5). Oosporein, a C2 symmetrical 2,5-dihydroxybenzoquinone derivative

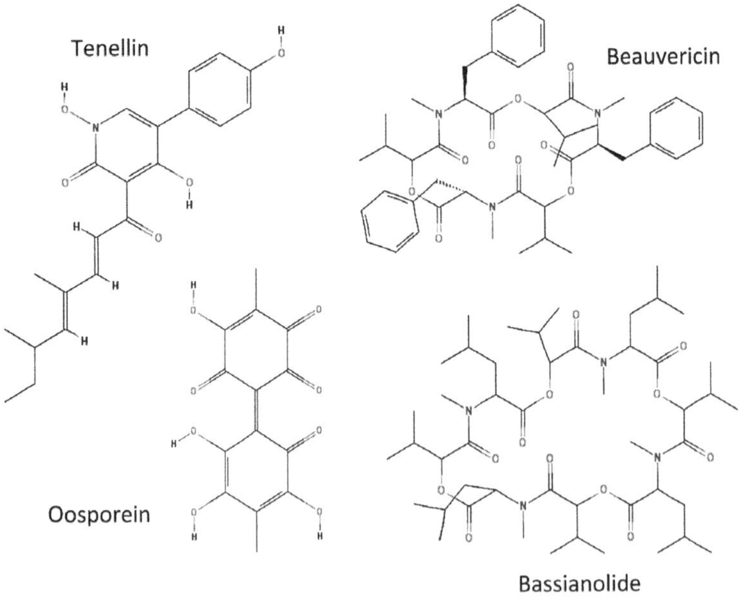

Figure 5 Chemical structures of select *Beauveria bassiana* secondary metabolites. Oosporein CID = 5,359,404, beauvericin CID = 105,014, tenellin CID = 54,704,235, bassianolide CID = 163,065. *Images obtained from NCBI PubChem Compound Database (https://pubchem.ncbi.nlm.nih.gov).*

whose chemical synthesis has been reported, was originally identified as a pigment from the basidiomycete, *Oospora colorans* (van Beyma) (Kogl & van Wessem, 1944; Love, Bonner-Stewart, & Forrest, 2009). Since then, it has been found in various soil fungi and is a major secondary metabolite in many species of the entomopathogenic *Beauveria* genus (Strasser, Abendstein, Stuppner, & Butt, 2000). Oosporein displays a wide range of biological activities; it has been reported to possess mild antibacterial activity (Brewer, Jen, Jones, & Taylor, 1984), antifungal and anti-oomycotic activity against important plant pathogens including *Rhizopus solani*, *Botrytis cinerea*, *Pythium ultimum*, and *Phytophthora infestans* (Mao, Huang, Yang, Chen, & Chen, 2010; Nagaoka, Nakata, Kouno, & Ando, 2004), the potential to inhibit proliferation of tumor cell lines (Mao et al., 2010), and preferential inhibition of Herpes simplex I DNA polymerase over those from HeLa and *Escherichia coli* (Terry et al., 1992). Furthermore, at high concentrations, oosporein has been reported to be toxic to various poultry including broiler chicks and turkeys, resulting in gout, kidney damage, and even death after prolonged and systemic exposure (Cole, Kirksey, Cutler, & Davis, 1974; Pegram, Wyatt, & Smith, 1981). A seven-gene oosporein biosynthetic cluster, *BbOps1*−7 has been identified in *B. bassiana* that included a polyketide synthase (PKS) responsible for orsellinic acid biosynthesis (*BbOps1*), a transporter (*BbOps2*), a transcriptional activator (*BbOps3*), a hydrolase (*BbOps4*), a laccase (*BbOps5*), a glutathione S-transferase (*BbOps6*), and a cupin family dioxygenase (*BbOps7*). Targeted gene knockouts of any of the *Ops* genes, with the exception of *Ops2* (transporter), resulted in loss of oosporein production; curiously, the *ΔBbOps2* mutant displayed increased oosporein production. Insect bioassays performed with the *ΔBbOps2* mutant indicated decreased virulence, presumably due to an impaired suppression of insect host immunity (Feng, Shang, Cen, & Wang, 2015). Molecular dissection of the TF Msn2 in *B. bassiana* suggests that *Msn2* can act as a repressor of oosporein biosynthesis under certain conditions, eg, targeted gene knockout of *Msn2* resulted in derepression of oosporein. *ΔBbMsn2* exhibited an earlier onset of oosporein production and was able to produce oosporein within a wider range of pH values than the WT parent. These data show that a pH-dependent regulation mechanism of oosporein exists, in which Msn2 TF apparently acts as a negative regulator (Luo, Li, et al., 2014).

Bassianolide is an octadepsipeptide with a 24-membered macrolactone ring that has been detected in *B. bassiana*-infected insect cadavers, and in its purified form can induce atony in insect larvae. Bassianolide also possesses acetylcholinesterase-mediated muscle contraction inhibitory activity, as well as antiplasmodial and antimycobacterial activities. Cyclooligomer depsipeptides are typically synthesized via nonribosomal iterative oligomerization by head to tail condensation of amino acid and hydroxycarboxylic acid precursors. A gene for a nonribosomal peptide synthetase (NRPS), *BbBsls*, has been identified and characterized as containing the enzymatic domains for (1) synthesis of an enzyme-bound dipeptidol monomer intermediate from D-2-hydroxyisovalerate and L-leucine and (2) subsequent formation of the cyclic tetramer ester (bassianolide) via recursive condensation of the dipeptidol monomer. Targeted gene inactivation of the *Bsls* gene resulted in significant reduction in virulence in topical bioassays against a range of lepidopteran hosts (Xu, Rozco et al., 2009).

Beauvericin is another *B. bassiana* produced cyclooligomer depsipeptide that displays antibiotic and insecticidal activities, as well as antiproliferative biological activities. Beauvericin is thought to act as an ionophore that can transport monovalent cations across membrane leading to uncoupling of oxidative phosphorylation. In *B. bassiana*, a *Beas* gene coding for a NRPS uses D-hydroxyisovalerate and L-phenylalanine for iterative synthesis of the dipeptidol intermediate which is then converted to the cyclic trimeric ester (beauvericin) in a recursive process. Targeted inactivation of the *Beas* gene resulted in loss of beauvericin production and significant reduction in virulence in topical insect bioassays (Xu et al., 2008, 2007). The results probing bassianolide and beauvericin production indicate that these compounds exert nonredundant activities critical for insect virulence in the fungus. This point was further illustrated by targeted inactivation of the *Kirv* gene. This encodes a ketoisovalerate reductase that produces a common precursor for synthesis of both depsipeptides (Xu, Wijeratne et al., 2009). The *ΔBbKirV* mutant also displayed greatly reduced virulence when tested against a number of lepidopteran hosts, however, as the product of the enzyme also feeds into other pathways; effects on other critical systems cannot be ruled out. The 2-pyridone tenellin, compound shown to be cytotoxic to mammalian erythrocytes, has also been shown to be synthesized via a fused type I PKS-NRPS termed *TenS*. Targeted inactivation of the *TenS* gene resulted in loss of tenellin production, however, unlike results seen for bassianolide and beauvericin, no loss of virulence was seen in insect bioassays using *G. mellonella* as the target host (Eley et al., 2007; Halo, Henegen, et al., 2008; Halo, Marshall, et al., 2008; Jirakkakul et al., 2015). Although it is possible that effects could be seen with other hosts (and the authors performed LT_{50} analysis at only a single spore concentration), these data suggest that tenellin does not substantially contribute to virulence per se, and its biological role(s) remains unknown.

5.10 Metabolic Pathways and Other Genes Examined

Mannitol and trehalose are compatible solutes that function as cell osmotic protectants, with the former also able to quench reactive oxygen species and act as a reservoir of reducing power (Corina & Munday, 1971; Eleutherio, Panek, De Mesquita, Trevisol, & Magalhaes, 2015; Hult, Veide, & Gatenbeck, 1980; Panek, Bernardes, Ortiz, Baker, & Mattoon, 1981). Both compounds have been linked to stress tolerance, spore viability and germination, and/or spore dispersal (Ni & Yu, 2007; Solomon et al., 2006). Two enzymes involved in mannitol metabolism, a mannitol-1-phosphate dehydrogenase (Mpd) and a mannitol dehydrogenase (Mtd), functioning in the intraconversion of mannitol-1 phosphate ($+NAD^+$) $\leftarrow \rightarrow$ fructose 6-phosphate ($+NADH + H^+$) and mannitol ($+NAD^+$) $\leftarrow \rightarrow$ mannose ($+NADH + H^+$), respectively, have been characterized in *B. bassiana* (Wang, Lu, & Feng, 2012). Targeted gene knockouts of either *Mpd* or *Mtd* resulted in reduced growth on several carbohydrate sources including glucose, mannitol, and fructose, with increased susceptibility to stress including oxidative, osmotic, heat, and UVB. Mutants also showed a decrease in the accumulation of mannitol, but increased levels of trehalose, both in the presence and absence of stress. However, a decrease in conidial yield was noted only for the *ΔBbMpd* mutant. Despite the notable reduction in stress resistance, only a slight decrease in virulence was seen in either mutant strain in topical insect bioassays.

A number of genes involved in several additional processes that include autophagy, lipid body homeostasis, and a LaeA-like methyltransferase have been examined in *B. bassiana* (Qin et al., 2014). Autophagy, the nonselective degradation of cellular molecules, is important for normal physiological processes that range from development, proliferation, and remodeling to those implicated in stress response and aging. Atg5 is a protein member of the "core" autophagosome machinery considered to be involved in all autophagy-related pathways. The *ΔBbAtg5* mutant displayed growth and conidial viability abnormalities on nutrient-limiting media, but unlike several plant and animal fungi, in which the gene was found to be required for infection, only a moderate decrease in virulence was noted (Zhang et al., 2013). Caleosins are proteins first characterized in plant seeds that are thought to be components of the proteinaceous coat of intracellular lipid bodies. A single caleosin gene, *Cal1,* was identified and characterized in

B. bassiana (Fan, Ortiz-Urquiza, Garrett, Pei, & Keyhani, 2015). Loss of the *Cal1* gene had little effect on growth, stress response, or conidial germination. However, abnormal conidial development was noted in which more compact assemblages of the sympodially produced conidia were noted and spore dispersal was affected. Although no obvious changes in lipid body formation were noted, alterations in vacuole and/or accumulations of ER/multilamellar-like structures were seen, and altered cellular lipid profiles in conidia as a function of growth on various lipid substrates was reported. A small to moderate decrease in virulence in topical insect bioassays was seen using conidia harvested from PDA plates (Fan et al., 2015). Thus far, a single methyltransferase, MtrA, out of a family of at least 17 other putative methyltransferases has been examined in *B. bassiana* (Qin et al., 2014). *MtrA* displayed homology (~30% identity) to the *A. nidulans* LaeA methyltransferase implicated in regulation of secondary metabolite production, as well as contributing to various developmental processes. Targeted gene knockout of *MtrA* resulted in reduced growth in select nutrient sources including glucose, mannitol, and trehalose, increased hyphal branching, impaired protein secretion, and a sharp decrease in conidial viability, ie, to <10% after 30 days. Virulence was markedly decreased in the *ΔBbMtrA* strain, and the mutant failed to sporulate on insect cadavers.

6. CONCLUSIONS AND FUTURE PROSPECTS

Significant progress has been made in the genetic characterization of the entomopathogenic fungus *B. bassiana*, most within the past 5 years. The development of robust protocols for genetic manipulation coupled with the availability of the fungal genome has led to a sharp increase in the genetic dissection of genes involved in fungal development, stress response, and virulence. With few exceptions, much of this work has been based upon identification of homologous genes in *B. bassiana* whose functions have been described in other organisms (mainly fungi). Although this has led to some unique insights, most results are confirmatory in that the broad functional outlines of the genes and their protein products analyzed were already known. In addition, as seen in this chapter, much of the work is fragmented with cohesive links often missing. Of the 98 genes described in this chapter, 32 were found to affect the trifecta of "conidiation, stress response, and virulence." It is the opinion of the authors that this suggests less that all of these genes are essential players in these processes, but that perhaps even minor perturbations in normal cellular physiology and homeostasis will likely have broad pleiotropic effects on the outputs examined. This is not to downplay that significant advances have been made, and such efforts are needed and should be continued. However, as additional information becomes available, it is hoped that a transition to the establishment of *B. bassiana* as a model system for functional genomics, with de novo identification of gene functions will occur, and mechanistic insights will be derived into how the genes examined affect development, stress response, and/or virulence. *Beauveria bassiana* affords unique aspects that can complement other fungal model systems; living as a saprophyte, fungal development, stress response, and environmental adaptations can be addressed, living as an insect pathogen, issues of virulence and host adaptation ranging from invasion to immune evasion can be investigated using a tractable system with readily available hosts (It is interesting to note that a number of researchers have proposed using insects (larvae) as hosts for animal pathogenic and even plant pathogenic fungi. While such experiments are certainly feasible, and may yield interesting results, the authors view such efforts with a bit of bemused skepticism. Maybe insights into *B. bassiana* insect virulence can be gained by using a mouse model!), and finally, living as a plant epiphyte and/or endophyte, its symbiotic relationships with plants can be examined. The latter, in particular, remains a nascent field, particularly at the genetic level.

ACKNOWLEDGMENTS

This work was supported in part by NSF grant IOS-1121392 to NOK. The authors also wish to thank their colleagues in field who have been generous with their advice, collaborations, sharing of resources, and congeniality.

REFERENCES

Behie, S. W., & Bidochka, M. J. (2014). Nutrient transfer in plant-fungal symbioses. *Trends in Plant Science, 19*, 734–740. http://dx.doi.org/10.1016/j.tplants.2014.06.007.

Behie, S. W., Zelisko, P. M., & Bidochka, M. J. (2012). Endophytic insect-parasitic fungi translocate nitrogen directly from insects to plants. *Science, 336*, 1576–1577. http://dx.doi.org/10.1126/Science.1222289.

Bidochka, M. J., Clark, D. C., Lewis, M. W., & Keyhani, N. O. (2010). Could insect phagocytic avoidance by entomogenous fungi have evolved via selection against soil amoeboid predators? *Microbiology-Sgm, 156*, 2164–2171.

Bidochka, M. J., & Khachatourians, G. G. (1987). Purification and properties of an extracellular protease produced by the entomopathogenic fungus *Beauveria bassiana*. *Applied and Environmental Microbiology, 53*, 1679–1684.

Bidochka, M. J., & Khachatourians, G. G. (1988). Regulation of extracellular protease in the entomopathogenic fungus *Beauveria bassiana*. *Experimental Mycology, 12*, 161–168. http://dx.doi.org/10.1016/0147-5975(88)90005-9.

Bidochka, M. J., & Khachatourians, G. G. (1990). Identification of *Beauveria bassiana* extracellular protease as a virulence factor in pathogenicity toward the migratory grasshopper, *Melanoplus sanguinipes*. *Journal of Invertebrate Pathology, 56*, 362–370.

Bidochka, M. J., & Khachatourians, G. G. (1994). Basic proteases of entomopathogenic fungi differ in their adsorption properties to insect cuticle. *Journal of Invertebrate Pathology, 64*, 26–32. http://dx.doi.org/10.1006/jipa.1994.1064.

Bidochka, M. J., Menzies, F. V., & Kamp, A. M. (2002). Genetic groups of the insect-pathogenic fungus *Beauveria bassiana* are associated with habitat and thermal growth preferences. *Archives of Microbiology, 178*, 531–537. http://dx.doi.org/10.1007/s00203-002-0490-7.

Biswas, C., Dey, P., Satpathy, S., & Satya, P. (2012). Establishment of the fungal entomopathogen *Beauveria bassiana* as a season long endophyte in jute (*Corchorus olitorius*) and its rapid detection using SCAR marker. *BioControl, 57*, 565–571. http://dx.doi.org/10.1007/S10526-011-9424-0.

Boomsma, J. J., Jensen, A. B., Meyling, N. V., & Eilenberg, J. (2014). Evolutionary interaction networks of insect pathogenic fungi. *Annual Review of Entomology, 59*, 467–485. http://dx.doi.org/10.1146/annurev-ento-011613-162054.

Boucias, D., & Pendland, J. (1991). Attachment of mycopathogens to cuticle. In G. T. Cole, & H. C. Hoch (Eds.), *The fungal spore and disease initiation in plants and animals* (pp. 101–127). New York, NY: Plenum Press.

Boucias, D. G., Mazet, I., Pendland, J., & Hung, S. Y. (1995). Comparative-analysis of the in-vivo and in-vitro metabolites produced by the entomopathogen *Beauveria-bassiana*. *Canadian Journal of Botany, 73*, S1092–S1099.

Boucias, D. G., Pendland, J. C., & Latge, J. P. (1988). Nonspecific factors involved in attachment of entomopathogenic deuteromycetes to host insect cuticle. *Applied and Environmental Microbiology, 54*, 1795–1805.

Brewer, D., Jen, W. C., Jones, G. A., & Taylor, A. (1984). The antibacterial activity of some naturally occurring 2,5-dihydroxy-1,4-benzoquinones. *Canadian Journal of Microbiology, 30*, 1068–1072.

Brownbridge, M., Costa, S., & Jaronski, S. T. (2001). Effects of in vitro passage of *Beauveria bassiana* on virulence to *Bemisia argentifolii*. *Journal of Invertebrate Pathology*, 77, 280–283.

Brownbridge, M., Reay, S. D., Nelson, T. L., & Glare, T. R. (2012). Persistence of *Beauveria bassiana* (Ascomycota: Hypocreales) as an endophyte following inoculation of radiata pine seed and seedlings. *Biological Control, 61*, 194–200. http://dx.doi.org/10.1016/J.Biocontrol.2012.01.002.

Chantasingh, D., Kitikhun, S., Keyhani, N. O., Boonyapakron, K., Thoetkiattikul, H., Pootanakit, K., et al. (2013). Identification of catalase as an early up-regulated gene in *Beauveria bassiana* and its role in entomopathogenic fungal virulence. *Biological Control, 67*, 85–93. http://dx.doi.org/10.1016/J.Biocontrol.2013.08.004.

Charnley, A. K. (2003). Fungal pathogens of insects: cuticle degrading enzymes and toxins. *Advances in Botanical Research, 40*, 241–321. http://dx.doi.org/10.1016/S0065-2296(05)40006-3.

Chen, Y., Zhu, J., Ying, S. H., & Feng, M. G. (2014a). The GPI-anchored protein Ecm33 is vital for conidiation, cell wall integrity, and multi-stress tolerance of two filamentous entomopathogens but not for virulence. *Applied Microbiology and Biotechnology*. http://dx.doi.org/10.1007/s00253-014-5577-y.

Chen, Y., Zhu, J., Ying, S. H., & Feng, M. G. (2014b). Three mitogen-activated protein kinases required for cell wall integrity contribute greatly to biocontrol potential of a fungal entomopathogen. *Plos One, 9*, e87948. http://dx.doi.org/10.1371/journal.pone.0087948.

Cho, E. M., Boucias, D., & Keyhani, N. O. (2006). EST analysis of cDNA libraries from the entomopathogenic fungus *Beauveria (Cordyceps) bassiana*. II. Fungal cells sporulating on chitin and producing oosporein. *Microbiology-Sgm, 152*, 2855–2864.

Cho, E. M., Kirkland, B. H., Holder, D. J., & Keyhani, N. O. (2007). Phage display cDNA cloning and expression analysis of hydrophobins from the entomopathogenic fungus *Beauveria (Cordyceps) bassiana*. *Microbiology-Sgm, 153*, 3438–3447.

Cho, E. M., Liu, L., Farmerie, W., & Keyhani, N. O. (2006). EST analysis of cDNA libraries from the entomopathogenic fungus *Beauveria (Cordyceps) bassiana*. I. Evidence for stage-specific gene expression in aerial conidia, in vitro blastospores and submerged conidia. *Microbiology-Sgm, 152*, 2843–2854.

Cole, R. J., Kirksey, J. W., Cutler, H. G., & Davis, E. E. (1974). Toxic effects of oosporein from Chaetomium-Trilaterale. *Journal of Agricultural and Food Chemistry, 22*, 517–520. http://dx.doi.org/10.1021/Jf60193a049.

Corina, D. L., & Munday, K. A. (1971). Studies on polyol function in *Aspergillus clavatus* — role for mannitol and ribitol. *Journal of General Microbiology, 69*, 221.

de Crecy, E., Jaronski, S., Lyons, B., Lyons, T. J., & Keyhani, N. O. (2009). Directed evolution of a filamentous fungus for thermotolerance. *BMC Biotechnology, 9*.

Crespo, R., Juarez, M. P., Dal Bello, G. M., Padin, S., Fernandez, G. C., & Pedrini, N. (2002). Increased mortality of *Acanthoscelides obtectus* by alkane-grown *Beauveria bassiana*. *BioControl, 47*, 685–696.

Cunningham, K. W., & Fink, G. R. (1996). Calcineurin inhibits VCX1-dependent H^+/Ca^{2+} exchange and induces Ca^{2+} ATPases in *Saccharomyces cerevisiae*. *Molecular and Cellular Biology, 16*, 2226–2237.

Dias, B. A., Neves, P. M. O. J., Furlaneto-Maia, L., & Furlaneto, M. C. (2008). Cuticle-degrading proteases produced by the entomopathogenic fungus *Beauveria bassiana* in the presence of coffee berry borer cuticle. *Brazilian Journal of Microbiology, 39*, 301–306.

Donatti, A. C., Furlaneto-Maia, L., Fungaro, M. H. P., & Furlaneto, M. C. (2008). Production and regulation of cuticle-degrading proteases from *Beauveria bassiana* in the presence of *Rhammatocerus schistocercoides* cuticle. *Current Microbiology, 56*, 256–260.

Eleutherio, E., Panek, A., De Mesquita, J. F., Trevisol, E., & Magalhaes, R. (2015). Revisiting yeast trehalose metabolism. *Current Genetics, 61*, 263–274. http://dx.doi.org/10.1007/s00294-014-0450-1.

Eley, K. L., Halo, L. M., Song, Z. S., Powles, H., Cox, R. J., Bailey, A. M., et al. (2007). Biosynthesis of the 2-pyridone tenellin in the insect pathogenic fungus *Beauveria bassiana*. *ChemBioChem, 8*, 289–297.

Fan, Y., Ortiz-Urquiza, A., Garrett, T., Pei, Y., & Keyhani, N. O. (2015). Involvement of a caleosin in lipid storage, spore dispersal, and virulence in the entomopathogenic filamentous fungus, *Beauveria bassiana*. *Environmental Microbiology*. http://dx.doi.org/10.1111/1462–2920.12990.

Fan, Y., Ortiz-Urquiza, A., Kudia, R. A., & Keyhani, N. O. (2012). A fungal homologue of neuronal calcium sensor-1, *Bbcsa1*, regulates extracellular acidification and contributes to virulence in the entomopathogenic fungus *Beauveria bassiana*. *Microbiology, 158*, 1843–1851. http://dx.doi.org/10.1099/mic.0.058867-0.

Fan, Y., Zhang, S., Kruer, N., & Keyhani, N. O. (2011). High-throughput insertion mutagenesis and functional screening in the entomopathogenic fungus *Beauveria bassiana*. *Journal of Invertebrate Pathology, 106*, 274–279.

Fan, Y. H., Fang, W. G., Guo, S. J., Pei, X. Q., Zhang, Y. J., Xiao, Y. H., et al. (2007). Increased insect virulence in *Beauveria bassiana* strains overexpressing an engineered chitinase. *Applied and Environmental Microbiology, 73*, 295–302. http://dx.doi.org/10.1128/Aem.01974-06.

Fan, Y. H., Fang, W. G., Xiao, Y. H., Yang, X. Y., Zhang, Y. J., Bidochka, M. J., et al. (2007). Directed evolution for increased chitinase activity. *Applied Microbiology and Biotechnology, 76*, 135–139. http://dx.doi.org/10.1007/s00253-007-0996-7.

Fang, W. G., Feng, J., Fan, Y. H., Zhang, Y. J., Bidochka, M. J., Leger, R. J. S., et al. (2009). Expressing a fusion protein with protease and chitinase activities increases the virulence of

the insect pathogen *Beauveria bassiana*. *Journal of Invertebrate Pathology, 102*, 155–159. http://dx.doi.org/10.1016/j.jip.2009.07.013.
Fang, W. G., Leng, B., Xiao, Y. H., Jin, K., Ma, J. C., Fan, Y. H., et al. (2005). Cloning of *Beauveria bassiana* chitinase gene *Bbchit1* and its application to improve fungal strain virulence. *Applied and Environmental Microbiology, 71*, 363–370.
Fang, W. G., Scully, L. R., Zhang, L., Pei, Y., & Bidochka, M. J. (2008). Implication of a regulator of G protein signalling (BbRGS1) in conidiation and conidial thermotolerance of the insect pathogenic fungus *Beauveria bassiana*. *FEMS Microbiology Letters, 279*, 146–156. http://dx.doi.org/10.1111/j.1574-6968.2007.00978.x.
Fang, W. G., Zhang, Y. J., Yang, X. Y., Zheng, X. L., Duan, H., Li, Y., et al. (2004). *Agrobacterium tumefaciens*-mediated transformation of *Beauveria bassiana* using an herbicide resistance gene as a selection marker. *Journal of Invertebrate Pathology, 85*, 18–24.
Feng, P., Shang, Y. F., Cen, K., & Wang, C. S. (2015). Fungal biosynthesis of the bibenzoquinone oosporein to evade insect immunity. *Proceedings of the National Academy of Sciences of the United States of America, 112*, 11365–11370. http://dx.doi.org/10.1073/pnas.1503200112.
Gibson, D. M., Donzelli, B. G., Krasnoff, S. B., & Keyhani, N. O. (2014). Discovering the secondary metabolite potential encoded within entomopathogenic fungi. *Natural Product Reports, 31*, 1287–1305. http://dx.doi.org/10.1039/c4np00054d.
Glare, T., Caradus, J., Gelernter, W., Jackson, T., Keyhani, N., Kohl, J., et al. (2012). Have biopesticides come of age? *Trends in Biotechnology, 30*, 250–258. http://dx.doi.org/10.1016/j.tibtech.2012.01.003.
Goettel, M. S., Hajek, A. E., Siegel, J. P., & Evans, H. C. (2001). Safety of fungal biocontrol agents. In T. M. Butt, C. Jackson, & N. Magan (Eds.), *Fungi as biocontrol agents, progress, problems, and potential* (pp. 347–375). Wallingford, UK: CAB International.
Guan, Y., Wang, D. Y., Ying, S. H., & Feng, M. G. (2015). A novel Ras GTPase (Ras3) regulates conidiation, multi-stress tolerance and virulence by acting upstream of Hog1 signaling pathway in *Beauveria bassiana*. *Fungal Genetics and Biology, 82*, 85–94. http://dx.doi.org/10.1016/j.fgb.2015.07.002.
Guerri-Agulloo, B., Gomez-Vidal, S., Asensio, L., Barranco, P., & Lopez-Llorca, L. V. (2010). Infection of the red palm weevil (*Rhynchophorus ferrugineus*) by the entomopathogenic fungus *Beauveria bassiana*: a SEM study. *Microscopy Research and Technique, 73*, 714–725. http://dx.doi.org/10.1002/jemt.20812.
Gupta, S. C., Leathers, T. D., Elsayed, G. N., & Ignoffo, C. M. (1992). Insect cuticle-degrading enzymes from the entomogenous fungus *Beauveria bassiana*. *Experimental Mycology, 16*, 132–137. http://dx.doi.org/10.1016/0147-5975(92)90019-N.
Gupta, S. C., Leathers, T. D., Elsayed, G. N., & Ignoffo, C. M. (1994). Relationships among enzyme-activities and virulence parameters in *Beauveria bassiana* infections of *Galleria mellonella* and *Trichoplusia ni*. *Journal of Invertebrate Pathology, 64*, 13–17. http://dx.doi.org/10.1006/jipa.1994.1062.
Hajek, A. E., & Delalibera, I. (2010). Fungal pathogens as classical biological control agents against arthropods. *BioControl, 55*, 147–158.
Halo, L. M., Heneghan, M. N., Yakasai, A. A., Song, Z., Williams, K., Bailey, A. M., et al. (2008). Late stage oxidations during the biosynthesis of the 2-pyridone tenellin in the entomopathogenic fungus *Beauveria bassiana*. *Journal of the American Chemical Society, 130*, 17988–17996.
Halo, L. M., Marshall, J. W., Yakasai, A. A., Song, Z., Butts, C. P., Crump, M. P., et al. (2008). Authentic heterologous expression of the tenellin iterative polyketide synthase nonribosomal peptide synthetase requires coexpression with an enoyl reductase. *ChemBioChem, 9*, 585–594.
He, P. H., Wang, X. X., Chu, X. L., Feng, M. G., & Ying, S. H. (2015). RNA sequencing analysis identifies the metabolic and developmental genes regulated by BbSNF1 during

conidiation of the entomopathogenic fungus *Beauveria bassiana*. *Current Genetics, 61*, 143–152. http://dx.doi.org/10.1007/s00294-014-0462-x.
He, Z., Zhang, S., Keyhani, N. O., Song, Y., Huang, S., Pei, Y., et al. (2014). A novel mitochondrial membrane protein, Ohmm, limits fungal oxidative stress resistance and virulence in the insect fungal pathogen *Beauveria bassiana*. *Environmental Microbiology*. http://dx.doi.org/10.1111/1462–2920.12713.
Hegedus, D. D., Pfeifer, T. A., MacPherson, J. M., & Khachatourians, G. G. (1991). Cloning and analysis of five mitochondrial tRNA-encoding genes from the fungus *Beauveria bassiana*. *Gene, 109*, 149–154.
Holder, D. J., & Keyhani, N. O. (2005). Adhesion of the entomopathogenic fungus *Beauveria (Cordyceps) bassiana* to substrata. *Applied and Environmental Microbiology, 71*, 5260–5266.
Holder, D. J., Kirkland, B. H., Lewis, M. W., & Keyhani, N. O. (2007). Surface characteristics of the entomopathogenic fungus *Beauveria (Cordyceps) bassiana*. *Microbiology-Sgm, 153*, 3448–3457.
Hu, Y., Wang, J., Ying, S. H., & Feng, M. G. (2014). Five vacuolar Ca^{2+} exchangers play different roles in calcineurin-dependent Ca^{2+}/Mn^{2+} tolerance, multistress responses and virulence of a filamentous entomopathogen. *Fungal Genetics and Biology, 73*, 12–19. http://dx.doi.org/10.1016/j.fgb.2014.09.005.
Huang, S., He, Z., Zhang, S., Keyhani, N. O., Song, Y., Yang, J., et al. (2015). Interplay between calcineurin and the Slt2 MAP-kinase in mediating cell wall integrity, conidiation, and virulence in the insect fungal pathogen *Beauveria bassiana*. *Fungal Genetic and Biology, 83*, 78–91.
Hult, K., Veide, A., & Gatenbeck, S. (1980). The distribution of the NADP(H)-regenerating mannitol cycle among fungal species. *Archives of Microbiology, 128*, 253–255. http://dx.doi.org/10.1007/Bf00406168.
Humber, R. A. C. (2015). Usda-ars collection of entomopathogenic fungal cultures (ARSEF). In *Emerging pests and pathogens research unit RWHCfAaH* (Ithaca, NY).
Inglis, G. D., Goettel, M. S., Butt, T. M., & Strasser, H. (2001). Use of hyphomycetous fungi for managing insect pests. In L. A. Lacey, & H. K. Kaya (Eds.), *Field manual of techniques in invertebrate pathology* (pp. 651–679). Netherlands: Kluwer Academic.
Inglis, G. D., Goettel, M. S., & Johnson, D. L. (1993). Persistence of the entomopathogenic fungus, *Beauveria bassiana*, on phylloplanes of crested wheatgrass and alfalfa. *Biological Control, 3*, 258–270.
Jackson, M. A., Dunlap, C. A., & Jaronski, S. T. (2010). Ecological considerations in producing and formulating fungal entomopathogens for use in insect biocontrol. *BioControl, 55*, 129–145.
Jaronski, S. T. (2010). Ecological factors in the inundative use of fungal entomopathogens. *BioControl, 55*, 159–185.
Jin, K., Zhang, Y. J., Fang, W. G., Luo, Z. B., Zhou, Y. H., & Pei, Y. (2010). Carboxylate transporter gene JEN1 from the entomopathogenic fungus *Beauveria bassiana* is involved in conidiation and virulence. *Applied and Environmental Microbiology, 76*, 254–263.
Jin, K., Zhang, Y. J., Luo, Z. B., Xiao, Y. H., Fan, Y. H., Wu, D., et al. (2008). An improved method for *Beauveria bassiana* transformation using phosphinothricin acetlytransferase and green fluorescent protein fusion gene as a selectable and visible marker. *Biotechnology Letters, 30*, 1379–1383.
Jin-Zhu, X., Bo, H., Chang-Sheng, Q., & Zeng-Zhi, L. (2009). Sequence and phylogenetic analysis of *Beauveria bassiana* with mitochondrial genome. *Mycosystema, 28*, 718–723.
Jirakkakul, J., Cheevadhanarak, S., Punya, J., Chutrakul, C., Senachak, J., Buajarern, T., et al. (2015). Tenellin acts as an iron chelator to prevent iron-generated reactive oxygen species toxicity in the entomopathogenic fungus *Beauveria bassiana*. *FEMS Microbiology Letters, 362*. http://dx.doi.org/10.1093/femsle/fnu032.

Joshi, L., St Leger, R. J., & Bidochka, M. J. (1995). Cloning of a cuticle degrading protease from the entomopathogenic fungus, *Beauveria bassiana*. *FEMS Microbiology Letters, 125*, 211–217.

Keyhani, N. O., Li, X. B., & Roseman, S. (2000). Chitin catabolism in the marine bacterium *Vibrio furnissii*. Identification and molecular cloning of a chitoporin. *Journal of Biological Chemistry, 275*, 33068–33076.

Keyhani, N. O., & Roseman, S. (1996a). The chitin catabolic cascade in the marine bacterium *Vibrio furnissii* — molecular cloning, isolation, and characterization of a periplasmic beta-N-acetylglucosaminidase. *Journal of Biological Chemistry, 271*, 33425–33432.

Keyhani, N. O., & Roseman, S. (1996b). The chitin catabolic cascade in the marine bacterium *Vibrio furnissii* — molecular cloning, isolation, and characterization of a periplasmic chitodextrinase. *Journal of Biological Chemistry, 271*, 33414–33424.

Keyhani, N. O., & Roseman, S. (1999). Physiological aspects of chitin catabolism in marine bacteria. *Biochimica et Biophysica Acta, 1473*, 108–122.

Keyhani, N. O., Wang, L. X., Lee, Y. C., & Roseman, S. (1996). The chitin catabolic cascade in the marine bacterium *Vibrio furnissii*. Characterization of an N,N′-diacetyl-chitobiose transport system. *Journal of Biological Chemistry, 271*, 33409–33413.

Kim, H. K., Hoe, H. S., Suh, D. S., Kang, S. C., Hwang, C., & Kwon, S. T. (1999). Gene structure and expression of the gene from *Beauveria bassiana* encoding bassiasin I, an insect cuticle-degrading serine protease. *Biotechnology Letters, 21*, 777–783. http://dx.doi.org/10.1023/A:1005583324654.

Kirkland, B. H., Cho, E. M., & Keyhani, N. O. (2004). Differential susceptibility of *Amblyomma maculatum* and *Amblyomma americanum* (Acari: Ixodidea) to the entomopathogenic fungi *Beauveria bassiana* and *Metarhizium anisopliae*. *Biological Control, 31*, 414–421.

Kirkland, B. H., Eisa, A., & Keyhani, N. O. (2005). Oxalic acid as a fungal acaracidal virulence factor. *Journal of Medical Entomology, 42*, 346–351.

Kirkland, B. H., & Keyhani, N. O. (2011). Expression and purification of a functionally active class I fungal hydrophobin from the entomopathogenic fungus *Beauveria bassiana* in *E. coli*. *Journal of Indusrtrial Microbiology and Biotechnology, 38*, 327–335.

Kirkland, B. H., Westwood, G. S., & Keyhani, N. O. (2004). Pathogenicity of entomopathogenic fungi *Beauveria bassiana* and *Metarhizium anisopliae* to Ixodidae tick species *Dermacentor variabilis, Rhipicephalus sanguineus*, and *Ixodes scapularis*. *Journal of Medical Entomology, 41*, 705–711.

Kogl, F., & van Wessem, G. C. (1944). Analysis concerning pigments of fungi XIV Concerning oosporein, the pigment of *Oospora colorans* van Beyma. *Recueil Des Travaux Chimiques Des Pays-Bas, 63*, 5–24.

Kolattukudy, P. E., Rogers, L. M., Li, D. X., Hwang, C. S., & Flaishman, M. A. (1995). Surface signaling in pathogenesis. *Proceedings of the National Academy of Sciences of the United States of America, 92*, 4080–4087.

Kucera, M. (1971). Toxins of entomophagous fungus *Beauveria bassiana*. 2. Effect of nitrogen sources on formation of toxic protease in submerged culture. *Journal of Invertebrate Pathology, 17*, 211.

Kurtti, T. J., & Keyhani, N. O. (2008). Intracellular infection of tick cell lines by the entomopathogenic fungus *Metarhizium anisopliae*. *Microbiology-Sgm, 154*, 1700–1709.

Leathers, T. D., Gupta, S. C., & Alexander, N. J. (1993). Mycopesticides: status, challenges, and potential. *Journal of Industrial Microbiology, 12*, 69–75.

Leopold, J., & Samsinakova, A. (1970). Quantitative estimation of chitinase and several other enzymes in fungus *Beauveria bassiana*. *Journal of Invertebrate Pathology, 15*, 34.

Lewis, M. W., Robalino, I. V., & Keyhani, N. O. (2009). Uptake of the fluorescent probe FM4-64 by hyphae and haemolymph-derived in vivo hyphal bodies of the entomopathogenic fungus *Beauveria bassiana*. *Microbiology-Sgm, 155*, 3110–3120.

Li, F., Shi, H. Q., Ying, S. H., & Feng, M. G. (2015a). Distinct contributions of one Fe- and two Cu/Zn-cofactored superoxide dismutases to antioxidation, UV tolerance and virulence of *Beauveria bassiana*. *Fungal Genetics and Biology, 81*, 160—171. http://dx.doi.org/10.1016/j.fgb.2014.09.006.

Li, F., Shi, H. Q., Ying, S. H., & Feng, M. G. (2015b). WetA and VosA are distinct regulators of conidiation capacity, conidial quality, and biological control potential of a fungal insect pathogen. *Applied Microbiology and Biotechnology*. http://dx.doi.org/10.1007/s00253-015-6823-7.

Li, F., Wang, Z. L., Zhang, L. B., Ying, S. H., & Feng, M. G. (2015). The role of three calcineurin subunits and a related transcription factor (Crz1) in conidiation, multistress tolerance and virulence in *Beauveria bassiana*. *Applied Microbiology and Biotechnology, 99*, 827—840. http://dx.doi.org/10.1007/s00253-014-6124-6.

Liao, X. G., Fang, W. G., Zhang, Y. J., Fan, Y. H., Wu, X. W., Zhou, Q., et al. (2008). Characterization of a highly active promoter, p*Bbgpd*, in *Beauveria bassiana*. *Current Microbiology, 57*, 121—126.

Liao, X. G., Zhang, Y. J., Fan, Y. H., Ma, J. C., Zhou, Y. H., Jin, D., et al. (2009). An ethanol inducible alc system for regulating gene expression in *Beauveria bassiana*. *World Journal of Microbiology & Biotechnology, 25*, 2065—2069.

Liu, Q., Li, J. G., Ying, S. H., Wang, J. J., Sun, W. L., Tian, C. G., et al. (2015). Unveiling equal importance of two 14-3-3 proteins for morphogenesis, conidiation, stress tolerance and virulence of an insect pathogen. *Environmental Microbiology, 17*, 1444—1462. http://dx.doi.org/10.1111/1462-2920.12634.

Liu, Q., Ying, S. H., & Feng, M. G. (2011). Characterization of *Beauveria bassiana* neutral trehalase (BbNTH1) and recognition of crucial stress-responsive elements to control its expression in response to multiple stresses. *Microbiological Research, 166*, 282—293. http://dx.doi.org/10.1016/J.Micres.2010.04.001.

Liu, Q., Ying, S. H., Li, J. G., Tian, C. G., & Feng, M. G. (2013). Insight into the transcriptional regulation of Msn2 required for conidiation, multi-stress responses and virulence of two entomopathogenic fungi. *Fungal Genetics and Biology, 54*, 42—51. http://dx.doi.org/10.1016/j.fgb.2013.02.008.

Love, B. E., Bonner-Stewart, J., & Forrest, L. A. (2009). An efficient synthesis of oosporein. *Tetrahedron Letters, 50*, 5050—5052. http://dx.doi.org/10.1016/J.Tetlet.2009.06.103.

Luo, X. D., Keyhani, N. O., Yu, X. D., He, Z. J., Luo, Z. B., Pei, Y., et al. (2012). The MAP kinase *Bbslt2* controls growth, conidiation, cell wall integrity, and virulence in the insect pathogenic fungus *Beauveria bassiana*. *Fungal Genetics and Biology, 49*, 544—555. http://dx.doi.org/10.1016/J.Fgb.2012.05.002.

Luo, Z., Li, Y., Mousa, J., Bruner, S., Zhang, Y., Pei, Y., et al. (2014). Bbmsn2 acts as a pH-dependent negative regulator of secondary metabolite production in the entomopathogenic fungus *Beauveria bassiana*. *Environmental Microbiology*. http://dx.doi.org/10.1111/1462—2920.12542.

Luo, Z., Qin, Y., Pei, Y., & Keyhani, N. O. (2014). Ablation of the *creA* regulator results in amino acid toxicity, temperature sensitivity, pleiotropic effects on cellular development and loss of virulence in the filamentous fungus *Beauveria bassiana*. *Environmental Microbiology, 16*, 1122—1136. http://dx.doi.org/10.1111/1462-2920.12352.

Ma, J. C., Zhou, Q., Zhou, Y. H., Liao, X. G., Zhang, Y. J., Jin, D., et al. (2009). The size and ratio of homologous sequence to non-homologous sequence in gene disruption cassette influences the gene targeting efficiency in *Beauveria bassiana*. *Applied Microbiology and Biotechnology, 82*, 891—898. http://dx.doi.org/10.1007/s00253-008-1844-0.

Mantilla, J. G., Galeano, N. F., Gaitan, A. L., Cristancho, M. A., Keyhani, N. O., & Gongora, C. E. (2012). Transcriptome analysis of the entomopathogenic fungus *Beauveria bassiana* grown on cuticular extracts of the coffee berry borer (*Hypothenemus hampei*). *Microbiology, 158*, 1826—1842. http://dx.doi.org/10.1099/mic.0.051664-0.

Mao, B. Z., Huang, C., Yang, G. M., Chen, Y. Z., & Chen, S. Y. (2010). Separation and determination of the bioactivity of oosporein from *Chaetomium cupreum*. *African Journal of Biotechnology, 9*, 5955–5961.

Maurer, P., Rejasse, A., Capy, P., Langin, T., & Riba, G. (1997). Isolation of the transposable element hupfer from the entomopathogenic fungus *Beauveria bassiana* by insertion mutagenesis of the nitrate reductase structural gene. *Molecular & General Genetics, 256*, 195–202.

Mazet, I., Hung, S. Y., & Boucias, D. G. (1994). Detection of toxic metabolites in the hemolymph of *Beauveria bassiana* infected *Spodoptera exigua* larvae. *Experientia, 50*, 142–147.

Meyling, N. V., & Eilenberg, J. (2006). Isolation and characterisation of *Beauveria bassiana* isolates from phylloplanes of hedgerow vegetation. *Mycological Research, 110*, 188–195. http://dx.doi.org/10.1016/j.mycres.2005.09.008.

Meyling, N. V., & Hajek, A. E. (2010). Principles from community and metapopulation ecology: application to fungal entomopathogens. *BioControl, 55*, 39–54.

Molnar, I., Gibson, D. M., & Krasnoff, S. B. (2010). Secondary metabolites from entomopathogenic hypocrealean fungi. *Natural Product Reports, 27*, 1241–1275.

Nagaoka, T., Nakata, K., Kouno, K., & Ando, T. (2004). Antifungal activity of oosporein from an antagonistic fungus against *Phytophthora infestans*. *Zeitschrift fuer Naturforschung, C, 59*, 302–304.

Ni, M., & Yu, J. H. (2007). A novel regulator couples sporogenesis and trehalose biogenesis in *Aspergillus nidulans*. *PLoS One, 2*, e970. http://dx.doi.org/10.1371/journal.pone.0000970.

Oh, J., Kong, W. S., & Sung, G. H. (2015). Complete mitochondrial genome of the entomopathogenic fungus *Beauveria pseudobassiana* (Ascomycota, Cordycipitaceae). *Mitochondrial DNA, 26*, 777–778. http://dx.doi.org/10.3109/19401736.2013.855747.

Ortiz-Urquiza, A., & Keyhani, N. O. (2013). Action on the surface: entomopathogenic fungi versus the insect cuticle. *Insects, 4*, 357–374.

Ortiz-Urquiza, A., & Keyhani, N. O. (2014). Stress response signaling and virulence: insights from entomopathogenic fungi. *Current Genetics*. http://dx.doi.org/10.1007/s00294-014-0439-9.

Ortiz-Urquiza, A., Keyhani, N. O., & Quesada-Moraga, E. (2013). Culture conditions affect virulence and production of insect toxic proteins in the entomopathogenic fungus *Metarhizium anisopliae*. *Biocontrol Science and Technology, 23*, 1199–1212. http://dx.doi.org/10.1080/09583157.2013.822474.

Ortiz-Urquiza, A., Luo, Z., & Keyhani, N. O. (2015). Improving mycoinsecticides for insect biological control. *Applied Microbiology and Biotechnology, 99*, 1057–1068. http://dx.doi.org/10.1007/s00253-014-6270-x.

Panek, A. D., Bernardes, E., Ortiz, C. H., Baker, S., & Mattoon, J. R. (1981). Biochemical genetics of trehalose metabolism in Saccharomyces-Cerevisiae. *Anais da Academia Brasileira de Ciencias, 53*, 165–172.

Pedrini, N., Ortiz-Urquiza, A., Huarte-Bonnet, C., Fan, Y., Juarez, M. P., & Keyhani, N. O. (2015). Tenebrionid secretions and a fungal benzoquinone oxidoreductase form competing components of an arms race between a host and pathogen. *Proceedings of the National Academy of Sciences of the United States of America, 112*, E3651–E3660. http://dx.doi.org/10.1073/pnas.1504552112.

Pedrini, N., Ortiz-Urquiza, A., Huarte-Bonnet, C., Zhang, S., & Keyhani, N. O. (2013). Targeting of insect epicuticular lipids by the entomopathogenic fungus *Beauveria bassiana*: hydrocarbon oxidation within the context of a host-pathogen interaction. *Frontiers in Microbiology, 4*, 24. http://dx.doi.org/10.3389/fmicb.2013.00024.

Pedrini, N., Zhang, S. Z., Juarez, M. P., & Keyhani, N. O. (2010). Molecular characterization and expression analysis of a suite of cytochrome P450 enzymes implicated in insect hydrocarbon degradation in the entomopathogenic fungus *Beauveria bassiana*. *Microbiology-Sgm, 156*, 2549–2557. http://dx.doi.org/10.1099/Mic.0.039735-0.

Pegram, R., Wyatt, R. D., & Smith, T. L. (1981). Oosporein-toxicosis in the turkey poult. *Avian Diseases, 26*, 47–59.

Pendland, J. C., Hung, S. Y., & Boucias, D. G. (1993). Evasion of host defense by in vivo-produced protoplast-like cells of the insect mycopathogen *Beauveria bassiana. Journal of Bacteriology, 175*, 5962–5969.

Pfeifer, T. A., Hegedus, D. D., & Khachatourians, G. G. (1993). The mitochondrial genome of the entomopathogenic fungus *Beauveria bassiana*: analysis of the ribosomal RNA region. *Canadian Journal of Microbiology, 39*, 25–31.

Pfeifer, T. A., & Khachatourians, G. G. (1989). Isolation and characterization of DNA from the entomopathogen *Beauveria bassiana. Experimental Mycology, 13*, 392–402. http://dx.doi.org/10.1016/0147-5975(89)90035-2.

Pfeifer, T. A., & Khachatourians, G. G. (1992). *Beauveria bassiana* protoplast regeneration and transformation using electroporation. *Applied Microbiology and Biotechnology, 38*, 376–381.

Pomerening, J. R., Ubersax, J. A., & Ferrell, J. E. (2008). Rapid cycling and precocious termination of G1 phase in cells expressing CDK1AF. *Molecular Biology of the Cell, 19*, 3426–3441. http://dx.doi.org/10.1091/mbc.E08-02-0172.

Qazi, S. S., & Khachatourians, G. G. (2008). Addition of exogenous carbon and nitrogen sources to aphid exuviae modulates synthesis of proteases and chitinase by germinating conidia of *Beauveria bassiana. Archives of Microbiology, 189*, 589–596.

Qin, Y., Ortiz-Urquiza, A., & Keyhani, N. O. (2014). A putative methyltransferase, *mtrA*, contributes to development, spore viability, protein secretion, and virulence in the entomopathogenic fungus *Beauveria bassiana. Microbiology*. http://dx.doi.org/10.1099/mic.0.078469–0.

Qiu, L., Wang, J. J., Chu, Z. J., Ying, S. H., & Feng, M. G. (2014). Phytochrome controls conidiation in response to red/far-red light and daylight length and regulates multistress tolerance in *Beauveria bassiana. Environmental Microbiology, 16*, 2316–2328. http://dx.doi.org/10.1111/1462-2920.12486.

Qiu, L., Wang, J. J., Ying, S. H., & Feng, M. G. (2015). Wee1 and Cdc25 control morphogenesis, virulence and multistress tolerance of *Beauveria bassiana* by balancing cell cycle-required cyclin-dependent kinase 1 activity. *Environmental Microbiology, 17*, 1119–1133. http://dx.doi.org/10.1111/1462-2920.12530.

Rao, Y. K., Lu, S. C., Liu, B. L., & Tzeng, Y. M. (2006). Enhanced production of an extracellular protease from *Beauveria bassiana* by optimization of cultivation processes. *Biochemical Engineering Journal, 28*, 57–66. http://dx.doi.org/10.1016/j.bej.2005.09.005.

Rehner, S. A., & Buckley, E. (2005). A *Beauveria* phylogeny inferred from nuclear ITS and EF1-alpha sequences: evidence for cryptic diversification and links to Cordyceps teleomorphs. *Mycologia, 97*, 84–98.

Rehner, S. A., Minnis, A. M., Sung, G. H., Luangsa-ard, J. J., Devotto, L., & Humber, R. A. (2011). Phylogeny and systematics of the anamorphic, entomopathogenic genus *Beauveria. Mycologia, 103*, 1055–1073. http://dx.doi.org/10.3852/10-302.

Roy, H. E., Steinkraus, D. C., Eilenberg, J., Hajek, A. E., & Pell, J. K. (2006). Bizarre interactions and endgames: entomopathogenic fungi and their arthropod hosts. *Annual Review of Entomology, 51*, 331–357.

Safavi, S. A. (2011). Successive subculturing alters spore-bound Pr1 activity, germination and virulence of the entomopathogenic fungus, *Beauveria bassiana. Biocontrol Science and Technology, 21*, 883–890. http://dx.doi.org/10.1080/09583157.2011.588317.

Samsinakova, A., & Kalalova, S. (1983). The influence of a single-spore isolate and repeated subculturing on the pathogenicity of conidia of the entomophagous fungus *Beauveria bassiana. Journal of Invertebrate Pathology, 42*, 156–161.

Shapiro-Ilan, D. I., Fuxa, J. R., Lacey, L. A., Onstad, D. W., & Kaya, H. K. (2005). Definitions of pathogenicity and virulence in invertebrate pathology. *Journal of Invertebrate Pathology, 88*, 1–7. http://dx.doi.org/10.1016/j.jip.2004.10.003.

Shimizu, S., Tsuchitani, Y., & Matsumoto, T. (1993). Production of an extracellular protease by *Beauveria bassiana* in the hemolymph of the silkworm, *Bombyx mori*. *Letters in Applied Microbiology, 16*, 291–294. http://dx.doi.org/10.1111/j.1472-765X.1993.tb00360.x.
Smith, R. J., & Grula, E. A. (1983). Chitinase is an inducible enzyme in *Beauveria bassiana*. *Journal of Invertebrate Pathology, 42*, 319–326.
Solomon, P. S., Waters, O. D. C., Jorgens, C. I., Lowe, R. G. T., Rechberger, J., Trengove, R. D., et al. (2006). Mannitol is required for asexual sporulation in the wheat pathogen *Stagonospora nodorum* (glume blotch). *Biochemical Journal, 399*, 231–239.
Song, T. T., & Feng, M. G. (2011). In vivo passages of heterologous *Beauveria bassiana* isolates improve conidial surface properties and pathogenicity to *Nilaparvata lugens* (Homoptera: Delphacidae). *Journal of Invertebrate Pathology, 106*, 211–216. http://dx.doi.org/10.1016/j.jip.2010.09.022.
Song, T. T., Zhao, J., Ying, S. H., & Feng, M. G. (2013). Differential contributions of five ABC transporters to mutidrug resistance, antioxidion and virulence of *Beauveria bassiana*, an entomopathogenic fungus. *PLoS One, 8*, e62179. http://dx.doi.org/10.1371/journal.pone.0062179.
St Leger, R. J., Cooper, R. M., & Charnley, A. K. (1987). Distribution of chymoelastases and trypsin-like enzymes in five species of entomopathogenic deuteromycetes. *Archives of Biochemistry and Biophysics, 258*, 123–131.
St Leger, R. J., Joshi, L., Bidochka, M. J., Rizzo, N. W., & Roberts, D. W. (1996). Characterization and ultrastructural localization of chitinases from *Metarhizium anisopliae*, M-flavoviride, and *Beauveria bassiana* during fungal invasion of host (*Manduca sexta*) cuticle. *Applied and Environmental Microbiology, 62*, 907–912.
St Leger, R., Joshi, L., Bidochka, M. J., & Roberts, D. W. (1996). Construction of an improved mycoinsecticide overexpressing a toxic protease. *Proceedings of the National Academy of Sciences of the United States of America, 93*, 6349–6354.
St Leger, R. J., Joshi, L., & Roberts, D. W. (1997). Adaptation of proteases and carbohydrates of saprophytic, phytopathogenic and entomopathogenic fungi to the requirements of their ecological niches. *Microbiology, 143*(Pt 6), 1983–1992.
Strasser, H., Abendstein, D., Stuppner, H., & Butt, T. M. (2000). Monitoring the distribution of secondary metabolites produced by the entomogenous fungus *Beauveria brongniartii* with particular reference to oosporein. *Mycological Research, 104*, 1227–1233.
Tartar, A., & Boucias, D. G. (2004). A pilot-scale expressed sequence tag analysis of *Beauveria bassiana* gene expression reveals a tripeptidyl peptidase that is differentially expressed in vivo. *Mycopathologia, 158*, 201–209.
Terry, B. J., Liu, W. C., Cianci, C. W., Proszynski, E., Fernandes, P., Bush, K., et al. (1992). Inhibition of herpes-simplex virus type-1 DNA-polymerase by the natural product oosporein. *Journal of Antibiotics, 45*, 286–288.
Thomas, S. R., & Elkinson, J. S. (2004). Pathogenicity and virulence. *Journal of Invertebrate Pathology, 85*, 146–151. http://dx.doi.org/10.1016/j.jip.2004.01.006.
Urtz, B. E., & Rice, W. C. (2000). Purification and characterization of a novel extracellular protease from *Beauveria bassiana*. *Mycological Research, 104*, 180–186. http://dx.doi.org/10.1017/S0953756299001215.
Vega, F. E., Goettel, M. S., Blackwell, M., Chandler, D., Jackson, M. A., Keller, S., et al. (2009). Fungal entomopathogens: new insights on their ecology. *Fungal Ecology, 2*, 149–159.
Vega, F. E., Posada, F., Aime, M. C., Pava-Ripoll, M., Infante, F., & Rehner, S. A. (2008). Entomopathogenic fungal endophytes. *Biological Control, 46*, 72–82.
Viaud, M., Couteaudier, Y., Levis, C., & Riba, G. (1996). Genome organization in *Beauveria bassiana*: electrophoretic karyotype, gene mapping, and telomeric fingerprints. *Fungal Genetics and Biology, 20*, 175–183.

Wanchoo, A., Lewis, M. W., & Keyhani, N. O. (2009). Lectin mapping reveals stage-specific display of surface carbohydrates in in vitro and haemolymph-derived cells of the entomopathogenic fungus *Beauveria bassiana*. *Microbiology-Sgm, 155*, 3121–3133.

Wang, C. S., & Leger, R. J. S. (2007). The *Metarhizium anisopliae* perilipin homolog *MPL1* regulates lipid metabolism, appressorial turgor pressure, and virulence. *Journal of Biological Chemistry, 282*, 21110–21115.

Wang, C. S., & St Leger, R. J. (2007). The MAD1 adhesin of *Metarhizium anisopliae* links adhesion with blastospore production and virulence to insects, and the MAD2 adhesin enables attachment to plants. *Eukaryotic Cell, 6*, 808–816.

Wang, J., Liu, J., Hu, Y., Ying, S. H., & Feng, M. G. (2013). Cytokinesis-required Cdc14 is a signaling hub of asexual development and multi-stress tolerance in *Beauveria bassiana*. *Scientific Reports-UK, 3*, 3086. http://dx.doi.org/10.1038/Srep03086.

Wang, J., Zhou, G., Ying, S. H., & Feng, M. G. (2013). P-type calcium ATPase functions as a core regulator of *Beauveria bassiana* growth, conidiation and responses to multiple stressful stimuli through cross-talk with signalling networks. *Environmental Microbiology, 15*, 967–979. http://dx.doi.org/10.1111/1462-2920.12044.

Wang, J., Zhou, G., Ying, S. H., & Feng, M. G. (2014). Adenylate cyclase orthologues in two filamentous entomopathogens contribute differentially to growth, conidiation, pathogenicity, and multistress responses. *Fungal Biology, 118*, 422–431. http://dx.doi.org/10.1016/j.funbio.2014.03.001.

Wang, J. J., Qiu, L., Cai, Q., Ying, S. H., & Feng, M. G. (2014). Three alpha-1,2-mannosyltransferases contribute differentially to conidiation, cell wall integrity, multistress tolerance and virulence of *Beauveria bassiana*. *Fungal Genetics and Biology, 70*, 1–10. http://dx.doi.org/10.1016/j.fgb.2014.06.010.

Wang, J. J., Qiu, L., Cai, Q., Ying, S. H., & Feng, M. G. (2015). Transcriptional control of fungal cell cycle and cellular events by Fkh2, a forkhead transcription factor in an insect pathogen. *Scientific Reports-UK, 5*, 10108. http://dx.doi.org/10.1038/srep10108.

Wang, J. J., Qiu, L., Chu, Z. J., Ying, S. H., & Feng, M. G. (2014). The connection of protein O-mannosyltransferase family to the biocontrol potential of *Beauveria bassiana*, a fungal entomopathogen. *Glycobiology, 24*, 638–648. http://dx.doi.org/10.1093/glycob/cwu028.

Wang, X. X., He, P. H., Feng, M. G., & Ying, S. H. (2014). BbSNF1 contributes to cell differentiation, extracellular acidification, and virulence in *Beauveria bassiana*, a filamentous entomopathogenic fungus. *Applied Microbiology and Biotechnology, 98*, 8657–8673. http://dx.doi.org/10.1007/s00253-014-5907-0.

Wang, X. X., Ji, X. P., Li, J. X., Keyhani, N. O., Feng, M. G., & Ying, S. H. (2013). The putative alpha-glucoside transporter gene *BbAGT1* contributed to carbohydrate utilization, growth, conidiation and virulence of filamentous entomopathogenic fungi *Beauveria bassiana*. *Research in Microbiology*. http://dx.doi.org/10.1016/j.resmic.2013.02.008.

Wang, Z. L., Li, F., Li, C., & Feng, M. G. (2014). Bbssk1, a response regulator required for conidiation, multi-stress tolerance, and virulence of *Beauveria bassiana*. *Applied Microbiology and Biotechnology, 98*, 5607–5618. http://dx.doi.org/10.1007/s00253-014-5644-4.

Wang, Z. L., Lu, J. D., & Feng, M. G. (2012). Primary roles of two dehydrogenases in the mannitol metabolism and multi-stress tolerance of entomopathogenic fungus *Beauveria bassiana*. *Environmental Microbiology, 14*, 2139–2150. http://dx.doi.org/10.1111/j.1462-2920.2011.02654.x.

Wang, Z. L., Zhang, L. B., Ying, S. H., & Feng, M. G. (2013). Catalases play differentiated roles in the adaptation of a fungal entomopathogen to environmental stresses. *Environmental Microbiology, 15*, 409–418. http://dx.doi.org/10.1111/j.1462-2920.2012.02848.x.

Wraight, S. P., Ramos, M. E., Avery, P. B., Jaronski, S. T., & Vandenberg, J. D. (2010). Comparative virulence of *Beauveria bassiana* isolates against lepidopteran pests of vegetable crops. *Journal of Invertebrate Pathology, 103*, 186–199. http://dx.doi.org/10.1016/j.jip.2010.01.001.

Wu, J., Ridgway, H., Carpenter, M., & Glare, T. (2008a). Efficient transformation of *Beauveria bassiana* by *Agrobacterium tumefaciens*-mediated insertional mutagenesis. *Australasian Plant Pathology, 37*, 537–542.

Wu, J., Ridgway, H. J., Carpenter, M. A., & Glare, T. R. (2008b). Identification of novel genes associated with conidiation in *Beauveria bassiana* with suppression subtractive hybridization. *Mycologia, 100*, 20–30.

Xiao, G., Ying, S. H., Zheng, P., Wang, Z. L., Zhang, S., Xie, X. Q., et al. (2012). Genomic perspectives on the evolution of fungal entomopathogenicity in *Beauveria bassiana. Scientific Reports, 2*, 483. http://dx.doi.org/10.1038/srep00483.

Xie, X. Q., Guan, Y., Ying, S. H., & Feng, M. G. (2013). Differentiated functions of Ras1 and Ras2 proteins in regulating the germination, growth, conidiation, multi-stress tolerance and virulence of *Beauveria bassiana. Environmental Microbiology, 15*, 447–462. http://dx.doi.org/10.1111/j.1462-2920.2012.02871.x.

Xie, X. Q., Li, F., Ying, S. H., & Feng, M. G. (2012). Additive contributions of two manganese-cored superoxide dismutases (MnSODs) to antioxidation, UV tolerance and virulence of *Beauveria bassiana. PLoS One, 7*, e30298. http://dx.doi.org/10.1371/journal.pone.0030298.

Xie, X. Q., Wang, J., Huang, B. F., Ying, S. H., & Feng, M. G. (2010). A new manganese superoxide dismutase identified from *Beauveria bassiana* enhances virulence and stress tolerance when overexpressed in the fungal pathogen. *Applied Microbiology and Biotechnology, 86*, 1543–1553. http://dx.doi.org/10.1007/S00253-010-2437-2.

Xu, Y., Orozco, R., Wijeratne, E. M. K., Gunatilaka, A. A. L., Stock, S. P., & Molnar, I. (2008). Biosynthesis of the cyclooligomer depsipeptide beauvericin, a virulence factor of the entomopathogenic fungus *Beauveria bassiana. Chemistry & Biology, 15*, 898–907.

Xu, Y., Zhan, J., Wijeratne, E. M. K., Burns, A. M., Gunatilaka, A. A. L., & Molnar, I. (2007). Cytotoxic and antihaptotactic beauvericin analogues from precursor-directed biosynthesis with the insect pathogen *Beauveria bassiana* ATCC 7159. *Journal of Natural Products, 70*, 1467–1471.

Xu, Y. Q., Rozco, R., Wijeratne, E. M. K., Espinosa-Artiles, P., Gunatilaka, A. A. L., Stock, S. P., et al. (2009). Biosynthesis of the cyclooligomer depsipeptide bassianolide, an insecticidal virulence factor of *Beauveria bassiana. Fungal Genetics and Biology, 46*, 353–364.

Xu, Y. Q., Wijeratne, E. M. K., Espinosa-Artiles, P., Gunatilaka, A. A. L., & Molnar, I. (2009). Combinatorial mutasynthesis of scrambled beauvericins, cyclooligomer depsipeptide cell migration inhibitors from *Beauveria bassiana. Chembiochem, 10*, 345–354.

Ying, S. H., & Feng, M. G. (2011). A conidial protein (CP15) of *Beauveria bassiana* contributes to the conidial tolerance of the entomopathogenic fungus to thermal and oxidative stresses. *Applied Microbiology and Biotechnology, 90*, 1711–1720. http://dx.doi.org/10.1007/s00253-011-3205-7.

Ying, S. H., Feng, M. G., & Keyhani, N. O. (2013a). A carbon responsive G-protein coupled receptor modulates broad developmental and genetic networks in the entomopathogenic fungus, *Beauveria bassiana. Environmental Microbiology, 15*, 2902–2921. http://dx.doi.org/10.1111/1462-2920.12169.

Ying, S. H., Feng, M. G., & Keyhani, N. O. (2013b). Use of uridine auxotrophy (*ura3*) for markerless transformation of the mycoinsecticide *Beauveria bassiana. Applied Microbiology and Biotechnology, 97*, 3017–3025. http://dx.doi.org/10.1007/s00253-012-4426-0.

Ying, S. H., Ji, X. P., Wang, X. X., Feng, M. G., & Keyhani, N. O. (2014). The transcriptional co-activator multiprotein bridging factor 1 from the fungal insect pathogen, *Beauveria bassiana*, mediates regulation of hyphal morphogenesis, stress tolerance and virulence. *Environmental Microbiology, 16*, 1879–1897. http://dx.doi.org/10.1111/1462-2920.12450.

Zhang, L., Wang, J., Xie, X. Q., Keyhani, N. O., Feng, M. G., & Ying, S. H. (2013). The autophagy gene *BbATG5*, involved in the formation of the autophagosome, contributes to cell differentiation and growth but is dispensable for pathogenesis in the entomopathogenic fungus *Beauveria bassiana*. *Microbiology, 159*, 243–252. http://dx.doi.org/10.1099/mic.0.062646-0.

Zhang, L. B., Tang, L., Ying, S. H., & Feng, M. G. (2015). Subcellular localization of six thioredoxins and their antioxidant activity and contributions to biological control potential in *Beauveria bassiana*. *Fungal Genetics and Biology, 76*, 1–9. http://dx.doi.org/10.1016/j.fgb.2015.01.008.

Zhang, S., Fan, Y., Xia, Y. X., & Keyhani, N. O. (2010). Sulfonylurea resistance as a new selectable marker for the entomopathogenic fungus *Beauveria bassiana*. *Applied Microbiology and Biotechnology, 87*, 1151–1156. http://dx.doi.org/10.1007/s00253-010-2636-x.

Zhang, S., Widemann, E., Bernard, G., Lesot, A., Pinot, F., Pedrini, N., et al. (2012). CYP52X1, representing new cytochrome P450 subfamily, displays fatty acid hydroxylase activity and contributes to virulence and growth on insect cuticular substrates in entomopathogenic fungus *Beauveria bassiana*. *Journal of Biological Chemistry, 287*, 13477–13486. http://dx.doi.org/10.1074/jbc.M111.338947.

Zhang, S. Z., Xia, Y. X., & Keyhani, N. O. (2011). Contribution of the *gas1* gene of the entomopathogenic fungus *Beauveria bassiana*, encoding a putative glycosylphosphatidylinositol-anchored beta-1,3-glucanosyltransferase, to conidial thermotolerance and virulence. *Applied and Environmental Microbiology, 77*, 2676–2684.

Zhang, S. Z., Xia, Y. X., Kim, B., & Keyhani, N. O. (2011). Two hydrophobins are involved in fungal spore coat rodlet layer assembly and each play distinct roles in surface interactions, development and pathogenesis in the entomopathogenic fungus, *Beauveria bassiana*. *Molecular Microbiology, 80*, 811–826.

Zhang, Y. J., Feng, M. G., Fan, Y. H., Luo, Z. B., Yang, X. Y., Wu, D., et al. (2008). A cuticle-degrading protease (CDEP-1) of *Beauveria bassiana* enhances virulence. *Biocontrol Science and Technology, 18*, 551–563. http://dx.doi.org/10.1080/09583150802082239.

Zhang, Y. J., Zhang, J. Q., Jiang, X. D., Wang, G. J., Luo, Z. B., Fan, Y. H., et al. (2010). Requirement of a mitogen activated protein kinase for appressorium formation and penetration of insect cuticle by the entomopathogenic fungus *Beauveria bassiana*. *Applied and Environmental Microbiology, 76*, 2262–2270. http://dx.doi.org/10.1128/Aem.02246-09.

Zhang, Y. J., Zhao, J. H., Fang, W. G., Zhang, J. Q., Luo, Z. B., Zhang, M., et al. (2009). Mitogen-activated protein kinase hog1 in the entomopathogenic fungus *beauveria bassiana* regulates environmental stress responses and virulence to insects. *Applied and Environmental Microbiology, 75*, 3787–3795. http://dx.doi.org/10.1128/Aem.01913-08.

Zibaee, A., & Bandani, A. R. (2009). Purification and characterization of the cuticle-degrading protease produced by the entomopathogenic fungus, *Beauveria bassiana* in the presence of Sunn pest, *Eurygaster integriceps* (Hemiptera: Scutelleridae) cuticle. *Biocontrol Science and Technology, 19*, 797–808. http://dx.doi.org/10.1080/09583150903132172. pii:913279411.

Zimmermann, G. (1986). The *Galleria* bait method for detection of entomopathogenic fungi in soil. *Journal of Applied Entomology-Zeitschrift Fur Angewandte Entomologie, 102*, 213–215.

CHAPTER SEVEN

Insect Immunity to Entomopathogenic Fungi

H.-L. Lu[1] and R.J. St. Leger
University of Maryland, College Park, MD, United States
[1]Corresponding author: E-mail: hllu@umd.edu

Contents

Abstract

The study of infection and immunity in insects has achieved considerable prominence with the appreciation that their host defense mechanisms share many fundamental characteristics with the innate immune system of vertebrates. Studies on the highly tractable model organism *Drosophila* in particular have led to a detailed understanding of conserved innate immunity networks, such as Toll. However, most of these studies have used opportunistic human pathogens and may not have revealed specialized immune strategies that have arisen through evolutionary arms races with natural insect pathogens. Fungi are the commonest natural insect pathogens, and in this review, we focus on studies using *Metarhizium* and *Beauveria* spp. that have addressed immune system function and pathogen virulence via behavioral avoidance, the use of physical barriers, and the activation of local and systemic immune responses. In particular, we highlight studies on the evolutionary genetics of insect immunity and discuss insect–pathogen coevolution.

Advances in Genetics, Volume 94
ISSN 0065-2660
http://dx.doi.org/10.1016/bs.adgen.2015.11.002

Throughout the course of their lives, insects are surrounded by a heterogeneous microbial population. Individually, most of these microbes will have an infinitesimal impact on the insect's health and welfare, but some will be symbionts that help keep the insect alive whereas others will be pathogens that would destroy them. Only a very small minority of microbes can successfully parasitize healthy insects. It is axiomatic that the survival of insects has depended on their ability to interpose physical and chemical barriers to surrounding organisms. The first, and most effective of these barriers is the cuticular integument: a physical barrier to infection composed of macromolecules, such as tanned proteins and chitin (St. Leger, 1991; St Leger, Goettel, Roberts, & Staples, 1991). The value of this barrier can be clearly seen when it is breached in some way. Physical damage to the cuticle allows microbial invasion of the underlying nutrient-rich tissues, and it is a commonplace observation that normally innocuous microbes can prove lethal if the cuticle is wounded or bypassed by direct injection of spores. The *Drosophila* model system for studying human pathogens, including the fungus *Aspergillus fumigatus*, has depended on this property. The discovery that the *Drosophila* Toll pathway was involved in host defense against injected *A. fumigatus* infections ushered in a new era for the study of immunity in this model organism, as well as in mammals (Lemaitre, Nicolas, Michaut, Reichhart, & Hoffmann, 1996; Lionakis, 2011). Since then, *Drosophila*'s high degree of molecular, cellular, and physiological conservation with humans has allowed the modeling of 43 human pathogens; much of what we know about innate immunity was first deciphered by injecting opportunistic human pathogens into immunocompromised flies (Panayidou, Ioannidou, & Apidianakis, 2014).

These studies have shown that the immune defense in *Drosophila melanogaster* is based on two main components: the humoral, or systemic, immunity and the cellular immunity (Lemaitre & Hoffmann, 2007) (Fig. 1). The fast-acting responses are largely mediated by hemocytes circulating in the hemolymph and include the coagulation or melanization of foreign objects, phagocytosis of microbes, and cellular encapsulation of parasites (Lemaitre & Hoffmann, 2007). The slow response is induced over the course of several hours following a systemic infection and is tailored to combat specific pathogen classes. The antifungal response in *Drosophila* is largely mediated by the evolutionarily conserved *Toll* (*Tl*) pathway (Lemaitre & Hoffmann, 2007), and leads to induction of antifungal peptides, mainly Drosomycin (Drs) and Metchnikowin, into the hemolymph. Many of these pathways are upregulated in response to natural insect pathogens

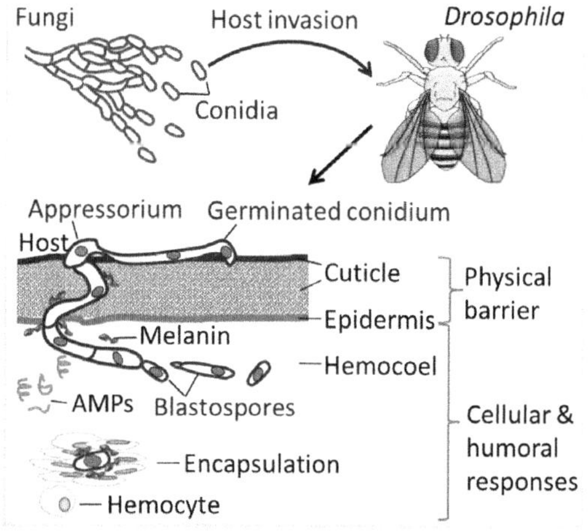

Figure 1 Schematic representation of *Drosophila melanogaster* immunity. Fungal invasion is initiated by the attachment and germination of a conidium (spore) on the host cuticle and the formation of an appressorium (a sticky holdfast). Appressoria produce hyphae that penetrate the insect cuticle via a combination of mechanical pressure and hydrolytic enzymes. Penetrant hyphae bud off blastospores, which multiply and circulate in the insect hemocoel. The cuticle acts as an initial physical barrier and first line of defense in insects. Fungi that succeed in breaching the cuticle induce a wide range of humoral (eg, production of antimicrobial peptides) and cellular responses (eg, encapsulation and melanization). *AMPs*, antimicrobial peptides.

such as *Metarhizium* and *Beauveria* although often with little effect (Lu, Wang, Brown, Euerle, & St Leger, 2015), suggesting that opportunistic human pathogens do not adequately model natural pathogens. In order to understand the natural dynamics of infection, natural pathogens and manipulations of pathogen loads via natural routes of infection are required (Keebaugh & Schlenke, 2014; Kounatidis & Ligoxygakis, 2012; Magwire, Bayer, Webster, Cao, & Jiggins, 2011; Rolff & Reynolds, 2009; Sibley et al., 2008). Furthermore, in order to understand the evolution of the insect immune system, we need to be able to quantify the strength of selection in wild populations (Kraaijeveld & Wertheim, 2009). Given the importance of *Drosophila* to our understanding of immunity, it is surprising that very little is known about the rate of infection of *Drosophila* by fungi, microsporidia, and bacteria in the field (Keebaugh & Schlenke, 2014; Kraaijeveld & Godfray, 2009), although genetic variation for resistance against *Beauveria* does exist within natural populations (Tinsley, Blanford, & Jiggins, 2006). As noted by Rolff and Reynolds (2009), there is likewise very little data quantifying the frequency of wounds in natural populations of insects, which at least would offer an estimate of the opportunities for opportunistic infection. This has important implications as septic infections imposed on experimental insects through injection or body piercing are often inherently unnatural. Moreover, insect immune genes must be assumed to have coevolved with the virulence genes of the most frequently encountered, often specialist pathogens. Thus, much of what we know about immune systems is based on how *Drosophila* responds to pathogens and infection modes they have rarely, if ever, encountered in nature during their evolutionary history (Keebaugh & Schlenke, 2014). Broad host range strains of *Metarhizium* and *Beauveria* have been used in several studies on *Drosophila* (Gottar et al., 2006; Matskevich, Quintin, & Ferrandon, 2010; Roxström-Lindquist, Terenius, & Faye, 2004; Tinsley et al., 2006), but given their broad host range, it seems unlikely that they are engaging in a strict coevolutionary arms race with a particular *Drosophila* population (Kawecki, 1998). Narrow host range entomopathogenic fungi exist by the hundreds of thousands, if not millions, but specific *Drosophila* pathogens that would facilitate understanding of host–pathogen evolution and identify specialized immune mechanisms have not been identified.

There are a lot of important review papers in the literature, which present, in detail, specific groups of signaling pathways or mechanisms of innate immunity in insects. This review will provide an overview of these mechanisms, many of which were described using opportunistic pathogens. We will, however, emphasize studies covering the genetic architecture of insect host interactions with natural fungal entomopathogens. Fungi cause the majority of insect disease (Roberts & St Leger, 2004), and both broad and narrow host range fungi play a crucial role in natural ecosystems and are being developed as alternatives to chemical insecticides (St Leger, Wang, & Fang, 2011).

1. BEHAVIORAL AVOIDANCE OF PATHOGENS

The most effective defense against disease may be behavioral avoidance of pathogens (De Roode & Lefèvre, 2012). Examples of this include social insects, such as the termite *Macrotermes michaelseni* that can ascertain the virulence of *Metarhizium* and *Beauveria* strains from a distance and is thus more strongly repelled by the more virulent strains (Mburu et al., 2009), and the bug *Anthocoris nemorum* that avoids foraging and ovipositing on plants contaminated with *Beauveria bassiana* spores (Meyling, Pell, & Eilenberg, 2006). Social insects have limited genetic diversity in their crowded colonies, and honeybees and ants have fewer genes responsible for innate immunity than some other insects possess (Evans et al., 2006; Libbrecht, Oxley, Kronauer, & Keller, 2013). *Prima facie*, these features should make them particularly prone to disease, but they are in practice highly resistant to most nonspecialist pathogens. Social insects have multiple collective behavioral defenses against pathogens, such as grooming other colony members (Schmid-Hempel, 1998) or the intake of tree resin with antipathogenic properties (Christe, Oppliger, Bancala, Castella, & Chapuisat, 2003). It has thus been proposed that such prophylactic behaviors might reduce the selective pressure for increasing the number of immune genes in the genome (C.D. Smith et al., 2011; C.R. Smith et al., 2011; Suen et al., 2011). Often called social immunity, grooming nest mates and other behaviors may have multiple functions and, thereby be more efficient than having individual genetic components for innate immunity (Evans & Spivak, 2010; Le Conte et al., 2011). An important example of hygienic behavior involves worker honeybees (*Apis mellifera*) being able to detect and remove from the nest larvae infected with the fungus *Ascosphaera apis* (chalkbrood). The chemical cue that bees use to recognize infected brood is phenethyl acetate (Swanson et al., 2009), and each step in the removal—uncapping a diseased brood cell, followed by removal of the contents—is controlled by a suite of genes, including four genes involved in olfaction, learning and social behavior, and one gene involved in circadian locomotion (Oxley, Spivak, & Oldroyd, 2010). Another example of a pleotropic social immune behavior is production of propolis, which is a resinous substance that honeybees encase their hives in that, is simultaneously waterproof and antimicrobial (Evans & Spivak, 2010).

Behavioral analyses in conjunction with RNA-seq of genes involved in physiological immune defenses in the leaf cutter ant *Acromyrmex echinatior* confirmed the existence of efficient prophylactic behaviors and showed that pathogenic challenges with the ant pathogen *Metarhizium brunneum* triggered an increase in immune gene expression. However, ants challenged with a fungus-garden pathogen showed a decrease in immune gene expression while displaying more prophylactic behaviors, suggesting that trade-offs might occur between physiological and behavioral immune responses (Yek, Boomsma, & Schiøtt, 2013). The concept of trade-offs is central to many evolutionary hypotheses for limited life span and optimal allocation of resources (Garland, 2014).

However, the presumption that a reduction in immunity genes in social insects is associated with social grooming has been recently challenged.

A comparative-genomic analysis showed that only 3 of 16 immune gene families differed significantly between social and solitary insect species (Simola et al., 2013). This finding, combined with the fact that the nonsocial parasitoid wasp *Nasonia vitripennis* also contains fewer immune genes than do flies and beetles (Werren et al., 2010), suggests that the depletion of immune genes in social insects is not as dramatic as initially proposed and might not be directly associated with sociality. This obviously does not detract from the existence and benefits of social immunity.

2. THE IMPACT OF PHYSIOLOGICAL STATE ON IMMUNE FUNCTIONS IN INSECTS

Insect immune systems are known to be dynamic, and their responsiveness fluctuates with changes in the native microbiota, energy availability, feeding, circadian rhythm, age, and even past exposure to microbes (Schneider, 2009). This plasticity exists because of a complex web of interconnections between the immune system and other physiological systems that can influence behavior. A well-studied example in insects is behavioral fever. Insects seek environments above their own optimum temperature when this can also allow insects to limit pathogen development. Insects like migratory locust, *Locusta migratoria*, can overcome *Metarhizium* infection by basking in the sun and raising their body temperature to fever levels (Elliot, Blanford, & Thomas, 2002; Ouedraogo, Cusson, Goettel, & Brodeur, 2003).

Changes in physiological state can alter immune system function directly via neural/neuroendocrine/immune connections that adapt the immune system to changing needs. Changes in physiological state can also alter immune system function indirectly by reducing the resources available for an immune response (Adamo, 2009).

Recently, mutant screens have focused on pathogen life history traits, such as within host growth, host survival, and sporulation (pathogen reproduction), whereas most earlier studies just monitored antimicrobial peptide (AMP) transcription (Lemaitre & Hoffmann, 2007). These studies cast a wider net than has been previously used to measure immunity to fungi in *Drosophila*, enabling investigations on how host genotypes differentially affect pathogen fitness and how defense is interconnected with other aspects of host physiology that can set the stage for trade-offs between immunity and other costly life history traits (Ayres, Freitag, & Schneider, 2008; Lu et al., 2015; Schmid-Hempel, 2003; Short & Lazzaro, 2013). Using *Metarhizium anisopliae* ARSEF 549 (Ma549) as a natural fungal pathogen of

Drosophila in a mutagenesis screen revealed that 9% of the Minos element insertions tested affected disease resistance to Ma549 (Lu et al., 2015). Thus a large fraction of the genome contributes to disease resistance predicting extensive pleiotropy. Lu et al. (2015) also reported that 87% of mutated genes in mutant fly lines more susceptible to *Metarhizium* infection are involved in a broad spectrum of biological functions not directly connected with the canonical immune system, including basic cellular processes (mitosis, transcription, translation, and protein modification), early development, muscle and nervous system development and function, chemosensation and vision, and metabolism (Fig. 2). Overall, 58 mutations affecting cellular processes and metabolism increase susceptibility (~40% of all susceptibility genes), and it is plausible that dysfunction in these processes could

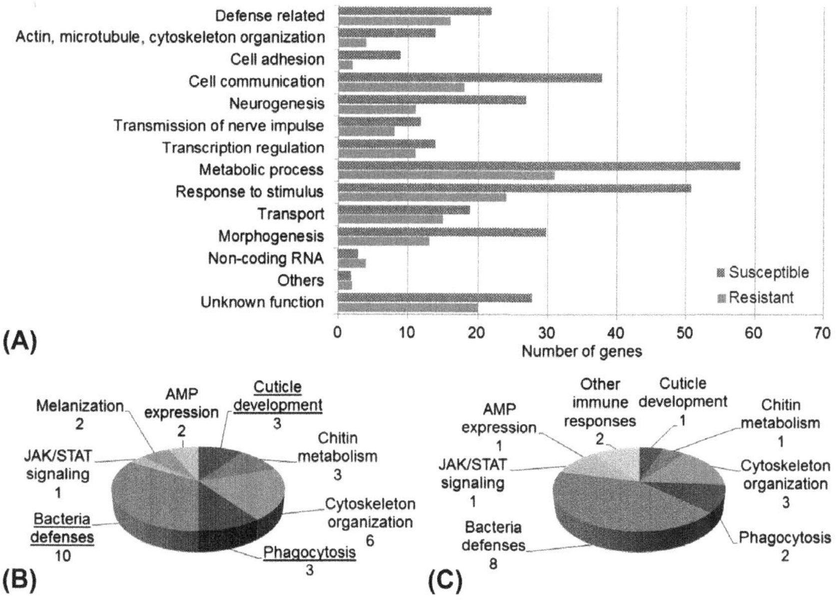

Figure 2 Classification of *Drosophila* genes that change susceptibility to *Metarhizium anisopliae* (Ma549) infection identified from a screen of 2613 mutant fly lines (Lu et al., 2015). Genes with mutations that increase (susceptible, blue (black in print versions) bar with pattern) or decrease (resistant, red (gray in print versions) bar) susceptibility to Ma549 infection were categorized based on gene ontology and published references (A). The x-axis indicates the number of genes in each category. Genes in the "defense-related" category in Fig. 2A subdivided by roles in susceptible (B) and resistant (C) lines (the number of genes in each subcategory is under each heading). Underlined subcategories contain genes that overlap with another subcategory. *AMP*, antimicrobial peptide.

specifically reduce expenditure of energy on immune responses. Approximately 25% of the mutations represent diverse pathways affecting morphogenesis and neurogenesis. These could be involved in tissue regeneration and counteract or repair damage from invasive fungal growth. Twenty-one genes affecting resistance have human orthologs that impact temperament or cognitive development including learning. Overall, a surprising number of mutations affect neuronal functioning implying that disease resistance is linked to behaviors, perhaps by affecting nutrient uptake or social interactions in a manner which alters disease resistance. Consistent with this, many *Drosophila* mutants affecting susceptibility to *M. anisopliae* have orthologs to human genes impacting nutrient uptake and obesity via sensory inputs and immunity, and nine mutated genes have human orthologs that also influence temperament, providing a plausible link between behavioral or cognitive traits and disease resistance. Three of the susceptible *Drosophila* lines were mutated in octopamine (*Octbeta2R, Oamb*) and dopamine (*Dop1R1*) receptors, which work together in *Drosophila* learning to form appetitive memories for sugar and other foods (Burke et al., 2012). Octopamine is also the key hormone involved in the acute stress response and prepares the insect for flight or fight behaviors. It mediates a connection between the nervous system and the immune system because both nerve cells and immune cells have octopamine receptors (Adamo, 2009). Octopamine seems to produce a mix of immunoenhancing and immunosuppressive effects. Thus, injecting octopamine into locusts increases their susceptibility to *M. anisopliae* (Goldsworthy, Opoku-Ware, & Mullen, 2005), but background levels of octopamine in nonstressed insects may help to maintain normal immune functions (immune homeostasis) (Adamo, 2009). The overall effect of stress is to reduce immune function presumably as a trade-off that reconfigures physiological networks to maximize the flight or fight response (Adamo, 2012). The details of this process are largely enigmatic in part due to the complexity of physiological networks and the dynamic nature of the biochemical networks underlying stress and immunity interactions (Adamo, 2012).

It has recently been established that insects can anticipate infections by upregulating immune genes when they find themselves in scenarios associated with increased disease risk. Mating is frequently associated with a heightened risk of disease, as sexually transmitted infections (STIs) are often highly prevalent and pathogenic. The heads of female *Drosophila* specifically upregulate two members of the Turandot family of immune and stress response genes (TotM and TotC) when they hear male courtship song, and TotM has been shown to provide protection following STIs (Zhong et al., 2013). Surprisingly, there is little evidence that courtship stimulates the upregulation of the Toll pathway, although TotM is also induced by fungal infections. STIs differ from other modes of transmission in that they tend to cause chronic low-level infections, which do not result in the rapid septicemia typically associated with acute immune challenges. Zhong et al. (2013) found that sexually transmitted *Metarhizium* causes weak fitness costs to female *Drosophila* that are ameliorated by TotM, but not the canonical antifungal Toll pathway gene Dif, whereas topical *Metarhizium* infections cause substantial fitness costs that were ameliorated by Dif, but not TotM. Significantly therefore, fruit flies have different mechanisms for immunity against low-dose STIs and against high-dose topical infections, even for the same pathogen (Zhong et al., 2013). The mechanisms through which TotM enhances immunity are unknown, but Zhong et al. (2013) speculate that it might help the fly tolerate persistent fungal infections. Clearly, STIs exert a selective pressure strong enough to affect insect immune systems, but there are likely to be many other biological scenarios where immune anticipation could be advantageous, such as during crowding of conspecifics and feeding (Bailey, Gray, & Zuk, 2011; Barnes & Siva-Jothy, 2000). Zhong et al. (2013) raise the interesting possibility that control of many immune genes, including TotM, by circadian clock genes might reflect "anticipation" of predictable fluctuations of disease risk over the course of 24 h. Thus, the courtship induced, preemptive upregulation of TotM might represent a general pattern of immune anticipation in insects, which underlines the intimate link between the brain, behavior, and immunity (Lu et al., 2015; Maier, Watkins, & Fleshner, 1994).

3. CUTICLE AS A BARRIER TO MICROBIAL INFECTIONS

Fungal entomopathogens gain access to the host via cuticle penetration (Goettel, St Leger, Rizzo, Staples, & Roberts, 1989; Vega & Kaya, 2012). As a first step, spores need to adhere to the host surface through mucilage and adhesive proteins (Wang & St Leger, 2007a). Penetration is then achieved both by enzyme secretion and the development of specialized structures (appressoria) that exert mechanical pressure (Wang & St Leger, 2007a). When the fungus reaches the body cavity of the host, it produces hydrolytic enzymes to assimilate nutrients and toxins with immunosuppressive activity (Freimoser, Screen, Bagga, Hu, & St Leger, 2003; Gillespie, Burnett, & Charnley, 2000; Pal, St Leger, & Wu, 2007). If the host fails to clear the infection, the fungus eventually kills the insect and transmission is achieved by sporulation from the cadaver (Fig. 1).

The insect cuticle is a physical barrier to infection composed of macromolecules such as tanned proteins and chitin relatively resistant to degradation and that provides an inhospitable environment for microbes by having a low water activity, a shortage of readily available nutrients, and antimicrobial compounds such as short-chain fatty acids (St. Leger, 1991). The intact cuticle is impervious to organisms that do not possess an active mechanism of cuticle penetration (eg, protozoa, bacteria, *rickettsia*, and viruses). With a few specialized exceptions, these organisms invade insects after being ingested with infected food. Direct penetration of intact cuticle is the normal mode of entry by most fungi adapted for entomopathogenicity. This stems in part from the fact that fungi can actively penetrate and move through host cuticle, forcibly breaching barriers by enzymatic and physical mechanisms. In addition, fungal pathogens have evolved a complex array of tropic responses which may contribute to infection processes. Most fungi are prevented from being entomopathogens because they lack one or more of the necessary morphological or biochemical cuticle-degrading mechanisms, or they have not evolved ways to overcome disease resistance mechanisms in the cuticle.

Comparative genomics has shed new light on the physiology and evolution of insect pathogens by comparing seven *Metarhizium* species (Gao et al., 2011; Hu et al., 2014), *Cordyceps militaris* (Zheng et al., 2011), *Cordyceps sinensis* (Hu et al., 2013), and *B. bassiana* (Xiao et al., 2012), with saprophytes and plant pathogens. Entomopathogens have more proteases than plant pathogenic fungi. This expansion is dramatic in *B. bassiana* and broad host range *Metarhizium* species, and less marked in *Metarhizium acridum* and *Cordyceps* species that are more specialized pathogens. Therefore, the proliferation of proteases may reflect an adaptation to infect insects via the cuticle and is possibly influenced by host range. Another common feature of insect—pathogenic fungi is the abundance of chitinases compared with plants pathogens. This is likely an adaptation to the amount of chitin in the insect cuticle. For entomopathogens, the necessity of crossing the protein—chitin procuticle has therefore had a major impact on their evolution; this is a testament to the critical role the cuticle plays in insect immunity against entomopathogenic fungi.

Other factors including gland secretions and the resident microbial community make the cuticle an even more formidable barrier to infection. Thus, *Bacillus pumilus* on the cuticles of the hemipteran herbivores *Delphacodes kuscheli* and *Dalbulus maidis* inhibits germination of *B. bassiana* (Toledo,

Alippi, & de Remes Lenicov, 2011). Some insects coat their cuticles with gland secretions containing antimicrobials. For instance, all 26 species of ant examined in one study wiped secretions produced by thoracic metapleural glands onto their bodies (Fernández-Marín, Zimmerman, Rehner, & Wcislo, 2006). Ants increase their use of this fungicidal secretion within 1-h postinfection by a fungal entomopathogen (Fernández-Marín et al., 2006). When these glands are covered, infected ants died within a few days. In a similar fashion, bedbugs inhibit *M. anisopliae* with glandular secretions of (E)-2-hexenal and (E)-2-octenal (Ulrich, Feldlaufer, Kramer, & St. Leger, 2015). It has long been recognized that closely related insect species can differ greatly in susceptibility to broad host range strains of entomopathogens. The tenebrionid beetle *Tribolium castaneum* is a resistant species due at least in part to its secreting benzoquinones that act as defensive compounds against *B. bassiana*. The fungus counters by producing a benzoquinone reductase (BqrA) that detoxifies benzoquinones. Knocking out the *BqrA* gene reduces virulence whereas strains overexpressing BqrA show increased virulence, suggesting a coevolutionary arms race between the beetle and the fungus in the production/detoxification of a cuticular antimicrobial compound (Pedrini, Ortiz-Urquiza, Huarte-Bonnet, Fan, Juarez, & Keyhani, 2015).

4. OVERVIEW OF INSECT IMMUNE DEFENSE MECHANISMS

Innate immunity is present in all multicellular forms of life, most of which, like insects, lack the antibody production that characterizes the adaptive immune response of higher vertebrates. The insect innate immune system is capable of recognition and subsequent eradication of some microbes and parasites through humoral and cellular mechanisms (Lemaitre & Hoffmann, 2007). Invading microbes are detected by recognition molecules performing surveillance and the signal is transduced through two primary signaling pathways, termed the Toll and Imd pathways, that approximately regulate antifungal and antibacterial defenses, respectively (Valanne, Wang, & Rämet, 2011). In humoral immunity, AMPs are produced by a specialized tissue, the fat body (functionally equivalent of the liver), and subsequently secreted into the hemolymph. Insect AMPs are small, cationic, membrane-active molecules that accumulate in the hemolymph, reaching high concentrations in response to infection; they exhibit a broad range of activities against different classes of pathogens and can still be detected in the hemolymph up to several weeks after challenge (Lemaitre & Hoffmann, 2007). Cellular immune responses involve the action of circulating hemocytes, which differ among insect species and are divided into certain classes based on morphological characteristics, antigenic properties, and functional features. Their number rapidly increases during an infection, and they are responsible for several cellular defenses including cell spreading, formation of cell aggregates, nodulation, phagocytosis, and encapsulation; they can also take part in humoral reactions (Strand, 2008). In addition, insects can activate complex proteolytic cascades that regulate coagulation and melanization of hemolymph, defenses associated with production of reactive oxygen and nitrogen species, and epithelial responses in the gut that also play important roles in fighting microbial infections (Cerenius, Kawabata, Lee, Nonaka, & Söderhäll, 2010; Charroux & Royet, 2010). The important role of the inducible antimicrobial defense against fungal infection is demonstrated by *Metarhizium* employing the cell wall-surface collagen-like Mcl1 protein against the activation of immune pathways in *Manduca* caterpillars (Wang & St Leger, 2006). Nevertheless, genes encoding different AMPs are overexpressed in *Drosophila* in response to either *Metarhizium* or *Beauveria* infection, so at least some of the antifungal responses of the insect host are deployed against infection by these pathogens (Gottar et al., 2006; Lu et al., 2015; Tzou, Reichhart, & Lemaitre, 2002). However, some *Metarhizium* and *Beauveria* strains are immune to drosomycin, an AMP with potent antifungal activity, suggesting that adaptation to being entomopathogenic involved evolving multiple overlapping countermeasures to avoid or mitigate the antagonistic effects of the immune response host defenses. Thus, *B. bassiana* at least is also able to inhibit the process of melanization in *Drosophila* (Matskevich et al., 2010).

5. IMMUNE RECOGNITION OF FUNGI

In the last decade, the importance of symbionts in host–parasite interaction has been increasingly realized (Gross et al., 2009). Each of the 900,000-plus species of insects has its own unique microbial population that can influence its health and welfare. A diverse range of invertebrate taxa, including many insect species, is infected with obligate *rickettsia*, maternally transmitted endosymbionts in the genus *Wolbachia* that have been shown to help protect their host from pathogens (Hedges, Brownlie, O'Neill, & Johnson, 2008; Jaenike & Brekke, 2011; Scarborough, Ferrari,

& Godfray, 2005). Host acquisition of endosymbionts influences the evolution of both host and parasite (Jaenike, Unckless, Cockburn, Boelio, & Perlman, 2010; Jiggins & Hurst, 2011) and a parallelism between endosymbionts acquisition and the occurrence of beneficial nuclear mutations has been proposed (Jaenike, 2012). Insects also maintain a population of commensal microbes that contribute to the homeostasis of the host by aiding digestion and facilitating nutritional assimilation (Vlisidou & Wood, 2015). Therefore, symbionts add an extra level of complexity to host–parasite interactions, and the immune system has had to evolve to distinguish between and react appropriately to pathogens that destroy their hosts and symbionts that keep them healthy (Vlisidou & Wood, 2015). To do so, insect hosts have developed control mechanisms that monitor and maintain the balance between nonself signals (ie, microbial molecules) and danger signals (ie, molecules released by host cells upon damage). This allows them to disregard recognition of beneficial microbes and maintain their communities, or activate a robust immune response when it recognizes pathogenic microbes that cause substantial cell damage (and consequently higher amounts of danger signals) (Lazzaro & Rolff, 2011; Vance, Isberg, & Portnoy, 2009). This general property of immune system activation occurs not only locally (eg, in the gut), but also systemically in vertebrates and invertebrates (Lazzaro & Rolff, 2011).

Aside distinguishing symbionts from pathogens, immune responses triggered by the presence of pathogens in the body cavity show some degree of genus specificity, as different genera of pathogens trigger the expression of different AMPs. Several studies have assessed *Drosophila* transcriptional response to different pathogens (De Gregorio, Spellman, Rubin, & Lemaitre, 2001; Lemaitre, Reichhart, & Hoffmann, 1997; Roxström-Lindquist et al., 2004; Wertheim, Kraaijeveld, Hopkins, Walther Boer, & Godfray, 2011; Wertheim et al., 2005). For example, Roxström-Lindquist and collaborators assessed gene expression in *Drosophila* 24 h after infection by *B. bassiana*, the protozoan parasite *Octosporea muscaedomesticae*, the Gram-negative bacterium *Serratia marcescens* and *Drosophila* C virus (Edwards, Chesnut, Satta, & Wakeland, 1997; Roxström-Lindquist et al., 2004). They found a high degree of microbe specificity, with the fungal infection generating the strongest response with 298 genes induced.

The insect response to invading microorganisms is initiated by recognition of certain microbial structures by host receptors of the innate immune system, and pathogen-induced processes that contribute to the progression of disease. In particular, recognition of microbial infection is controlled by pattern recognition proteins that bind to conserved microbe-associated molecular pattern molecules produced by microbes, such as bacteria and fungi (Müller, Vogel, Alber, & Schaub, 2008). Bacteria recognition is achieved through peptidoglycan recognition proteins (PGRPs) that sense peptidoglycan (PGN), an essential component of bacteria cell wall. PGRP-LC isoforms and PGRP-LE are specialized in recognizing Gram-negative PGN, while PGRP-SA, PGRP-SD, and GNBP1 are specialized in sensing Gram-positive PGN. In *Drosophila*, the *Toll* signaling pathway controls the systemic antifungal host response, and it can be activated by either Gram-negative binding protein 3 (GNBP3) or the Persephone protease. GNBP3 binds beta(1,3)-glucan, a component of fungal cell wall (Matskevich et al., 2010), and the Persephone protease is activated by fungal proteases (Gottar et al., 2006). As mentioned previously, protease gene families are enormously expanded in *Metarhizium* compared to nonentomopathogenic fungi (Gao et al., 2011); many are expressed in order to cross the insect cuticle, but, crucially, their expression stops in the hemolymph (Freimoser, Hu, & St Leger, 2005). Engineering constitutive expression of the virulence factor Pr1 subtilisin protease triggered a massive melanization response in the hemolymph of caterpillars which killed the fungus as well as the insect (St Leger, Joshi, Bidochka, Rizzo, & Roberts, 1996). This unforeseen activation presumably results from the caterpillar equivalent of Persephone. GNBP3 also triggers antifungal defenses that are not dependent on activation of the *Toll* pathway by assembling "attack complexes," which comprise phenoloxidase (PO), the main melanization enzyme, and the necrotic serpin (Matskevich et al., 2010).

6. CELLULAR IMMUNE RESPONSES TO FUNGI

Although insect humoral immune defenses are still better understood than cellular responses, the gap is rapidly closing, as significant progress has been made recently with the aid of genetic techniques that have identified molecules required for regulating cellular defenses. In *Drosophila*, it has been shown that the cellular and humoral responses act in concert to combat infection (Elrod-Erickson, Mishra, & Schneider, 2000; Nehme et al., 2011), with cellular defenses playing the bigger role. It has been estimated that systemically infected larvae containing 3000 nonpathogenic bacteria can eliminate almost 95% of the bacterial load in 30 min by phagocytosis (Shia et al., 2009). This reduction in bacterial load facilitates microbe clearance by AMPs in the later stages of the infection and, at the same time, minimizes the risk of antimicrobial resistance development by the pathogen (Haine, Moret, Siva-Jothy, & Rolff, 2008).

The cellular response relies on blood cells (hemocytes) present in the body cavity. Hemocytes are the professional phagocytes of the immune system and are able to recognize the particle to be ingested and engulf and destroy it. Phagocytosis involves extensive membrane reorganization, cytoskeletal remodeling, and intracellular vesicle trafficking. *Drosophila* blood cells comprise undifferentiated hemocyte progenitors that exist during the early embryonic stages, and at least three differentiated blood cell types that function in larval and adult stages (Evans & Banerjee, 2003; Hartenstein, 2006; Lanot, Zachary, Holder, & Meister, 2001; Lemaitre et al., 1996): plasmatocytes, crystal cells, and lamellocytes. Plasmatocytes display phagocytic activity and represent the functional equivalents of mammalian monocytes/macrophages. Plasmatocytes represent 90–95% of all hemocytes in *Drosophila* and are responsible for the phagocytosis of microorganisms and apoptotic cells. Phagocytosis initiation can be either direct, via specific cell-surface receptors, or indirect, via opsonins that mark a surface so that it can be detected by phagocytic receptors. Several receptors have been shown to be involved in phagocytosis, the most studied are Eater and Dscam. Flies lacking Eater showed impaired phagocytosis and decreased survival after bacterial infection (Kocks et al., 2005), but to date Eater has not been shown to play a role in resisting fungi. Through alternative splicing, insects have the capacity to express thousands of Dscam isoforms potentially providing precise recognition of thousands of pathogens, though this has not yet been demonstrated. Hemocyte-specific loss of *Drosophila Dscam* impaired the efficiency of phagocytic uptake of bacteria (Watson et al., 2005). Similarly, silencing of mosquito (*Anopheles gambiae) Dscam* compromises the mosquito's resistance to bacteria and the malaria parasite *Plasmodium* (Dong, Taylor, & Dimopoulos, 2006), indicating a broad spectrum defense. A group of six thioester-containing proteins (TEPs) has an important role in this defense mechanism. TEPs are secreted proteins and three of them are upregulated upon infection with bacteria. There is evidence that they bind to pathogens and promote phagocytosis in a similar fashion as the complement proteins in vertebrates (Blandin et al., 2004; Stroschein-Stevenson, Foley, O'Farrell, & Johnson, 2006). Hemocytes also mediate the humoral response. Plasmatocytes are known to play roles in triggering AMP production, as eliminating hemocytes abolishes AMP expression in the larval fat body (Shia et al., 2009).

Lamellocytes, another category of hemocytes, are specialized for encapsulating foreign bodies too large to be phagocytosed, such as parasitoid wasp eggs laid in *Drosophila* larvae. The molecular mechanisms of encapsulation are poorly understood, but N-glycosylation of lamellocyte membrane components has a key role in the encapsulation response (Mortimer, Kacsoh, Keebaugh, & Schlenke, 2012). Lamellocytes are stress induced; mutations in the gene *cactus* (the *Drosophila* homologue of the mammalian IκB) or the constitutive expression of the gene *dorsal* (NF-κB transcription factor) can induce their differentiation (Lemaitre & Hoffmann, 2007).

Crystal cells, the last group of hemocytes, contain the key enzyme in melanin biosysnthesis called prophenoloxidase (ProPO). The enzyme is stored in the form of crystalline inclusions and released upon rupture of the crystal cells.

The general assumption is that insects depend solely on an innate immune response to fight invading microorganisms; by definition, innate immunity lacks adaptive characteristics. However, there are some reports showing that *Drosophila* can be primed with a sublethal dose of *Streptococcus pneumoniae* to protect against an otherwise-lethal second challenge of *S. pneumonia* via a phagocyte-dependent mechanism (Pham, Dionne, Shirasu-Hiza, & Schneider, 2007). This protective effect has loose specificity for *S. pneumonia* and persists for the life of the fly. While not all microbial challenges induced this specific, primed response, a similar specific protection could be elicited by *B. bassiana*, a natural fly pathogen (Pham et al., 2007). These results indicate that insect immune responses can indeed adapt, and suggest that insect hemocytes may also present an activation response similar to the one known in mammalian leukocytes.

7. INTERACTION OF FUNGI WITH THE PHENOLOXIDASE AND COAGULATION RESPONSES

The PO immune response involves the formation of short-lasting toxic substances (eg, chemically reactive quinones) and long-lasting products such as melanin (a brown–black pigment) that is rapidly deposited at wound sites and participates in encapsulation and killing of microbial pathogens (Cerenius, Lee, & Söderhäll, 2008). The PO enzyme initially exists as the inactive precursor proPO. This is synthesized and released by specialized circulating cells into the hemolymph (eg, the lamellocytes in *Drosophila*) following microbial challenge and is promptly activated by plasma serine proteases (Cerenius & Söderhäll, 2004). The melanization cascade is tightly regulated, as it generates toxic by-products, including reactive oxygen species. It is positively and negatively regulated by a network of specific clip domain serine proteases (CLIPB's), enzymatically incompetent CLIPB homologues (CLIPA's) and serine protease inhibitors (Kafatos, Waterhouse, Zdobnov, & Christophides, 2009). The importance of melanization in the fight against entomopathogenic fungi is consistent with the tight regulation that fungal pathogens have over their own proteases so as not to activate this pathway (St Leger, Joshi, Bidochka, Rizzo, et al., 1996), and with the observation that many entomopathogenic fungi have evolved anti-PO activity, as illustrated by active PO suppression by *B. bassiana* (Matskevich et al., 2010, Feng, Shang, Cen, & Wang, 2015). In spite of their countermeasures, *Metarhizium* and *Beauveria* are more virulent to *Drosophila* and mosquitoes carrying deletions in POs, suggesting that melanization is likely an effective defense against some fungal infections (Binggeli, Neyen, Poidevin, & Lemaitre, 2014; Yassine, Kamareddine, & Osta, 2012).

Recent evidence implies that the melanization cascade (or proPO activating system), overlaps with the humoral and cellular insect immune defenses. Plasmatocytes are strongly associated with the coagulation response (or hemolymph clotting), which acts to seal wounds and prevent microorganisms from entering and spreading throughout the insect body cavity (Eleftherianos & Revenis, 2011; Loof et al., 2011). Hemolectin, a plasmatocyte-specific gene, codes for the main component of the clotting fibers and is required for coagulation in *D. melanogaster* (Lemaitre & Hoffmann, 2007). Clots composed of melanized hemolectin fibers are rapidly generated at sites of injury and act to trap microbes. It is an integral part of the insect immune response to bacteria and nematodes (Scherfer et al., 2004; Wang et al., 2010). The potential immune function of clotting in response to infection by *M. anisopliae* was demonstrated by increased virulence against *Drosophila* carrying an insertion in the hemolectin gene (Lu et al., 2015). Infection by *Metarhizium* often produces a melanization response in the cuticle at the site of entry (St Leger, Charnley, & Cooper, 1988; St Leger, Cooper, & Charnley, 1988), and the involvement of hemolectin suggests that penetrant fungal hyphae are responded to as septic wounding or that clotting can occur in the absence of significant injury. Melanization increases insect resistance to *Metarhizium* spp., mainly due to toxic effects from L-DOPA oxidation products (St Leger, Charnley, et al., 1988; St Leger, Cooper, et al., 1988).

8. HUMORAL IMMUNE RESPONSES TO FUNGI

In humoral immunity, AMPs are produced by a specialized tissue, the fat body, and are subsequently secreted into the body cavity in response to an infection. As discussed above in section five, microbial detection relies on direct contact between a host pattern recognition receptor protein and a pathogen molecule. Fungi and Gram-positive bacteria recognition triggers a serine protease cascade that activates the Toll signaling pathway (Lemaitre & Hoffmann, 2007). In *Drosophila*, this results in the expression of AMPs, like defensin, active against Gram-positive bacteria, and drosomycin and metchnikowin, active against fungi (Fehlbaum et al., 1994; Levashina et al., 1995). Mutations in certain genes of the Toll pathway cause an immune-deficient phenotype characterized by the lack of expression of drosomycin and metchnikowin, and a marked sensitivity to infections by opportunistic fungi (Govind, 2008); although, this phenotype does not alter susceptibility to *M. anisopliae*, as the fungus is resistant to known *Drosophila* AMPs (Lu et al., 2015). On the other hand, Gram-negative bacteria recognition triggers the Imd signaling pathway with consequent expression of AMPs like cecropins, diptericin, attacin, and drosocin. Other signaling cascades that activate immune genes in the *Drosophila* fat body have been recognized, including the JAK/STAT and the JNK pathways, although their precise contribution to immune defense is not clear.

Owing to the constant threat of attack from fungal pathogens and to their lack of an adaptive immune activity, other insects besides *Drosophila* express a large number of AMPs for protection against fungal diseases (Faruck, Yusof, & Chowdhury, 2015). Nevertheless, in comparison to antibacterial peptides, insect antifungals have not been systematically screened for and only a few are well characterized: termicin from termites (Da Silva et al., 2003), heliomicin from the tobacco budworm *Heliothis virescens* (Lamberty et al., 1999), and the gallerimycin peptide from the greater wax moth *Galleria mellonella* larvae (Schuhmann, Seitz, Vilcinskas, & Podsiadlowski, 2003). Determining the disparate structures of antifungal peptides is daunting with incomplete knowledge of action and classification. Drosomycin is a cysteine-rich peptide containing 44 amino acids with a twisted three-stranded sheet structure steadied by disulfide bonds. The sheet structure means drosomycin has charged (hydrophilic) and uncharged (hydrophobic) elements (ie, it is ampipathic), which allow it to penetrate and disrupt cell membranes. A larger number of peptides that are active against both bacteria and fungi have been studied, the most renowned of which are the cecropins. The giant silk moth *Hyalophora cecropia* produces two types of cecropins, Cecropin A and Cecropin B, which are lytic linear peptides that are active against bacteria and opportunistic, fungal human pathogens *Fusarium oxysporum* and *A. fumigatus* (De Lucca et al., 2000). However, cecropins are inactive against *Metarhizium* (Lu and St. Leger, unpublished data). Thanatin is a small peptide, with only 21 residues. In water, it is able to assume the form of an antiparallel sheet structure that is well defined with a disulfide bridge (Vigers et al., 1992). Thanatin is nonhemolytic and can inhibit the growth of *F. oxysporum* and *A. fumigatus* (Mandard et al., 1998).

9. THE EVOLUTIONARY GENETICS OF INSECT IMMUNITY

Throughout the evolutionary history of complex life forms, the burden of infectious diseases has been massive. A rich diversity of pathogens and parasites threaten insects with high fitness and immunity costs, which fuel the dynamic coevolutionary arms races between hosts and pathogens (Schmid-Hempel, 2003; Wertheim, 2015). It is therefore hardly surprising that immunity-related genes are among those most strongly targeted by natural selection in eukaryote genomes. For example, many studies have found that *Drosophila* immunity genes have a significantly higher rate of adaptive substitution compared to control genes (Clark & Wang, 1997; Jiggins & Hurst, 2003; Jiggins & Kim, 2006, 2007; Lazzaro, 2008; Obbard, Jiggins, Bradshaw, & Little, 2011; Sackton et al., 2007; Schlenke & Begun, 2005). This confirms that parasites indeed exercise a strong selection pressure on *Drosophila*.

Interestingly, the recognition, signal transduction, and immune effectors (AMPs, etc.) that comprise the immune system differ not only in how fast they evolve, but also in the ways in which they can change (Juneja & Lazzaro, 2009). The complete genome sequencing of 12 species of *Drosophila* showed that Toll and Imd signaling pathways, for example, show a high degree of amino acid divergence. This suggests that natural, currently unknown pathogens of *Drosophila* are evolving to circumvent the *Drosophila* immune system by avoiding recognition. Yet across flies, bees, and beetles, the main proteins in these signaling pathways are highly conserved; they are orthologs (Juneja & Lazzaro, 2009). By contrast, the 12 species of *Drosophila* show high levels of gene duplication and turnover for AMP genes, but no evidence of positive selection at the amino acid level (Obbard, Welch, Kim, & Jiggins, 2009; Sackton et al., 2007; Salazar-Jaramillo et al., 2014).

The most likely explanation is that AMPs are constrained as they bind to highly conserved targets; though, again, there is a problem of interpreting these results, as the identity of the natural parasites and their interacting virulence proteins have not been determined. Furthermore, *Drosophila* AMP genes have no identifiable homologues in the genomes of mosquitoes, bees, or *Tribolium* beetles. Instead, these insects each have their own distinct AMP gene families (Juneja & Lazzaro, 2009). This is usually interpreted as resulting from red queen dynamics. These molecules operate at the interface between the host and the pathogen, and are therefore crucial for recognition by the host of an invading organism, and to mediating the targeting and antagonistic effects of the immune response on the pathogen. At the same time, the parasite is under selection to go undetected, in order to avoid or mitigate the antagonistic effects of the immune response. Therefore, the interaction between host and parasite genotypes, through these interfacing molecules, and the selection pressure they exert upon one another set the scene for antagonistic coevolution: the host and parasite continuously evolve in response to each other (Wertheim, 2015).

AMPs are quite efficient at resisting opportunistic infections that occasionally gain access to a host via injuries. Thus, many human pathogens such as *A. fumigatus* will not infect wild-type *Drosophila*, but are highly virulent when injected into Toll knockout flies (Lemaitre et al., 1996). Generalist insect pathogens, including *Beauveria* and *Metarhizium*, have evolved resistance to drosomycin (Lu et al., 2015; Tzou et al., 2002), the principal antifungal *Drosophila* AMP, presumably because they were under strong selective pressure to resist this form of defense. Consequently, there must also be selective pressure on insect hosts to adapt their AMPs at the amino acid level over modest evolutionary timescales. However, given the broad host range of some *Metarhizium* species, it seems unlikely that they are engaging in a strict coevolutionary arms race with a particular host species or population (Kawecki, 1998). *Prima facie*, such arms race coevolution will most likely occur with specialized pathogens. Although it has not been confirmed that generalist and specialist pathogens differ in terms of the selection pressure they impose on host immunity (Keebaugh & Schlenke, 2014), it can be assumed that they do given that their pathogenic strategies are different. As described in Wang, St. Leger, & Wang, (2016), generalists such as *M. anisopliae* kill hosts quickly via toxins and grow saprophytically in the cadaver. In contrast, like many specialists, *M. acridum* (a specialist locust pathogen) causes a systemic infection of host tissues before the host dies, which suggests a much smaller role for toxins.

Using *M. anisopliae* in infection experiments gives us the possibility to study how hosts respond to a generalist fungal pathogen and to assess if variability among host populations is present, possibly due to divergent life histories. In contrast, *M. acridum*, as well as many species of *Cordyceps* and Entomophthoromycota, has very narrow host ranges, eg, *Entomophaga maimaga*, a biocontrol agent that exclusively infects gypsy moths. These natural host–pathogen pairs are understudied opportunities to investigate the genetic basis of antagonistic coevolution and uncover any specialized immune systems, besides those used to combat opportunistic or generalist pathogens. Parasites are usually considered to have a shorter generation time and therefore to be ahead in the coevolutionary race. Though this might not be true for entomopathogenic fungi, they have to kill their host in order to reproduce, thus pathogen and host generation times may be synchronized. It is postulated that the 10–14 days it takes some *Metarhizium* strains to kill mosquitoes could result in an evolution proof insecticide as the infected mosquitoes will likely be able to reproduce in that amount of time, reducing selective pressures to evolve resistance to the fungus (Thomas & Read, 2007).

Adaptation of pathogens to their hosts depends critically on factors affecting pathogen reproductive rate, but the extent to which varying host genotype might affect the evolution of pathogen life history is unclear. *Metarhizium* life stages show apparent density-dependent effects; following penetration of the cuticle, greater growth within the host insect (*Drosophila*) was correlated with shorter life spans and earlier onset of reproduction (spore production on cadavers) (Lu et al., 2015). The most commonly accepted model for animal and plant pathogens assumes that within host growth is a critical parameter in the evolution of virulence, but that high virulence is traded off against increased host mortality reducing the infectious period for transmission between hosts (Bruns, Carson, & May, 2012; Staves & Knell, 2010). As a consequence, natural selection should favor pathogens with intermediate exploitation, and hence intermediate virulence, at which level lifetime transmission is maintained. This model pertains to the great majority of host–pathogen interactions whereby the living infected animal disseminates the pathogen. In contrast, most insect–pathogenic fungi only propagate on hosts after death. Thus, variation in fungal growth between *Drosophila* lines with different susceptibilities to *M. anisopliae* does not affect total spore production. Instead, trade-offs affect latent periods (the interval between infection and sporulation), which are longer on resistant host flies as they survive for longer, so a *Metarhizium* strain that evolves to kill resistant flies as quickly as susceptible flies should have higher lifetime fitness. However, transgenic *Metarhizium* strains that kill very quickly come at a cost of decreased spore production (Pava-Ripoll, Posada, Momen, Wang, & St Leger, 2008; St Leger, Joshi, Bidochka, & Roberts, 1996); slow killing speed is strategic as it allows the fungus to fully exploit host tissues to maximize yields of conidia (Wang & St Leger, 2007b). This will limit the apparent benefit of a very short latent period and thereby constrain the evolution of *Metarhizium* virulence.

Differing host—pathogen interactions will produce widely varied costs that impact where trade-offs occur among life history traits, but trade-offs cannot be avoided somewhere during the pathogen life cycle. One of the central principals of life history theory is that because organisms are constrained by resource limitations they cannot simultaneously optimize all aspects of fitness (Zuk & Stoehr, 2002). This premise has been challenged by studies documenting hormesis, the beneficial effects of low-level toxins and other stressors (Gems & Partridge, 2008). The question of potential immunological trade-offs due to hormesis was addressed with topical application of dead *Metarhizium robertsii* spores to *Drosophila*. This increases the longevity and fecundity of *Drosophila*, but reduces ability to fight off live infections, showing that, at least in this case, hormesis is manifested as stress-induced trade-offs with immunity, not as cost-free benefits (McClure et al., 2014). The authors conclude that the consequences of hormetic treatments for infected humans could be dire.

Sporulating postmortem has many other consequences for entomopathogen life-cycle traits. Sporulation capacity in plant pathogens is a measure of host resistance (Bruns et al., 2012), but that would not apply to an insect pathogen that sporulates postmortem. Unlike some other *Metarhizium* strains, Ma549 deployed by Lu et al. (2015) does not employ toxins and kills principally by invasive growth and depletion of host nutrients (Samuels, Charnley, & Reynolds, 1988). This pathogen strategy would likely engender selection on the host to reduce fungal infectivity and growth, as these are the traits most strongly affected by host genotype. However, the costs of launching an immune response against fungal infection are unknown, and it cannot be ruled out that the direction of resources toward a powerful immune response could contribute to an early death, or at least reduced fecundity (Kraaijeveld & Wertheim, 2009).

10. FUNGAL COUNTERMEASURES TO HOST IMMUNITY

As described above, once inside the insect, entomopathogenic fungi face a highly potent immune response by which the host attempts to eliminate or mitigate an infection. However, the immune response of a host to a professional entomopathogen may be ineffective because of the ability of the pathogen to survive immune defenses. There is evidence that pathogenic fungi are capable of interacting with the innate immune system to successfully infect their hosts. This involves immune evasion strategies that interfere with, disrupt or manipulate constitutive or induced immune defenses. *Metarhizium* destruxins (cyclic peptide toxins) play a specific role in this context, as they are able to suppress AMP gene expression (Pal et al., 2007) and also block phagocytosis by various actions that include inhibiting V-ATPase, which maintains an acidic environment within lysosomes and vacuoles to help digest ingested materials (Chen, Hu, Yu, & Ren, 2014). Likewise, oosporein, a quinone derivative produced by *B. bassiana*, inhibits ProPO activity and downregulates expression of an antifungal peptide gallerimycin in larvae of the wax moth *Galleria* (Feng et al., 2015).

At least some *Metarhizium* and *Beauveria* strains are resistant to products of the *Drosophila* Toll pathway including the antifungal peptide drosomycin (Lu et al., 2015; Tzou et al., 2002). Evolving resistance against AMPs was probably a prerequisite for *Metarhizium* spp. to function as specialized entomopathogens. Tolerance to drosomycin will prevent clearance of Ma549 from the hemocoel, which likely dooms the fly (Lu et al., 2015). *Metarhizium* can also adopt a camouflage strategy as its blastospores can evade hemocytes by producing a hydrophilic collagen (Mcl1) coat (Wang & St Leger, 2006). Furthermore, conidia can be internalized and grow within arthropod phagocytic cells, where they can avoid additional immune reactions ("anatomical seclusion" strategy), while dispersing through the insect body (Kurtti & Keyhani, 2008). As mentioned above, entomopathogenic fungi also reduce their hosts PO activity (Matskevich et al., 2010), which again highlights the dynamic interference between entomopathogenic fungi and the insect innate immune mechanism. A mutant screen showed that the consequence of disrupting many *Drosophila* genes that influence cellular immunity was an increase in susceptibility to *Metarhizium* (Lu et al., 2015), consistent with a dynamic evolutionary arms race between pathogen and host.

11. TOLERANCE VERSUS RESISTANCE

Recent studies have highlighted that successful immunity to infection with bacteria requires two facets: first, targeting the microbe for elimination, termed resistance; and second, inducing tissue protective responses, termed disease tolerance (Medzhitov, Schneider, & Soares, 2012; Råberg, Graham, & Read, 2009). Mechanistically, they are defined by their effects on both organism survival and pathogen load. Resistance mechanisms promote survival by decreasing microbial numbers, whereas tolerance mechanisms improve the ability to survive very high pathogen loads. The concept of tolerance was only recently introduced into the field of animal immunity, and the mechanisms are poorly understood, but active repair mechanisms and metabolic adaptations are likely key to surviving infection. The few available studies suggest that in both mice and *Drosophila*, a set of conditions that might favor tolerance to one bacterial pathogen can promote resistance against another (Ayres et al., 2008; Cunnington, de Souza, Walther, & Riley, 2012; Igboin, Griffen, & Leys, 2012).

It may be more difficult to evolve tolerance to a filamentous fungal pathogen that unlike bacteria actively penetrates and colonizes infected tissues. Lu et al. (2015) reported that mutant fly lines that succumbed quickly to *M. anisopliae* all had higher fungal loads than wild-type flies, suggesting that the principal defect in these flies is that they are less able to restrain *Metarhizium* growth, consistent with a reduction in resistance rather than tolerance. Likewise, none of the long-lived lines appeared to tolerate *M. anisopliae* better than the wild type because their delayed mortality coincided with rapid proliferation of *M. anisopliae* in the hemolymph (Lu et al., 2015). However, experiments where flies were artificially selected for increased resistance to *Beauveria* showed evolved tolerance may be possible. Although there were no differences in life span after infection between selected and control flies, selected flies continued laying eggs for longer than infected control flies so their lifetime fecundity was greater (Kraaijeveld & Godfray, 2008). Kraaijeveld and Godfray (2008) suggested that the selection regime selected for improved tolerance to infection rather than resistance. They also, however, suggested a cost of tolerance selection as selection flies have lower lifetime fecundity when uninfected than control flies.

12. CONCLUDING REMARKS AND FUTURE PERSPECTIVES

Insects are excellent models for studying host—fungal interactions and immune responses. Identification of host immune reactions or immune-related molecules in *Drosophila* has already greatly benefited human health by assisting strategies for controlling human pathogens. Early studies on fungal diseases focused on molecules involved in sensing and signaling, whereas more recent studies have started to investigate immunity using a more holistic approach to provide a more complete understanding of disease. These studies can reveal how immunity and its dysregulation can affect whole-body pathophysiology. They have also shown that innate immune genes are under strong positive selection, which suggests that parasites impose strong evolutionary pressures on their hosts. Undoubtedly, the power of *Drosophila* genetics, including functional genomics analysis, in combination with genetic and RNAi screens will continue to offer a powerful approach for elucidating the regulation of insect immune genes responsible for antifungal reactions. This will in turn help us to investigate and understand the evolutionary conservation of antifungal immune responses in vertebrate animals. It is likely that susceptibility to disease will involve alleles with various degrees of penetrance and in which many common or rare variants are involved, each having a modest effect. Association studies based on whole-genome sequencing of hundreds of wild *Drosophila* strains have identified the genetic underpinnings of considerable heterogeneity in *Drosophila* populations regarding their responses to various stresses, including starvation (Harbison et al., 2009; Ivanov et al., 2015; Shorter et al., 2015). Similar approaches integrating whole-genome sequencing data, proteomic profiles, RNA-seq, mass spectrometry, epigenetic markers (e.g., genome-wide methylation profiles), and evolutionary and population genetics data should facilitate the identification of functionally important genes and variants that contribute to heterogeneity between individuals and populations of insects in terms of susceptibility to infection and disease progression. These studies will provide new insights into the physiological consequences of microbial colonization and how immunity interacts with behavior, metabolism, physiology, and hormonal regulation. They will continue to further our understanding of the genetic basis for any tolerance mechanisms and the means by which a host adapts to infection and associated damage. They will also open up new routes for translational research into human and animal diseases and new strategies for insect pest control.

As mentioned in the introduction, in order to understand the evolution of the insect immune system we need to be able to quantify the strength of selection, and very little information is available on natural pathogens and their infection rates in wild populations of *Drosophila*. To fully leverage the *Drosophila* model system, this needs to be remedied by observing and assaying wild-type populations over time to characterize the process of adaptation to different environments and pathogens (Kaltz & Shykoff, 1998). Alternatively, additional insect models can be adopted for which there is greater background ecological information. Investigations of *Metarhizium* infections in *Manduca sexta* and other physiological insect models have identified gene regulatory mechanisms used by fungi to selectively switch on pathogenic genes in insect blood (Wang & St Leger, 2006). These studies have led to development of transgenic fungi as biological control agents against insect vectors of disease and agricultural insect pests (Fang et al., 2011; Pava-Ripoll et al., 2008; Wang & St Leger, 2007b).

The evolution of host life history strategies, such as earlier reproduction in response to greater pathogen prevalence, has been explored, at least in laboratory conditions (Kraaijeveld & Godfray, 2008). In contrast, the evolution of pathogen life histories in response to host genotypic variation is poorly understood. Studies with plant pathogenic fungi are consistent with the idea that host resistance diversity is the driving force behind variation in pathogenicity (Tack, Thrall, Barrett, Burdon, & Laine, 2012), but the relative contributions of host and pathogen to variation in pathogen life history within the host remain unclear. For entomopathogens, this challenge is being met by whole-genome sequencing, which is also beginning to play an important role in the development of insect–fungal model systems for studying host defense mechanisms. Comparative genomics of both pathogen and host genomes is a powerful approach allowing scrutiny of the cross talk between pathogen and host genomes throughout infection processes. This will potentially reveal the total number and nature of genes that are involved in parasite–host interactions, as well as molecules that could participate in the invasion of the host and evasion or abrogation of host immune responses. Untangling these contributions will allow us to identify traits with sufficient variation for selection to act upon and, therefore, help identify mechanisms of coevolution between pathogens and their hosts. The current focus on generalist pathogens such as *B. bassiana* and *M. robertsii* hinders any strong generalizations across the notoriously diverse array of host and parasite life histories. It is likely that additional specialized immune mechanisms have evolved to combat specialized pathogens, but so far these have been overlooked. New data across a broad suite of study systems will be necessary to unravel the contributions of different host and pathogen life history traits responsible for generating patterns of coevolution.

ACKNOWLEDGMENTS

This work was supported by the Biotechnology Risk Assessment Program competitive grant number 2015-33522-24107 from the USDA, National Institute of Food and Agriculture, National Science Foundation grant 1257685 and by the National Institute of Allergy and Infectious Diseases of the National Institutes of Health under award number RO1 AI106998.

REFERENCES

Adamo, S. A. (2009). The impact of physiological state on immune function in insects. In J. Rolff, & S. E. Reynolds (Eds.), *Insect infection and immunity: Evolution, ecology, and mechanisms* (pp. 173–186). Oxford: Oxford University Press.

Adamo, S. A. (2012). The effects of the stress response on immune function in invertebrates: an evolutionary perspective on an ancient connection. *Hormones and Behavior, 62*, 324–330.

Ayres, J. S., Freitag, N., & Schneider, D. S. (2008). Identification of *Drosophila* mutants altering defense of and endurance to *Listeria monocytogenes* infection. *Genetics, 178*, 1807–1815.

Bailey, N. W., Gray, B., & Zuk, M. (2011). Exposure to sexual signals during rearing increases immune defence in adult field crickets. *Biology Letters, 7*, 217–220.

Barnes, A. I., & Siva-Jothy, M. T. (2000). Density-dependent prophylaxis in the mealworm beetle *Tenebrio molitor* L. (Coleoptera: Tenebrionidae): cuticular melanization is an indicator of investment in immunity. *Proceedings of the Royal Society B: Biological Sciences, 267*, 177–182.

Binggeli, O., Neyen, C., Poidevin, M., & Lemaitre, B. (2014). Prophenoloxidase activation is required for survival to microbial infections in *Drosophila*. *PLoS Pathogens, 10*, e1004067.

Blandin, S., Shiao, S. H., Moita, L. F., Janse, C. J., Waters, A. P., Kafatos, F. C., & Levashina, E. A. (2004). Complement-like protein TEP1 is a determinant of vectorial capacity in the malaria vector *Anopheles gambiae*. *Cell, 116*, 661–670.

Bruns, E., Carson, M., & May, G. (2012). Pathogen and host genotype differently affect pathogen fitness through their effects on different life-history stages. *BMC Evolutionary Biology, 12*, 135.

Burke, C. J., Huetteroth, W., Owald, D., Perisse, E., Krashes, M. J., Das, G. … Waddell, S. (2012). Layered reward signalling through octopamine and dopamine in *Drosophila*. *Nature, 492*, 433–437.

Cerenius, L., Kawabata, S., Lee, B. L., Nonaka, M., & Söderhäll, K. (2010). Proteolytic cascades and their involvement in invertebrate immunity. *Trends in Biochemical Sciences, 35*, 575–583.

Cerenius, L., Lee, B. L., & Söderhäll, K. (2008). The proPO-system: pros and cons for its role in invertebrate immunity. *Trends in Immunology, 29*, 263–271.

Cerenius, L., & Söderhäll, K. (2004). The prophenoloxidase-activating system in invertebrates. *Immunological Reviews, 198*, 116–126.

Charroux, B., & Royet, J. (2010). *Drosophila* immune response: from systemic antimicrobial peptide production in fat body cells to local defense in the intestinal tract. *Fly (Austin), 4*, 40–47.

Chen, X. R., Hu, Q. B., Yu, X. Q., & Ren, S. X. (2014). Effects of destruxins on free calcium and hydrogen ions in insect hemocytes. *Insect Science, 21*, 31–38.

Christe, P., Oppliger, A., Bancala, F., Castella, G., & Chapuisat, M. (2003). Evidence for collective medication in ants. *Ecology Letters, 6*, 3.

Clark, A. G., & Wang, L. (1997). Molecular population genetics of *Drosophila* immune system genes. *Genetics, 147*, 713–724.

Cunnington, A. J., de Souza, J. B., Walther, M., & Riley, E. M. (2012). Malaria impairs resistance to Salmonella through heme- and heme oxygenase-dependent dysfunctional granulocyte mobilization. *Nature Medicine, 18*, 120–127.

Da Silva, P., Jouvensal, L., Lamberty, M., Bulet, P., Caille, A., & Vovelle, F. (2003). Solution structure of termicin, an antimicrobial peptide from the termite *Pseudacanthotermes spiniger*. *Protein Science, 12*, 438–446.

De Gregorio, E., Spellman, P. T., Rubin, G. M., & Lemaitre, B. (2001). Genome-wide analysis of the *Drosophila* immune response by using oligonucleotide microarrays. *Proceedings of the National Academy of Sciences of the United States of America, 98*, 12590–12595.

De Lucca, A. J., Bland, J. M., Vigo, C. B., Jacks, T. J., Peter, J., & Walsh, T. J. (2000). D-Cecropin B: proteolytic resistance, lethality for pathogenic fungi and binding properties. *Medical Mycology, 38*, 301–308.

De Roode, J. C., & Lefèvre, T. (2012). *Behavioral Immunity in Insects, 3*, 789–820.

Dong, Y., Taylor, H. E., & Dimopoulos, G. (2006). AgDscam, a hypervariable immunoglobulin domain-containing receptor of the *Anopheles gambiae* innate immune system. *PLoS Biology, 4*, e229.

Edwards, S. V., Chesnut, K., Satta, Y., & Wakeland, E. K. (1997). Ancestral polymorphism of Mhc class II genes in mice: implications for balancing selection and the mammalian molecular clock. *Genetics, 146*, 655–668.

Eleftherianos, I., & Revenis, C. (2011). Role and importance of phenoloxidase in insect hemostasis. *Journal of Innate Immunity, 3*, 28–33.

Elliot, S. L., Blanford, S., & Thomas, M. B. (2002). Host-pathogen interactions in a varying environment: temperature, behavioural fever and fitness. *Proceedings of the Royal Society B: Biological Sciences, 269*, 1599–1607.

Elrod-Erickson, M., Mishra, S., & Schneider, D. (2000). Interactions between the cellular and humoral immune responses in *Drosophila*. *Current Biology, 10*, 781–784.

Evans, C. J., & Banerjee, U. (2003). Transcriptional regulation of hematopoiesis in *Drosophila*. *Blood Cells, Molecules & Diseases, 30*, 223–228.

Evans, J. D., Aronstein, K., Chen, Y. P., Hetru, C., Imler, J. L., Jiang, H. … Hultmark, D. (2006). Immune pathways and defence mechanisms in honey bees *Apis mellifera*. *Insect Molecular Biology, 15*, 645–656.

Evans, J. D., & Spivak, M. (2010). Socialized medicine: individual and communal disease barriers in honey bees. *Journal of Invertebrate Pathology, 103*(Suppl. 1), S62–S72.

Fang, W., Vega-Rodríguez, J., Ghosh, A. K., Jacobs-Lorena, M., Kang, A., & St Leger, R. J. (2011). Development of transgenic fungi that kill human malaria parasites in mosquitoes. *Science, 331*, 1074–1077.

Faruck, M. O., Yusof, F., & Chowdhury, S. (2015). An overview of antifungal peptides derived from insect. *Peptides*. Epub ahead of print.

Fehlbaum, P., Bulet, P., Michaut, L., Lagueux, M., Broekaert, W. F., Hetru, C., & Hoffmann, J. A. (1994). Insect immunity. Septic injury of *Drosophila* induces the synthesis of a potent antifungal peptide with sequence homology to plant antifungal peptides. *Journal of Biological Chemistry, 269*, 33159–33163.

Feng, P., Shang, Y., Cen, K., & Wang, C. S. (2015). Fungal biosynthesis of the bibenzoquinone oosporein to evade insect immunity. *Proceedings of the National Academy of Sciences of the United States of America, 112*(36), 11365–11370.

Fernández-Marín, H., Zimmerman, J. K., Rehner, S. A., & Wcislo, W. T. (2006). Active use of the metapleural glands by ants in controlling fungal infection. *Proceedings of the Royal Society B: Biological Sciences, 273*, 1689–1695.

Freimoser, F. M., Hu, G., & St Leger, R. J. (2005). Variation in gene expression patterns as the insect pathogen *Metarhizium anisopliae* adapts to different host cuticles or nutrient deprivation in vitro. *Microbiology, 151*, 361–371.

Freimoser, F. M., Screen, S., Bagga, S., Hu, G., & St Leger, R. J. (2003). Expressed sequence tag (EST) analysis of two subspecies of *Metarhizium anisopliae* reveals a plethora of secreted proteins with potential activity in insect hosts. *Microbiology, 149*, 239–247.

Gao, Q., Jin, K., Ying, S. H., Zhang, Y., Xiao, G., Shang, Y. ... Wang, C. (2011). Genome sequencing and comparative transcriptomics of the model entomopathogenic fungi *Metarhizium anisopliae* and *M. acridum*. *PLoS Genetics, 7*, e1001264.

Garland, T. (2014). Trade-offs. *Current Biology, 24*, R60–R61.

Gems, D., & Partridge, L. (2008). Stress-response hormesis and aging: 'that which does not kill us makes us stronger'. *Cell Metabolism, 7*, 200–203.

Gillespie, J. P., Burnett, C., & Charnley, A. K. (2000). The immune response of the desert locust *Schistocerca gregaria* during mycosis of the entomopathogenic fungus, *Metarhizium anisopliae* var *acridum*. *Journal of Insect Physiology, 46*, 429–437.

Goettel, M. S., St Leger, R. J., Rizzo, N. W., Staples, R. C., & Roberts, D. W. (1989). Ultrastructural localization of a cuticledegrading protease produced by the entomopathogenic fungus *Metarhizium anisopliae* during penetration of host (*Manduca sexta*) cuticle. *Microbiology, 135*, 2233–2239.

Goldsworthy, G. J., Opoku-Ware, K., & Mullen, L. M. (2005). Adipokinetic hormone and the immune responses of locusts to infection. *Annals of the New York Academy of Sciences, 1040*, 106–113.

Gottar, M., Gobert, V., Matskevich, A. A., Reichhart, J. M., Wang, C., Butt, T. M. ... Ferrandon, D. (2006). Dual detection of fungal infections in *Drosophila* via recognition of glucans and sensing of virulence factors. *Cell, 127*, 1425–1437.

Govind, S. (2008). Innate immunity in *Drosophila*: pathogens and pathways. *Insect Science, 15*, 29–43.

Gross, R., Vavre, F., Heddi, A., Hurst, G. D., Zchori-Fein, E., & Bourtzis, K. (2009). Immunity and symbiosis. *Molecular Microbiology, 73*, 751–759.

Haine, E. R., Moret, Y., Siva-Jothy, M. T., & Rolff, J. (2008). Antimicrobial defense and persistent infection in insects. *Science, 322*, 1257–1259.

Harbison, S. T., Carbone, M. A., Ayroles, J. F., Stone, E. A., Lyman, R. F., & Mackay, T. F. (2009). Co-regulated transcriptional networks contribute to natural genetic variation in *Drosophila* sleep. *Nature Genetics, 41*, 371–375.

Hartenstein, V. (2006). Blood cells and blood cell development in the animal kingdom. *Annual Review of Cell and Developmental Biology, 22*, 677–712.

Hedges, L. M., Brownlie, J. C., O'Neill, S. L., & Johnson, K. N. (2008). *Wolbachia* and virus protection in insects. *Science, 322*, 702.

Hu, X., Xiao, G., Zheng, P., Shang, Y., Su, Y., Zhang, X. ... Wang, C. (2014). Trajectory and genomic determinants of fungal-pathogen speciation and host adaptation. *Proceedings of the National Academy of Sciences of the United States of America, 111*, 16796–16801.

Hu, X., Zhang, Y., Xiao, G., Zhang, P., Xia, Y., Zhang, X. ... Wang, C. (2013). Genome survey uncovers the secrets of sex and lifestyle in caterpillar fungus. *Chinese Science Bulletin, 58*(23), 2846–2854.

Igboin, C. O., Griffen, A. L., & Leys, E. J. (2012). The *Drosophila melanogaster* host model. *Journal of Oral Microbiology, 4*.

Ivanov, D. K., Escott-Price, V., Ziehm, M., Magwire, M. M., Mackay, T. F., Partridge, L., & Thornton, J. M. (2015). Longevity GWAS using the *Drosophila* genetic reference panel. *Journals of Gerontology. Series A, Biological Sciences and Medical Sciences, 70*(12), 1470–1478.

Jaenike, J. (2012). Population genetics of beneficial heritable symbionts. *Trends in Ecology and Evolution, 27*, 226–232.

Jaenike, J., & Brekke, T. D. (2011). Defensive endosymbionts: a cryptic trophic level in community ecology. *Ecology Letters, 14*, 150–155.

Jaenike, J., Unckless, R., Cockburn, S. N., Boelio, L. M., & Perlman, S. J. (2010). Adaptation via symbiosis: recent spread of a *Drosophila* defensive symbiont. *Science, 329*, 212–215.

Jiggins, F. M., & Hurst, G. D. (2003). The evolution of parasite recognition genes in the innate immune system: purifying selection on *Drosophila melanogaster* peptidoglycan recognition proteins. *Journal of Molecular Evolution, 57*, 598–605.

Jiggins, F. M., & Hurst, G. D. (2011). Microbiology. Rapid insect evolution by symbiont transfer. *Science, 332*, 185–186.

Jiggins, F. M., & Kim, K. W. (2006). Contrasting evolutionary patterns in *Drosophila* immune receptors. *Journal of Molecular Evolution, 63*, 769–780.

Jiggins, F. M., & Kim, K. W. (2007). A screen for immunity genes evolving under positive selection in *Drosophila*. *Journal of Evolutionary Biology, 20*, 965–970.

Juneja, P., & Lazzaro, B. P. (2009). *Providencia sneebia* sp. nov. and *Providencia burhodogranariea* sp. nov., isolated from wild *Drosophila melanogaster*. *International Journal of Systematic and Evolutionary Microbiology, 59*, 1108–1111.

Kafatos, F., Waterhouse, R., Zdobnov, E., & Christophides, G. (2009). Comparative genomics of insect immunity. In J. Rolff, & S. E. Reynolds (Eds.), *Insect infection and immunity: Evolution, ecology, and mechanisms*. Oxford: Oxford University Press.

Kaltz, O., & Shykoff, J. A. (1998). Local adaptation in host–parasite systems. *Heredity, 81*, 361–370.

Kawecki, T. J. (1998). Red queen meets Santa Rosalia: arms races and the evolution of host specialization in organisms with parasitic lifestyles. *American Naturalist, 152*, 635–651.

Keebaugh, E. S., & Schlenke, T. A. (2014). Insights from natural host-parasite interactions: the *Drosophila* model. *Developmental and Comparative Immunology, 42*, 111–123.

Kocks, C., Cho, J. H., Nehme, N., Ulvila, J., Pearson, A. M., Meister, M. ... Ezekowitz, R. A. (2005). Eater, a transmembrane protein mediating phagocytosis of bacterial pathogens in *Drosophila*. *Cell, 123*, 335–346.

Kounatidis, I., & Ligoxygakis, P. (2012). *Drosophila* as a model system to unravel the layers of innate immunity to infection. *Open Biology, 2*, 120075.

Kraaijeveld, A. R., & Godfray, H. C. (2008). Selection for resistance to a fungal pathogen in *Drosophila melanogaster*. *Heredity (Edinb), 100*, 400–406.

Kraaijeveld, A. R., & Godfray, H. C. (2009). Evolution of host resistance and parasitoid counter-resistance. *Advances in Parasitology, 70*, 257–280.

Kraaijeveld, A., & Wertheim, B. (2009). Costs and genomic aspects of *Drosophila* immunity to parasites and pathogens. In J. Rolff, & S. Reynolds (Eds.), *Insect infection and immunity* (pp. 187–205). Oxford: Oxford University Press.

Kurtti, T. J., & Keyhani, N. O. (2008). Intracellular infection of tick cell lines by the entomopathogenic fungus *Metarhizium anisopliae*. *Microbiology, 154*, 1700–1709.

Lamberty, M., Ades, S., Uttenweiler-Joseph, S., Brookhart, G., Bushey, D., Hoffmann, J. A., & Bulet, P. (1999). Insect immunity. Isolation from the lepidopteran *Heliothis virescens* of a novel insect defensin with potent antifungal activity. *Journal of Biological Chemistry, 274*, 9320–9326.

Lanot, R., Zachary, D., Holder, F., & Meister, M. (2001). Postembryonic hematopoiesis in *Drosophila*. *Developmental Biology, 230*, 243–257.

Lazzaro, B. P. (2008). Natural selection on the *Drosophila* antimicrobial immune system. *Current Opinion in Microbiology, 11*, 284–289.

Lazzaro, B. P., & Rolff, J. (2011). Immunology. Danger, microbes, and homeostasis. *Science, 332*, 43–44.

Le Conte, Y., Alaux, C., Martin, J. F., Harbo, J. R., Harris, J. W., Dantec, C. … Navajas, M. (2011). Social immunity in honeybees (*Apis mellifera*): transcriptome analysis of varroa-hygienic behaviour. *Insect Molecular Biology, 20*, 399–408.

Lemaitre, B., & Hoffmann, J. (2007). The host defense of *Drosophila melanogaster*. *Annual Review of Immunology, 25*, 697–743.

Lemaitre, B., Nicolas, E., Michaut, L., Reichhart, J. M., & Hoffmann, J. A. (1996). The dorsoventral regulatory gene cassette spätzle/Toll/cactus controls the potent antifungal response in *Drosophila* adults. *Cell, 86*, 973–983.

Lemaitre, B., Reichhart, J. M., & Hoffmann, J. A. (1997). *Drosophila* host defense: differential induction of antimicrobial peptide genes after infection by various classes of microorganisms. *Proceedings of the National Academy of Sciences of the United States of America, 94*, 14614–14619.

Levashina, E. A., Ohresser, S., Bulet, P., Reichhart, J. M., Hetru, C., & Hoffmann, J. A. (1995). Metchnikowin, a novel immune-inducible proline-rich peptide from *Drosophila* with antibacterial and antifungal properties. *European Journal of Biochemistry, 233*, 694–700.

Libbrecht, R., Oxley, P. R., Kronauer, D. J., & Keller, L. (2013). Ant genomics sheds light on the molecular regulation of social organization. *Genome Biology, 14*, 212.

Lionakis, M. S. (2011). *Drosophila* and *Galleria* insect model hosts: new tools for the study of fungal virulence, pharmacology and immunology. *Virulence, 2*, 521–527.

Loof, T. G., Mörgelin, M., Johansson, L., Oehmcke, S., Olin, A. I., Dickneite, G. … Herwald, H. (2011). Coagulation, an ancestral serine protease cascade, exerts a novel function in early immune defense. *Blood, 118*, 2589–2598.

Lu, H. L., Wang, J. B., Brown, M. A., Euerle, C., & St Leger, R. J. (2015). Identification of *Drosophila* mutants affecting defense to an entomopathogenic fungus. *Scientific Reports, 5*, 12350.

Magwire, M. M., Bayer, F., Webster, C. L., Cao, C., & Jiggins, F. M. (2011). Successive increases in the resistance of *Drosophila* to viral infection through a transposon insertion followed by a duplication. *PLoS Genetics, 7*, e1002337.

Maier, S. F., Watkins, L. R., & Fleshner, M. (1994). Psychoneuroimmunology. The interface between behavior, brain, and immunity. *American Psychologist, 49*, 1004–1017.

Mandard, N., Sodano, P., Labbe, H., Bonmatin, J. M., Bulet, P., Hetru, C. … Vovelle, F. (1998). Solution structure of thanatin, a potent bactericidal and fungicidal insect peptide, determined from proton two-dimensional nuclear magnetic resonance data. *European Journal of Biochemistry, 256*, 404–410.

Matskevich, A. A., Quintin, J., & Ferrandon, D. (2010). The *Drosophila* PRR GNBP3 assembles effector complexes involved in antifungal defenses independently of its Toll-pathway activation function. *European Journal of Immunology, 40*, 1244–1254.

Mburu, D. M., Ochola, L., Maniania, N. K., Njagi, P. G., Gitonga, L. M., Ndung'u, M. W. … Hassanali, A. (2009). Relationship between virulence and repellency of entomopathogenic isolates of *Metarhizium anisopliae* and *Beauveria bassiana* to the termite *Macrotermes michaelseni*. *Journal of Insect Physiology, 55*, 774–780.

McClure, C. D., Zhong, W., Hunt, V. L., Chapman, F. M., Hill, F. V., & Priest, N. K. (2014). Hormesis results in trade-offs with immunity. *Evolution, 68*, 2225–2233.

Medzhitov, R., Schneider, D. S., & Soares, M. P. (2012). Disease tolerance as a defense strategy. *Science, 335*, 936–941.

Meyling, N. V., Pell, J. K., & Eilenberg, J. (2006). Dispersal of *Beauveria bassiana* by the activity of nettle insects. *Journal of Invertebrate Pathology, 93*, 121–126.

Mortimer, N. T., Kacsoh, B. Z., Keebaugh, E. S., & Schlenke, T. A. (2012). Mgat1-dependent N-glycosylation of membrane components primes *Drosophila melanogaster* blood cells for the cellular encapsulation response. *PLoS Pathogens, 8*, e1002819.

Müller, U., Vogel, P., Alber, G., & Schaub, G. A. (2008). The innate immune system of mammals and insects. *Contributions to Microbiology, 15*, 21–44.
Nehme, N. T., Quintin, J., Cho, J. H., Lee, J., Lafarge, M. C., Kocks, C., & Ferrandon, D. (2011). Relative roles of the cellular and humoral responses in the *Drosophila* host defense against three gram-positive bacterial infections. *PLoS One, 6*, e14743.
Obbard, D. J., Jiggins, F. M., Bradshaw, N. J., & Little, T. J. (2011). Recent and recurrent selective sweeps of the antiviral RNAi gene Argonaute-2 in three species of *Drosophila. Molecular Biology and Evolution, 28*, 1043–1056.
Obbard, D. J., Welch, J. J., Kim, K. W., & Jiggins, F. M. (2009). Quantifying adaptive evolution in the *Drosophila* immune system. *PLoS Genetics, 5*, e1000698.
Ouedraogo, R. M., Cusson, M., Goettel, M. S., & Brodeur, J. (2003). Inhibition of fungal growth in thermoregulating locusts, *Locusta migratoria*, infected by the fungus *Metarhizium anisopliae* var *acridum. Journal of Invertebrate Pathology, 82*, 103–109.
Oxley, P. R., Spivak, M., & Oldroyd, B. P. (2010). Six quantitative trait loci influence task thresholds for hygienic behaviour in honeybees (*Apis mellifera*). *Molecular Ecology, 19*, 1452–1461.
Pal, S., St Leger, R. J., & Wu, L. P. (2007). Fungal peptide Destruxin A plays a specific role in suppressing the innate immune response in *Drosophila melanogaster. Journal of Biological Chemistry, 282*, 8969–8977.
Panayidou, S., Ioannidou, E., & Apidianakis, Y. (2014). Human pathogenic bacteria, fungi, and viruses in *Drosophila*: disease modeling, lessons, and shortcomings. *Virulence, 5*, 253–269.
Pava-Ripoll, M., Posada, F. J., Momen, B., Wang, C., & St Leger, R. (2008). Increased pathogenicity against coffee berry borer, *Hypothenemus hampei* (Coleoptera: Curculionidae) by *Metarhizium anisopliae* expressing the scorpion toxin (AaIT) gene. *Journal of Invertebrate Pathology, 99*, 220–226.
Pedrini, N., Ortiz-Urquiza, A., Huarte-Bonnet, C., Fan, Y., Juarez, M. P., & Keyhani, N. O. (2015). Tenebrionid secretions and a fungal benzoquinone oxidoreductase form competing components of an arms race between a host and pathogen. *Proceedings of the National Academy of Sciences of the United States of America, 112*, E3651–E3660.
Pham, L. N., Dionne, M. S., Shirasu-Hiza, M., & Schneider, D. S. (2007). A specific primed immune response in *Drosophila* is dependent on phagocytes. *PLoS Pathogens, 3*, e26.
Råberg, L., Graham, A. L., & Read, A. F. (2009). Decomposing health: tolerance and resistance to parasites in animals. *Philosophical Transactions of the Royal Society of London. Series B, Biological Sciences, 364*, 37–49.
Roberts, D. W., & St Leger, R. J. (2004). *Metarhizium* spp., cosmopolitan insect-pathogenic fungi: mycological aspects. *Advances in Applied Microbiology, 54*, 1–70.
Rolff, J., & Reynolds, S. E. (2009). *Insect infection and immunity: Evolution, ecology and mechanisms*. Oxford: Oxford University Press.
Roxström-Lindquist, K., Terenius, O., & Faye, I. (2004). Parasite-specific immune response in adult *Drosophila melanogaster*: a genomic study. *EMBO Reports, 5*, 207–212.
Sackton, T. B., Lazzaro, B. P., Schlenke, T. A., Evans, J. D., Hultmark, D., & Clark, A. G. (2007). Dynamic evolution of the innate immune system in *Drosophila. Nature Genetics, 39*, 1461–1468.
Salazar-Jaramillo, L., Paspati, A., van de Zande, L., Vermeulen, C. J., Schwander, T., & Wertheim, B. (2014). Evolution of a cellular immune response in *Drosophila*: a phenotypic and genomic comparative analysis. *Genome Biology and Evolution, 6*, 273–289.
Samuels, R. I., Charnley, A. K., & Reynolds, S. E. (1988). The role of destruxins in the pathogenicity of 3 strains of *Metarhizium anisopliae* for the tobacco hornworm *Manduca sexta. Mycopathologia, 104*, 51–58.
Scarborough, C. L., Ferrari, J., & Godfray, H. C. (2005). Aphid protected from pathogen by endosymbiont. *Science, 310*, 1781.

Scherfer, C., Karlsson, C., Loseva, O., Bidla, G., Goto, A., Havemann, J. ... Theopold, U. (2004). Isolation and characterization of hemolymph clotting factors in *Drosophila melanogaster* by a pullout method. *Current Biology, 14*, 625–629.

Schlenke, T. A., & Begun, D. J. (2005). Linkage disequilibrium and recent selection at three immunity receptor loci in *Drosophila* simulans. *Genetics, 169*, 2013–2022.

Schmid-Hempel, P. (1998). *Parasites in social insects*. Princeton University Press.

Schmid-Hempel, P. (2003). Variation in immune defence as a question of evolutionary ecology. *Proceedings of the Royal Society B: Biological Sciences, 270*, 357–366.

Schneider, D. (2009). Physiological integration of innate immunity. In J. Rolff, & S. E. Reynolds (Eds.), *Insect infection and immunity: Evolution, ecology, and mechanisms*. Oxford: Oxford University Press.

Schuhmann, B., Seitz, V., Vilcinskas, A., & Podsiadlowski, L. (2003). Cloning and expression of gallerimycin, an antifungal peptide expressed in immune response of greater wax moth larvae, *Galleria mellonella*. *Archives of Insect Biochemistry and Physiology, 53*, 125–133.

Shia, A. K., Glittenberg, M., Thompson, G., Weber, A. N., Reichhart, J. M., & Ligoxygakis, P. (2009). Toll-dependent antimicrobial responses in *Drosophila* larval fat body require Spätzle secreted by haemocytes. *Journal of Cell Science, 122*, 4505–4515.

Short, S. M., & Lazzaro, B. P. (2013). Reproductive status alters transcriptomic response to infection in female *Drosophila melanogaster*. *G3 (Bethesda), 3*, 827–840.

Shorter, J., Couch, C., Huang, W., Carbone, M. A., Peiffer, J., Anholt, R. R., & Mackay, T. F. (2015). Genetic architecture of natural variation in *Drosophila melanogaster* aggressive behavior. *Proceedings of the National Academy of Sciences of the United States of America, 112*, E3555–E3563.

Sibley, C. D., Duan, K., Fischer, C., Parkins, M. D., Storey, D. G., Rabin, H. R., & Surette, M. G. (2008). Discerning the complexity of community interactions using a *Drosophila* model of polymicrobial infections. *PLoS Pathogens, 4*, e1000184.

Simola, D. F., Wissler, L., Donahue, G., Waterhouse, R. M., Helmkampf, M., Roux, J. ... Gadau, J. (2013). Social insect genomes exhibit dramatic evolution in gene composition and regulation while preserving regulatory features linked to sociality. *Genome Research, 23*, 1235–1247.

Smith, C. D., Zimin, A., Holt, C., Abouheif, E., Benton, R., Cash, E. ... Tsutsui, N. D. (2011). Draft genome of the globally widespread and invasive Argentine ant (*Linepithema humile*). *Proceedings of the National Academy of Sciences of the United States of America, 108*, 5673–5678.

Smith, C. R., Smith, C. D., Robertson, H. M., Helmkampf, M., Zimin, A., Yandell, M. ... Gadau, J. (2011). Draft genome of the red harvester ant *Pogonomyrmex barbatus*. *Proceedings of the National Academy of Sciences of the United States of America, 108*, 5667–5672.

St. Leger, R. J. (1991). The integument as a barrier to microbial infections. In A. Retnakaran, & K. Binnington (Eds.), *The Physiology of Insect Epidermis* (pp. 284–306). Australia: CSIRO.

St. Leger, R. J., Charnley, A. K., & Cooper, R. M. (1988). Production of polyphenol pigments and phenoloxidase by the entomopathogen, *Metarhizium anisopliae*. *Journal of Invertebrate Pathology, 52*, 215–220.

St Leger, R. J., Cooper, R. M., & Charnley, A. K. (1988). The effect of melanization of *Manduca sexta* cuticle on growth and infection by *Metarhizium anisopliae*. *Journal of Invertebrate Pathology, 52*, 459–470.

St Leger, R. J., Goettel, M., Roberts, D. W., & Staples, R. C. (1991). Pre-penetration events during infection of host cuticle by *Metarhizium anisopliae*. *Journal of Invertebrate Pathology, 58*, 168–179.

St Leger, R. J., Joshi, L., Bidochka, M. J., Rizzo, N. W., & Roberts, D. W. (1996). Biochemical characterization and ultrastructural localization of two extracellular trypsins produced

by *Metarhizium anisopliae* in infected insect cuticles. *Applied and Environmental Microbiology, 62*, 1257–1264.
St Leger, R. J., Joshi, L., Bidochka, M. J., & Roberts, D. W. (1996). Construction of an improved mycoinsecticide overexpressing a toxic protease. *Proceedings of the National Academy of Sciences of the United States of America, 93*, 6349–6354.
St Leger, R. J., Wang, C., & Fang, W. (2011). New perspectives on insect pathogens. *Fungal Biology Reviews, 25*, 84–88.
Staves, P. A., & Knell, R. J. (2010). Virulence and competitiveness: testing the relationship during inter- and intraspecific mixed infections. *Evolution, 64*, 2643–2652.
Strand, M. R. (2008). The insect cellular immune response. *Insect Science, 15*, 1–14.
Stroschein-Stevenson, S. L., Foley, E., O'Farrell, P. H., & Johnson, A. D. (2006). Identification of *Drosophila* gene products required for phagocytosis of *Candida albicans*. *PLoS Biology, 4*, e4.
Suen, G., Teiling, C., Li, L., Holt, C., Abouheif, E., Bornberg-Bauer, E. … Currie, C. R. (2011). The genome sequence of the leaf-cutter ant *Atta cephalotes* reveals insights into its obligate symbiotic lifestyle. *PLoS Genetics*, 7, e1002007.
Swanson, J. A., Torto, B., Kells, S. A., Mesce, K. A., Tumlinson, J. H., & Spivak, M. (2009). Odorants that induce hygienic behavior in honeybees: identification of volatile compounds in chalkbrood-infected honeybee larvae. *Journal of Chemical Ecology, 35*, 1108–1116.
Tack, A. J., Thrall, P. H., Barrett, L. G., Burdon, J. J., & Laine, A. L. (2012). Variation in infectivity and aggressiveness in space and time in wild host-pathogen systems: causes and consequences. *Journal of Evolutionary Biology, 25*, 1918–1936.
Thomas, M. B., & Read, A. F. (2007). Can fungal biopesticides control malaria? *Nature Reviews Microbiology, 5*, 377–383.
Tinsley, M. C., Blanford, S., & Jiggins, F. M. (2006). Genetic variation in *Drosophila melanogaster* pathogen susceptibility. *Parasitology, 132*, 767–773.
Toledo, A. V., Alippi, A. M., & de Remes Lenicov, A. M. (2011). Growth inhibition of *Beauveria bassiana* by bacteria isolated from the cuticular surface of the corn leafhopper, *Dalbulus maidis* and the planthopper, *Delphacodes kuscheli*, two important vectors of maize pathogens. *Journal of Insect Science, 11*, 29.
Tzou, P., Reichhart, J. M., & Lemaitre, B. (2002). Constitutive expression of a single antimicrobial peptide can restore wild-type resistance to infection in immunodeficient *Drosophila* mutants. *Proceedings of the National Academy of Sciences of the United States of America, 99*, 2152–2157.
Ulrich, K. R., Feldlaufer, M. F., Kramer, M., & St. Leger, R. J. (2015). Inhibition of the entomopathogenic fungus *Metarhizium anisopliae* sensu lato in vitro by the bed bug defensive secretions (E)-2-hexenal and (E)-2-octenal. *BioControl*, 1–10.
Valanne, S., Wang, J. H., & Rämet, M. (2011). The *Drosophila* Toll signaling pathway. *Journal of Immunology, 186*, 649–656.
Vance, R. E., Isberg, R. R., & Portnoy, D. A. (2009). Patterns of pathogenesis: discrimination of pathogenic and nonpathogenic microbes by the innate immune system. *Cell Host and Microbe, 6*, 10–21.
Vega, F. E., & Kaya, H. F. (2012). *Insect pathology*. Academic Press.
Vigers, A. J., Wiedemann, S., Roberts, W. K., Legrand, M., Selitrennikoff, C. P., & Fritig, B. (1992). Thaumatin-like pathogenesis-related proteins are antifungal. *Plant Science, 83*, 155–161.
Vlisidou, I., & Wood, W. (2015). *Drosophila* blood cells and their role in immune responses. *FEBS Journal, 282*, 1368–1382.
Wang, C., & St Leger, R. J. (2006). A collagenous protective coat enables *Metarhizium anisopliae* to evade insect immune responses. *Proceedings of the National Academy of Sciences of the United States of America, 103*, 6647–6652.

Wang, C., & St Leger, R. J. (2007a). The MAD1 adhesin of *Metarhizium anisopliae* links adhesion with blastospore production and virulence to insects, and the MAD2 adhesin enables attachment to plants. *Eukaryotic Cell, 6*, 808–816.

Wang, C., & St Leger, R. J. (2007b). A scorpion neurotoxin increases the potency of a fungal insecticide. *Nature Biotechnology, 25*, 1455–1456.

Wang, Z., Wilhelmsson, C., Hyrsl, P., Loof, T. G., Dobes, P., Klupp, M. ... Theopold, U. (2010). Pathogen entrapment by transglutaminase—a conserved early innate immune mechanism. *PLoS Pathogens, 6*, e1000763.

Wang, J. B., St. Leger, R. J., & Wang, C. (2016). Advances in Genomics of Insect Pathogenic Fungi. *Advances in Genetics, 94*, 67–105.

Watson, F. L., Püttmann-Holgado, R., Thomas, F., Lamar, D. L., Hughes, M., Kondo, M. ... Schmucker, D. (2005). Extensive diversity of Ig-superfamily proteins in the immune system of insects. *Science, 309*, 1874–1878.

Werren, J. H., Richards, S., Desjardins, C. A., Niehuis, O., Gadau, J., Colbourne, J. K. ... Gibbs, R. A. (2010). Functional and evolutionary insights from the genomes of three parasitoid *Nasonia* species. *Science, 327*, 343–348.

Wertheim, B. (2015). Genomic basis of evolutionary change: evolving immunity. *Frontiers in Genetics, 6*, 222.

Wertheim, B., Kraaijeveld, A. R., Hopkins, M. G., Walther Boer, M., & Godfray, H. C. (2011). Functional genomics of the evolution of increased resistance to parasitism in *Drosophila*. *Molecular Ecology, 20*, 932–949.

Wertheim, B., Kraaijeveld, A. R., Schuster, E., Blanc, E., Hopkins, M., Pletcher, S. D. ... Godfray, H. C. (2005). Genome-wide gene expression in response to parasitoid attack in *Drosophila*. *Genome Biology, 6*, R94.

Xiao, G., Ying, S. H., Zheng, P., Wang, Z. L., Zhang, S., Xie, X. Q. ... Feng, M. G. (2012). Genomic perspectives on the evolution of fungal entomopathogenicity in *Beauveria bassiana*. *Scientific Reports, 2*, 483.

Yassine, H., Kamareddine, L., & Osta, M. A. (2012). The mosquito melanization response is implicated in defense against the entomopathogenic fungus *Beauveria bassiana*. *PLoS Pathogens, 8*, e1003029.

Yek, S. H., Boomsma, J. J., & Schiøtt, M. (2013). Differential gene expression in Acromyrmex leaf-cutting ants after challenges with two fungal pathogens. *Molecular Ecology, 22*, 2173–2187.

Zheng, P., Xia, Y., Xiao, G., Xiong, C., Hu, X., Zhang, S. ... Wang, C. (2011). Genome sequence of the insect pathogenic fungus *Cordyceps militaris*, a valued traditional Chinese medicine. *Genome Biology, 12*, R116.

Zhong, W., McClure, C. D., Evans, C. R., Mlynski, D. T., Immonen, E., Ritchie, M. G., & Priest, N. K. (2013). Immune anticipation of mating in *Drosophila*: Turandot M promotes immunity against sexually transmitted fungal infections. *Proceedings of the Royal Society B: Biological Sciences, 280*, 20132018.

Zuk, M., & Stoehr, A. M. (2002). Immune defense and host life history. *American Naturalist, 160*(Suppl. 4), S9–S22.

CHAPTER EIGHT

Disease Dynamics in Ants: A Critical Review of the Ecological Relevance of Using Generalist Fungi to Study Infections in Insect Societies

R.G. Loreto[*,§,1] **and D.P. Hughes**[*,1]
[*]Pennsylvania State University, University Park, PA, United States
[§]CAPES Foundation, Ministry of Education of Brazil, Brasília, DF, Brazil
[1]Corresponding authors: E-mail: raquelgloreto@gmail.com; dhughes@psu.edu

Contents

Abstract

It is assumed that social life can lead to the rapid spread of infectious diseases and outbreaks. In ants, disease outbreaks are rare and the expression of collective behaviors is invoked to explain the absence of epidemics in natural populations. Here, we address the ecological approach employed by many studies that have notably focused (89% of the studies) on two genera of generalist fungal parasites (*Beauveria* and *Metarhizium*). We ask whether these are the most representative models to study the evolutionary ecology of ant—fungal parasite interactions. To assess this, we critically examine the literature on ants and their interactions with fungal parasites from the past 114 years (1900—2014). We discuss how current evolutionary ecology approaches emerged from studies focused on the biological control of pest ants. We also analyzed the ecological relevance of the laboratory protocols used in evolutionary ecology studies employing generalist parasites, as well as the rare natural occurrence of these parasites

Advances in Genetics, Volume 94
ISSN 0065-2660
http://dx.doi.org/10.1016/bs.adgen.2015.12.005

on ants. After a detailed consideration of all the publications, we suggest that using generalist pathogens such as *Beauveria* and *Metarhizium* is not an optimal approach if the goal is to study the evolutionary ecology of disease in ants. We conclude by advocating for approaches that incorporate greater realism.

1. INTRODUCTION

The emergence of the insect societies is one of the great evolutionary transitions in organic life (Maynard Smith & Szathmáry, 1995), and the notable dominance of ants, and social insects in general, implies advantages of group living over solitary life. One advantage is the extended life span of reproductive castes as a result of the protective nature of a colony (Keller & Genoud, 1997). As a counterpoint, major costs may exist for social life. Notably, group living presents conditions that are assumed to be ideal for parasite development and transmission. The organization and dynamics of the colony depend on the exchange of information and resources among nestmates, resulting in high rates of contact within the confined environment that is the nest. This close contact is compounded by elevated relatedness, an overlap of generations with reproducing and sterile adults, pupae, larvae, and eggs in the same space, and the controlled microclimate within the nest. Thus, it is assumed that social life can lead to the rapid spread of infectious diseases and outbreaks (Cremer, Armitage, & Schmid-Hempel, 2007).

Social insects, like all insects, are host to a diverse array of macroparasites and microparasites. Macroparasites include parasitic insects (eg, flies, wasps, and strepsipterans), worms (nematodes, trematodes, and cestodes) as well as mites, beetles, lepidopterans, and other arthropods that enter the nest to sequester resources (Schmid-Hempel, 1998). However, we know from standard epidemiological theory that macroparasitic infections tend not to lead to major disease outbreaks that strongly affect the host population (Anderson & May, 1979). On the other hand, microparasites (bacteria, fungi, protozoa and viruses) can cause epidemics that in turn slow down population growth. In ants, the abundance of entomopathogenic fungi in tropical forests has been suggested to act as an important control of ant populations (Evans, 1974). The potential for these microparasites to cause infectious diseases has been hypothesized as a major break on the evolution of sociality (Alexander, 1974; Arneberg, Skorping, Grenfell, & Read, 1998; Côté & Poulin, 1995).

A second hypothesis emerged from studies motivated by the use of infectious disease agents as biological control tools of ants and termites (Chouvenc, Su, & Grace, 2011), as well as the efforts aimed at reducing infectious disease spread in bees (Bailey & Ball, 1991). Overall, these studies indicate that those same conditions that could facilitate disease spread would also result in complex collective behaviors that could mitigate the parasite threat, a phenomena described as "social immunity" (Cremer et al., 2007). This second hypothesis (the social immunity hypothesis) predicts that the emergent properties of sociality, such as collective behavior, in fact lead to disease control and not necessarily an increased rate of transmission, as predicted by the previous hypothesis (ie, that social living promotes epidemics). This socially mediated dampening down of disease inside the nest may be particularly true for ants (discussed in this chapter) and termites (Chouvenc et al., 2011), where a summary of the literature would conclude that they are very efficient at collectively avoiding and controlling disease spread. For these two important groups of insects, social immunity is often used to explain the absence of disease outbreaks in natural populations, despite the apparently ideal ecological conditions for the rapid spread of infectious diseases within the colony.

Recently in the literature on disease in ant societies, we have encountered many studies supporting the social immunity hypothesis. The majority of them use fungal parasites as a source of disease (Fig. 1A and Boomsma, Schmid-Hempel, and Hughes (2005)). Overall, these studies show changes in the behavior of the colony and conclude that social immunity in ant societies is very efficient. To examine this in more detail, we analyzed 114 years of literature (1900–2014) on ant–fungal parasite interactions. Most studies used the generalist entomopathogenic fungi *Beauveria* and *Metarhizium* (89% of studies on ant–fungal parasite interactions, Fig. 1B). This is likely due to the enthusiasm and then subsequent failure of the attempt to use these two genera of fungi to control pest ants (this chapter), which mirrors a review of 50 years of similar studies in termites (Chouvenc et al., 2011). Although these fungi did not result in successful biological control in the field, they were found to kill ant workers, thus elevating these fungi to important tools for laboratory-based assays of infection. From there the approach has shifted to a framework that draws general conclusions about the evolutionary ecology of ant–parasite interactions.

It appears to us that our current knowledge on the ecology and evolution of ant–fungal parasite interactions is heavily drawn from studies using

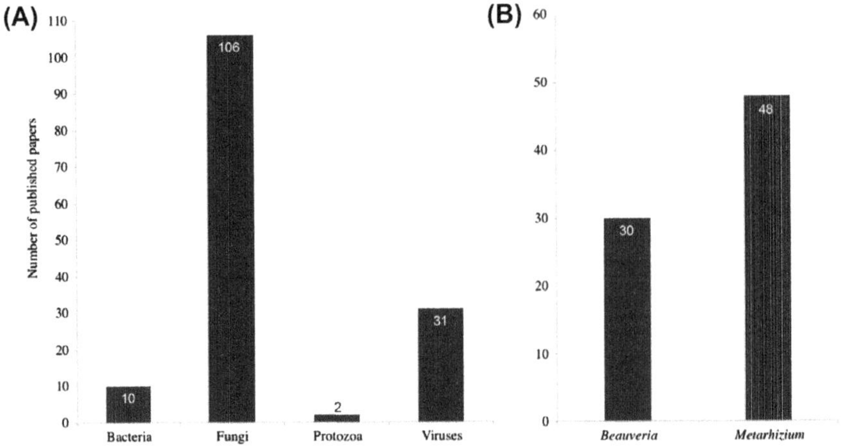

Figure 1 Number of publications exploring the different group of microparasites of ants over last 14 years (2001–14). (A) Number of published papers exploring each of the four categories of microparasites: bacteria (10), fungi (106), protozoa (2), and viruses (31). When more than one parasite was used in the same publication, the publication was listed once for each parasite. (B) Number of publications on ant–parasite interactions using *Beauveria* or *Metarhizium* as infectious disease agents.

simplistic conditions, using far higher doses of parasite propagules than those experienced in nature, and often presented to ants in unrealistic routes (Table S1). We suggest that employing a methodology that lacks ecological relevance may lead to incorrect conclusions. This, in turn, possibly prevents general insights into the evolution and ecology of ant–fungal parasite interactions. The equivocal assumption that such studies illuminate our understanding of ant–fungal coevolution is now being compounded by the adoption of formal mathematical models (Novak & Cremer, 2015; Theis, Ugelvig, Marr, & Cremer, 2015). However, if the biology is not realistic to begin with, then such approaches have limited utility. In addition, we also observed few records of natural infections of ants by generalist fungi. In line with that, it has been suggested that *Beauveria* and *Metarhizium* are not obligate insect parasites, but rather facultative/opportunistic associates that also interact with plants (Gao et al., 2011; Vega, 2008). We also raise the issue for how the sublethal effect of these infections has been considered in ant–generalist parasite studies. Taken together, we argue that the use of *Beauveria* and *Metarhizium* as a tool to understand infectious diseases in ants will better serve us when implemented in conditions that reflect natural conditions. Additionally, studies on parasites known to be specialized on ants are encouraged, as they would significantly complement our current knowledge.

2. ORIGIN AND TRENDS OF USING GENERALIST FUNGAL PARASITES TO STUDY ANT–FUNGAL PARASITE INTERACTIONS

During the second half of 1980s, the economic importance of leaf-cutting ants (*Atta* and *Acromyrmex*) and fire ants (*Solenopsis*) led to efforts to develop entomopathogenic fungi as biological control agents of these ants. Special attention was devoted to fungi from the genera *Beauveria* and *Metarhizium* due to their successful application on other groups of insects (Goettel, Eilenberg, & Glare, 2010). The initial tests for pathogenicity and host susceptibility to different strains of *Beauveria* and *Metarhizium* were promising, with high mortality and sporulation rates on individual workers (Alves, Stimac, & Camargo, 1988; Sanchěz-penã & Thorvilson, 1992; Silva & Diehl-Fleig, 1988; Stimac, Alves, & Camargo, 1987). However, the application of these fungi at the colony level both in laboratory and field conditions had inconsistent results that were not sufficiently robust for effective pest control (Diehl-Fleig, Silva, Specht, & Valim-Labres, 1993; Kermarrec, Febvay, & Decharme, 1986; Pereira & Stimac, 1992; Stimac, Alves, & Camargo, 1989). For example, Stimac et al. (1989) reported an average of 80% of control in *Solenopsis* spp. nests by *Beauveria bassiana*, while Oi, Pereira, Stimac, and Wood (1994) reported 24% of nest mortality using a similar technique of fungal application.

In the biological control of pest insects, the social life of ants can add difficulties for controlling these organisms compared to solitary insects. In general, only a small portion of the colony explores the outside environment at one time, limiting the exposure of the colony to the fungi. Compounding the problem of using such fungi as biological control tools is that the queen, the reproductive unit of the colony, tends to be well protected, especially in mature colonies, making it unlikely that an application of fungi would eliminate her. Thus, the application of the fungus may decrease the number of ants temporarily, but it does not necessarily lead to control. Additionally, the behavioral ecology of the ants could be a barrier for the establishment of the onward infection chain needed for successful biological control. From the earliest reports of field applications of entomopathogenic fungi in ant colonies, researchers found that the ants would abandon areas of the nest where the inoculation with the fungi occurred (Machado, Diehl-Fleig, Silva, & Lucchese, 1988), as well as removing the inoculated baits (Diehl-Fleig & Lucchese, 1991) and infected cadavers (Pereira & Stimac, 1992) away from their nest.

In an attempt to link the inconsistency around ant control following the application of entomopathogenic fungi with general insights into colony defenses, Jaccoud, Hughes, and Jackson (1999) studied the epizootiology of *Metarhizium* infections in leaf-cutting ants. Although they seemed to be motived by biological control, these authors stepped away from the traditional framework, focusing more on the ecology of the interaction. A few years later, Hughes, Eilenberg, and Boomsma (2002) published the first paper on ant–fungal parasite interactions that explicitly considered the ability of ants to survive fungal infections, making clear that the framework of the paper was not biological control. Before this, biological control focused studies used ant mortality, never survival, as the most relevant measurement (Table S1). It was the beginning of a new trend, which would shift the focus of the ant–fungal parasite studies from biological control to the ecology and evolution of infectious diseases in ants.

3. THE ECOLOGICAL RELEVANCE OF LABORATORY EXPERIMENTATION WITH *BEAUVERIA* AND *METARHIZIUM* IN ANTS

The difficulty of working with ant colonies in the wild and the convenience of generalist parasites likely explain why the majority of ant–fungal parasite studies have been performed in the laboratory. These studies have shaped our understanding of ant–fungal parasite interactions and, more broadly, the ecology and evolution of disease threats in ant societies. The controlled conditions of laboratory experiments, achieved by simplifying setups, do provide valuable information on how general defenses are organized. However, basing our view only on studies performed in such contexts can be problematic because they may not translate to defenses implemented in more complex environments. Environmental complexity, such as daily temperature fluctuation (Murdock, Paaijmans, Cox-Foster, Read, & Thomas, 2012), spatial heterogeneity (Brockhurst & Koskella, 2013), and community diversity (Orlofske, Jadin, Preston, & Johnson, 2012; Thrall, Hochberg, Burdon, & Bever, 2007), is known to play an important role in host–parasite interactions. We could reasonably assume that is also true where ants are the host.

One major issue with laboratory setups is how the conidia (ie, infective propagule of the fungal pathogen) encounter ants. This is the first step of the infection, and is, therefore, ecologically relevant for the interaction. As in other insects, ants can get infected when they encounter fungal conidia

(Hajek & St. Leger, 1994; Kermarrec et al., 1986). These conidia would be scattered on the soil and other surfaces where the ants nest and forage for food. Therefore, a realistic protocol for infection would be mixing fungi into the soil and exposing the ants to it. This was often done in biological control research (Pereira & Stimac, 1992; Pereira, Stimac, & Alves, 1993), where it was established that applying spores in soil considerably decreased the infection rate (Fuxa & Richter, 2004; Pereira & Stimac, 1992; Pereira et al., 1993; Stimac, Pereira, Alves, & Wood, 1993). In studies on behavioral immunity of ants, the experimental arenas are often sterile petri dishes (Konrad et al., 2012; Tragust, Mitteregger, et al., 2013).

Despite the appeal of naturalistic conditions such as placing spores in soil or having ants walk over conidia applied on filter paper (Castella, Chapuisat, & Christe, 2008; Chapuisat, Oppliger, Magliano, & Christe, 2007; Mattoso, Moreira, & Samuels, 2012; Reber, Castella, Christe, & Chapuisat, 2008; Schmidt, Linksvayer, Boomsma, & Pedersen, 2011), most researchers have adopted a peculiar approach that is not natural. Out of the 57 papers published in the last 12 years on ant ecological immunology, 31 used a topical application of conidia in a suspension applied directly onto the ant cuticle (Fig. 2 and Table S1). These conidia suspensions are often very concentrated, with the number of spores ranging from 10^4 to 10^9 conidia

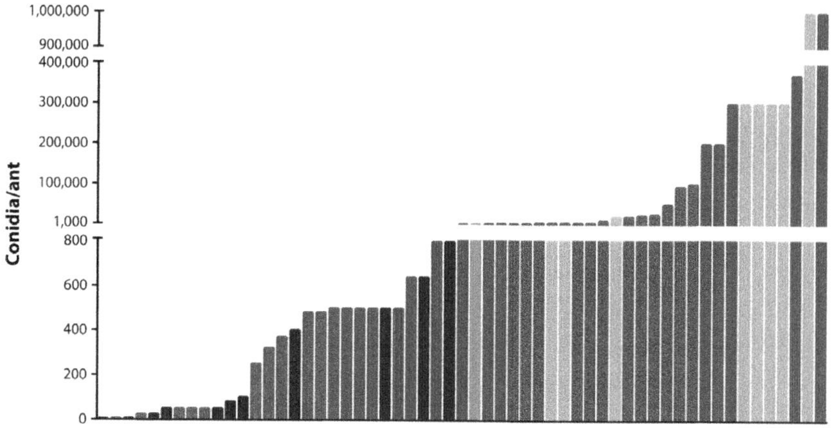

Figure 2 Number of conidia applied per ant in experiments using topical application of conidial suspension as exposure method to *Beauveria* or *Metarhizium*. The blue (black in print versions) bars represent the experiments where survival was not decreased by the parasite. The red (dark gray in print versions) bars represent the experiments where survival was decreased by the parasite. Gray bars are studies where survival was not measured. The doses in the first five studies were below or equal to those found in nature (see Table S2).

per mL (Fig. 2). Such concentrations are unlikely to be found in nature. The highest reported concentration of conidia in the soil for either *Beauveria* or *Metarhizium* is 6.8×10^3 and 5×10^4 colony-forming units (CFU) per gram, respectively (Hughes, Thomsen, Eilenberg, & Boomsma, 2004; Keller, Kessler, & Schweizer, 2003). It is difficult to estimate how many spores an ant would come in contact with based on the number of CFU found in 1 g of soil. It would also be challenging to translate the number of CFU per gram into a solution of conidia (per mL). However, for the majority of studies, the experimental dosages are orders of magnitude higher than what can be found in natural situations (Fig. 2). In those studies, justifications for the applied dosages are rare. When provided, they are often related to experimental need to ensure the infection process (Reber, Purcell, Buechel, Buri, & Chapuisat, 2011) or to elicit a behavioral response (Pull, Hughes, & Brown, 2013). Moreover, the topical application of conidia solution does not reflect the natural route of infection, which is via ants brushing against a source of spores or walking over them. The application method in the laboratory setup eliminates ecological features of what is an important step of the ant–fungal parasite interactions.

Fungal conidia could also potentially be transmitted inside the nest by contacting a nestmate that had direct contact with the fungi. As discussed above, the sibling–sibling transmission is assumed to be an important constraint of social living. Konrad et al. (2012) showed that ants infected with a high dosage of conidia (10^9 mL^{-1}) subsequently transmitted conidia to 17 ants out of 45 (37%) they had contact with. However, in the same study, the mortality of these secondarily exposed nestmates due to the fungus was only 3 out of 150 worker ants (2%). A study by Reber et al. (2011) had shown that in two different experiments, using a total of 1555 ants, only one was infected and died by secondary contact with conidia via nestmates. Therefore, if realistic spore loads were applied, it is unlikely that secondary infections and subsequent deaths would have occurred.

We suggest, based on our review of the literature, that ants might not naturally encounter the high amounts of conidia typically used for experimental infections and certainly not as highly concentrated solutions of conidia directly applied to their bodies. However, it could be possible that worker ants come in contact with a large amount of conidia if they are handling infectious cadavers (either of an ant or some other insect). The encounter with an infectious cadaver may happen when the ants are digging the nest galleries or foraging directly on the forest floor. The amount of conidia produced by a cadaver varies according to the ant species (Walker &

Hughes, 2011), but it was estimated in the leaf-cutter ant *Acromyrmex echinatior* to be 1.2×10^7 ($\pm 1.6 \times 10^6$) conidia per cadaver (Hughes et al., 2002). Studies on biological control have investigated the mortality of workers following the introduction of cadavers in the nest, showing that it correlates with the number of cadavers (Pereira & Stimac, 1992). However, it remains to be investigated how ants react to situations where they encounter infectious cadavers.

Since the initial infection is a key point in any host—parasite interaction, we looked at 31 studies that topically applied conidia of *Beauveria* or *Metarhizium*. For those that compared infected and control ants, we found, not surprisingly, that ant survival is less where researchers expose the ants to higher loads of conidia (Fig. 2). The lower doses of fungi (10^4 mL^{-1}) used in the studies also did not affect grooming levels (Reber et al., 2011). Grooming is the most common behavior used to measure social immunity. In some cases, the higher doses of conidia (10^7 mL^{-1}) resulted in decreased grooming levels (Okuno, Tsuji, Sato, & Fujisaki, 2012). In some cases, the approach taken is perhaps not as objective as it should be. For example, in a study of founding *Lasius niger* queens, Pull et al. (2013) state that queens were "exposed to a higher dosage of conidia (2 μL of conidia solution at a concentration of 5×10^8 conidia mL^{-1}) in order to elicit a greater antiseptic grooming response from the queens." In Fig. 3 of the same study (Pull et al., 2013), the researchers report an increase in self-grooming behavior as a response to the parasite. These are possible examples of how our ecologically unrealistic setups could produce misleading conclusions into the evolutionary ecology of diseases in ant societies.

During our literature research we came across very few negative results. This implies that most of the publications discovered an effect of fungal application on the tested subject. This could be due to the robustness of our current knowledge or, after evaluating the implemented protocols, a bias toward positive results, likely driven by the nonrealistic applications of conidia. Only one publication was entirely based on a negative result (Reber & Chapuisat, 2012b), while the other ones had the negative results buried among the highlighted, positive, results (Okuno et al., 2012; Tragust, Ugelvig, Chapuisat, Heinze, & Cremer, 2013). As such, the negative results, when present, did not receive their due attention. In studies of disease dynamics, reporting the failure of the disease agent to transmit (ie, a negative result) is of fundamental interest because it reveals aspects of context-dependent virulence, which is a property of both host and parasite (Ebert, 1994; Read, 1994).

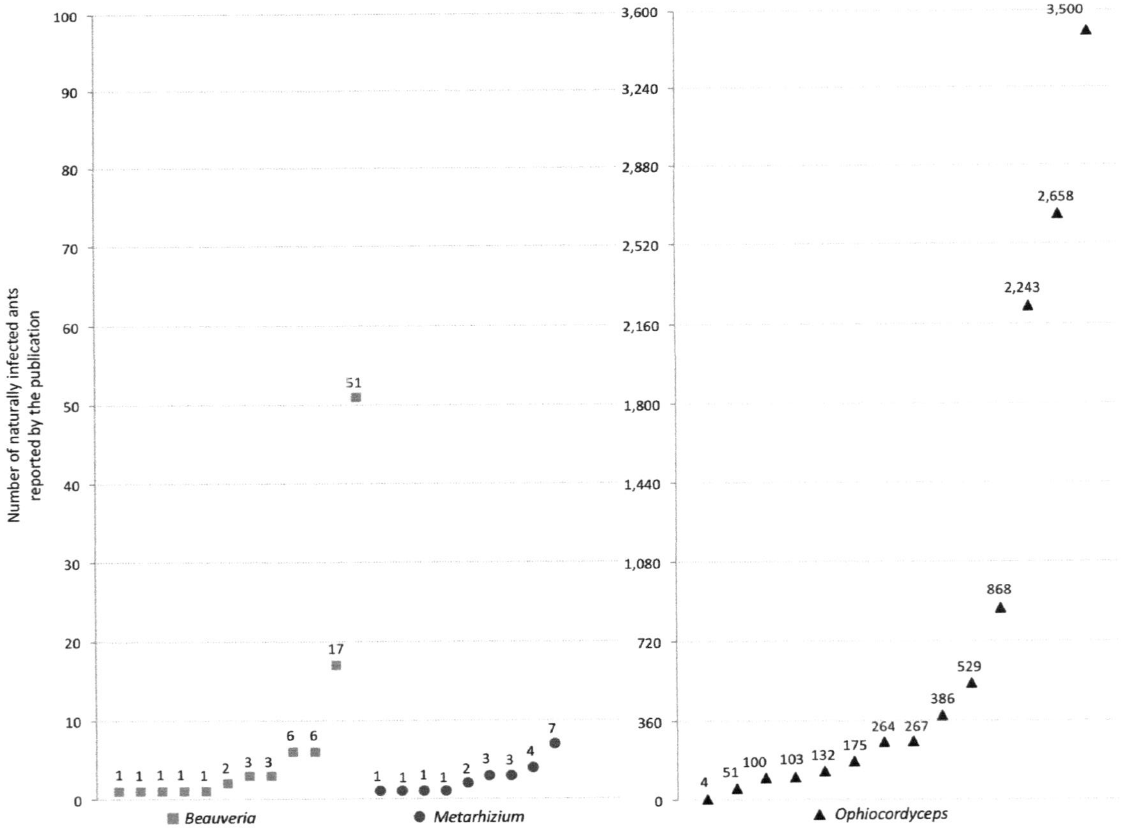

Figure 3 Number of natural infections of ants by generalist and specialist fungi reported in publications. All the publications that reported the number of isolates or cadavers for *Beauveria* (*red (light gray in print versions) squares*), *Metarhizium* (*blue (gray in print versions) circle*) or *Ophiocordyceps* (*black triangles*) were included. Note that it was necessary to include a second x-axis for reporting the number of ants naturally infected by *Ophiocordyceps* sp. The publications used for this figure are also reported in Table S3.

Very recently, researchers started adopting mathematical models to explore disease transmission within ant societies (Novak & Cremer, 2015; Theis et al., 2015). These models are based on assumptions from empirical data, collected following laboratory studies, such as those that elicited grooming (Theis et al., 2015) or those which observed cross infections among nestmates (Novak & Cremer, 2015). We have already discussed that such results might be due to the nature of the laboratory setups, and likely do not represent the real ecology of ant–fungal interactions. Consequently, the theoretical models based on these results may also be only applicable to these nonnatural conditions. Mathematical explorations are certainly illuminating, but we encourage them to be parameterized based on a realistic ecological basis (Andersen et al., 2012).

Although we highlight the limitation of the laboratory studies that do not reflect natural conditions, we do not suggest that laboratory studies themselves are the problem. The simplification of the system is necessary to establish cause and effect. We suggest that it would be fairly easy to use more realistic conidia loads and/or exposure methods, and to change some other aspects to incorporate more biological relevance into laboratory setups. Despite the limitations of the protocols in the published studies, the generally held view that an ant nest is a "fortress" (Schmid-Hempel, 1998), well defended against diseases, seems to be corroborated by the absence of outbreaks of *Beauveria* or *Metarhizium* in their societies. However, as other fungi are commonly found infecting ants (Araujo, Evans, Geiser, Mackay, & Hughes, 2015; Csata et al., 2013; Evans, Elliot, & Hughes, 2011), we question if the absence of *Beauveria* or *Metarhizium* is in fact due to the efficient social defenses.

4. NATURAL OCCURRENCE OF *BEAUVERIA* AND *METARHIZIUM* IN ANTS: OPPORTUNISTIC PARASITES?

The use of *Beauveria* and *Metarhizium* to study ant diseases is often justified by the assumption that they are common parasites of ants (Reber et al., 2011; Ribeiro et al., 2012; Tragust, Mitteregger, et al., 2013; Tranter, LeFevre, Evison, & Hughes, 2015; Walker & Hughes, 2011). The reasons for this assumption are (1) these fungi are soil-borne parasites, commonly associated with the soil of ant nests, and (2) a few records of natural infections of these parasites in ants do exist. Although *Beauveria* and *Metarhizium* do cause epizootics in other insects, such as coleopterans (Townsend, Glare, & Willoughby, 1995), hemipterans (Shimazu, 1989), and lepidopterans (Tefera & Pringle, 2003; Townsend et al., 1995), there is no record of a natural epizootic events in ant colonies and natural infections are rather isolated and very rare (Fig. 3 and Table S3).

Beauveria and *Metarhizium* incidence is relatively higher on ground nesting founding queens—up to 10% (Cardoso, 2010) (Fig. 3 and Table S3). Following the nuptial flight, the reproductive female attempts to found a colony. This is a critical and vulnerable stage for the future queen, where she has limited resources until the first batch of workers emerge and start foraging. At this stage of the colony, generalist and opportunistic parasites may represent a bottleneck for the next generation of colonies, since the infected, nonresistant queens will die. Consequently, the workers may be potentially preselected (via their mother) for resistance, since only the queen lays eggs in the colony. Because the parasites are not equally distributed on soil (Hughes, Thomsen, et al., 2004; Keller et al., 2003), the variation in the level of exposure of the queens may be an important source of selection, removing from the population queens that are more susceptible to generalist parasites or priming the immune system of the queen. It might be the reason for the strong "colony effect," where workers from different colonies have different responses under laboratory conditions (Baer, Krug, Boomsma, & Hughes, 2005; Fountain & Hughes, 2011; Hughes, Petersen, et al., 2004; Ribeiro et al., 2012). Nonetheless, the overall success of the founder queens is low and more than 95% of them die before laying the first batch of eggs (Hölldobler & Wilson, 1990).

The bottleneck the queens go through in the founding stage of the colony could, in part, explain why there are few records of worker cadavers infected by these generalist fungi in nature. Hughes, Thomsen, et al. (2004) actively searched for ant cadavers around the nests of leaf-cutting ants and found zero cadavers of *Beauveria* or *Metarhizium* infected ants. They also collected 3300 live ants over a year to verify infections from which only 3 (0.09%) were infected by *Metarhizium anisopliae* (Fig. 3 and Table S3). The fungus *Beauveria* was not detected. Reber and Chapuisat (2012a) collected 4050 ants from 81 colonies of *Formica selysi*. They found that *Beauveria* prevalence at the colony level was 17% (14 colonies out of 81), but at individual level, it was only 0.42% (17 ants out of 4050). In that study, zero workers were infected by *Metarhizium*. A few other publications have reported that live ants collected in field and kept in the laboratory until death showed fungal growth following few days of postmortem incubation of the cadaver (Baird, Woolfolk, & Watson, 2007; Hughes, Thomsen, et al., 2004; Sanchéz-penã & Thorvilson, 1992) (Fig. 3 and Table S3).

These sampling methodologies, which collect ants and establish infection status following their death and the subsequent growth of the fungus following incubation do not account for sublethal infections. Sublethal effects of parasites, such as reduced fertility and competitiveness, are considered important ecological and evolutionary impacts of diseases (Boots et al., 2003; Boots & Norman, 2000). This important aspect of host–parasite ecology has not been considered in the ant studies we examined. Interestingly, in our opinion, this would be one of the aspects to be explored and better characterized in laboratory setups. Considering that ant workers are sterile, a reduction in fertility or competitiveness might not be the most appropriate parameter for studying sublethal effect. On the other hand, their social nature can open new avenues to study sublethal effects of parasite. For example, do worker ants infected by either *Metarhizium* or *Beauveria* forage less or generally alter their expression of altruism to nestmates?

Other fungi in the genus *Aspergillus* are also used (Table S1) because they are considered opportunistic parasites, while species in the genera *Beauveria* and *Metarhizium*, are considered to be obligate pathogens in ant studies (Novak & Cremer, 2015; Tranter et al., 2015). Ironically, the fungus *Aspergillus*, claimed as an opportunistic parasite, is often present in ants collected both in the field and those reared in the laboratory (Lacerda et al., 2014; Lacerda, Della Lucia, Pereira, Peternelli, & Totola, 2010; Rodrigues, Silva, Bacci, Forti, & Pagnocca, 2010; Tranter, Graystock, Shaw, Lopes, & Hughes, 2014). If *Beauveria* and *Metarhizium* were common parasites in ants, as claimed, we would expect to find them as frequently or more often than the opportunistic parasite, *Aspergillus*. It is worth noting that *Beauveria* and *Metarhizium* play a role in other ecological interactions, such as their role as endophytes, or as associates of roots in the rhizosphere and as mycoparasites (Vega et al., 2009). For example, *Metarhizium* spp. are known to colonize roots and promote plant growth (St. Leger, 2008). Genomes of *Metarhizium* species reveal many genes involved in the colonization of plant tissue (Gao et al., 2011). In fact, an emerging view among mycologists is that they are facultative parasites of insects (Bidochka, Clark, Lewis, & Keyhani, 2010; Oulevey, Widmer, Koelliker, & Enkerli, 2009; Padilla-Guerrero, Barelli, Gonzalez-Hernandez, Torres-Guzman, & Bidochka, 2011). Additionally, these two fungi have been treated as opportunistic parasites in the biological control arena (Brodeur, 2012).

The notion that the fungi *Beauveria* and *Metarhizium* are a common parasite of ants is widespread, but most likely not true. For *Metarhizium*, there are some extreme cases in the ant literature where it is even described as a "specialist parasite" (Tranter et al., 2014, 2015). The rare natural infection of ants by *Beauveria* and *Metarhizium* (discussed in this chapter), their alternative lifestyles in plants (Vega et al., 2009), and their opportunistic nature recognized by both mycology and biological control disciplines (Brodeur, 2012) indicate that these fungi are not necessarily a potent threat for ants. Therefore, we suggest that there is no evidence for assuming that *Beauveria* and *Metarhizium* are common parasites of ants in natural populations, and the designation of any of them as "specialized parasites" is wrong.

5. FUTURE PERSPECTIVES

Beauveria and *Metarhizium* have been and continue to be important tools to understand how ants organize their defenses against opportunistic threats. However, we consider the unjustified methodologies implemented by researchers as a barrier to knowledge collection. It is important to reconsider the unrealistic doses of conidia and exposure routes, a reality also discussed in other host–parasite systems (Poulin, 2010a, 2010b). High concentrations of microbial parasites will always affect ant mortality or behavior, but they do not allow us to objectively understand the link between social behavior and evolved behaviors that function to mitigate diseases. We are not implying that high doses cannot be used, when, for example, the goal is simply to kill, study within host growth or better understand the terminal stages of the infection (ie, prior to death) (Heinze & Walter, 2010). However, where the goal is to understand hygienic behaviors, ant-to-ant cross infection and social network disruption following exposure, we strongly recommend future research efforts incorporate ecologically relevant conditions. Recently, a few studies have used more realistic setups to investigate ants and diseases (Loreto, Elliot, Freitas, Pereira, & Hughes, 2014; Quevillon, Hanks, Bansal, & Hughes, 2015), which prove that it is feasible and fairly simple to implement. We hope to see more ant–parasite studies continue in this direction.

Although we highlight that the two focal parasites on ant studies are rather rare and likely opportunistic in natural populations, we understand the importance of these models. However, we were surprised by the lack of information on sublethal doses and its effects, both at individual and colony levels. Perhaps, it is due to the experimental ease by which high doses of conidia can be used to produce high mortality rates (Fig. 3). Additionally, there were no studies on the detection and response to infectious cadavers, a likely real scenario. *Beauveria* and *Metarhizium* may be useful models to investigate these unknown aspects of ant–parasites interactions, but we would also argue that our view could be biased by the limitation of these models.

While the current studies on ant–parasite interactions support the view that they possess a robust social immunity, there are many other parasites, including other fungi, which are specialized in infecting ant societies (Csata et al., 2013; Loreto et al., 2014; Marikovsky, 1962). It is worth noting that studies investigating immune defenses in honeybees and bumblebees predominantly focus on specialized parasites (Bailey & Ball, 1991), which can contribute to the loss of these beneficial insects. These fungi have been overlooked in studies focused on the ecology and evolution of ant–parasite interactions. They are very common in the natural habitat of ants (Table S3), have a prevalence as high as 100% at the colony level (Konrad, Grasse, Tragust, & Cremer, 2015; Loreto et al., 2014), and allow for both field and laboratory experimentation (de Bekker et al., 2014). We see a lot of potential in exploring these specialized interactions: they could lead us to a broader understanding of social immunity and how parasites can overcome collective defenses.

Finally, models are often chosen for particular reasons. Due to ethical and logistical reasons, most of what we know about the molecular mechanisms of diseases in humans comes from artificial models, such as *Drosophila melanogaster* (Keebaugh & Schlenke, 2014; Lemaitre & Hoffmann, 2007), *Caenorhabditis elegans* (Markaki & Tavernarakis, 2010), and the mouse (Rosenthal & Brown, 2007). However, these models often lack the social component of disease spread, which is an important factor in human societies. In addition, a disease model that reflects the low genetic diversity and high density of our agricultural and livestock groups would be beneficial. Social insects have the social properties that reflect high-density living of closely related individuals. The effort to prevent colony collapse disorder in bees has advanced our knowledge on diseases in insect societies. However, we suggest that ants may be a better model to study the ecology and evolution of group living disease dynamics due to the larger sizes of their colonies of the nonwinged workers. If we study these social models in the proper way, many valuable lessons could be learned from them.

ACKNOWLEDGMENTS

We are thankful to Dr. Rebeca Rosengaus, Dr. Klaus Reinhardt, and Dr. Andrew Read for helpful discussions and to three anonymous reviewers for their constructive comments. We are also thankful to Djoshkun Shengjuler for assisting us in the making of the figures for this manuscript. RGL is funded by CAPES-Brazil (grant 6203-10-8). DPH is funded by Pennsylvania State University. Part of this work was funded by NSF (grant 1414296) as part of the joint NSF-NIH-USDA Ecology and Evolution of Infectious Diseases program.

SUPPLEMENTARY DATA

Supplementary data related to this article can be found online at http://dx.doi.org/10.1016/bs.adgen.2015.12.005.

REFERENCES

Alexander, R. D. (1974). The evolution of social behavior. *Annual Review of Ecology and Systematics, 5*, 325–383.

Alves, S. B., Stimac, J. L., & Camargo, M. T. V. (1988). Susceptibility of *Solenopsis invicta* Buren and *Solenopsis saevissima* fr. Smith to isolates of *Beauveria bassiana* Bals. vuill. *Anais da Sociedade Entomologica do Brasil, 17*, 379–388.

Andersen, S. B., Ferrari, M., Evans, H. C., Elliot, S. L., Boomsma, J. J., & Hughes, D. P. (2012). Disease dynamics in a specialized parasite of ant societies. *PLoS One*, 7, e36352.

Anderson, R. M., & May, R. M. (1979). Population biology of infectious-diseases: part I. *Nature, 280*, 361–367.

Araujo, J. P. M., Evans, H. C., Geiser, D. M., Mackay, W. P., & Hughes, D. P. (2015). Unravelling the diversity behind the *Ophiocordyceps unilateralis* (Ophiocordycipitaceae) complex: three new species of zombie-ant fungi from the Brazilian Amazon. *Phytotaxa, 220*, 224–238.

Arneberg, P., Skorping, A., Grenfell, B., & Read, A. F. (1998). Host densities as determinants of abundance in parasite communities. *Proceedings of the Royal Society B, 265*, 1283–1289.

Baer, B., Krug, A., Boomsma, J. J., & Hughes, W. O. H. (2005). Examination of the immune responses of males and workers of the leaf-cutting ant *Acromyrmex echinatior* and the effect of infection. *Insectes Sociaux, 52*, 298–303.

Bailey, L., & Ball, B. (1991). Honey bee pathology. *Honey Bee Pathology*, 1–200.

Baird, R., Woolfolk, S., & Watson, C. E. (2007). Survey of bacterial and fungal associates of black/hybrid imported fire ants from mounds in Mississippi. *Southeastern Naturalist, 6*, 615–632.

de Bekker, C., Quevillon, L. E., Smith, P. B., Fleming, K. R., Ghosh, D., Patterson, A. D., & Hughes, D. P. (2014). Species-specific ant brain manipulation by a specialized fungal parasite. *BMC Evolutionary Biology, 14*.

Bidochka, M. J., Clark, D. C., Lewis, M. W., & Keyhani, N. O. (2010). Could insect phagocytic avoidance by entomogenous fungi have evolved via selection against soil amoeboid predators? *Microbiology-Sgm, 156*, 2164–2171.

Boomsma, J. J., Schmid-Hempel, P., & Hughes, W. O. H. (2005). Life histories and parasite pressure across the major groups of social insects. Insect evolutionary ecology. In *Proceedings of the Royal Entomological Society's 22nd symposium* (pp. 139–175).

Boots, M., Greenman, J., Ross, D., Norman, R., Hails, R., & Sait, S. (2003). The population dynamical implications of covert infections in host-microparasite interactions. *Journal of Animal Ecology, 72*, 1064–1072.

Boots, M., & Norman, R. (2000). Sublethal infection and the population dynamics of host-microparasite interactions. *Journal of Animal Ecology, 69*, 517–524.

Brockhurst, M. A., & Koskella, B. (2013). Experimental coevolution of species interactions. *Trends in Ecology & Evolution, 28*, 367–375.

Brodeur, J. (2012). Host specificity in biological control: insights from opportunistic pathogens. *Evolutionary Applications, 5*, 470–480.

Cardoso, S. R. S. (2010). *Morfogênese de ninhos iniciais de* atta *spp. (hymenoptera: Formicidae), mortalidade em condições naturais e avaliação da ação de fungos entomopatogênicos* (Ph.D. UNESP, Botucatú).

Castella, G., Chapuisat, M., & Christe, P. (2008). Prophylaxis with resin in wood ants. *Animal Behaviour, 75*, 1591–1596.

Chapuisat, M., Oppliger, A., Magliano, P., & Christe, P. (2007). Wood ants use resin to protect themselves against pathogens. *Proceedings of the Royal Society B, 274*, 2013–2017.

Chouvenc, T., Su, N.-Y., & Grace, J. K. (2011). Fifty years of attempted biological control of termites - analysis of a failure. *Biological Control, 59*, 69–82.

Côté, I. M., & Poulin, R. B. (1995). Parasitism and group-size in social animals: a meta-analysis. *Behavioral Ecology, 6*, 159–165.

Cremer, S., Armitage, S. A. O., & Schmid-Hempel, P. (2007). Social immunity. *Current Biology, 17*, R693–R702.

Csata, E., Czekes, Z., Eros, K., Nemet, E., Hughes, M., Csosz, S., & Marko, B. (2013). Comprehensive survey of Romanian myrmecoparasitic fungi: new species, biology and distribution. *North-Western Journal of Zoology, 9*, 23–29.

Diehl-Fleig, E., & Lucchese, M. E. D. P. (1991). Behavioral responses of *Acromyrmex striatus* workers (Hymenoptera: Formicidae) in the presence of entomopathogenic fungi. *Revista Brasileira de Entomologia, 35*, 101–108.

Diehl-Fleig, E., Silva, M. E. D., Specht, A., & Valim-Labres, M. E. (1993). Efficiency of *Beauveria bassiana* for *Acromyrmex* spp. control (Hymenoptera: Formicidae). *Anais da Sociedade Entomologica do Brasil, 22*, 281–285.

Ebert, D. (1994). Virulence and local adaptation of a horizontally transmitted parasite. *Science, 265*, 1084–1086.

Evans, H. C. (1974). Natural control of arthropods, with special reference to ants (Formicidae), by fungi in the tropical high forest of Ghana. *Journal of Applied Ecology, 11*, 37–49.

Evans, H. C., Elliot, S. L., & Hughes, D. P. (2011). Hidden diversity behind the zombie-ant fungus *Ophiocordyceps unilateralis*: four new species described from carpenter ants in Minas Gerais, Brazil. *PLoS One, 6*.

Fountain, T., & Hughes, W. O. H. (2011). Weaving resistance: silk and disease resistance in the weaver ant *Polyrhachis dives*. *Insectes Sociaux, 58*, 453–458.

Fuxa, J. R., & Richter, A. R. (2004). Effects of soil moisture and composition and fungal isolate on prevalence of *Beauveria bassiana* in laboratory colonies of the red imported fire ant (Hymenoptera: Formicidae). *Environmental Entomology, 33*, 975–981.

Gao, Q. A., Jin, K., Ying, S. H., Zhang, Y. J., Xiao, G. H., Shang, Y. F.… Wang, C. S. (2011). Genome sequencing and comparative transcriptomics of the model entomopathogenic fungi *Metarhizium anisopliae* and *M. acridum*. *PLoS Genetics, 7*.

Goettel, M. S., Eilenberg, J., & Glare, T. (2010). Entomopathogenic fungi and their role in regulation of insect populations. In L. I. Gilbert, & S. S. Gill (Eds.), *Insect control: Biological and synthetic agents*. London: Academic Press.

Hajek, A. E., & St. Leger, R. J. (1994). Interactions between fungal pathogens and insect hosts. *Annual Review of Entomology, 39*, 293–322.

Heinze, J., & Walter, B. (2010). Moribund ants leave their nests to die in social isolation. *Current Biology, 20*, 249–252.

Hölldobler, B., & Wilson, E. O. (1990). *The ants*. Cambridge: Harvard University Press.

Hughes, W. O. H., Eilenberg, J., & Boomsma, J. J. (2002). Trade-offs in group living: transmission and disease resistance in leaf-cutting ants. *Proceedings of the Royal Society B, 269*, 1811–1819.

Hughes, W. O. H., Petersen, K. S., Ugelvig, L. V., Pedersen, D., Thomsen, L., Poulsen, M., & Boomsma, J. J. (2004). Density-dependence and within-host competition in a semelparous parasite of leaf-cutting ants. *BMC Evolutionary Biology, 4*.

Hughes, W. O. H., Thomsen, L., Eilenberg, J., & Boomsma, J. J. (2004). Diversity of entomopathogenic fungi near leaf-cutting ant nests in a neotropical forest, with particular reference to *Metarhizium anisopliae* var. anisopliae. *Journal of Invertebrate Pathology, 85*, 46–53. e1001300.

Jaccoud, B. D., Hughes, W. O. H., & Jackson, C. W. (1999). The epizootiology of a *Metarhizium* infection in mini-nests of the leaf-cutting ant *Atta sexdens rubropilosa*. *Entomologia Experimentalis et Applicata, 93*, 51–61.

Keebaugh, E. S., & Schlenke, T. A. (2014). Insights from natural host-parasite interactions: the *Drosophila* model. *Developmental and Comparative Immunology, 42*, 111–123.

Keller, L., & Genoud, M. (1997). Extraordinary lifespans in ants: a test of evolutionary theories of ageing. *Nature, 389*, 958–960.

Keller, S., Kessler, P., & Schweizer, C. (2003). Distribution of insect pathogenic soil fungi in Switzerland with special reference to *Beauveria brongniartii* and *Metharhizium anisopliae*. *BioControl, 48*, 307–319.

Kermarrec, A., Febvay, G., & Decharme, M. (1986). Protection of leaf-cutting ants from biohazards: is there a future for microbiological control? In C. S. Lofgren, & R. K. Meer (Eds.), *Fire ants and leaf-cutting ants: Biology and management* (p. 433). Boulder and London: Westview Press.

Konrad, M., Grasse, A. V., Tragust, S., & Cremer, S. (2015). Anti-pathogen protection versus survival costs mediated by an ectosymbiont in an ant host. *Proceedings of the Royal Society B, 282*, 20141976.

Konrad, M., Vyleta, M. L., Theis, F. J., Stock, M., Tragust, S., Klatt, M.... Cremer, S. (2012). Social transfer of pathogenic fungus promotes active immunisation in ant colonies. *PLoS Biology, 10.*

Lacerda, F. G., Della Lucia, T. M. C., Desouza, O., Pereira, O. L., Kasuya, M. C. M., De Souza, L. M. ... De Souza, D. J. (2014). Social interactions between fungus garden and external workers of *Atta sexdens* (Linnaeus) (Hymenoptera: Formicidae). *Italian Journal of Zoology, 81*, 298–303.

Lacerda, F. G., Della Lucia, T. M. C., Pereira, O. L., Peternelli, L. A., & Totola, M. R. (2010). Mortality of *Atta sexdens rubropilosa* (Hymenoptera: Formicidae) workers in contact with colony waste from different plant sources. *Bulletin of Entomological Research, 100*, 99–103.

Lemaitre, B., & Hoffmann, J. (2007). The host defense of *Drosophila melanogaster*. *Annual Review of Immunology, 25*, 697–743.

Loreto, R. G., Elliot, S. L., Freitas, M. L. R., Pereira, T. M., & Hughes, D. P. (2014). Long-term disease dynamics for a specialized parasite of ant societies: a field study. *PLoS One, 9*, e103516.

Machado, V., Diehl-Fleig, E., Silva, M. E. D., & Lucchese, M. E. D. P. (1988). Observed reactions in colonies of some species of *Acromyrmex* (Hymenoptera: Formicidae) when inoculated with entomopathogenic fungi. *Ciencia e Cultura (Sao Paulo), 40*, 1106–1108.

Marikovsky, P. I. (1962). On some features of behaviour of the ant *Formica rufa* L. infected with fungous disease. *Insectes Sociaux, 9*, 173–179.

Markaki, M., & Tavernarakis, N. (2010). Modeling human diseases in *Caenorhabditis elegans*. *Biotechnology Journal, 5*, 1261–1276.

Mattoso, T. C., Moreira, D. D. O., & Samuels, R. I. (2012). Symbiotic bacteria on the cuticle of the leaf-cutting ant *Acromyrmex subterraneus subterraneus* protect workers from attack by entomopathogenic fungi. *Biology Letters, 8*, 461–464.

Maynard Smith, J., & Szathmáry, E. (1995). *The major transitions in evolution* (p. 346). Oxford: Oxford University Press.

Murdock, C. C., Paaijmans, K. P., Cox-Foster, D., Read, A. F., & Thomas, M. B. (2012). Rethinking vector immunology: the role of environmental temperature in shaping resistance. *Nature Reviews Microbiology, 10*, 869–876.

Novak, S., & Cremer, S. (2015). Fungal disease dynamics in insect societies: optimal killing rates and the ambivalent effect of high social interaction rates. *Journal of Theoretical Biology, 372*, 54–64.

Oi, D. H., Pereira, R. M., Stimac, J. L., & Wood, L. A. (1994). Field applications of *Beauveria bassiana* for control of the red imported fire ant (Hymenoptera: Formicidae). *Journal of Economic Entomology, 87*, 623–630.

Okuno, M., Tsuji, K., Sato, H., & Fujisaki, K. (2012). Plasticity of grooming behavior against entomopathogenic fungus *Metarhizium anisopliae* in the ant *Lasius japonicus*. *Journal of Ethology, 30*, 23–27.

Orlofske, S. A., Jadin, R. C., Preston, D. L., & Johnson, P. T. J. (2012). Parasite transmission in complex communities: predators and alternative hosts alter pathogenic infections in amphibians. *Ecology, 93*, 1247–1253.

Oulevey, C., Widmer, F., Koelliker, R., & Enkerli, J. (2009). An optimized microsatellite marker set for detection of *Metarhizium anisopliae* genotype diversity on field and regional scales. *Mycological Research, 113*, 1016–1024.

Padilla-Guerrero, E. I., Barelli, L., Gonzalez-Hernandez, G. A., Torres-Guzman, J. C., & Bidochka, M. J. (2011). Flexible metabolism in *Metarhizium anisopliae* and *Beauveria bassiana*: role of the glyoxylate cycle during insect pathogenesis. *Microbiology-Sgm, 157*, 199–208.

Pereira, R. M., & Stimac, J. L. (1992). Transmission of *Beauveria bassiana* within nests of *Solenopsis invicta* (Hymenoptera: Formicidae) in the laboratory. *Environmental Entomology, 21*, 1427–1432.

Pereira, R. M., Stimac, J. L., & Alves, S. B. (1993). Soil antagonism affecting the dose-response of workers of the red imported fire ant, *Solenopsis invicta*, to *Beauveria bassiana* conidia. *Journal of Invertebrate Pathology, 61*, 156–161.

Poulin, R. (2010a). The scaling of dose with host body mass and the determinants of success in experimental cercarial infections. *International Journal for Parasitology, 40*, 371–377.

Poulin, R. (2010b). The selection of experimental doses and their importance for parasite success in metacercarial infection studies. *Parasitology, 137*, 889–898.

Pull, C. D., Hughes, W. O. H., & Brown, M. J. F. (2013). Tolerating an infection: an indirect benefit of co-founding queen associations in the ant *Lasius niger*. *Naturwissenschaften, 100*, 1125–1136.

Quevillon, L. E., Hanks, E. M., Bansal, S., & Hughes, D. P. (2015). Social, spatial, and temporal organization in a complex insect society. *Scientific Reports, 5*.

Read, A. F. (1994). The evolution of virulence. *Trends in Microbiology, 2*, 73–76.

Reber, A., Castella, G., Christe, P., & Chapuisat, M. (2008). Experimentally increased group diversity improves disease resistance in an ant species. *Ecology Letters, 11*, 682–689.

Reber, A., & Chapuisat, M. (2012a). Diversity, prevalence and virulence of fungal entomopathogens in colonies of the ant *Formica selysi*. *Insectes Sociaux, 59*, 231–239.

Reber, A., & Chapuisat, M. (2012b). No evidence for immune priming in ants exposed to a fungal pathogen. *PLoS One, 7*.

Reber, A., Purcell, J., Buechel, S. D., Buri, P., & Chapuisat, M. (2011). The expression and impact of antifungal grooming in ants. *Journal of Evolutionary Biology, 24*, 954–964.

Ribeiro, M. M. R., Amaral, K. D., Seide, V. E., Souza, B. M. R., Lucia, T. M. C. D., Kasuya, M. C. M., & de Souza, D. J. (2012). Diversity of fungi associated with *Atta bisphaerica* (Hymenoptera: Formicidae): the activity of *Aspergillus ochraceus* and *Beauveria bassiana*. *Psyche, 2012*, 1–6.

Rodrigues, A., Silva, A., Bacci, M., Jr., Forti, L. C., & Pagnocca, F. C. (2010). Filamentous fungi found on foundress queens of leaf-cutting ants (Hymenoptera: Formicidae). *Journal of Applied Entomology, 134*, 342–345.

Rosenthal, N., & Brown, S. (2007). The mouse ascending: perspectives for human-disease models. *Nature Cell Biology, 9*, 993–999.

Sanchéz-penã, S. R., & Thorvilson, H. G. (1992). 2 Fungi infecting red imported fire ant (Hymenoptera: Formicidae) founding queens from Texas. *Southwestern Entomologist, 17*, 181–182.

Schmid-Hempel, P. (1998). *Parasites in social insects*. Princeton: Princeton University Press.

Schmidt, A. M., Linksvayer, T. A., Boomsma, J. J., & Pedersen, J. S. (2011). No benefit in diversity? The effect of genetic variation on survival and disease resistance in a polygynous social insect. *Ecological Entomology, 36*, 751–759.

Shimazu, M. (1989). *Metarhizium cylindrosporae* chen et guo (Deuteromycotina, Hyphomycetes), the causative agent of an epizootic on *Graptopsaltria nigrofuscata* Motchulski (Homoptera, Cicadidae). *Applied Entomology and Zoology, 24*, 430–434.

Silva, M. E. D., & Diehl-Fleig, E. (1988). Evaluation of different strains of entomopathogenic fungi to control the ant *Atta sexdens piriventris* Santschi 1919 (Hymenoptera: Formicidae). *Anais da Sociedade Entomologica do Brasil, 17*, 263–270.

St. Leger, R. J. S. (2008). Studies on adaptations of *Metarhizium anisopliae* to life in the soil. *Journal of Invertebrate Pathology, 98*, 271–276.

Stimac, J. L., Alves, S. B., & Camargo, M. T. V. (1987). Susceptibility of *Solenopsis* spp. to different species of entomopathogenic fungi. *Anais da Sociedade Entomologica do Brasil, 16*, 377–388.

Stimac, J. L., Alves, S. B., & Camargo, M. T. V. (1989). Control of *Solenopsis* spp. (Hymenoptera: Formicidae) with *Beauveria bassiana* bals. Vuill. in laboratory and field conditions. *Anais da Sociedade Entomologica do Brasil, 18*, 95–104.

Stimac, J. L., Pereira, R. M., Alves, S. B., & Wood, L. A. (1993). Mortality in laboratory colonies of *Solenopsis invicta* (Hymenoptera: Formicidae) treated with *Beauveria bassiana (Deuteromycetes)*. *Journal of Economic Entomology, 86*, 1083–1087.

Tefera, T., & Pringle, K. L. (2003). Effect of exposure method to *Beauveria bassiana* and conidia concentration on mortality, mycosis, and sporulation in cadavers of Chilo partellus (Lepidoptera: Pyralidae). *Journal of Invertebrate Pathology, 84*, 90–95.

Theis, F. J., Ugelvig, L. V., Marr, C., & Cremer, S. (2015). Opposing effects of allogrooming on disease transmission in ant societies. *Philosophical Transactions of the Royal Society B-Biological Sciences, 370*.

Thrall, P. H., Hochberg, M. E., Burdon, J. J., & Bever, J. D. (2007). Coevolution of symbiotic mutualists and parasites in a community context. *Trends in Ecology & Evolution, 22*, 120–126.

Townsend, R. J., Glare, T. R., & Willoughby, B. E. (1995). The fungi *Beauvera* spp. cause epizootics in grass grub populations in Waikato. In *48th New Zealand plant protection conference* (Vol. 48, pp. 237–241) (Hastings, New Zealand).

Tragust, S., Mitteregger, B., Barone, V., Konrad, M., Ugelvig, L. V., & Cremer, S. (2013). Ants disinfect fungus-exposed brood by oral uptake and spread of their poison. *Current Biology, 23*, 76–82.

Tragust, S., Ugelvig, L. V., Chapuisat, M., Heinze, J., & Cremer, S. (2013). Pupal cocoons affect sanitary brood care and limit fungal infections in ant colonies. *BMC Evolutionary Biology, 13*.

Tranter, C., Graystock, P., Shaw, C., Lopes, J. F. S., & Hughes, W. O. H. (2014). Sanitizing the fortress: protection of ant brood and nest material by worker antibiotics. *Behavioral Ecology and Sociobiology, 68*, 499–507.

Tranter, C., LeFevre, L., Evison, S. E. F., & Hughes, W. O. H. (2015). Threat detection: contextual recognition and response to parasites by ants. *Behavioral Ecology, 26*, 396–405.

Vega, F. E. (2008). Insect pathology and fungal endophytes. *Journal of Invertebrate Pathology, 98*, 277–279.

Vega, F. E., Goettel, M. S., Blackwell, M., Chandler, D., Jackson, M. A., Keller, S.… Roy, H. E. (2009). Fungal entomopathogens: new insights on their ecology. *Fungal Ecology, 2*, 149–159.

Walker, T. N., & Hughes, W. O. H. (2011). Arboreality and the evolution of disease resistance in ants. *Ecological Entomology, 36*, 588–595.

CHAPTER NINE

Entomopathogenic Fungi: New Insights into Host–Pathogen Interactions

T.M. Butt*,[1], C.J. Coates*, I.M. Dubovskiy[§] and N.A. Ratcliffe*,[¶]
*Swansea University, Swansea, Wales, United Kingdom
[§]SB Russian Academy of Sciences, Novosibirsk, Russia
[¶]Universidade Federal Fluminense, Niteroi, Rio de Janeiro, Brazil
[1]Corresponding author: E-mail: t.butt@swansea.ac.uk

Contents

Advances in Genetics, Volume 94
ISSN 0065-2660
http://dx.doi.org/10.1016/bs.adgen.2016.01.006

Abstract

Although many insects successfully live in dangerous environments exposed to diverse communities of microbes, they are often exploited and killed by specialist pathogens. Studies of host–pathogen interactions (HPI) provide valuable insights into the dynamics of the highly aggressive coevolutionary arms race between entomopathogenic fungi (EPF) and their arthropod hosts. The host defenses are designed to exclude the pathogen or mitigate the damage inflicted while the pathogen responds with immune evasion and utilization of host resources. EPF neutralize their immediate surroundings on the insect integument and benefit from the physiochemical properties of the cuticle and its compounds that exclude competing microbes. EPF also exhibit adaptations aimed at minimizing trauma that can be deleterious to both host and pathogen (eg, melanization of hemolymph), form narrow penetration pegs that alleviate host dehydration and produce blastospores that lack immunogenic sugars/enzymes but facilitate rapid assimilation of hemolymph nutrients. In response, insects deploy an extensive armory of hemocytes and macromolecules, such as lectins and phenoloxidase, that repel, immobilize, and kill EPF. New evidence suggests that immune bioactives work synergistically (eg, lysozyme with antimicrobial peptides) to combat infections. Some proteins, including transferrin and apolipophorin III, also demonstrate multifunctional properties, participating in metabolism, homeostasis, and pathogen recognition. This review discusses the molecular intricacies of these HPI, highlighting the interplay between immunity, stress management, and metabolism. Increased knowledge in this area could enhance the efficacy of EPF, ensuring their future in integrated pest management programs.

1. INTRODUCTION

Entomopathogenic fungi (EPF) offer environmentally friendly alternatives to conventional synthetic chemicals for arthropod pest control. Of the over 750 different species of EPF, attention has focused on development of species belonging to the order Hypocreales as they have a relatively wide host range and are amenable to mass production (Butt, Jackson, & Magan, 2001). About 80% of the commercially available EPF products are based on the genera *Metarhizium* and *Beauveria* (de Faria & Wraight, 2007). The use of EPF is likely to increase because of the withdrawal of many chemical pesticides due to perceived environmental risks, legislation encouraging use of EPF and other biopesticides, the consumer's desire for zero residues on food, and the development of pesticide resistance by many pests of socioeconomic importance (Hemingway & Ranson, 2000; Whalon, Mota-Sanchez, & Hollingworth, 2008).

EPF occur naturally in soils from disparate habitats worldwide (Scheepmaker & Butt, 2010; Zimmermann, 2007). EPF are rhizosphere competent, exist as endophytes, and even exhibit "mycorrhizal" characteristics transferring nitrogen from infected insects to host plants (Behie & Bidochka, 2014; Behie, Jones, & Bidochka, 2015; reviewed by Moonjely, Barelli, & Bidochka, 2016). Exactly how EPF evolved to kill arthropods (eg, insects, ticks, spiders, and mites) is unclear, but strains have clearly evolved different degrees of specificity and virulence. Phylogenetic studies suggest that, although the invasive and developmental processes of EPF are similar, the fungal virulence toward insects arose independently several times by convergent evolution (Zheng, Xia, Zhang, & Wang, 2013). Analysis of genomic data reveals that EPF differ in key aspects of the molecular mechanisms mediating virulence (Xiao et al., 2012). Overall, both arthropod hosts and fungal pathogens display much plasticity in their coevolution by adapting to tolerate each other's weaponry and concomitantly evolving counterattack measures.

The key stages in the invasive and developmental processes of EPF are spore adhesion, germination, differentiation of infection structures, penetration, colonization of the hemocoel, and sporulation following emergence from mycosed cadavers (Fig. 1; Table 1). The host attempts to resist infection and colonization through preformed and induced defenses (Figs. 2–4; Tables 2 and 3). This review outlines the major interactions between EPF and their arthropod hosts, drawing attention to the coevolutionary arms race between these organisms, and implications for the use of EPF in pest management programs.

2. PRE-ADHESION AND COMMUNITY-LEVEL IMMUNITY

The ability of arthropods to detect and avoid EPF depends on species and development stage (Baverstock, Roy, & Pell, 2010). For example, common flower-bugs avoid *Beauveria bassiana* on leaves but not in soils (Meyling & Pell, 2006), whereas Japanese beetle larvae shun soil-containing *Metarhizium anisopliae* (Villani et al., 1994). Social insects live in dense groups and often nest in soil, which directly exposes them to EPF. These insects (eg, termites and ants) can detect the presence of EPF, such as *Metarhizium*, and infected members of the colony using olfactory cues (Ugelvig & Cremer, 2007; Yanagawa & Shimzu, 2007). The termite, *Macrotermes michaelseni* can discriminate between virulent and avirulent strains of *M. anisopliae* and *B. bassiana* based on the volatile organic compound (VOC) profiles of these fungi (Mburu, Maniania, & Hassanali, 2013; Mburu, Ndung'u, Maniania, &

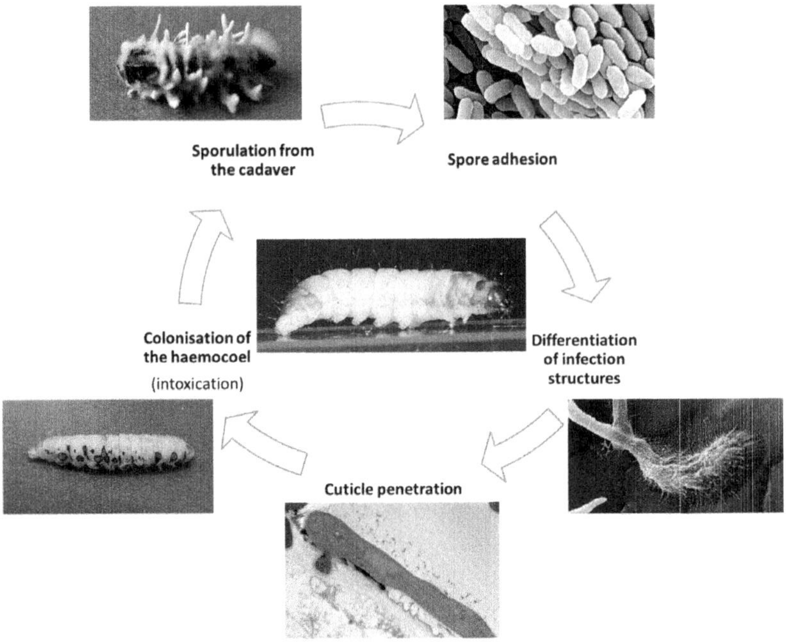

Figure 1 ***General overview of the invasive and developmental processes of entomopathogenic fungi in an insect host.*** The key stages include adhesion of infection propagules (eg, conidium and blastospore) to the surface of the host cuticle, differentiation of infection structures (appressoria and penetration pegs), penetration of the cuticle using enzymes and mechanical force, colonization of the hemocoel and emergence of the conidiophores for external sporulation on the cadaver.

Hassanali, 2011). Ants can detect EPF and alter their behavior accordingly. For example, *Formica selysi* workers enhance the rate of self-grooming when exposed to *Metarhizium brunneum*, while *Lasius neglectus* workers increase brood care and sanitary behavior in the presence of contaminated workers (Tragust et al., 2013; Ugelvig & Cremer, 2007). Some insects groom conspecifics in a bid to improve hygiene and to prevent pathogens from spreading. Grooming can be performed concurrently with "disinfectants" such as formic acid, and antimicrobial peptides (AMPs) and proteinaceous salivary deposits in ants and termites, (Bulmer, Bachelet, Raman, Rosengaus, & Sasisekharan, 2009; Tragust et al., 2013). Ants actively spread these secretions from a cuticular metathoracic gland to protect themselves and their nestmates. Salivary gland secretions appear to provide a similar antimicrobial role in termites. Indeed, two antifungal proteins, termicin and Gram-negative bacteria–binding protein 2 (GNBP2 = β-1,3-glucanase), have

Table 1 Key fungal virulence genes of *Beauveria* and *Metarhizium* species

Functional group	Genes	Description	Key references
Adhesion to cuticle	*Mad 1, Mad 2*	Adhesin-like proteins	Wang and St. Leger (2007a), Barelli et al. (2011)
	Hyd 1, Hyd 2, Hyd 3	Hydrophobins	Cho, Kirkland, Holder, and Keyhani (2007), Zhang et al. (2011)
	SsgA	Hydrophobin-like protein	St. Leger, Staples, and Roberts (1992), Bidochka, De Koning, and St. Leger (2001)
	cwp10	Increasing net conidial hydrophobicity	Li et al. (2010)
Cuticle degrading	*Pr1, Pr2, Pr4*	Subtilisin, trypsin, and cysteine proteases	St. Leger, Bidochka, and Roberts (1994), St. Leger, Joshi, Bidochkam, and Roberts (1995), Bagga et al. (2004)
	chit1, chi2, chi3, chi4, Bbchit1, Bbchit2	Chitinases	Kang, Park, and Lee (1999), Baratto et al. (2003), Duo-Chuan (2006), Fan et al. (2007)
	MrCYP52, cyp52x1	Cytochrome P450, fatty acid hydroxylase activity	Lin et al. (2011), Zhang et al. (2012)
Stress management	*HSP25, HSP30, HSP70, HSP90*	Heat shock proteins (chaperones)	Barelli et al. (2011), Liao et al. (2014)
	Hog1, Pmr1	Mitogen-activated protein kinase (MAPK)	Jin, Ming, and Xia (2012), Wang, Yang, et al. (2013), Wang, Zhou, et al. (2013)
Adaptation to hemolymph/ immunomodulation	*Mos1*	Osmosensor	Wang, Duan, and St. Leger (2003)
	Mcl1	Collagen-like protein	Wang and St. Leger (2006)
	Mr-npc2a	Sterol carrier	Zhao et al. (2014)
	dtxS1-dtxS4	Nonribosomal destruxins biosynthesis	Wang, Kang, Lu, Bai, and Wang (2012)

(*Continued*)

Table 1 Key fungal virulence genes of *Beauveric* and *Metarhizium* species—cont'd

Functional group	Genes	Description	Key references
Multifactorial (transcription factors)	*MrpacC*	Cuticle degrading Mycosis of insect cadavers Evasion of host immunity	Huang, Shang, Chen, Gao, and Wang (2015)
	MrSkn7	Appressorium formation Cell wall biosynthesis Evade host immunity	Shang, Chen, Chen, Lu, and Wang (2015)
	cag8	Hydrophobin synthesis Mycelial growth Blastospores production Regulation of conidiation	Fang, Pei, and Bidochka (2007)
	BbMBF1	Tolerance to oxidative, osmotic, and heat stress Hyphal morphogenesis	Ying, Ji, Wang, Feng, and Keyhani (2014)
	MaPKA1	Differentiation of appressoria Cuticle degradation Nutrient acquisition	Fang, Feng, et al. (2009), Fang, Pava-ripoll, et al. (2009)
Nutrient assimilation	*nrr1*	Nitrogen regulator response	Screen, Bailey, Charnley, Cooper, and Clarkson (1998)
	Crr1	Carbon regulator	Screen et al. (1997)
	Mest1	Hydrolysis of stored lipids	Wang and St. Leger (2007b), Wang et al. (2011)
	MPL1	Prelipin–lipid metabolism	
	ATM1	Trehalose-hydrolyzing enzyme	Zhao et al. (2006)
	mrGAT	Triacylglycerol biosynthesis	Gao, Shang, Huang, and Wang (2013)

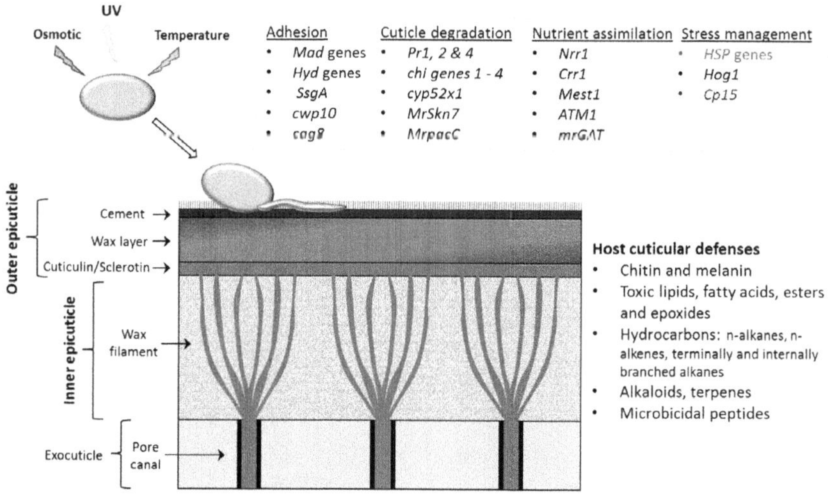

Figure 2 ***Adhesion and pre-penetration events.*** Interactions among entomopathogens, the host epicuticular defenses, and environmental factors. Inocula (conidia and blastospores) of EPF are exposed to a wide range of stressors (UV light, fungistatic fatty acids, and antifungal allelochemicals) and have evolved mechanisms in order to cope with these (eg, HSP, cell wall–bound enzymes). See Table 1 for more details of EPF virulence genes and Table 2 for fungistatic compounds associated with the cuticle.

been identified in termite salivary glands (Bulmer et al., 2009; Lamberty et al., 2001). More examples of arthropod antifungal glandular secretions are listed in Table 2. Finally, some insects, such as locusts (*Schistocerca gregaria*) infected with *Metarhizium acridum*, can raise their body temperature by basking in the sun to eliminate EPF (Blanford & Thomas, 1999).

Population-level immunity or density-dependent prophylaxis is a sophisticated defense strategy against infectious microbes, notably EPF (Wang, Yang, Cui, & Kang, 2013; Wilson-Rich, Spivak, Fefferman, & Starks, 2009). This has significant implications as regards use of EPF as biopesticides. In situ dilution of inocula can lead to unintended immune stimulation of arthropod targets rather than death, inadvertently promoting more EPF-resistant communities. This phenomenon would be more influential among social or gregarious insects rather than solitary species.

Studies suggest that some arthropods are attracted to EPF, with collembolans showing a preference for substrates containing conidia of several EPF (Dromph & Vestergaard, 2002). Female *Anopheles stephensi* mosquitoes are also attracted to spores of *B. bassiana* and *M. anisopliae*, as well as to fungal-infected larvae of the moths *Manduca sexta* and *Heliothis subflexa*

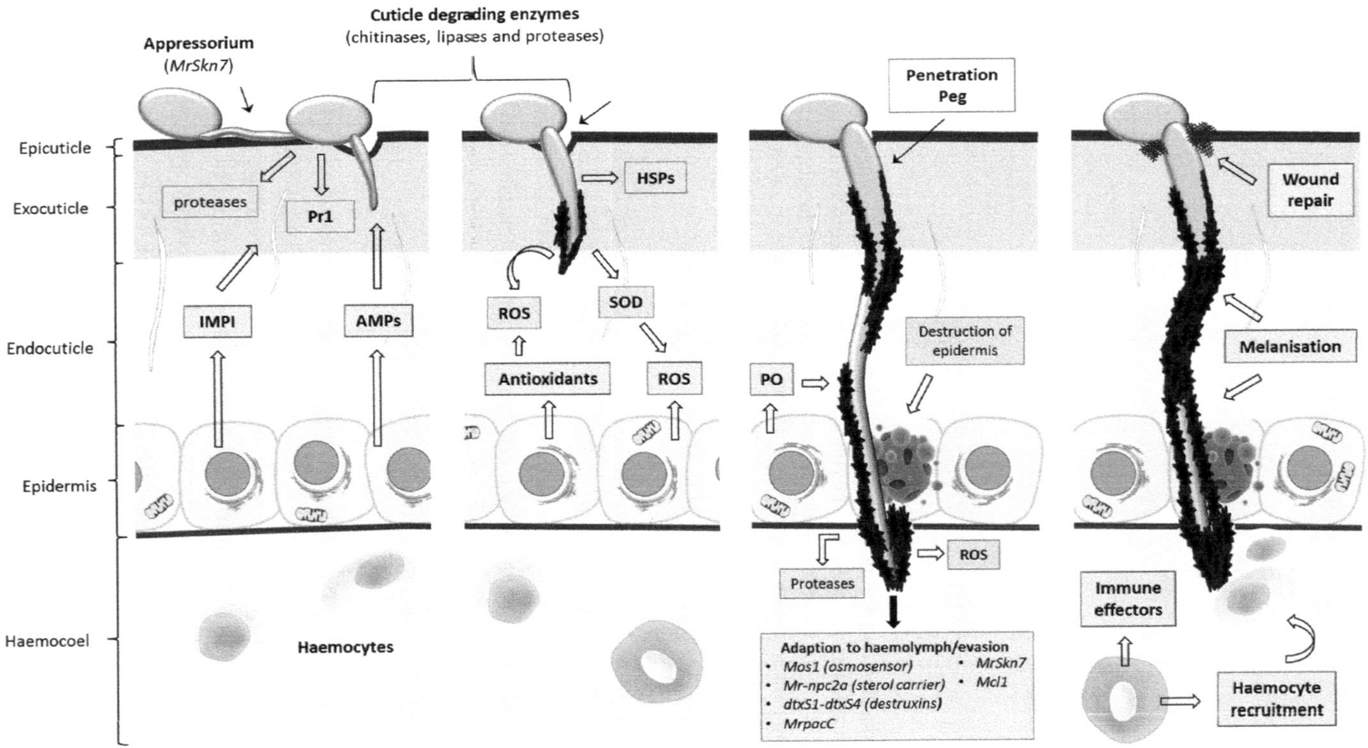

Figure 3 ***Post-penetration events.*** Entomopathogenic fungi release a cocktail of cuticle degrading enzymes and use mechanical force to penetrate the insect integument. The invading hyphae counteract the host's defenses, sequester nutrients, and exhibit physiological and phenotypic adaptations in preparation for reproduction and evasion of the immune system. Prior to hemocoel invasion, the insect hemocytes are mobilized to attack the pathogen.

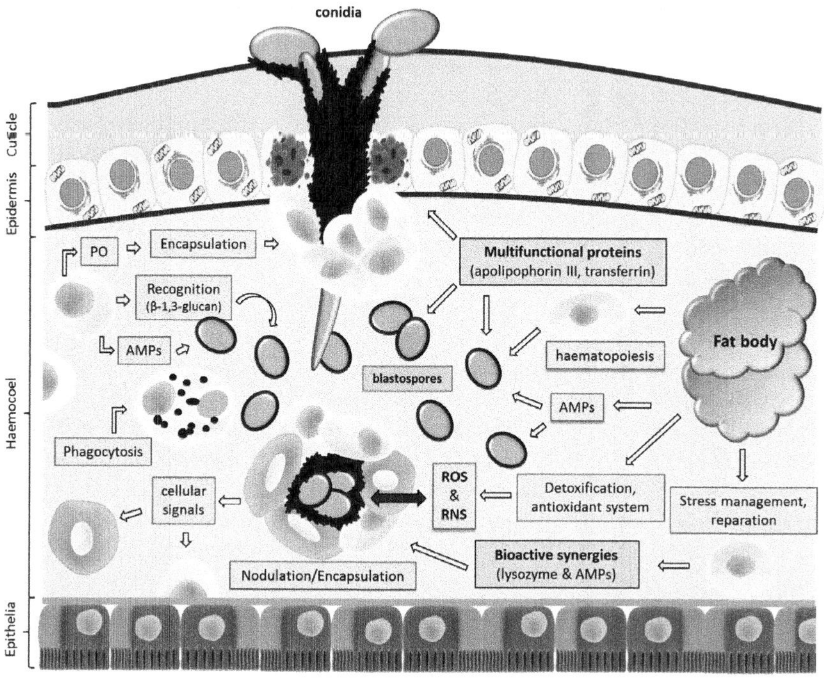

Figure 4 ***Insect cellular and humoral responses to entomopathogenic fungi.*** EPF in the hemolymph are recognized by hemocytes and soluble receptors. Besides cellular phagocytosis and encapsulation reactions, the host exploits multifunctional proteins and synergies between bioactive compounds to immobilize and kill the fungus. Pathogens activate many response networks, including stress management and antioxidants. The immune tactics represented in blue (black in print versions) boxes are recent discoveries or often overlooked by immunologists and mycologists. See Tables 3 and 4 for more details on AMPs and multifunctional proteins.

(George, Jenkins, Blanford, Thomas, & Baker, 2013). Whether EPF use attractants to lure and infect new hosts, or, whether the insect exposes itself to a low dose of EPF in a bid to "self-vaccinate" is unknown. However, luring arthropods (host or nonhost) would help in dispersal of EPF and increase chances of contacting fresh hosts.

The phenomenon of immune memory or vaccination in invertebrates is a contentious issue (Rowley & Powell, 2007). Pre-exposure of *Galleria mellonella* larvae to nonlethal doses of *Aspergillus fumigatus* conidia or increasing concentrations of β-1,3-glucan primes/protects the insect from a subsequent lethal dose of *A. fumigatus* or *Candida albicans*, respectively (Fallon, Troy, & Kavanagh, 2011; Mowlds, Coates, Renwick, & Kavanagh, 2010). Enhanced *Galleria* survival was linked to the de novo synthesis of immune-related

Table 2 Cuticle and defensive secretions as barriers to fungal infection

Arthropod	Fungistatic factor(s)	Key references
Brassy willow leaf beetle (*Phratora vitellinae*)	Salicylaldehyde	Gross, Schumacher, Schmidtberg, and Vilcinskas (2008)
Mustard beetle (*Phaedon cochleariae*)	Epi-chysomelidial (iridoid monoterpene)	Gross et al. (2008)
Mustard beetle (*P. cochleariae*)	Isothiocyanates	Inyang, Butt, Beckett, et al. (1999), Inyang, Butt, Doughty, et al. (1999)
Striated white butterfly (*Pieris melete*)	Allyl isothiocyanate	Atsumi and Saito (2015)
Bed bug (*Cimex lectularius*)	Aldehydes (*E*)-2-hexenal and (*E*)-2-octenal	Ulrich, Feldlaufer, Kramer, and Leger (2015)
Fire ant (*Solenopsis invicta*)	Venom alkaloids	Storey, Vandermeer, Boucias, and McCoy (1991)
Woodroach (*Cryptocercus punctulatus*)	Gram-negative bacteria binding proteins (GNBPs)—β-1,3-glucanase	Bulmer, Denier, Velenovsky, and Hamilton (2012)
Termites (*Reticulitermes* sp.)	Termicins and GNBPs	Hamilton and Bulmer (2012)
Termites (*Nasutitermes* sp.)	Terpenoids	Rosengaus, Lefebvre, and Traniello (2000)
Biting midge (*Forcipomyia nigra*)	Fatty acids (several: eg, pelargonic acid, capric acid, and palmitoleic acid)	Urbanek et al. (2012)
Corn-earworm (*Heliothis zea*), fall armyworm, *Spodoptera fragiterda*	Fatty acids (eg, caprylic acid, valeric acid, and nonanoic acid)	Smith and Grula (1982)
Silkworm (*B. mori*), fall webworm (*Hyphantria cunea*)	Fatty acids (eg, caprylic acid and capric acid)	Saito and Aoki (1983)
Blowfly (*Calliphora vicina*), pine moth (*Dendrolimus pini*), waxmoth (*G. mellonella*)	Fatty acids	Golebiowski, Malinski, Bogus, Kumirska, and Stepnowski (2008)
Silkworm (*B. mori*), rice stemborer (*Chilo simplex*)	Fatty acids (eg, caprylic acid and capric acid)	Koidsumi (1957)

Table 2 Cuticle and defensive secretions as barriers to fungal infection—cont'd

Arthropod	Fungistatic factor(s)	Key references
Aphids (*Sitobion avenae, Hyalopterus pruni, Brevicoryne brassicae*)	Fatty acids (eg, dodecanoic acid eicosanoic acid, pentanoic acid, and E,E-2,4-hexadienoic acid)	Szafranek et al. (2001)
Booklouse (*Liposcelis bostrychophila*)	Fatty amides	Lord and Howard (2004)
Stink bugs (*Dichelops melacanthus, Nezara viridula, Chinavia ubica,* and *Euschistus heros*)	Various volatile compounds	Lopes, Laumann, Blassioli-Moraes, Borges, and Faria (2015)
Tick (*Hyalomma excavatum*)	Fatty acids (C16:0, C18:0, C18:1ω9C, C20:0)	Ment et al. (2013)

proteins (phenoloxidase and apolipophorin III) and higher numbers of circulating hemocytes. Irrespective of whether the agent used was biotic (EPF) or abiotic (thermal stress), the priming effects are transient and seem to last only 24–48 h (Browne, Surlis, & Kavanagh, 2014).

3. ADHESION AND PRE-PENETRATION EVENTS

3.1 Adhesion of Infection Propagules to Host Surface

The first step in the EPF infection process is adhesion of conidia to the host surface (Figs. 2 and 5). Since mortality is dose dependent, it is vital that as many conidia adhere to the cuticle as possible (Butt & Goettel, 2000). The surface of aerial conidia of most hypocrealean EPF is covered with a rodlet layer (Fig. 5) composed of hydrophobin proteins that confer a hydrophobic charge facilitating the passive attachment of conidia to hydrophobic surfaces (Holder & Keyhani, 2005). In contrast to blastospores, conidia attach poorly to hydrophilic surfaces especially in aquatic environments (Greenfield, Lord, Dudley, & Butt, 2014; Holder, Kirkland, Lewis, & Keyhani, 2007).

EPF are known to possess several hydrophobin genes with three (*hyd1*, *hyd2*, and *hyd3*) present in *M. brunneum* (Sevim et al., 2012). These are differentially expressed (depending on development stage) with *hyd*-deficient mutants exhibiting decreased virulence (Sevim et al., 2012). Targeted gene

Table 3 Sensitivity of fungi (EPF and non-EPF) to AMPs

AMP	Source	Test fungi	Susceptible (S) or Resistant (R)	References
Cecropin A	*D. melanogaster*	*M. anisopliae*, *Geotrichum candidum*, *S. cerevisiae*	S	Ekengren and Hultmark (1999)
Cecropin B	*D. melanogaster*	*M. anisopliae*, *G. candidum*, *S. cerevisiae*	S	Ekengren and Hultmark (1999)
Cecropin A	*D. melanogaster*	*B. bassiana*	R	Ekengren and Hultmark (1999)
Cecropin B	*D. melanogaster*	*B. bassiana*	R	Ekengren and Hultmark (1999)
Cecropin	*Hyalaphora cecropia*	*Fusarium* sp., *Aspergillus* sp.	S	De Lucca et al. (1997), De Lucca, Bland, Jacks, Grimm, and Walsh (1998)
Cecropin	*Hyphantrea*	*Candida tropicalis*	S	Park et al. (1997)
Drosomycin (defensin)	*D. melanogaster*	*B. bassiana*, *M. anisopliae*	R	Fehlbaum et al. (1994), Zhang and Zhu (2009)
Heliomicin (defensin)	*H. virescens*	*C. albicans*, *Pichia pastoris*, *Cryptococcus neoformans*, *N. crassa*, *Fusarium* sp., *A. fumigatus*, *Trichoderma viridae*	S	Lamberty et al. (1999)
Metchnikowin	*D. melanogaster*	*N. crassa*	S	Levashina et al. (1995)
Gallerimycin (defensin)	*G. mellonella*	*M. anisopliae*, *T. viridae*, *Pyricularia grisea*, *yeasts*	S	Lee et al. (2004), Schumann, Seitz, Vilcinskas, and Podsiadlowski (2003)
Moricin like peptides	*G. mellonella*	*Fusarium*, yeasts	S	Brown, Howard, Kasprzak, Gordon, and East (2008)
Termicin (defensin)	*Pseudocanthothermes spiniger*, *Reticulitermes flavipes*	*F. culmorum*, *F. oxysporum*, *N. crassa*, *N. haematococca*	S	Lamberty et al. (2001)
Termicin (defensin)	*P. spiniger*, *R. flavipes*	*B. bassiana*	R	Lamberty et al. (2001)

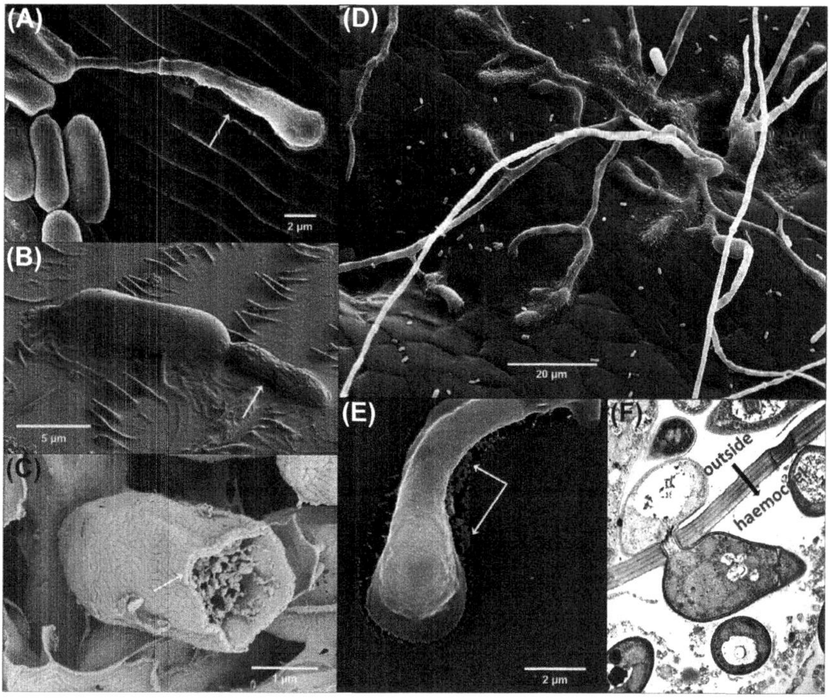

Figure 5 ***Key morphological features of entomopathogenic fungi interacting with the insect cuticle.*** (A) Blastospores develop infection structures covered in mucilage. (B) Appressorium formation. The different surface characteristics of the germinating blastospore can be distinguished. (C) The hydrophobin rodlet layer of an aerial conidium. (D) Differential hyphae colonize the insect cuticle. Some rod-shaped bacteria can also be seen in the image. (E) Removing the mucilage of an invading *Metarhizium* hypha revealed distinct enzymatic erosion of the integument. (F) EPF penetration of the insect cuticle layers. Key features are indicated using *arrows. Modified from Schreiter, G., Butt, T. M., Beckett, A., Moritz, G. & Vestergaard, S. (1994). Invasion and development of Verticillium lecanii in the Western Flower Thrips, Frankliniella occidentalis.* Mycological Research, 98, *1025–1034.* (See color plate)

inactivation of *hyd1* did not alter adhesive properties of *B. bassiana* conidia, but altered surface carbohydrate epitopes and reduced hydrophobicity and virulence (Zhang, Xia, Kim, & Keyhani, 2011). In contrast, *hyd2* mutants retained distorted surface bundles, but the conidia displayed both decreased cell surface hydrophobicity and adhesion without affecting virulence (Zhang et al., 2011). We postulate the fungus up-regulates other virulence factors to compensate, further demonstrating EPF plasticity.

Several other cell wall proteins play important roles in adhesion and stress management (Table 1). CWP10 is a non-hydrophobic cell wall protein that enhances conidial hydrophobicity and adhesion (Li, Ying, Shan, & Feng, 2010), while CP15 helps conidia cope with thermal and oxidative stresses (Fig. 2; Ying & Feng, 2011). Mad1 and Mad2 facilitate adhesion to arthropod and plant roots, respectively (Barelli, Padilla-Guerrero, & Bidochka, 2011; Wang & St. Leger, 2007a). Disrupting *mad1* delays germination and reduces virulence while disruption of *mad2* blocks adhesion to plant surfaces without affecting infection of arthropod hosts (Wang & St. Leger, 2007a). Molecular evolutionary analyses revealed that the *mad1* gene is highly conserved among *Metarhizium* lineages, whereas the *mad2* gene is divergent. This suggests that early plant-fungal interactions have influenced the infectious capacity of EPF, enabling them to exist both as entomopathogens and endophytes (Wyrebek & Bidochka, 2013). Some cell wall–bound proteins appear to have dual or multifunctional properties. For example, the glycolytic enzyme glyceraldehyde-3-phosphate dehydrogenase (GAPDH) is produced by *M. anisopliae* in response to different carbon sources, and also has adhesion-like properties (Broetto et al., 2010).

Several enzymes are present in the cell wall of EPF conidia (Schrank & Vainstein, 2010). Virulent conidia possess high levels of spore-bound Pr1 (Shah, Wang, & Butt, 2005), which, together with other preformed enzymes, may degrade fungistatic compounds at the host surface (Table 2) and concomitantly release nutrients that stimulate germination (Ortiz-Urquiza & Keyhani, 2013; Santi et al., 2010). Rapid germination is an attribute of virulent strains; it facilitates infection while climatic conditions are favorable. Multiple penetration sites enable the pathogen to rapidly colonize the host, overpower the host's defenses, and prevent infection by opportunistic saprophytes (Altre & Vandenberg, 2001). Enzyme activity occurs prior to germ tube formation, reflected in etching of the host cuticle beneath the spore (Fig. 5E; Schreiter, Butt, Beckett, Moritz, & Vestergaard, 1994). Enzyme activity is usually restricted to the mucilage that anchors the germ tube and appressorium to the host cuticle (Fig. 5).

The cuticle is the first and most important barrier to EPF (Figs. 2 and 3; Table 2). Dynamic interactions at the cuticle surface influence the pathogens ability to infect its host. Conidia of *Metarhizium* species fail to remain attached to the surface of mosquitoes (eg, *Aedes*, *Culex*, and *Anopheles*) larvae due to weaker adhesion forces compared with those between the conidia and terrestrial hosts (Greenfield et al., 2014). The hydrophilic nature of the larval cuticle and possibly certain chemicals prevent conidia from adhering (Greenfield et al., 2014). Blastospores produce mucilage that enables their establishment on the larval cuticle and appears to be relatively stable underwater (Alkhabairi & Butt, unpublished observations).

A variety of other mechanisms prevent spore adhesion to and/or germination on insect cuticles. While some of the chemical barriers are intrinsic to the host, others may be derived from the environment (Table 2). Fungistatic compounds include plant-derived chemicals or bioactives produced by bacteria on the cuticle surface (Ortiz-Urquiza & Keyhani, 2013; Toledo, Alippi, & de Remes Lenicov, 2011). Some social insects use microbicidal secretions alongside grooming to cleanse the exoskeleton (as discussed in Section 2). Plant volatiles can also deter infection due to their sequestration on the insect cuticle. For example, crucifer isothiocyanates inhibit germination of *M. anisopliae* resulting in reduced infection of mustard beetle (*Phaedon cochleariae*) and striated white butterfly (*Pieris melete*) larvae (Atsumi & Saito, 2015; Inyang, Butt, Beckett, & Archer, 1999; Inyang, Butt, Doughty, Todd, & Archer, 1999). Many plant allelochemicals have been identified that inhibit germination and growth both in vitro and in vivo (Table 2). Other ways plants influence host—pathogen interactions (HPI) have been reviewed extensively by Cory and Hoover (2006).

3.2 EPF Have Excellent Stress Management and Detoxification Systems

The cuticle surface is a harsh environment for EPF; conidia are exposed to harmful UV, fluctuating humidity and temperature, antagonistic microbes and the host innate defenses (Fig. 2). EPF have evolved various mechanisms to cope with these biotic and abiotic stresses, often involving genes linked with virulence, osmolyte balance, cell wall integrity, and signal transduction (Fig. 2, Table 1; Ortiz-Urquiza & Keyhani, 2014). Heat shock proteins (HSPs), as well as specific cell wall proteins mentioned earlier, help with stress management (Liao, Lu, Fang, & Leger, 2014). Besides preformed enzymes that degrade fungistatic fatty acids in the epicuticular waxes (see Section 3.1) many other genes are up-regulated such as *MrCYP52*, a cytochrome P450 monoxygenase family 52 gene (Lin et al., 2011). *MrCYP52* is expressed on insect cuticle and in artificial media containing extracted cuticle hydrocarbons (Lin et al., 2011). Extraction of alkanes from the cuticle prevents induction of *MrCYP52* and limits growth. Disrupting *MrCYP52* reduces in vitro growth on epicuticular hydrocarbons, delays germination and appressorium differentiation on the cuticle and reduces virulence for wax moth larvae (Lin et al., 2011). Although the genome of *Metarhizium robertsii* encodes 123 highly divergent *CYP* genes, surprisingly this generalist species has only a single *CYP52* compared with four in the more specialized *M. acridum* (Gao et al., 2011), suggesting that generalists are dependent on other genes for detoxification.

Hog1 is a mitogen-activated protein kinase (MAPK; Table 1) that enables EPF to respond to abiotic stresses as illustrated by *B. bassiana hog1* null mutants which were more sensitive to hyperosmotic stress, high temperature, and oxidative stress compared to wild-type controls (Zhang et al., 2009). These mutants were also less pathogenic due to decreased spore viability as well as fewer *hyd1/hyd2* transcripts corresponding to poor adhesion and fewer appressoria (Zhang et al., 2009).

Continuous cultivation on artificial media results in EPF degeneration or attenuation with cells making very little Pr1 or destruxins, and producing sterile sectors (Wang, Butt, & St. Leger, 2005). A *M. anisopliae* mutant lacking a dispensable chromosome lost the ability to produce Pr1 and destruxins and was weakly pathogenic to *T. molitor* (Wang, Skrobek, & Butt, 2003; Wang, Typas, & Butt, 2002). These observations suggest that the fungus possesses other pathogenicity determinants so that it is not solely dependent on Pr1 and destruxins. EPF degeneration is linked to their inability to cope with stress (Wang et al., 2005). However, overexpression of glutathione peroxidase can help increase fungal tolerance to oxidative stress and stabilize growth during subculturing (Xiong, Xia, Zheng, & Wang, 2013).

3.3 Differentiation of Infection Structures

3.3.1 Influence of Arthropod Host on Appressorium Differentiation

Differentiation of appressoria is a prerequisite for infection by most EPF. The physicochemical and nutritional cues at the host surface influence both differentiation of infection structures and specificity (Fig. 2; Lin et al., 2011; St. Leger, Butt, Goettel, Staples, & Roberts, 1989; St. Leger, Butt, Staples, & Roberts, 1989). Thus, conidia of *M. acridum* will germinate and differentiate appressoria on host (*S. gregaria*) cuticle, but fail to germinate on a nonhost (*Leptinotarsa decimlineata*). Interestingly, germination appeared normal on the hemipteran bug, *Magicicada septendecim*, but differentiation of appressoria was reduced (Wang & St. Leger, 2005). Altogether these studies illustrate the complexity of the interactions and suggest that infection can be blocked at different stages. Wang and St. Leger (2005) noted that although *M. acridum* transcript profiles were similar in locust and beetle extracts, the latter was enriched in genes for detoxification and redox processes, while the locust extract up-regulated more genes involved in cell division and biomass.

Appressoria may be produced directly from conidia or at the end of germ tubes or even laterally from hyphae (Figs. 3 and 5). Some species, such as *Nomuraea rileyi*, do not produce appressoria, but can penetrate directly (Srisukchayakul, Wiwat, & Pantuwatana, 2005). On occasion, *Beauveria* and *Metarhizium* germlings will also penetrate directly. Cuticular physicochemical cues influence appressorium differentiation and phenotype. Appressoria morphologies range from clavate, spherical, or cup shaped. EPF differentiate appressoria on hard, nutrient poor, hydrophobic surfaces (St. Leger, Butt, Goettel, et al., 1989). Lipids, polyols, trehalose, and glycogen are the main nutrient reserves in fungal spores (Hallsworth & Magan, 1996; Wang & St. Leger, 2005). Lipids are transported to the developing appressorium and degraded to release glycerol, which increases hydrostatic pressure and provides a driving force for mechanical penetration (Wang & St. Leger, 2007b). Perilipin (MPL1) plays a pivotal role in lipid metabolism and is considered a pathogenicity determinant (Table 1). Expression of the perilipin gene, *mpl1*, is easily detectable when *M. anisopliae* is engaged in accumulating and breaking down lipids. Mutants lacking MPL1 have reduced levels of total lipids, thinner hyphae, and fewer lipid droplets especially in appressoria (Wang & St. Leger, 2007b).

3.3.2 Signaling and Appressorium Differentiation

Complex links exist between adhesion, appressorium differentiation, and infection of the host cuticle by EPF. This requires cross-talk between a plethora of genes whose expression must be choreographed to ensure successful identification and infection of the host. The pathogen concomitantly has to cope with environmental stresses and the hosts defense responses (Figs. 2–4). This choreography is facilitated partly by a suite of proteins (eg, transpanins, P-type Ca^{2+}-ATPases, calcineurin, and kinases) and secondary messengers constituting the signaling apparatus (Fang, Feng, et al., 2009; Fang, Pava-ripoll, Wang, & St. Leger, 2009; Wang, Yang, et al., 2013; Wang, Zhou, Ying, & Feng, 2013).

Protein kinase A (PKA) plays a pivotal role in sensing host stimuli and transduction of these signals to regulate infection processes. By disrupting the class I PKA catalytic subunit gene *MaPKA1*, Fang, Feng, et al. (2009) and Fang, Pava-ripoll, et al. (2009) showed that growth and virulence of *M. ansiopliae* was greatly reduced with nearly all key stages of the infection process being affected including adhesion, speed of germination and differentiation of appressoria. Disruption of the *B. bassiana* mitogen—activated protease kinase gene *BbMPK1* also results in loss of virulence, partly due to poor adhesion and differentiation of appressoria (Zhang et al., 2010). When injected into the hemocoel, mutant conidia produced hyphae that grew but were unable to reemerge and sporulate at the host surface. This suggests that *BbMPK1* is essential for penetration of the cuticle and reemergence from mycosed cadavers (Zhang et al., 2010). Other kinases also influence sporulation, cell wall integrity, stress tolerance, and virulence including the MAP kinase Bbslt2 (Luo et al., 2012) and Group III histidine kinase mhk1 (Zhou, Wang, Qiu, & Feng, 2012; see Table 1).

3.3.3 Differentiation of Appressoria on Diverse Substrates

Appressoria are produced on the arthropod cuticle, artificial substrates, and plants, suggesting that EPF respond to a wide range of cues. *Metarhizium flavoviride* and *M. anisopliae* will produce appressoria on synthetic, hard, hydrophobic substrates supplemented with traces of nutrients (St. Leger, Butt, Goettel, et al., 1989; St. Leger, Butt, Staples, et al., 1989; Xavier-Santos, Magalhães, & Lima, 1999). Plant waxes can entrap EPF conidia or inhibit their germination, but removal of the wax layer will stimulate germination, differentiation of appressoria and even penetration of the underlying epidermal layer (Inyang, Butt, Beckett, et al., 1999; Inyang, Butt, Doughty, et al., 1999). EPF can exist as endophytes in a wide range of plant species entering through the leaves, seeds, and roots. *Metarhizium* species are almost exclusively found on and around roots while *B. bassiana* are found throughout the plant (Behie et al., 2015).

4. PENETRATION OF THE INTEGUMENT

4.1 Role of EPF Enzymes in the Infection Process

EPF produce an extensive cocktail of extracellular hydrolytic enzymes including lipases, proteases, chitinases, phospholipase C, and catalase (Santi et al., 2010; Schrank & Vainstein, 2010). Some cuticle-degrading enzymes (CDEs) are virulence determinants (Nunes, Martins, Furlaneto, & Barros, 2010; St. Leger, Joshi, & Roberts, 1998). The importance of these enzymes is validated through gene knockout, enzyme-deficient mutants and overexpression studies. Genetically modified EPF overexpressing proteases and chitinases display increased virulence compared with the wild-type strains (Fan et al., 2007; St. Leger, Joshi, Bidochka, & Roberts, 1996; Zhang et al., 2008).

Most EPF species possess multiple copies of chitinases and proteases. Generalist strains of *M. anisopliae* possess six chitinase genes and 11 *Pr1 (Pr1A−K)* genes (Bagga, Hu, Screen, & Leger, 2004; Schrank & Vainstein, 2010), with specialist species possessing fewer copies of the respective genes (Bagga et al., 2004). Constructs overexpressing CHI2 endochitinase are far more efficient at killing the host, while knockout constructs are less virulent than the wild type (Boldo et al., 2009). Pr1A accounts for over 90% of the Pr1 activity, it is produced by both specialists and generalists. Genes encoding for Pr1A and the cyclic peptide destruxin appear to be located on the dispensable chromosome (Wang et al., 2003; Wang, Sykes, et al., 2002; Wang, Typas, et al., 2002). Pr1 works in concert with other enzymes (eg, Pr2 and chitinases) in digesting the protein−chitin cuticle to enable the fungus to gain access to the nutrient-rich hemocoel and to emerge for external sporulation (Table 1; Small & Bidochka, 2005).

Each appressorium produces a narrow penetration peg (Figs. 2, 3 and 5) and anchors the fungus to the cuticle to counteract the downward pressure generated by the peg. Since pressure = force × area, the force generated must be considerable since the area is relatively small. The cuticle is occasionally distorted at the penetration site suggesting that substantial force is being applied during infection (Butt, Ibrahim, Clark, & Beckett, 1995). Blastospores can produce penetration pegs from appressoria along the length or tip of the blastospore (Fig. 5; Schreiter et al., 1994).

4.2 Insect Responses to Counter EPF during Penetration

Preventing the loss of hemolymph is essential to insect survival. Whether the cuticle is compromised by predators, pathogens, parasites, or herbivores, the hemostatic response is initiated alongside immune defenses (Theopold, Li, Fabbri, Scherfer, & Schmidt, 2002). Narrow penetration pegs (see previous section) minimize hemolymph leakage during infection. Insect hemostasis involves the polymerization of lipophorins and vitellogenin-like proteins via the activity of calcium-dependent transglutaminases (E.C. 2.3.2.13). These clotting proteins contain a cysteine-rich domain homologous to the mammalian von-Willebrand factor blood clotting proteins (Vilmos & Kurucz, 1998). In 2010, Wang et al. documented a novel function for transglutaminases in *Drosophila*. During microbial invasion, micro clots form (similar to hemocyte-free nodules) and act to sequester microbes in the hemolymph. Transglutaminase anchors the microbes to these microclot matrices for killing by AMPs and hemocytes. Insect survival can be greatly improved due to ecdysis when conidia are shed with the old cuticle, and providing the fungus has not penetrated the underlying fresh cuticle (Vestergaard, Gillespie, Butt, Schreiter, & Eilenberg, 1995). Besides fungistatic compounds in the epicuticular wax layer (Table 2) pathogens have to overcome other biological barriers such as cuticle thickness, mineralization, sclerotization, melanization, protease inhibitors, and AMPs (Figs. 2–4; Dubovskiy, Whitten, Kryukov, et al., 2013; Dubovskiy, Whitten, Yaroslavtseva, et al., 2013).

Susceptibility to infection can depend heavily on the insect developmental stage. Molted insects are particularly vulnerable since the new cuticle is soft and not fully sclerotized (Liu et al., 2014). Cuticle hardness is attributed directly to mineralization and/or sclerotization, making it recalcitrant to enzyme degradation (Andersen, 2010; Bogus et al., 2007). Calcium, magnesium, and phosphorus are the major mineral elements in puparial exuviae of the face fly, *Musca autumnalis*; house fly, *Musca domestica*; and stable fly, *Stomoxys calcitrans* (Roseland, Grodowitz, Kramer, Hopkins, & Broce, 1985). These dipterans use both catecholamines and minerals for stabilization of the puparial cuticle.

The actual thickness of the cuticle does not correlate broadly with resistance since EPF can infect a wide range of insects, some of which have relatively thick cuticles (Lacey et al., 2015). A combination of body surface structure, cuticle thickness, and wax secretions of the pine scale, *Matsucoccus matsumurae* influences its susceptibility to EPF (Liu et al., 2014). Delaying entry to the hemocoel allows the host to mobilize its cellular and humoral defenses (Dubovskiy, Whitten, Kryukov, et al., 2013; Figs. 3 and 4).

Besides playing a major role in ecdysis and wound healing, the epidermis also contributes to host immunity (Dubovskiy, Whitten, Yaroslavtseva, et al., 2013). During infection, regeneration factors are up-regulated in insect epidermal cells augmenting the integumentary defenses (Dubovskiy, Whitten, Yaroslavtseva, et al., 2013). Similarly, AMPs have also been detected in the integument (Brey et al., 1993). The extent to which these AMPs contribute to insect resistance to EPF or innate immunity remains unclear. They are likely involved with the early stages of recognition and destruction of fungal structures in the integument and may also prevent secondary infections by opportunistic saprophytes (Table 3).

Insects possess an array of protease inhibitors that are involved with development (eg, IMPI = insect metalloprotease inhibitor) and regulation of the prophenoloxidase (proPO) cascade (eg, serpins). Culture filtrates of *M. anisopliae* containing Pr1 and other proteases are inactivated by cell-free hemolymph (unpublished observations). According to Vilcinskas (2010), insects have been selected to evolve inhibitors to EPF proteolytic enzymes. One notable example is IMPI which has probably evolved in response to the EPF-specific metalloproteases (Qazi & Khachatourians, 2007; Vilcinskas, 2010).

Damage to the cuticle layers or fungal penetration of the epidermis leads to synthesis and accumulation of melanin within these tissues (Figs. 3 and 4). This process is initiated by the proPO activation cascade. Melanin pigments and their precursors are important structural and protective components of the cuticle. Melanogenesis involves the processing of phenolic substrates (eg, L-tyrosine and L-dihydroxphenylalaine) into red/brown/black pigments, consequently producing ROS and toxic intermediates. Phenoloxidases (PO; EC 1.10.3.1 and EC 1.14.18.1) activity occurs primarily in the (epi)cuticular structures, barrier epithelia (midgut and gonads) and hemolymph where it performs dual roles in cuticle hardening/tanning (Andersen, 2010) and immune defense (Cerenius, Lee, & Söderhäll, 2008). Such defenses include nonself-recognition and both humoral and hemocyte-mediated encapsulation of invading organisms (Butt et al., 1988; Ratcliffe, Rowley, Fitzgerald, & Rhodes, 1985). Accumulation of melanin and semiquinone intermediates in the insect cuticle not only limits the growth of certain fungi, possibly by acting as a physical barrier, but also suppresses synthesis of fungal CDEs, to further impede cuticle penetration (Söderhäll & Ajaxon, 1982; St. Leger et al., 1988). Dark, or melanic, insect morphs have an unusually high concentration of cuticular melanin, so possibly a positive correlation exists between melanism, PO activity, and resistance to parasites and pathogens such as *B. bassiana* and *M. anisopliae* (Dubovskiy, Whitten, Kryukov, et al., 2013; Dubovskiy, Whitten, Yaroslavtseva, et al., 2013). Melanism often occurs in crowded insect populations as a form of density-dependent prophylaxis due to the high risk of infection (see Section 2).

The fact that hemocytes will migrate to penetration sites prior to entry by the pathogen suggest that signals are being relayed from the site of injury (Gunnarsson, 1988). Hemocytes attach to and penetrate the basement membrane, mingle with epidermal cells (extravasation) and even form multilayered capsules at the infection site (Figs. 3 and 4) (Gunnarsson, 1988). The hemocytes migrate to repair the wound and release AMPs. Insects with few circulating hemocytes, such as hemipterans (eg, *Aphis fabae* and *Empoasca fabae*), form nonhemocytic, melanotic capsules around invading fungal hyphae at the penetration site (Butt et al., 1988; Gotz & Vey, 1974).

5. POST-PENETRATION HPI

5.1 Adaptation to the Hemolymph

Once EPF invade the hemolymph they multiply as yeast-like, thin-walled blastospores or hyphal bodies. Similar structures are also produced in liquid media, especially under conditions of elevated glucose and dissolved oxygen (Mascarin, Jackson, Kobori, Behle, Dunlap, et al., 2015; Mascarin, Jackson, Kobori, Behle, & Júnior, 2015), so their production may be triggered by an oxygenated, nutrient-rich, liquid environment. Blastospores offer a large surface area to volume ratio for the uptake of nutrients and will multiply in this form until nutrients are exhausted, after which, they transform into hyphae both in vitro and in vivo. Blastospores rapidly colonize the hemocoel, but will encounter host cellular and humoral defenses (Figs. 4 and 6) and high osmotic pressure (300–500 mOsmol/l). The osmosensor Mos1 helps the fungus respond to high osmotic pressure and adapt to the hemolymph (Wang et al., 2008). Repressing *Mos1* triggers pleiotropic effects, presumably because Mos1 acts as a scaffold protein in MAP kinase pathways that influence many biological processes, including: development, stress management, and virulence (Roman, Arana, Nombela, Alonso-Monge, & Pla, 2007; Wang & St. Leger, 2007a). Knockdown of *Mos1* does not affect expression of the immune coat protein gene *Mcl1* that is expressed exclusively in the hemolymph (Wang & St. Leger, 2006), but it does down-regulate the adhesion *Mad1* gene, which is important in spore "stickiness" and hyphal body differentiation (Wang & St. Leger, 2007a).

Carbohydrate heterogeneity (including quantity) at the fungal surface plays an important role in nonself recognition by arthropod hosts (Wanchoo, Lewis, & Keyhani, 2009). Blastospores appear to possess fewer carbohydrate epitopes than propagules such as conidia, submerged conidia, and hyphae (Pendland & Boucias, 1993; Wanchoo et al., 2009). *B. bassiana* cells isolated from the hemolymph of infected *M. sexta* and *Heliothis virescens* lacked many carbohydrate epitopes, but expressed more when transferred to an artificial culture medium (Wanchoo et al., 2009). These observations may explain why conidia and hyphae elicit stronger immune responses than

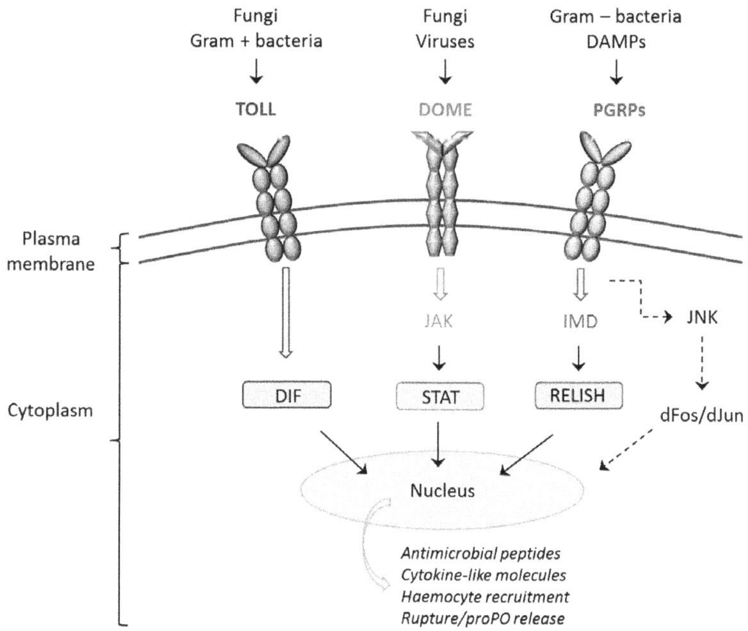

Figure 6 ***Hemocyte signaling cascades involved in pathogen recognition.*** Transmembrane pathogen recognition receptors (DOME, Toll, and PGRPs) interact with microbial ligands called pathogen-associated molecular patterns (PAMPs) or damage-associated molecular patterns (DAMPs) to activate the corresponding pathway. Fungal ligands (β-1,3-glucan, mannan, and chitin) are recognized by DOME and/or Toll leading to the activation of cytosolic transcription factors (DIF and/or STAT), which translocate to the nucleus to initiate the production of immune effectors.

blastospores or protoplasts (Enríquez-Vara et al., 2014; Pendland, Hung, & Boucias, 1993). Protoplasts are a rare occurrence in hypocrealean fungi, but are common in the entomophthorales (Butt, Beckett, & Wilding, 1981; Butt, Hoch, Staples, & St. Leger, 1989; Beauvais, Latgé, Vey, & Prevost, 1989). Some of the strategies EPF deploy to escape host immune responses are outlined in Section 6.

5.2 Role of PAMPs and PRRs in the Activation of Host Defenses

Invertebrates lack the ability to produce clonally derived immunoglobulins, yet they have developed the capacity to differentiate between major groups of microorganisms using pathogen recognition receptors (PRRs). PRRs are expressed constitutively and secreted into the hemolymph or presented on hemocyte, fat body, and epidermal cell plasma membranes. Some of the well-characterized PRRs include peptidoglycan recognition proteins (PGRPs), Gram-negative–binding proteins (GNBPs), hemolin (a member of the Ig superfamily) and β-glucan–binding proteins (βGBPs), among others (reviewed by Stokes, Yadav, Shokal, Smith, & Eleftherianos, 2015). PRRs function by binding to differential ligands and sugar moieties found on microbial cell surfaces, collectively known as PAMPs that include bacterial endotoxins (lipopolysaccharides, LPS), lipoteichoic acids (LTA), mannan, and β-1,3-glucans from fungi, and viral RNA (Uvell & Engstrom, 2007). Opsonization of foreign invaders in the hemolymph allows hemocyte-derived PRRs to interact with the immobilized pathogens and initiate defense reactions (Fig. 4).

There are three major signaling pathways involved in detecting microbes, viruses, and parasites: Toll, JAK-STAT, and Imd (see Fig. 6 and Lu & St. Leger, 2016). Recognition of fungal ligands (β-glucans, chitins, and mannans) by the Toll and JAK-STAT receptors initiates signal transduction cascades that culminate in the synthesis of immune effectors. For example, binding of β-1,3-glucan to Toll induces the expression of the fungicidal peptide, drosomycin, in wild-type *Drosophila melanogaster* (Lemaitre, Nicolas, Michaut, Reichhart, & Hoffmann, 1996).

Souza-Neto, Sim, and Dimopoulos (2009) demonstrated unequivocally that *Aedes aegypti* use the JAK-STAT pathway to launch antiviral defenses against dengue virus. Independent of the Toll pathway, RNA inhibition of the JAK-STAT receptor Domeless and the Janus kinase Hop increased insect susceptibility to the virus. Interestingly, *B. bassiana* infected *A. aegypti* demonstrate an enhanced ability to combat dengue virus, restricting viral replication to the midgut of the mosquito. It appears that EPF such as *Beauveria* can influence (perhaps indirectly) components of both the Toll and JAK-STAT pathways (Dong, Morton, Ramirez, Souza-Neto, & Dimopoulos, 2012).

5.3 Host Cellular Responses and the proPO Cascade

Insects possess different types of hemocytes (eg, coagulocytes, granular cells, plasmatocytes, oencoytoids, pro-hemocytes, and adipohemocytes) that participate in metamorphosis, wound repair, enzyme synthesis, differentiation of tissues, and immunity (Lu & St. Leger, 2016; Price & Ratcliffe, 1974; Strand, 2008). Activated hemocytes release bioactives and carry out phagocytosis, encapsulation, or nodulation to combat mycosis (Fig. 4).

Hemocyte death plays a substantial role in innate immunity having long been associated with encapsulation and nodulation reactions in insects (reviewed by Smith, 2010). More recently, insect hemocytes have been shown to release extracellular chromatin traps as an antimicrobial response, termed ETosis (Altincicek, Stotzel, Wygrecka, Preissner, & Vilcinskas, 2008; Robb, Dyrynda, Gray, Rossi, & Smith, 2014). During sepsis or mycosis, populations of microorganisms are opsonized via hemolymph and hemocyte-derived lectins, becoming immobilized. These immobilized microbes can either be nodulated or encapsulated. Typically, nodules consist of an amalgamation of viable and nonviable hemocytes, nonself- particles, and melanized debris (Ratcliffe & Gagen, 1977; Stanley-Samuelson et al., 1991). Miller, Nguyen, and Stanley-Samuelson (1994) demonstrated that microbial populations present in the hemolymph of *M. sexta* larvae lead to the biosynthesis of eicosanoids, resulting in hemocytes undergoing a conformational change allowing them to adhere to microorganisms. Circulating granular cells adhere to foreign bodies (using peroxinectin or integrin) and release plasmatocyte spreading proteins that attract plasmatocytes to the site of infection. Recruited plasmatocytes secrete cytoplasmic stored adhesion molecules onto their outer membrane surface resulting in multiple layers of hemocytes attaching to the invaders (Strand & Clark, 1999). Finally, a monolayer of granular cells adhere to and apoptose on the periphery, thus ending encapsulation. Asphyxiation, production of ROS, AMP release, and deposition of melanin collectively lead to the destruction of the encapsulated intruder (Lavine & Strand, 2002; Walters & Ratcliffe, 1983).

The intricate mechanisms of proPO activation and its regulation by serpins have been studied extensively in insects (reviewed by Cerenius, Kawabata, Lee, Nonaka, & Söderhäll, 2010). The cascade is initiated via the interaction of fungal PAMPs with soluble and/or hemocyte-bound PRRs present in the hemocoel (Section 5.2; Fig. 6). A succession of distinct protein zymogens are converted into their active conformations, ultimately leading to cellular rupture and the liberation of activated PO. PO converts phenolic substrates (in a two-step enzymatic reaction) into quinone species that facilitate the biogenesis of the microbicidal pigment, melanin (Nappi & Christensen, 2005). In a few insect species, the coordinated rupture of oenocytoid hemocytes (ie, cell death) in response to wounding/infection leads to the release of PO alongside various cytosolic and membrane components (Bidla, Dushay, & Theopold, 2007). In *Drosophila*, extracellular PO appears to form an activation complex with proteases and amino-phospholipids (phosphatidylserine (PS) and phosphatidylinositol; Bidla, Hauling, Dushay, & Theopold, 2009). Since PS interacts directly with PO (Bidla et al., 2009) and invertebrate hemocyanin is converted into a PO-like enzyme upon binding to PS liposomes and PS on apoptotic hemocytes (post-phagocytosis of *B. bassiana* spores; Coates, Kelly, & Nairn, 2011; Coates, Whalley, Wyman, & Nairn, 2013), it is possible that insect PO may interact with the PS-expressing apoptotic hemocytes on capsules/nodules, thereby aiding melanin deposition.

5.4 Humoral Responses and Multifunctional Proteins

Insects can produce a wide range of humoral defenses to resist fungal infection (Figs. 3 and 4; Tables 3 and 4). These include lectins, protease inhibitors, PO, AMPs, and reactive oxygen and nitrogen radicals (Jiang, Vilcinskas, & Kanost, 2010). In addition, increasing evidence suggests that a number of proteins involved in invertebrate innate immunity display functionally diverse roles (see Table 4). It is essential to note that PO activities are crucial for early host responses to EPF on the cuticle (Section 4.2; Figs. 2 and 3), during cellular and noncellular encapsulation responses to EPF in the hemocoel (Section 5.3; Fig. 4), and for the production of melanin and reactive oxygen and nitrogen species (ROS and RNS) (Dubovskiy et al., 2010). These oxygenic and nitrogenous volatiles can be damaging to both the host and fungal pathogen, affecting nucleic acids, peptides, and oxidation of lipids (Lyakhovich, Vavilin, Zenkov, & Menshchikova, 2006).

Both the fungus and insect possess antioxidant systems (detoxifying enzymes) aimed at neutralizing ROS and RNS. In insects, these are broadly divided into enzymatic antioxidants (superoxide dismutase (SOD), catalase, glutathione-S-transferase (GST), peroxidase) and nonenzymatic antioxidants (phenol-containing compounds, α-tocopherol, ascorbic acid, high and low molecular thiols) (Felton & Summers, 1995). EPF are able to grow faster than capsule formation and presumably use their own antioxidant machinery to limit the impact of host ROS. Both *B. bassiana* and *M. anisopliae* can survive and even break out of cellular or melanotic capsules (Bidochka, Clark, Lewis, & Keyhani, 2010).

AMPs are the foundation of the humoral immune response (Table 3) and are typically cationic, demonstrating broad-spectrum activities (Bulet & Stocklin, 2005). Most can be classified based on their amino acid composition, molecular structure, and mode of action. AMPs are produced in relatively small amounts and often target specific microbes (Lemaitre, Reichhart, & Hoffmann, 1997), yet some have multifunctional properties and work synergistically with established immune molecules including lysozyme and apolipophorin, thus leading to enhanced efficacy (see, Table 4).

Table 4 Examples of multifunctional immune proteins in insects

Bioactives	Functional properties	Key references
Apolipophorin III (homologous to mammalian apolipoprotein E)	• Pattern recognition • Binds to, and detoxifies, LPS and LTA • Stimulates AMP release • Enhances hemocyte-led phagocytosis and encapsulation • Binds directly to β-glucan on entomopathogenic conidia (*M. anisopliae*) • Synergistic effect with lysozyme • Lipid transport, an enzyme cofactor and lipoprotein metabolism	Whitten et al. (2004), Zdybicka-Barabas et al. (2011, 2012, 2013)
Hemocyanin?	• Oxygen transport and plasma osmolyte in larval insects • Embryonic development in migratory locusts • Hemocyanin-derived phenoloxidase activity?	Hagner-Holler et al. (2004), Pick, Schneuer, and Burmester (2010), Chen et al. (2015)
(pro)**Phenoloxidase**	• Cuticle sclerotization and developmental morphogenesis • Antimicrobial pigment (melanin) production • Generation of ROS and RNS • Aids hemocyte-mediated defenses, encapsulation	Cerenius et al. (2008)
Serpins (**ser**ine **p**rotease **in**hibitor**s**)	• Deactivation of microbe-derived proteases • Modulation of the proPO cascade • Protein storage • Hormone transport	Kanost (1999), Cerenius et al. (2008), Reichhart, Gubb, and Leclerc (2011)
Transferrin	• Iron binding, and transport from the gut • Ameliorating oxidative damage (Stress response?) • Iron deprivation from pathogens • Antimicrobial properties • Vitellogenesis	Wang, Kim, et al. (2009), Wang, Leclerque, Pava-Ripoll, Fang, and St. Leger (2009), Geiser and Winzerling (2012)

AMP, antimicrobial peptide; *LPS*, lipopolysaccharides; *LTA*, lipoteichoic acid; *RNS*, reactive nitrogen species; *ROS*, reactive oxygen species.

AMPs are synthesized and secreted by hemocytes, barrier epithelia, the fat body, and the reproductive organs of insects (Nehme et al., 2007). Not only do AMPs form membranous pores on microbes, they can also translocate intracellularly and inhibit cell wall, nucleic acid, and protein synthesis along with the disruption of cytoplasmic membrane integrity (Brogden, 2005). Insects possess species-specific families of AMPs, for example, the European honey bee, *Apis mellifera*, produces a family of AMPs known as jelleines (Fontana et al., 2004).

Surprisingly, relatively few studies have focused on antifungal peptides in insects, when compared with antibacterials (see Lu & St. Leger, 2016). According to Faruck, Yusof, and Chowdhury (2015), there are only eight recognized antifungal peptides characterized from insects (some are listed in Table 3). Fungi present in the hemocoel of *D. melanogaster* elicit the expression of drosomycin and metchnikowin via the insect Toll pathway (Kurata, Ariki, & Kawabata, 2006). Moreover, drosomycin was induced in *D. melanogaster* exposed to conidia of pathogenic strains of *B. bassiana* or *M. anisopliae*, but not to conidia of nonpathogenic strains of *P. fumoroseus* and *A. fumigatus* (Lemaitre et al., 1997).

Only modest information exists on the impact of AMPs on EPF with most antifungal studies focusing on non-EPF (Table 3). EPF have evolved mechanisms to resist, tolerate or ameliorate (eg, via proteolysis) the effects of AMPs (Vilcinskas, 2010). AMPs often lyze fungal hyphae by permeabilizing cell membranes. Heliomicin binds directly to fungal plasma membrane ceramides, such as glucosylceramides, while drosomycin interacts with sphingolipids leading to pore formation of susceptible fungi (Fehlbaum et al., 1994; Gao & Zhu, 2008).

Lipophorins are noncovalent, structurally adaptable, assemblages of proteins and lipids with hydrophobic cores (Wang, Sykes, et al., 2002; Wang, Typas, et al., 2002). Aside from protein metabolism and lipid transport, apolipophorin III is known to contribute to multiple aspects of insect innate immunity (reviewed by Zdybicka-Barabas & Cytrynska, 2013). Depending on the condition of an insect, either infected or healthy, concentrations of apolipophorin III in the hemolymph vary greatly (Zdybicka-Barabas & Cytrynska, 2011). Apolipoprotein III from *Bombyx mori* and *G. mellonella* is capable of binding to, and detoxifying, bacterial endotoxins (LPS) and lipoteichoic acids (Halwani, Niven, & Dunphy, 2000; Kato et al., 1994). Neutralization of LPS by lipophorin was linked to a reduction in mRNA abundance of the antibacterial peptide, cecropin. The lipid A moiety of LPS interacts directly with apolipophorin III, thus compromising its immunogenicity. In a thorough study carried out by Whitten, Tew, Lee, and Ratcliffe (2004), it was demonstrated that apolipophorin III from *G. mellonella* larvae participates in the detection of fungi (via β-1,3-glucan) and hemocyte-mediated encapsulation reactions. Rates of phagocytosis, hemocyte-directed production of oxygen volatiles, and nodule formation all seem to involve apolipophorin in some capacity (Niere et al., 1999; Whitten et al., 2004; Wiesner, Losen, Kopacek, Weise, & Gotz, 1997). Apolipophorins can adhere to a wide variety of yeasts, fungi, Gram-positive and Gram-negative bacteria. Adherence of *Galleria* apolipophorin III to microbial cell surfaces resulted in damage and membrane perturbations, assessed using confocal and atomic absorption microscopy. Incubation of microbes in the presence of apolipophorin III alone caused growth inhibition, metabolic disruption and vacuolization (Zdybicka-Barabas, Januszanis, Mak, & Cytrynska, 2011; Zdybicka-Barabas et al., 2012).

Recent observations suggest that insect bioactives produced in response to microbes tend to be more potent in combination. It is well known that Gram-positive bacteria are highly susceptible to the effects of lysozyme, whereas Gram-negative bacteria appear recalcitrant (Yu et al., 2002). That said, the muramidase activity and permeabilizing effects of *Galleria* lysozyme toward *Escherichia coli* were enhanced in the presence of apolipophorin III (Zdybicka-Barabas et al., 2013). The authors found no evidence to suggest that these proteins formed intermolecular complexes; therefore, they appear to work in synergy. Previous reports also state that bacteria incubated in the presence of lysozyme were more susceptible to the effects of AMPs: attacins, cecropins, insect defensins, and various *Galleria* (poly)peptides (Chalk, Townson, Natori, Desmond, & Ham, 1994; Cytrynska et al., 2001; Engstrom, Engstrom, Tao, Carlsson, & Bennich, 1984). Sowa-Jasilek et al. (2014) observed an increase in the antifungal activity of lysozyme toward *C. albicans* when co-incubated with *G. mellonella* anionic peptide 2. The production of immune effectors is a costly endeavor, therefore producing bioactives in lower concentrations, yet having them work in synergy has many advantages. Recently, in blowflies the synergistic interactions of AMP combinations have been demonstrated (Chernysh, Gordyam, & Suborova, 2015; Pöppel, Vogel, Wiesner, & Vilcinskas, 2015). This is most important as, to date, the properties of AMPs have been studied individually. Chernysh et al. (2015) showed that microorganisms failed to develop resistance to AMP complexes in vitro, while exposure to individual (conventional) antimicrobials led to resistance. Also in bumblebees (*Bombus pascuorum* and *Bombus terrestris*), the linear peptides hymenoptaecin and abaecin are more efficacious when produced in lower concentrations together, rather than individually at a much higher concentration (Rahnamaeian et al., 2015). *E. coli* cells were unaffected by 200 μM abaecin, but were highly susceptible to hymenoptaecin at concentrations >2 μM. Administering 160-fold less abaecin (~1.25) alongside hymenoptaecin led to a substantial increase in microbicidal activity. This was attributed to destabilization of the bacterial cell wall by hymenoptaecin, allowing abaecin to access the bacterial chaperone network regulator, DnaK. These results are supported by previous genetic analyses showing clearly that expression and release of both peptides occurs simultaneously in response to parasitic and microbial challenges (Erler, Popp, & Lattorff, 2011; Riddell, Sumner, Adams, & Mallon, 2011). In addition, numerous immune-related molecules probably act synergistically, and there may even be species-specific mechanisms. These antibacterial activities could be highly beneficial to EPF as they would exclude opportunistic saprophytes and the occurrences of sepsis both of which would be deleterious to the fungus.

A key example of a versatile immune protein is hemocyanin, a large, multi-subunit, copper-containing protein present within the hemolymph of invertebrates, which transports oxygen in a manner similar to vertebrate hemoglobin (Decker & van-Holde, 2010). Interestingly, hemocyanin-derived peptides have antifungal and antibacterial activity (Zhuang et al., 2015). Shrimp hemocyanin exhibits antifungal activity against *Fusarium* spp., *Botrytis cinerea*, *Neurospora crassa*, *Pythium ultimum*, and *C. albicans* (Destoumieux-Garzon et al., 2001; Lee, Lee, & Söderhäll, 2003). Since insects possess a gaseous tracheal system, the presence of hemocyanin was considered unnecessary. Hemocyanin has now been detected in several insect species including cockroaches (*Blaptica dubia*), stoneflies (*Perla marginata*), and grasshoppers (Chen et al., 2015; Hagner-Holler et al., 2004). Considering the well-established roles of hemocyanin in immune defenses of numerous noninsect invertebrates (reviewed by Coates & Nairn, 2014), a question remains: does hemocyanin play a role in insect immunity? Sequence analyses of insect hemocyanins highlight the presence of the six highly conserved, copper-binding, histidine residues found in all known invertebrate hemocyanins (Burmester, 2015). Therefore, it is possible that insect hemocyanin could oxidize phenolic substrates in a manner similar to its crustacean, chelicerate, and gastropod counterparts, thereby contributing to innate immunity. The role(s) of hemocyanin in insects requires substantial experimentation to evaluate its apparent physiological significance.

The glycoprotein transferrin is found in insect hemolymph and can be characterized by a metal-binding site with a particularly strong affinity for iron. Insect transferrin is similar to mammalian transferrin, both structurally and functionally (reviewed by Geiser & Winzerling, 2012). Hexapod transferrins seemingly transport iron from the gut to other tissues, such as the fat body (Nichol, Law, & Winzerling, 2002). In vivo studies suggest that mRNA abundance and transferrin synthesis correlates broadly to concentrations of iron injected directly into the hemolymph of many insect species, for example, *Bombus ignitus* (Wang, Kim, et al., 2009; Wang, Leclerque, et al., 2009). Recent genomic analyses suggest that there may be up to five transferrin-like proteins in insects.

Upon exposure to the entomopathogens, *B. bassiana* and *M. anisopliae*, *Tsf1* expression (the gene-encoding transferrin) was up-regulated in the fire ant *Solenopsis invicta* (Valles & Pereira, 2005), wax moth *G. mellonella* larvae (Dubovskiy, Whitten, Yaroslavtseva, et al., 2013), and the giant northern termite *Mastotermes darwiniensis* (Thompson, Crozier, & Crozier, 2003). Likewise, mycosis in *Apriona germari* and *Locusta migratoria* led to *Tsf1* expression and the de novo synthesis of transferrin (Lee et al., 2006; Wang et al., 2007). Using RNA interference to silence *Tsf1*, Kim et al. (2008) observed apoptotic cell death in the fat body and increases in iron and H_2O_2 concentrations in the hemolymph of beetle larvae (*Protaetia brevitarsis*) in response to fungal challenge. Excess iron and H_2O_2 in the hemolymph of insects is very dangerous, as ROS are likely to form through Fenton's/Harber—Weiss reactions and cause damage to tissues bathing in the hemolymph. Transcription and translation of transferrin are initiated in response to bacterial targets (Kim & Kim, 2010; Yoshiga et al., 1999), blood parasites (Dong et al., 2006), and viruses (Luplertlop et al., 2011), suggesting it is pivotal to insect immunity.

5.5 Stress Management

Both EPF and insects are exposed to a wide range of environmental stresses and have evolved strategies to mitigate their harmful effects which are often mediated by ROS (eg, H_2O_2 and OH^-) (Komarov et al., 2009; Komarov, Slepneva, Glupov, & Khramtsov, 2005; Nappi, Vass, Frey, & Carton, 1995; Slepneva, Komarov, Glupov, Serebrov, & Khramtsov, 2003). Some of the EPF stress responses are mentioned in Section 3.2 and reviewed by several researchers (Lovett & St. Leger, 2015; Rangel, Alder-Rangel, et al., 2015; Rangel, Braga, et al., 2015). Insect ROS plays important roles as antimicrobials and signaling molecules in response to EPF (Butt et al., 2013; Chen et al., 2013; Dubovskiy, Yaroslavtseva, Kryukov, Benkovskaya, & Glupov, 2013). ROS can be damaging to both the host and fungal pathogen leading to lipid peroxidation, DNA breaks, and protein degradation (Lyakhovich et al., 2006). To mitigate ROS damage, insects possess antioxidant and detoxifying enzymes, as outlined earlier.

Most insects infected with EPF upregulate antioxidant genes. Elevated activity of nonspecific esterases and GST were observed in the hemolymph and fat body of *L. migratoria* and *L. decemlineata* larvae during early stages of *M. anisopliae* infection (Dubovskiy et al., 2010, 2012). Catalase and peroxidase genes are upregulated in *Rhynchophorus ferrugineus* and *Periplaneta americana* during infection by *B. bassiana* and *Hirsutella thompsonii*, respectively (Chaurasia, Lone, & Gupta, 2016; Hussain, Rizwan-ul-Haq, Al-Ayedh, Ahmed, & Al-Jabr, 2015). Enhanced expression of hydroperoxide glutathione peroxidase was noted in the integuments of wax moth larvae exposed to EPF (Dubovskiy, Whitten, Yaroslavtseva, et al., 2013). To what extent the stress management apparatus limits EPF development has not been quantified but, independent of the pathogen, overexpression of catalase can extend the life span of insects (Orr & Sohal, 1994) and other animals (Schriner et al., 2005).

Arthropods contend with EPF and their noxious secretory products (see Section 6). Generally, antioxidant enzyme activity is much higher in insects exposed to fungal toxins. Responses tend to vary between insect species. For example, *Spodoptera litura* larvae cope with destruxin-induced oxidative stress by enhancing antioxidant enzyme activity and alter their metabolic use of glutathione and ascrobate (Sree & Padmaja, 2008a, 2008b), whereas *Plutella xylostella* up-regulate GST and cytochrome P450 (Han et al., 2013).

Host- directed repair mechanisms that act in response to damage caused by EPF pathogenesis or as a consequence of the harmful host-mediated immune responses (ie, collateral damage) are poorly understood (Chakrabarti, Liehl, Buchon, & Lemaitre, 2012; Sadd & Siva-Jothy, 2006; Wojda & Jakubowicz, 2007). Heat shock proteins (HSPs) have been mostly associated with the management of a wide range of stress factors (King & MacRae, 2015). Most HPI studies show elevated expression of stress management genes (eg, HSP 90) in response to *B. bassiana* and *M. anisopliae* infection (Dubovskiy, Whitten, Yaroslavtseva, et al., 2013). However, exposure to destruxin A results in down-regulation of one 19.5 kDa HSP and three 70 kDa HSPs, but significant up-regulation of a 20.4 kDa HSP in *B. mori* (Gong, Chen, Liu, Jin, & Hu, 2014), suggesting that the response will depend on the stress factor and the host. Thiol peroxidase expression is up-regulated in *A. aegypti* larvae exposed to proteases produced by conidia of *M. anisopliae*;

however, HSP70-regulated caspase activity leads to cell death and ultimately larval mortality (Butt et al., 2013).

6. FUNGAL STRATEGIES TO EVADE AND/OR TOLERATE THE HOST'S IMMUNE RESPONSE

EPF have evolved disparate strategies to negate or minimize the impact of the hosts immune defenses, including the repression of proteases that activate phenoloxidase (PO), shedding immunogenic surface carbohydrates, and collagen deposition in order to avoid immune stimulation, secreting immune modulators, and the ability to tolerate host AMPs.

During the invasion phase, production of many CDEs is repressed in nutrient-rich environments (Screen, Bailey, Charnley, Cooper, & Clarkson, 1997). These enzymes are produced again immediately before emergence from the insect corpse (Leao et al., 2015; Small & Bidochka, 2005). Most notable is Pr1 subtilisin protease, as this can activate POs. Overexpression of Pr1 leads to melanization of the hemolymph, which is catastrophic for both the host and pathogen (St. Leger et al., 1996). It is not in the interest of the pathogen to evoke responses which impede its growth since the success of the pathogen is dependent on generating biomass that it can convert to conidia. Equally, the host wants to prevent activation of enzymes that are regulated strictly to prevent self-harm. Insects employ different safety mechanisms such as protease inhibitors (serpins) which may be involved with various developmental processes as well as influencing PO activation (Kanost, 1999; Vilcinskas, 2010).

The surface carbohydrate profile of EPF depends on the fungal species and culture conditions. Monoclonal antibodies (MAbs) generated against epitopes on yeast-like hyphal bodies and hyphae of *N. rileyi* cross-react with antigens on blastospores and hyphal surfaces of EPF (*B. bassiana*, *Paecilomyces farinosus*), but not non-EPF (*C. albicans*, *Saccharomyces cerevisiae*) (Pendland & Boucias, 1998). Specific *N. rileyi* MAbs also cross-react with immunocompetent granular hemocytes from *Spodoptera exigua* and *Trichoplusia ni* larvae and with *S. exigua* plasmatocytes suggesting that fungal surface epitopes can mimic host surface molecules. This may explain why *N. rileyi* hyphal bodies are able to circulate freely in the hemolymph, seemingly unrecognized by granular cells (Pendland & Boucias, 1998).

M. anisopliae secretes a collagen-like immune evasion protein, MCL1, which is produced within 20 min of the pathogen contacting hemolymph (Wang & St. Leger, 2006). Ordinarily, hemocytes will recognize and ingest

M. anisopliae conidia, but not blastospores/hyphal bodies. Disrupting MCL1 increases attack of hyphal bodies by hemocytes and reduces virulence (Wang & St. Leger, 2006). Studies with staining reagents and hemocyte monolayers showed that MCL1 functions as an antiadhesive protective coat because it masks antigenic cell wall components (β-glucans). Blastospores produced in liquid media stain poorly with the vital fluorochrome calcofluor which binds to β-glucans.

EPF secrete a wide range of metabolites, at least in vitro (Gibson, Donzelli, Krasnoff, & Keyhani, 2014; Molnár, Gibson, & Krasnoff, 2010). The most notable metabolites of *Beauveria* species include bassianin, bassiacridin, bassianolid, tenellin, and oosporein, whereas *Metarhizium* species produce cyclosporine, swainsonine, and 39 congeners of the cyclic peptide destruxin (Gibson et al., 2014; Molnár et al., 2010; Wang et al., 2010; see Ortiz-Urquizza & Keyhani, 2016; Donzelli & Krasnoff, 2016). Some of these compounds are detected in vivo and have been linked with virulence and specificity (Amiri-Besheli, Khambay, Cameron, Deadman, & Butt, 2000; Kershaw, Moorhouse, Bateman, Reynolds, & Charnley, 1999). Disruption of the bassianolide synthetase gene showed that this toxin was a highly significant virulence factor against *G. mellonella*, *S. exigua*, and *Helicoverpa zea* (Xu et al., 2009). However, neither beauvericin nor tenellin appear to contribute to *B. bassiana* virulence against lepidopteran larvae (Eley et al., 2007; Xu et al., 2008). Similarly, disruption of destruxin synthesis did not greatly alter *M. roberstii* virulence (Donzelli, Krasnoff, Sun-Moon, Churchill, & Gibson, 2012).

Some bioactive metabolites are immune modulators and have been implicated in suppressing the host immune response (Pal, Leger, & Wu, 2007; Vey, Matha, & Dumas, 2002). Modulating the immune system is critical for most EPF; however, any activity that is injurious to the host could result in opportunistic microbes initiating sepsis and displacement of the EPF. Immune modulators would limit the impact of the hosts defenses and allow the fungus to colonize and generate biomass with relative ease leading to the production of copious conidia needed for dispersal and survival. Interestingly, several leading, human immune suppressant drugs were developed from EPF bioactives including cyclosporine and myriocin (El Enshasy, 2010; Molnár et al., 2010). Suppressing the host immune defenses facilitates energy savings that EPF can invest in growth and colonization rather than fighting the host.

EPF secrete bioactives, which exhibit antimicrobial and antifungal activity in vitro (Ravindran, Chitra, Wilson, & Sivaramakrishnan, 2014; Roy,

Brown, Rothery, Ware, & Majerus, 2008). It is possible these compounds are produced in vivo to exclude opportunistic saprophytic bacteria and fungi. As stated earlier, the host's immune response peaks during the first 48 h after which it declines, by this time the fungus will have generated enough biomass to produce significant quantities of antimicrobials. Most often, the quantities of biaoctives recovered in vivo are extremely low; therefore, it is possible that these compounds work synergistically in a similar manner to host AMPs (Singh & Kaur, 2014; see Section 5.4). Some EPF-secreted compounds are multifunctional. For example, hydroxyfungerins produced by *Metarhizium* species have both antibiotic and insecticidal activity (Uchida et al., 2005). Similarly, myriocin produced by *Isaria sinclairii*, functions as an antibiotic and immune suppressant (de Melo et al., 2013). Producing compounds with dual or multiple functions ensures the efficient colonization of the host.

7. USING KNOWLEDGE OF HPI IN PEST CONTROL PROGRAMS

Knowledge of HPI has and will continue to contribute significantly to pest control programs. Some aspects of HPI are receiving much attention, such as (1) strain improvement, (2) increasing host susceptibility to pathogens, (3) monitoring resistance to EPF, (4) host responses to EPF volatiles, and (5) risk assessment.

7.1 Strain Improvement

Commercial companies want strains that kill quickly, have a wide host range, and function in diverse environmental conditions. Traditional approaches to improving EPF efficacy were based on physiological manipulation of the fungus during production or development of better formulation and application strategies (Butt et al., 2001). Indeed, progress continues to be made in these areas. For example, new soaps, oils, and diatomaceous earths have been developed to remove or abrade the epicuticular waxes resulting in dehydration and death or mitigating fungistasis and improving infection even under conditions normally considered to be hostile to EPF (Luz, Rodrigues, & Rocha, 2012; Steenberg & Kilpinen, 2014). Knowledge of specificity facilitates use of nonhosts (eg, bees) to vector/carry EPF for efficacious pest control (Mommaerts & Smagghe, 2011).

Genetic engineering is also being used to improve promising candidate strains (Ortiz-Urquiza, Luo, & Keyhani, 2015; St. Leger & Wang, 2010;

reviewed by Zhao, Lovett, & Fang, 2016). Many EPF-derived genes have been identified that could be exploited to improve virulence, specificity, and ecological fitness (see Table 1). Incorporation of Pr1 and chitinase genes was used to increase the virulence of several EPF species (Fan et al., 2007; St. Leger et al., 1996), whereas insertion of esterase and HSP25 genes increased specificity and stress tolerance, respectively (Liao et al., 2014; Wang, Fang, Wang, & St. Leger, 2011). An engineered *B. bassiana* strain overexpressing both Pr1-like protease BbCDPE1 and chitinase Bbchit1 increased virulence and reduced the number of conidia needed to kill by 67% (Fan et al., 2007). EPF strains expressing venom genes from scorpions and spiders are proving to be highly potent against a wide range of pest species (Pava-Ripoll, Posada, Momen, Wang, & St. Leger, 2008; Peng & Xia, 2015; Xie et al., 2015). Wang & St. Leger (2007) showed that *M. anisopliae* expression of the scorpion toxin increased fungal toxicity 22-fold against *M. sexta* larvae and 9-fold against adult *A. aegypti*. Overexpressing virulence/venom genes reduces the time it takes to kill as well as application rates that have cost benefits for end users. Furthermore, improving stress management in EPF could improve shelf life or field persistence, which are major obstacles in their commercialization.

7.2 Efficacy-Enhancing Agents

Certain agents are known to enhance the efficacy of EPF in pest control programs. These efficacy-enhancing agents (EEAs) include botanicals, other microbes (*Bacillus thuringiensis*, *Bt*), entomopathogenic nematodes (EPN), and low-dose insecticides (Ansari, Shah, Tirry, & Moens, 2006; Ansari, Shah, & Butt, 2010; Kruykov et al., 2009; Shapiro-Ilan et al., 2014). The combinations are often additive or synergistic. The latter allows for both agents to be used at considerably reduced application rates (1−10% recommended rate), thus making significant savings for end users. Little or no use of insecticides means extremely low (<90−99%) or no residues, which allows for continuous cropping thereby generating more income for growers. This is in addition to the environmental and health benefits (especially reduced operator exposure).

Synergies have been observed in the control of disparate pest orders (eg, Diptera, Orthoptera, Hemiptera, and Coleoptera) with various mechanisms outlined to explain this phenomenon. Some workers report that the low-dose insecticides/botanicals create an unfavorable environment that increases insect movement resulting in enhanced acquisition of conidia and thereby faster and higher mortality. EEAs also elicit changes in the epicuticular waxes resulting in improved adhesion and germination and, consequently, higher mortality (Shah, Ansari, Prasad, & Butt, 2007; Shah, Gaffney, Ansari, Prasad, & Butt, 2008). EEAs can also weaken the host's immune system (Hiromori & Nishigaki, 2001). Larvae of *Anomola cuprea* exposed to synergistic combinations of insecticides (fenitrothion and teflubenzuron) and *M. anisopliae* had lower numbers of granular cells and reduced PO activity (Hiromori & Nishigaki, 2001). Chemical pesticides also cause oxidative stress (James & Xu, 2012) weakening the insect and leading to enhanced susceptibility to EPF.

Synergies have been observed between EPF and *Bt* in the control of Colorado potato beetle and other pests (Gao, Oppert, Lord, Liu, & Lei, 2012; Kryukov et al., 2009; Wraight & Ramos, 2005). The enhanced efficacy of *B. bassiana* and *M. anisopliae* was due to a weakening of the host's immune system by *Bt* (Yaroslavtseva, Kryukov, & Dubovskiy, unpublished data). Synergies have also been reported between EPF and EPN (Ansari, Shah, & Butt, 2008; Ansari et al., 2010). The symbiotic bacteria released by EPN suppress the host's immune system and produce proteases that indirectly target AMPs (Caldas, Cherqui, Pereira, & Simoes, 2002; Ji & Kim, 2004). Debilitating the host's defenses would enable the EPF to colonize with relative ease and explains why the host is killed quickly even though these biocontrol agents are used at reduced rates (Ansari et al., 2008). Park and Kim (2011) found that benzylideneacetone produced by *Xenorhabdus nematophila*, a bacterial symbiont of EPN, has immunosuppressive effects on both cellular and humoral responses of *S. exigua* to *B. bassiana* infection.

Studies of HPI could also help identify biochemical/molecular markers which could be used to monitor stress in insects and help accelerate identification of synergistic combinations of controlling agents. Post-identification of synergistic combinations could lead to more effective pest control, reducing application rates and costs to end users.

7.3 Monitoring Resistance

One of the frequently asked questions is "will insects develop resistance to EPF?" Dubovskiy, Whitten, Yaroslavtseva, et al. (2013) investigated this by comparing the defense and stress genes of selected (tolerant) and nonselected wax moth larvae. These workers observed that the selected line prioritized and reallocated pathogen species—specific augmentations to the integument, the primary barrier to EPF. The response to *B. bassiana* infection included enhanced expression of IMPI and cuticular POs, as well as stress-management factors (eg, antioxidants). The response was specific to *B. bassiana* as it did not impede *M. anisopliae* (Dubovskiy, Whitten, Yaroslavtseva, et al., 2013). Identification of host genes for use as markers to monitor resistance will be of immense value in future pest control programs. Concomitantly, it will be important to monitor any trade-offs, such as a change in fecundity.

7.4 EPF Volatile Organic Compounds (VOCs): New Pest Control Products

Insects (eg, anthocorid bug, coccinellids, mole crickets, parasitoid wasps, and termites) and mites (*Phytoseiulus*) are able to detect and avoid EPF belonging to the genera *Metarhizium* and *Beauveria* (Mburu et al., 2009; Meyling & Pell, 2006; Ormond, Thomas, Pell, Freeman, & Roy, 2011; Rannback, Cotes, Anderson, Rämert, & Meyling, 2015; Seiedy, Saboori, & Zahedi-Golpayegani, 2013; Thompson & Brandenburg, 2005). The arthropods are clearly responding to EPF VOCs, as they avoid soil or leaves treated with conidia of these pathogens. The fact that some beneficial predators and parasitoids avoid EPF suggests that the risk posed by EPF to beneficials will be greatly reduced and that they can be used to develop highly effective IPM programs. Such data is also valuable for registration purposes, as risk assessment is a major hurdle when commercializing EPF. However, the opportunity exists to develop the VOCs as pesticides. This approach avoids the need to treat large areas and thereby reduces costs and minimizes risks to nontarget invertebrates. At present, very few VOCs have been identified and, of the few studies where they have been characterized, the activity is due to a blend and not a single compound (Mburu et al., 2013). One of the major pests to target would be termites which cause billions of dollars of damage to structures and crops each year (Verma, Sharma, & Prasad, 2009). Knowledge of the repellent compounds could be used to formulate or genetically modify EPF so the repellent VOCs are masked or suppressed so target pests do not avoid the pathogen. An understanding of EPF VOCs will be critical in IPM programs to ensure repellents do not negate "lure and kill" strategies that use attractant semiochemicals (eg, pheromones and kairomones) to lure the target pest to the control agent.

7.5 Risk Assessment

EPF are far more specific in their host range than conventional chemical pesticides, but still evidence needs to be provided that EPF will not harm nontarget invertebrates such as pollinators, arthropod predators and parasitoids, and aquatic arthropods. Knowledge of HPI will not only provide much insight into factors determining specificity but also identify biochemical–molecular markers to determine if EPF pose a risk to nontarget invertebrates. As outlined earlier, EPF conidia and their metabolites induce stress in host insects, but there is growing evidence that they can also induce stress in nontarget invertebrates (Garrido-Jurado et al., 2015). Stress can lead to apoptosis, host death, or predispose the host to infections by opportunistic saprophytes (Butt et al., 2013). Caspases are proving excellent indicators as their levels are often elevated within a few hours of been exposed to EPF (Garrido-Jurado et al., 2015). They are also indicators of a wide range of stress factors (eg, insecticides, pollutants, pathogens, and UV radiation) linked with apoptosis and other immune responses (Accorsi, Zibaee, & Malagoli, 2015; Cooper, Granville, & Lowenberger, 2009).

Screening EPF against a range of invertebrates not only provides information on risks, but also reveals much about their plasticity and potential utility against a wider range of pests or developmental stages than first realized. For example, EPF are usually used to control adult and larval stages of pests, and have also been shown to have ovicidal activity or influence fecundity in red spider mites and mosquitoes (Rocha et al., 2015; Wekesa, Knapp, Maniania, & Boga, 2006). EPF can prevent maturation of egg masses of nonhosts such as the snail *Biomphalaria glabrata* (Duarte, Rodrigues, Fernandes, Humber, & Luz, 2015). More surprising is the tolerance of predatory mites to EPF allowing these to be used together for the efficacious control of thrips and red spider mites (Jacobson, Chandler, Fenlon, & Russell, 2001; Seiedy, Saboori, Allahyari, Talaei-Hassanloui, & Tork, 2012). These studies not only demonstrate the need to discriminate between perceived and actual risk, but also show that EPF could be used against pests that would have been overlooked due to speculative assumptions.

ACKNOWLEDGMENTS

T.M.B. was supported by a grant funded jointly by the Biotechnology and Biological Sciences Research Council, the Department for Environment, Food and Rural affairs, the Economic and Social Research Council, the Forestry Commission, the Natural Environment Research Council and the Scottish Government, under the Tree Health and Plant Biosecurity Initiative. I.M.D. gratefully acknowledges funding from the Russian Science Foundation (Grant Number 15-14-10014) and RFBR (Grant Number 15-34-20488 mol_a_ved).

REFERENCES

Accorsi, A., Zibaee, A., & Malagoli, D. (2015). The multifaceted activity of insect caspases. *Journal of Insect Physiology, 76*, 17–23.

Altincicek, B., Stotzel, S., Wygrecka, M., Preissner, K. T., & Vilcinskas, A. (2008). Host-derived extracellular nucleic acids enhance innate immune responses, induce coagulation, and prolong survival upon infection in insects. *Journal of Immunology, 181*, 2705–2712.

Altre, J. A., & Vandenberg, J. D. (2001). Penetration of cuticle and proliferation in hemolymph by *Paecilomyces fumosoroseus* isolates that differ in virulence against *Lepidopteran* larvae. *Journal of Invertebrate Pathology, 78*, 81–86.

Amiri-Besheli, B., Khambay, B., Cameron, S., Deadman, M. L., & Butt, T. M. (2000). Inter- and intra-specific variation in destruxin production by insect pathogenic *Metarhizium* spp., and its significance to pathogenesis. *Mycological Research, 104*(04), 447–452.

Andersen, S. O. (2010). Insect cuticular sclerotization: a review. *Insect Biochemistry and Molecular Biology, 40*(3), 166–178.

Ansari, M. A., Shah, F. A., & Butt, T. M. (2008). Combined use of entomopathogenic nematodes and *Metarhizium anisopliae* as a new approach for black vine weevil, *Otiorhynchus sulcatus* (Coleoptera: Curculionidae) control. *Entomologia Experimentalis et Applicata, 129*, 340–347.

Ansari, M. A., Shah, F. A., & Butt, T. M. (2010). The entomopathogenic nematode *Steinernema kraussei* and *Metarhizium anisopliae* work synergistically in controlling overwintering larvae of the black vine weevil, *Otiorhynchus sulcatus*, in strawberries growbags. *Biocontrol Science and Technology, 20*, 99–105.

Ansari, M. A., Shah, F. A., Tirry, L., & Moens, M. (2006). Field trials against *Hoplia philanthus* (Coleoptera: Scarabaeidae) with a combination of an entomopathogenic nematode and the fungus *Metarhizium anisopliae* CLO 53. *Biological Control, 39*(3), 453–459.

Atsumi, A., & Saito, T. (2015). Volatiles from wasabi inhibit entomopathogenic fungi: implications for tritrophic interactions and biological control. *Journal of Plant Interactions, 10*(1), 152–157.

Bagga, S., Hu, G., Screen, S. E., & Leger, R. J. S. (2004). Reconstructing the diversification of subtilisins in the pathogenic fungus *Metarhizium anisopliae. Gene, 324*, 159–169.

Baratto, U. M., da Silva, M. V., Santi, L., Passaglia, L., Schrank, I. S., Vainstein, M. H., & Schrank, A. (2003). Expression and characterization of the 42 kDa chitinase of the biocontrol fungus *Metarhizium anisopliae* in *Escherichia coli. Canadian Journal of Microbiology, 49*(11), 723–726.

Barelli, L., Padilla-Guerrero, I. E., & Bidochka, M. J. (2011). Differential expression of insect and plant specific adhesin genes, Mad1 and Mad2, in *Metarhizium robertsii. Fungal Biology, 115*(11), 1174–1185.

Baverstock, J., Roy, H. E., & Pell, J. K. (2010). Entomopathogenic fungi and insect behaviour: from unsuspecting hosts to targeted vectors. In *The ecology of fungal entomopathogens* (pp. 89–102). Netherlands: Springer.

Beauvais, A., Latgé, J. P., Vey, A., & Prevost, M. C. (1989). The role of surface components of the entomopathogenic fungus *Entomophaga aulicae* in the cellular immune response of *Galleria mellonella* (Lepidoptera). *Journal of General Microbiology, 135*(3), 489–498.

Behie, S. W., & Bidochka, M. J. (2014). Ubiquity of insect-derived nitrogen transfer to plants by endophytic insect-pathogenic fungi: an additional branch of the soil nitrogen cycle. *Applied and Environmental Microbiology, 80*(5), 1553–1560.

Behie, S. W., Jones, S. J., & Bidochka, M. J. (2015). Plant tissue localization of the endophytic insect pathogenic fungi *Metarhizium* and *Beauveria. Fungal Ecology, 13*, 112–119.

Bidla, G., Dushay, M. S., & Theopold, U. (2007). Crystal cell rupture after injury in *Drosophila* requires the JNK pathway, small GTPases and the TNF homolog Eiger. *Journal of Cell Science, 120*, 1209–1215.

Bidla, G., Hauling, T., Dushay, M. S., & Theopold, U. (2009). Activation of insect phenoloxidase after injury: endogenous versus foreign elicitors. *Journal of Innate Immunity, 1*, 301–308.

Bidochka, M. J., Clark, D. C., Lewis, M. W., & Keyhani, N. O. (2010). Could insect phagocytic avoidance by entomogenous fungi have evolved via selection against soil amoeboid predators? *Microbiology, 156*, 2164–2171.
Bidochka, M. J., De Koning, J., & St. Leger, R. J. (2001). Analysis of a genomic clone of hydrophobin (ssgA) from the entomopathogenic fungus *Metarhizium anisopliae. Mycological Research, 105*, 360–364.
Blanford, S., & Thomas, M. B. (1999). Host thermal biology: the key to understanding host-pathogen interactions and microbial pest control? *Agriculture and Forest Entomology, 1*, 195–202.
Boguś, M. I., Kędra, E., Bania, J., Szczepanik, M., Czygier, M., Jabłoński, M. P.... Polanowski, A. (2007). Different defense trategies of *Dendrolimus pini, Galleria mellonella*, and *Calliphora vicina* against fungal infection. *Journal of Insect Physiology, 53*(9), 909–922.
Boldo, J. T., Junges, A., Do Amaral, K. B., Staats, C. C., Vainstein, M. H., & Schrank, A. (2009). Endochitinase CHI2 of the biocontrol fungus *Metarhizium anisopliae* affects its virulence toward the cotton stainer bug *Dysdercus peruvianus. Current Genetics, 55*(5), 551–560.
Brey, P. T., Lee, W. J., Yamakawa, M., Koizumi, Y., Perrot, S., François, M., & Ashida, M. (1993). Role of the integument in insect immunity: epicuticular abrasion and induction of cecropin synthesis in cuticular epithelial cells. *Proceedings of the National Academy of Sciences of the United States of America, 90*, 6275–6279.
Broetto, L., Da Silva, W. O. B., Bailão, A. M., Soares, C. D. A., Vainstein, M. H., & Schrank, A. (2010). Glyceraldehyde-3-phosphate dehydrogenase of the entomopathogenic fungus *Metarhizium anisopliae*: cell-surface localization and role in host adhesion. *FEMS Microbiology Letters, 312*(2), 101–109.
Brogden, K. A. (2005). Antimicrobial peptides: pore formers or metabolic inhibitors in bacteria? *Nature Reviews Microbiology, 3*(3), 238–250.
Brown, S. E., Howard, A., Kasprzak, A. B., Gordon, K. H., & East, P. D. (2008). The discovery and analysis of a diverged family of novel antifungal moricin-like peptides in the wax moth *Galleria mellonella. Insect Biochemistry and Molecular Biology, 38*(2), 201–212.
Browne, N., Surlis, C., & Kavanagh, K. (2014). Thermal and physical stresses induce a short-term immune priming effect in *Galleria mellonella* larvae. *Journal of Insect Physiology, 63*, 21–26.
Bulet, P., & Stocklin, R. (2005). Insect antimicrobial peptides: structures, properties and gene regulation. *Protein and Peptide Letters, 12*(1), 3–11.
Bulmer, M. S., Bachelet, I., Raman, R., Rosengaus, R. B., & Sasisekharan, R. (2009). Targeting an antimicrobial effector function in insect immunity as a pest control strategy. *Proceedings of the National Academy of Sciences of the United States of America, 106*, 12652–12657.
Bulmer, M. S., Denier, D., Velenovsky, J., & Hamilton, C. (2012). A common antifungal defense strategy in *Cryptocercus* woodroaches and termites. *Insectes Sociaux, 59*(4), 469–478.
Burmester, T. (2015). Evolution of respiratory proteins across the Pancrustacea. *Integrative and Comparative Biology*, 1–10. http://dx.doi.org/10.1093/icb/icv079.
Butt, T. M., Beckett, A., & Wilding, N. (1981). Protoplasts in the in vivo life cycle of *Erynia neoaphidis. Journal of General Microbiology, 127*(2), 417–421.
Butt, T. M., & Goettel, M. (2000). Bioassays of entomogenous fungi. In A. Navon, & K. R. S. Ascher (Eds.), *Bioassays of entomopathogenic microbes and nematodes* (pp. 141–195). Wallingford, Oxon, U.K: CAB International.
Butt, T. M., Greenfield, B. P. J., Greig, C., Maffeis, T. G. G., Taylor, J. W. D., Piasecka, J.... Eastwood, D. C. (2013). *Metarhizium anisopliae* pathogenesis of mosquito larvae: a verdict of accidental death. *PLoS One, 8*(12), e81686.

Butt, T. M., Hoch, H. C., Staples, R. C., & St. Leger, R. J. (1989). Use of fluorochromes in the study of fungal cytology and differentiation. *Experimental Mycology, 13*(4), 303–320.

Butt, T. M., Ibrahim, L., Clark, S. J., & Beckett, A. (1995). The germination behaviour of *Metarhizium anisopliae* on the surface of aphid and flea beetle cuticles. *Mycological Research, 99*(8), 945–950.

Butt, T. M., Jackson, C. W., & Magan, N. (2001). Introduction-fungal biological control agents: progress, problems and potential. In T. M. Butt, C. W. Jackson, & N. Magan (Eds.), *Fungi as biocontrol agents: Progress, problems and potential*. Wallingford, UK: CABI Publishing.

Butt, T. M., Wraight, S. P., Galaini-Wraight, S., Humber, R. A., Roberts, D. W., & Soper, R. S. (1988). Humoral encapsulation of the fungus *Erynia radicans* (Entomophthorales) by the potato leafhopper, *Empoasca fabae* (Homoptera: Cicadellidae). *Journal of Invertebrate Pathology, 52*(1), 49–56.

Caldas, C., Cherqui, A., Pereira, A., & Simoes, N. (2002). Purification and characterization of an extracellular protease from *Xenorhabdus nematophila* involved in insect immunosuppression. *Applied and Environmental Microbiology, 68*(3), 1297–1304.

Cerenius, L., Kawabata, S., Lee, B. L., Nonaka, M., & Söderhäll, K. (2010). Proteolytic cascades and their involvement in invertebrate immunity. *Trends in Biochemical Sciences, 35*(10), 575–583.

Cerenius, L., Lee, B. L., & Söderhäll, K. (2008). The proPO-system: pros and cons for its role in invertebrate immunity. *Trends in Immunology, 29*, 263–271.

Chakrabarti, S., Liehl, P., Buchon, N., & Lemaitre, B. (2012). Infection-induced host translational blockage inhibits immune responses and epithelial renewal in the *Drosophila* gut. *Cell Host and Microbe, 12*, 60–70.

Chalk, R., Townson, H., Natori, S., Desmond, H., & Ham, P. J. (1994). Purification of an insect defensin from the mosquito, *Aedes aegypti*. *Insect Biochemistry and Molecular Biology, 24*, 403–410.

Chaurasia, A., Lone, Y., & Gupta, U. S. (2016). Effect of entomopathogenic fungi, *Hirsutella thompsonii* on mortality and detoxification enzyme activity in *Periplaneta americana*. *Journal of Entomology and Zoology Studies, 4*(1), 234–239.

Chen, B., Ma, R., Ma, G., Guo, X., Tong, X., Tang, G., & Kang, L. (2015). Haemocyanin is essential for embryonic development and survival in the migratory locust. *Insect Molecular Biology, 24*, 517–527.

Chen, X., Fu, S., Zhang, P., Gu, Z., Liu, J., Qian, Q., & Ma, B. (2013). Proteomic analysis of a disease-resistance enhanced lesion mimic mutant spotted leaf 5 in rice. *Rice, 6*, 1–10. http://www.thericejournal.com/content/6/1/1.

Chernysh, S., Gordyam, N., & Suborova, T. (2015). Insect antimicrobial peptide complexes prevent resistance development in bacteria. *PLoS One, 10*(7), e0130788. http://dx.doi.org/10.1371/journal.pone.0130788.

Cho, E. M., Kirkland, B. H., Holder, D. J., & Keyhani, N. O. (2007). Phage display cDNA cloning and expression analysis of hydrophobins from the entomopathogenic fungus *Beauveria (Cordyceps) bassiana*. *Microbiology, 153*(10), 3438–3447.

Coates, C. J., Kelly, S. M., & Nairn, J. (2011). Possible role of phosphatidylserine-hemocyanin interaction in the innate immune response of *Limulus polyphemus*. *Developmental and Comparative Immunology, 35*, 155–163.

Coates, C. J., & Nairn, J. (2014). Diverse immune functions of hemocyanins. *Developmental and Comparative Immunology, 45*, 43–55.

Coates, C. J., Whalley, T., Wyman, M., & Nairn, J. (2013). A putative link between phagocytosis-induced apoptosis and hemocyanin-derived phenoloxidase activation. *Apoptosis, 18*, 1319–1331.

Cooper, D. M., Granville, D. J., & Lowenberger, C. (2009). The insect caspases. *Apoptosis, 14*(3), 247–256.

Cory, J. S., & Hoover, K. (2006). Plant-mediated effects in insect–pathogen interactions. *Trends in Ecology and Evolution, 21*, 278–286. http://www.ncbi.nlm.nih.gov/entrez/query.fcgi?cmd=Retrieve&db=PubMed&dopt=Abstract&list_uids=16697914.
Cytryńska, M., Zdybicka-Barabas, A., Jabłoński, P., & Jakubowicz, T. (2001). Detection of antibacterial polypeptide activity in situ after sodium dodecyl sulphate polyacrylamide gel electrophoresis. *Analytical Biochemistry, 299*, 274–276.
De Lucca, A. J., Bland, J. M., Jacks, T. J., Grimm, C., Cleveland, T. E., & Walsh, T. J. (1997). Fungicidal activity of Cecropin A. *Antimicrobial Agents and Chemotherapy, 41*(2), 481–483.
De Lucca, A. J., Bland, J. M., Jacks, T. J., Grimm, C., & Walsh, T. J. (1998). Fungicidal and binding properties of the natural peptides cecropin B and dermaseptin. *Medical Mycology, 36*(5), 291–298.
Decker, H., & van-Holde, K. E. (2010). *Oxygen and the evolution of life*. New York: Springer.
Destoumieux-Garzon, D., Saulnier, D., Garnier, J., Jouffrey, C., Bulet, P., & Bachere, E. (2001). Crustacean immunity-antifungal peptides are generated from the C-terminus of shrimp hemocyanin in response to microbial challenge. *Journal of Biological Chemistry, 276*, 47070–47077.
Dong, Y., Aguilar, R., Xi, Z., Warr, E., Mongin, E., & Dimopolous, G. (2006). *Anopheles gambiae* immune responses to human and rodent *Plasmodium* parasite species. *PLoS Pathogens, 2*, e52.
Dong, Y., Morton, J. C., Jr., Ramirez, J. L., Souza-Neto, J. A., & Dimopoulos, G. (2012). The entomopathogenic fungus *Beauveria bassiana* activate toll and JAK-STAT pathway-controlled effector genes and anti-dengue activity in *Aedes aegypti*. *Insect Biochemistry and Molecular Biology, 42*, 126–132.
Donzelli, B. G. G., Krasnoff, S. B., Sun-Moon, Y., Churchill, A. C., & Gibson, D. M. (2012). Genetic basis of destruxin production in the entomopathogen *Metarhizium robertsii*. *Current Genetics, 58*(2), 105–116.
Donzelli, B. G. G., & Krasnoff, S. B. (2016). Molecular Genetics of Secondary Chemistry in Metarhizium Fungi. *Advances in Genetics, 94*, 365–436.
Dromph, K. M., & Vestergaard, S. (2002). Pathogenicity and attractiveness of entomopathogenic hyphomycete fungi to collembolans. *Applied Soil Ecology, 21*(3), 197–210.
Duarte, G. F., Rodrigues, J., Fernandes, É. K., Humber, R. A., & Luz, C. (2015). New insights into the amphibious life of *Biomphalaria glabrata* and susceptibility of its egg masses to fungal infection. *Journal of Invertebrate Pathology, 125*, 31–36.
Dubovskiy, I. M., Kryukov, V. Yu, Benkovskaya, G. V., Yaroslavtseva, O. N., Surina, E. V., & Glupov, V. V. (2010). Activity of the detoxificative enzyme system and encapsulation rate in the Colorado potato beetle *Leptinotarsa decemlineata* (Say) larvae under organophosphorus insecticide treatment and entomopathogenic fungus *Metharizium anisopliae* (Metsch.) infection. *Euroasian Entomological Journal, 9*(4), 577–582.
Dubovskiy, I. M., Slyamova, N. D., Kryukov, V. Yu, Yaroslavtseva, O. N., Levchenko, M. V., Belgibaeva, A. B. … Glupov, V. V. (2012). The activity of nonspecific esterases and glutathione-S-transferase in *Locusta migratoria* larvae infected with the fungus *Metarhizium anisopliae* (Ascomycota, Hypocreales). *Entomological Review, 92*(1), 27–32.
Dubovskiy, I. M., Whitten, M. M. A., Kryukov, V. Y., Yaroslavtseva, O. N., Grizanova, E. V., Greig, C. … Butt, T. M. (2013). More than a colour change: insect melanism, disease resistance and fecundity. *Proceedings of the Royal Society B: Biological Sciences, 280*, 20130584.
Dubovskiy, I. M., Whitten, M. M. A., Yaroslavtseva, O. N., Greig, C., Kryukov, V. Y., Grizanova, E. V. … Butt, T. M. (2013). Can insects develop resistance to insect pathogenic fungi? *PLoS One, 8*(4), e60248.
Dubovskiy, I. M., Yaroslavtseva, O., Kryukov, V., Benkovskaya, G., & Glupov, V. (2013). An increase in the immune system activity of the wax moth *Galleria mellonella* and of the

colorado potato beetle *Leptinotarsa decemlineata* under effect of organophosphorus insecticide. *Journal of Evolutionary Biochemistry and Physiology, 49*(6), 592–596.
Duo-Chuan, L. (2006). Review of fungal chitinases. *Mycopathologia, 161*(6), 345–360.
Ekengren, S., & Hultmark, D. (1999). Drosophila cecropin as an antifungal agent. *Insect Biochemistry and Molecular Biology, 29*(11), 965–972.
El Enshasy, H. (2010). Immunomodulation. In M. Hoffrichter (Ed.), *The mycota: Industrial applications* (2nd ed., pp. 165–194). Springer Verlag.
Eley, K. L., Halo, L. M., Song, Z., Powles, H., Cox, R. J., Bailey, A. M. … Simpson, T. J. (2007). Biosynthesis of the 2-Pyridone tenellin in the insect pathogenic fungus *Beauveria bassiana. ChemBioChem, 8*, 289–297.
Engstrom, A., Engstrom, P., Tao, Z. J., Carlsson, A., & Bennich, H. (1984). Insect immunity. The primary structure of the antibacterial protein attacin F and its relation to two native attacins from *Hyalophora cecropia. EMBO Journal, 3*, 2065–2070.
Enríquez-Vara, J. N., Guzmán-Franco1, A. W., Alatorre-Rosas, R., González-Hernández, H., Córdoba-Aguilar, A., & Contreras-Garduño, J. (2014). Immune response of *Phyllophaga polyphylla* larvae is not an effective barrier against *Metarhizium pingshaense. Invertebrate Survival Journal, 11*, 240–246.
Erler, S., Popp, M., & Lattorff, H. M. G. (2011). Dynamics of immune system gene expression upon bacterial challenge and wounding in a social insect (*Bombus terrestris*). *PLoS One, 6*, e18126.
Fallon, J., Troy, N., & Kavanagh, K. (2011). Pre-exposure of *Galleria mellonella* larvae to different doses of *Aspergillus fumigatus* conidia causes differential activation of cellular and humoral immune responses. *Virulence, 2*, 413–421.
Fan, Y., Fang, W., Guo, S., Pei, X., Zhang, Y., Xiao, Y. … Pei, Y. (2007). Increased insect virulence in *Beauveria bassiana* strains overexpressing an engineered chitinase. *Applied Environmental Microbiology, 73*(1), 295–302.
Fang, W., Pava-ripoll, M., Wang, S., & St. Leger, R. J. (2009). Protein kinase A regulates production of virulence determinants by the entomopathogenic fungus, *Metarhizium anisopliae. Fungal Genetics and Biology, 46*(3), 277–285.
Fang, W., Pei, Y., & Bidochka, M. J. (2007). A regulator of a G protein signalling (RGS) gene, cag8, from the insect-pathogenic fungus *Metarhizium anisopliae* is involved in conidiation, virulence and hydrophobin synthesis. *Microbiology, 153*(Pt 4), 1017–1025.
Fang, W. G., Feng, J., Fan, Y. H., Zhang, Y. J., Bidochka, M. J., Leger, R. J. S., & Pei, Y. (2009). Expressing a fusion protein with protease and chitinase activities increases the virulence of the insect pathogen *Beauveria bassiana. Journal of Invertebrate Pathology, 102*, 155–159.
de Faria, M. R., & Wraight, S.,P. (2007). Mycoinsecticides and Mycoacaricides: a comprehensive list with worldwide coverage and international classification of formulation types. *Biological Control, 43*, 237–256.
Faruck, M. O., Yusof, F., & Chowdhury, S. (2015). An overview of antifungal peptides derived from insect. *Peptides*. http://dx.doi.org/10.1016/j.peptides.2015.06.001.
Fehlbaum, P., Bulet, P., Michaut, L., Lagueux, M., Broekaert, W. F., Hetru, C., & Hoffmann, J. A. (1994). Insect immunity. Septic injury of *Drosophila* induces the synthesis of a potent antifungal peptide with sequence homology to plant antifungal peptides. *Journal of Biological Chemistry, 269*, 33159–33163.
Felton, G. W., & Summers, C. B. (1995). Antioxidant systems in insects. *Archives of Insect Biochemistry and Physiology, 29*, 187–197.
Fontana, R., Mendes, M. A., de Souza, B. M., Konno, K., César, L. M. M., Malaspina, O., & Palma, M. S. (2004). Jelleines: a family of antimicrobial peptides from the Royal Jelly of honeybees (*Apis mellifera*). *Peptides, 25*, 919–928.

Gao, B., & Zhu, S. Y. (2008). Differential potency of drosomycin to *Neurospora crassa* and its mutant: implications for evolutionary relationship between defensins from insects and plants. *Insect Molecular Biology, 17*, 405–411.
Gao, Q., Jin, K., Yingm, S.-H., Zhang, Y., Xiao, G., Shang, Y.... Wang, C. (2011). Genome sequencing and comparative transcriptomics of the model entomopathogenic fungi *Metarhizium anisopliae* and *M. acridum*. *PLoS Genetics*, 7(1), e1001264. http://dx.doi.org/10.1371/journal.pgen.1001264.
Gao, Q., Shang, Y., Huang, W., & Wang, C. (2013). Glycerol-3-phosphate acyltransferase contributes to triacylglycerol biosynthesis, lipid droplet formation, and host invasion in *Metarhizium robertsii*. *Applied Environmental Microbiology, 79*(24), 7646–7653.
Gao, Y., Oppert, B., Lord, J. C., Liu, C., & Lei, Z. (2012). *Bacillus thuringiensis* Cry3Aa toxin increases the susceptibility of *Crioceris quatuordecimpunctata* to *Beauveria bassiana* infection. *Journal of Invertebrate Pathology, 109*, 260–263.
Garrido-Jurado, I., Alkhaibari, A., Williams, S. R., Oatley-Radcliffe, D. L., Quesada-Moraga, E., & Butt, T. M. (2015). Toxicity testing of *Metarhizium* conidia and toxins against aquatic invertebrates. *Journal of Pest Science*. http://dx.doi.org/10.1007/s10340-015-0700-0.
Geiser, D. L., & Winzerling, J. J. (2012). Insect transferrins: multifunctional proteins. *Biochimica et Biophysica Acta, 1820*, 437–451.
George, J., Jenkins, N. E., Blanford, S., Thomas, M. B., & Baker, T. C. (2013). Malaria mosquitoes attracted by fatal fungus. *PLoS One, 8*(5), e62632.
Gibson, D. M., Donzelli, B. G., Krasnoff, S. B., & Keyhani, N. O. (2014). Discovering the secondary metabolite potential encoded within entomopathogenic fungi. *Natural Product Reports, 31*(10), 1287–1305.
Golebiowski, M., Malinski, E., Bogus, M. I., Kumirska, J., & Stepnowski, P. (2008). The cuticular fatty acids of *Calliphora vicina, Dendrolimus pini* and *Galleria mellonella* larvae and their role in resistance to fungal infection. *Insect Biochemistry and Molecular Biology, 38*(6), 619–627.
Gong, L., Chen, X., Liu, C., Jin, F., & Hu, Q. (2014). Gene expression profile of *Bombyx mori* hemocyte under the stress of destruxin A. *PLoS One, 9*(5), e96170.
Gotz, P., & Vey, A. (1974). Humoral encapsulation in Diptera (Insecta): defence reactions of *Chironomus* larvae against fungi. *Parasitology, 68*, 193–205.
Greenfield, B. P., Lord, A. M., Dudley, E., & Butt, T. M. (2014). Conidia of the insect pathogenic fungus, *Metarhizium anisopliae*, fail to adhere to mosquito larval cuticle. *Royal Society Open Science*. http://dx.doi.org/10.1098/rsos. 140193.
Gross, J., Schumacher, K., Schmidtberg, H., & Vilcinskas, A. (2008). Protected by fumigants: beetle perfumes in antimicrobial defense. *Journal of Chemical Ecology, 34*, 179–188.
Gunnarsson, S. G. S. (1988). Infection of *Schistocerca gregaria* by the fungus *Metarhizium anisopliae*: cellular reactions in the integument studied by scanning electron and light microscopy. *Journal of Invertebrate Pathology, 52*(1), 9–17.
Hagner-Holler, S., Schoen, A., Erker, W., Marden, J. H., Rupprecht, R., Decker, H., & Burmester, T. (2004). A respiratory hemocyanin in insects. *Proceedings of the National Academy of Sciences of the United States of America, 101*, 871–874.
Hallsworth, J. E., & Magan, N. (1996). Culture age, temperature, and pH affect the polyol and trehalose contents of fungal propagules. *Applied and Environmental Microbiology, 62*(7), 2435–2442.
Halwani, A. E., Niven, D. F., & Dunphy, G. B. (2000). Apolipophorin III and the interactions of lipoteichoic acids with immediate immune responses of *Galleria mellonella*. *Journal of Invertebrate Pathology, 76*, 233–241.
Hamilton, C., & Bulmer, M. S. (2012). Molecular antifungal defenses in subterranean termites: RNA interference reveals in vivo roles of termicins and GNBPs against a naturally encountered pathogen. *Developmental & Comparative Immunology, 36*(2), 372–377.

Han, P. F., Jin, F. L., Dong, X. L., Fan, J. Q., Qiu, B. L., Qiu, B., & Baek, S. R. K.-H. (2013). Transcript and protein profiling analysis of the destruxin a-induced response in larvae of *Plutella xylostella*. *PLoS One, 8*, e60771.

Hemingway, J., & Ranson, H. (2000). Insecticide resistance in insect vectors of human disease. *Annual Review of Entomology, 45*(1), 371–391.

Hiromori, H., & Nishigaki, J. (2001). Factor analysis of synergistic effect between the entomopathogenic fungus *Metarhizium anisopliae* and synthetic insecticides. *Applied Entomology and Zoology, 36*(2), 231–236.

Holder, D. J., & Keyhani, N. O. (2005). Adhesion of the entomopathogenic fungus *Beauveria (Cordyceps) bassiana* to substrata. *Appied and Environmental Microbiology, 2005*(71), 5260–5266.

Holder, D. J., Kirkland, B. H., Lewis, M. W., & Keyhani, N. O. (2007). Surface characteristics of the entomopathogenic fungus *Beauveria (Cordyceps) bassiana*. *Microbiology, 153*, 3448–3457.

Huang, W., Shang, Y., Chen, P., Gao, Q., & Wang, C. (2015). MrpacC regulates sporulation, insect cuticle penetration and immune evasion in *Metarhizium robertsii*. *Environmental Microbioliology, 17*(4), 994–1008.

Hussain, A., Rizwan-ul-Haq, M., Al-Ayedh, H., Ahmed, S., & Al-Jabr, A. M. (2015). Effect of *Beauveria bassiana* infection on the feeding performance and ntioxidant defence of red palm weevil, *Rhynchophorus ferrugineus*. *BioControl*, 1–11.

Inyang, E. N., Butt, T. M., Beckett, A., & Archer, S. (1999). The effect of crucifer epicuticular waxes and leaf extracts on the germination and virulence of *Metarhizium anisopliae* conidia. *Mycological Research, 103*(4), 419–426.

Inyang, E. N., Butt, T. M., Doughty, K. J., Todd, A. D., & Archer, S. (1999). The effects of isothiocyanates on the growth of the entomopathogenic fungus *Metarhizium anisopliae* and its infection of the mustard beetle. *Mycological Research, 103*(08), 974–980.

Jacobson, R. J., Chandler, D., Fenlon, J., & Russell, K. M. (2001). Compatibility of *Beauveria bassiana* (balsamo) vuillemin with *Amblyseius cucumeris oudemans* (acarina: Phytoseiidae) to control *Frankliniella occidentalis pergande* (thysanoptera: Thripidae) on cucumber plants. *Biocontrol Science and Technology, 11*(3), 391–400.

James, R. R., & Xu, J. (2012). Mechanisms by which pesticides affect insect immunity. *Journal of Invertebrate Pathology, 109*(2), 175–182.

Ji, D., & Kim, Y. (2004). An entomopathogenic bacterium, *Xenorhabdus nematophila*, inhibits the expression of an antibacterial peptide, cecropin, of the beet armyworm, *Spodoptera exigua*. *Journal of Insect Physiology, 50*(6), 489–496.

Jiang, H., Vilcinskas, A., & Kanost, M. R. (2010). Immunity in lepidopteran insects. In *Invertebrate immunity* (pp. 181–204). US: Springer.

Jin, K., Ming, Y., & Xia, Y. X. (2012). MaHog1, a Hog1-type mitogen-activated protein kinase gene, contributes to stress tolerance and virulence of the entomopathogenic fungus *Metarhizium acridum*. *Microbiology, 158*(12), 2987–2996.

Kang, S. C., Park, S., & Lee, D. G. (1999). Purification and characterization of a novel chitinase from the entomopathogenic fungus, *Metarhizium anisopliae*. *Journal of Invertebrate Pathology, 73*, 276–281. http://dx.doi.org/10.1006/jipa.1999.4843.

Kanost, M. R. (1999). Serine proteinase inhibitors in arthropod immunity. *Developmental and Comparative Immunology, 23*, 291–301.

Kato, Y., Motoi, Y., Taniai, K., Kadono-Okuda, K., Yamamoto, M., Higashino, Y. … Yamakawa, M. (1994). Lipopolysaccharide-lipophorin complex formation in insect hemolymph: a common pathway of lipopolysaccharide detoxification both in insects and mammals. *Insect Biochemistry and Molecular Biology, 24*, 547–555.

Kershaw, M. J., Moorhouse, E. R., Bateman, R. P., Reynolds, S. E., & Charnley, A. K. (1999). The role of destruxins in the pathogenicity of *Metarhizium anisopliae* for three species of insect. *Journal Invertebrate Pathology, 74*, 213–223.

Kim, B. Y., Lee, K. S., Choo, Y. M., Kim, I., Je, Y. H., Woo, S. D. ... Jin, B. R. (2008). Insect transferrin functions as an antioxidant protein in a beetle larva. *Comparative Biochemistry and Physiology B Biochemistry and Molecular Biology, 150*(2), 161–169. http://dx.doi.org/10.1016/j.cbpb.2008.02.009. Epub March 4, 2008.

Kim, J., & Kim, Y. (2010). A viral histone H4 supresses expression of a transferrin that plays a role in the immune response of the diamondback moth, *Plutella xylostella. Insect Molecular Biology, 19*, 567–574.

King, A. M., & MacRae, T. H. (2015). Insect heat shock proteins during stress and diapause. *Annual Review of Entomology, 60*, 59–75.

Koidsumi, K. (1957). Antifungal action of cuticular lipids in insects. *Journal of Insect Physiology, 1*, 40–51.

Komarov, D. A., Ryazanova, A. D., Slepneva, I. A., Khramtsov, V. V., Dubovskiy, I. M., & Glupov, V. V. (2009). Pathogen-targeted hydroxyl radical generation during melanization in insect hemolymph: EPR study of a probable cytotoxicity mechanism. *Applied Magnetic Resonance, 35*, 495–501.

Komarov, D. A., Slepneva, I. A., Glupov, V. V., & Khramtsov, V. V. (2005). Superoxide and hydrogen peroxide formation during enzymatic oxidation of DOPA by phenoloxidase. *Free Radical Research, 39*, 853–858.

Kryukov, V. Y., Khodyrev, V. P., Yaroslavtseva, O. N., Kamenova, A. S., Duisembekov, B. A., & Glupov, V. V. (2009). Synergistic action of entomopathogenic hyphomycetes and the bacteria *Bacillus thuringiensis* ssp. morrisoni in the infection of Colorado potato beetle *Leptinotarsa decemlineata. Applied Biochemistry and Microbiology, 45*(5), 511–516.

Kurata, S., Ariki, S., & Kawabata, S. (2006). Recognition of pathogens and activation of immune responses in *Drosophila* and horseshoe crab innate immunity. *Immunobiology, 211*, 237–249.

Lacey, L. A., Grzywacz, D., Shapiro-Ilan, D. I., Frutos, R., Brownbridge, M., & Goettel, M. S. (2015). Insect pathogens as biological control agents: back to the future. *Journal of Invertebrate Pathology, 132*, 1–41.

Lamberty, M., Ades, S., Uttenweiler-Joseph, S., Brookhart, G., Bushey, D., Hoffmann, J. A., & Bulet, P. (1999). Insect immunity. Isolation from the lepidopteran *Heliothis virescens* of a novel insect defensin with potent antifungal activity. *Journal of Biogical Chemistry, 274*, 9320–9326.

Lamberty, M., Zachary, D., Lanot, R., Bordereau, C., Robert, A., Hoffmann, J. A., & Bulet, P. (2001). Insect immunity – constitutive expression of a cysteine-rich antifungal and a linear antibacterial peptide in a termite insect. *Journal of Biological Chemistry, 276*(6), 4085–4092.

Lavine, M. D., & Strand, M. R. (2002). Insect hemocytes and their role in immunity. *Insect Biochemistry and Molecular Biology, 32*, 1295–1309.

Leão, M. P. C., Tiago, P. V., Andreote, F. D., de Araújo, W. L., & de Oliveira, N. T. (2015). Differential expression of the pr1A gene in *Metarhizium anisopliae* and *Metarhizium acridum* across different culture conditions and during pathogenesis. *Genetics and Molecular Biology, 38*(1), 86–92.

Lee, K. S., Kim, B. Y., Kim, H. J., Seo, S. J., Yoon, H. J., Choi, Y. S. ... Jin, B. R. (2006). Transferrin inhibits stress-associated apoptosis in a beetle. *Free Radical Biology and Medicine, 41*, 1151–1161.

Lee, S. Y., Lee, B. L., & Söderhäll, K. (2003). Processing of an antimicrobial peptide from hemocyanin of the freshwater crayfish *Pacifastacus leniusculus. Journal of Biological Chemistry, 278*, 7927–7933.

Lee, Y. S., Yun, E. K., Jang, W. S., Kim, I., Lee, J. H., Park, S. Y. ... Lee, I. H. (2004). Purfication, cDNA cloning and expression of an insect defensin from the great wax moth, *Galleria mellonella. Insect Moleclar Biology, 13*, 65–72.

Lemaitre, B., Nicolas, E., Michaut, L., Reichhart, J., & Hoffmann, J. (1996). The dorsoventral regulatory gene cassette spatzle/Toll/cactus controls the potent antifungal response in *Drosophila* adults. *Cell, 86*, 973–983.

Lemaitre, B., Reichhart, J. M., & Hoffmann, J. A. (1997). *Drosophila* host defense: differential induction of antimicrobial peptide genes after infection by various classes of microorganisms. *Proceedings of the National Academy of Sciences of the United States of America, 94*(26), 14614–14619.

Levashina, E. A., Ohresser, S., Bulet, P., Reichhart, J.-M., Hetru, C., & Hoffmann, J. A. (1995). Metchnikowin, a novel immune-inducible proline-rich peptide from *Drosophila* with antibacterial and antifungal properties. *European Journal of Biochemistry, 233*, 694–700.

Li, J., Ying, S. H., Shan, L. T., & Feng, M. G. (2010). A new non-hydrophobic cell wall protein (CWP10) of *Metarhizium anisopliae* enhances conidial hydrophobicity when expressed in *Beauveria bassiana*. *Applied Microbiology and Biotechnology, 85*(4), 975–984.

Liao, X., Lu, H. L., Fang, W., & Leger, R. J. S. (2014). Overexpression of a *Metarhizium robertsii* HSP25 gene increases thermotolerance and survival in soil. *Applied Microbiology and Biotechnology, 98*(2), 777–783.

Lin, L. C., Fang, W. G., Liao, X. G., Wang, F. Q., Wei, D. Z., & Leger, R. J. S. (2011). The MrCYP52 cytochrome P450 monoxygenase gene of *Metarhizium robertsii* is important for utilizing insect epicuticular hydrocarbons. *PLoS One, 6*(12).

Liu, W., Xie, Y., Dong, J., Xue, J., Zhang, Y., Lu, Y., & Wu, J. (2014). Pathogenicity of three entomopathogenic fungi to *Matsucoccus matsumurae*. *PLoS One, 9*(7), e103350. http://dx.doi.org/10.1371/journal.pone.0103350.

Lopes, R. B., Laumann, R. A., Blassioli-Moraes, M. C., Borges, M., & Faria, M. (2015). The fungistatic and fungicidal effects of volatiles from metathoracic glands of soybean-attacking stink bugs (Heteroptera: Pentatomidae) on the entomopathogen *Beauveria bassiana*. *Journal of Invertebrate Pathology, 132*, 77–85.

Lord, J. C., & Howard, R. W. (2004). A proposed role for the cuticular fatty amides of *Liposcelis bostrychophila* (Psocoptera: Liposcelidae) in preventing adhesion of entomopathogenic fungi with dry-conidia. *Mycopathologia, 158*, 211–217.

Lovett, B., & St. Leger, R. J. (2015). Stress is the rule rather than the exception for *Metarhizium*. *Current Genetics, 61*, 253–261.

Lu, H.-L., & St. Leger, R. J. (2016). Insect Immunity to Entomopathogenic Fungi. *Advances in Genetics, 94*, 251–286.

Luo, X., Keyhani, N. O., Yu, X., He, Z., Luo, Z., Pei, Y., & Zhang, Y. (2012). The MAP kinase Bbslt2 controls growth, conidiation, cell wall integrity, and virulence in the insect pathogenic fungus *Beauveria bassiana*. *Fungal Genetics and Biology, 49*(7), 544–555.

Luplertlop, N., Surasombatpattana, P., Patramool, S., Dumas, E., Wasinpiyamongkol, L., Saune, L. … Missé, D. (2011). Induction of a peptide with activity against a broad spectrum of pathogens in the *Aedes aegypti* salivary gland, following infection with dengue virus. *PLoS Pathogens, 7*, e1001252.

Luz, C., Rodrigues, J., & Rocha, L. F. N. (2012). Diatomaceous earth and oil enhance effectiveness of *Metarhizium anisopliae* against *Triatoma infestans*. *Acta Tropica, 122*, 29–35.

Lyakhovich, V. V., Vavilin, V. A., Zenkov, N. K., & Menshchikova, E. B. (2006). Active defense under oxidative stress. The antioxidant responsive element. *Biochemistry (Moscow), 71*, 962–974.

Mascarin, G. M., Jackson, M. A., Kobori, N. N., Behle, R. W., Dunlap, C. A., & Júnior, Í. D. (2015). Glucose concentration alters dissolved oxygen levels in liquid cultures of *Beauveria bassiana* and affects formation and bioefficacy of blastospores. *Applied Microbiology and Biotechnology*, 1–13.

Mascarin, G. M., Jackson, M. A., Kobori, N. N., Behle, R. W., & Júnior, Í. D. (2015). Liquid culture fermentation for rapid production of desiccation tolerant blastospores of *Beauveria bassiana* and *Isaria fumosorosea* strains. *Journal of Invertebrate Pathology, 127*, 11–20.

Mburu, D. M., Maniania, N. K., & Hassanali, A. (2013). Comparison of volatile blends and nucleotide sequences of two *Beauveria bassiana* isolates of different virulence and repellency towards the termite *Macrotermes michaelseni*. *Journal of Chemical Ecology, 39*, 101–108. http://dx.doi.org/10.1007/s10886-012-0207-6.
Mburu, D. M., Ndung'u, M. W., Maniania, N. K., & Hassanali, A. (2011). Comparison of volatile blends and gene sequences of two isolates of *Metarhizium anisopliae* of different virulence and repellency toward the termite *Macrotermes michaelseni*. *The Journal of Experimental Biology, 214*(6), 956–962.
Mburu, D. M., Ochola, L., Maniania, N. K., Njagi, P. G. N., Gitonga, L. M., Ndung'u, M. W.... Hassanali, A. (2009). Relationship between virulence and repellency of entomopathogenic isolates of *Metarhizium anisopliae* and *Beauveria bassiana* to the termite *Macrotermes michaelseni*. *Journal of Insect Physiology, 55*(9), 774–780.
de Melo, N. R., Abdrahman, A., Greig, C., Mukherjee, K., Thornton, C., Ratcliffe, N. A.... Butt, T. M. (2013). Myriocin significantly increases the mortality of a non-mammalian model host during *Candida* pathogenesis. *PLoS One, 8*, e78905. http://dx.doi.org/10.1371/journal.pone.0078905.
Ment, D., Gindin, G., Rot, A., Eshel, D., Teper-Bamnolker, P., Ben-Ze'ev, I.... Samish, M. (2013). Role of cuticular lipids and water-soluble compounds in tick susceptibility to *Metarhizium* infection. *Biocontrol Science and Technology, 23*(8), 956–967.
Meyling, N. V., & Pell, J. K. (2006). Detection and avoidance of an entomopathogenic fungus by a generalist insect predator. *Ecological Entomology, 31*(2), 162–171.
Miller, J. S., Nguyen, T., & Stanley-Samuelson, D. W. (1994). Eicosanoids mediate insect nodulation in response to bacterial infection. *Proceedings of the National Academy of Sciences of the United States of America, 91*, 12418–12422.
Moonjely, S., Barrelli, R., & Bidochka, M. J. (2016). Insect Pathogenic Fungi as Endophytes. *Advances in Genetics, 94*, 107–136.
Molnár, I., Gibson, D. M., & Krasnoff, S. B. (2010). Secondary metabolites from entomopathogenic hypocrealean fungi. *Natural Product Reports, 27*(9), 1241–1275.
Mommaerts, V., & Smagghe, G. (2011). Entomovectoring in plant protection. *Arthropod-Plant Interactions, 5*(2), 81–95.
Mowlds, P., Coates, C., Renwick, J., & Kavanagh, K. (2010). Dose-dependent cellular and humoral responses in *Galleria mellonella* larvae following β-glucan inoculation. *Microbes and Infection, 12*, 146–153.
Nappi, A. J., & Christensen, B. M. (2005). Melanogenesis and associated cytotoxic reactions: applications to insect innate immunity. *Insect Biochemistry and Molecular Biology, 35*, 443–459.
Nappi, A. J., Vass, E., Frey, F., & Carton, Y. (1995). Superoxide anion generation in *Drosophila* during melanotic encapsulation of parasites. *European Journal of Cell Biology, 68*, 450–456.
Nehme, N. T., Liégeois, S., Kele, B., Giammarinaro, P., Pradel, E., Hoffmann, J. A.... Dominique Ferrandon, D. (2007). A model of bacterial intestinal infections in *Drosophila melanogaster*. *PLoS Pathogens, 3*, e173. http://dx.doi.org/10.1371/journal.ppat.0030173.
Nichol, H., Law, J. H., & Winzerling, J. J. (2002). Iron metabolism in insects. *Annual Review of Entomology, 47*, 535–559.
Niere, M., Meisslitzer, C., Dettloff, M., Weise, C., Ziegler, M., & Wiesner, A. (1999). Insect immune activation by recombinant *Galleria mellonella* apolipophorin III. *Biochimica et Biophysica Acta, 1433*, 16–26.
Nunes, A. R. F., Martins, J. N., Furlaneto, M. C., & Barros, N. M. D. (2010). Production of cuticle-degrading proteases by *Nomuraea rileyi* and its virulence against *Anticarsia gemmatalis*. *Ciência Rural, 40*(9), 1853–1859.
Ormond, E. L., Thomas, A. P. M., Pell, J. K., Freeman, S. N., & Roy, H. E. (2011). Avoidance of a generalist entomopathogenic fungus by the ladybird, *Coccinella septempunctata*. *FEMS Microbiol Ecology, 77*, 229–237. http://dx.doi.org/10.1111/j.1574-6941.2011.01100.x.

Orr, W. C., & Sohal, R. S. (1994). Extension of life-span by overexpression of superoxide dismutase and catalase in *Drosophila melanogaster*. *Science, 263*(5150), 1128–1130.

Ortiz-Urquiza, A., & Keyhani, N. O. (2013). Action on the surface: entomopathogenic fungi versus the insect cuticle. *Insects, 4*, 357–374.

Ortiz-Urquiza, A., & Keyhani, N. O. (2014). Stress response signalling and virulence: insights from entomopathogenic fungi. *Current Genetics*, 1–11.

Ortiz-Urquiza, A., Luo, Z., & Keyhani, N. O. (2015). Improving mycoinsecticides for insect biological control. *Applied Microbiology and Biotechnology, 99*(3), 1057–1068.

Ortiz-Urquiza, A., & Keyhani, N. O. (2016). Molecular Genetics of Beauveria bassiana Infection of Insects. *Advances in Genetics, 94*, 165–250.

Pal, S., Leger, R. J. S., & Wu, L. P. (2007). Fungal peptide Destruxin A plays a specific role in suppressing the innate immune response in *Drosophila melanogaster*. *Journal of Biological Chemistry, 282*(12), 8969–8977.

Park, J., & Kim, Y. (2011). Benzylideneacetone suppresses both cellular and humoral immune responses of *Spodoptera exigua* and enhances fungal pathogenicity. *Journal of Asia-Pacific Entomology, 14*(4), 423–427.

Park, S. S., Shin, S. W., Park, D.-S., Oh, H. W., Boo, K. S., & Park, H.-Y. (1997). Protein purification and cDNA cloning of a cecropin-like peptide from the larvae of fall webworm (*Hyphantria cunea*). *Insect Biochemistry and Molecular Biology, 27*(8–9), 711–720.

Pava-Ripoll, M., Posada, F. J., Momen, B., Wang, C., & St. Leger, R. (2008). Increased pathogenicity against coffee berry borer, *Hypothenemus hampei* (Coleoptera: Curculionidae) by *Metarhizium anisopliae* expressing the scorpion toxin (AaIT) gene. *Journal Invertebrate Pathology, 99*, 220–226.

Pendland, J. C., & Boucias, D. G. (1993). Variations in the ability of galactose and mannose-specific lectins to bind to cell-wall surfaces during growth of the insect pathogenic fungus *Paecilomyces farinosus*. *European Journal of Cell Biology, 60*, 322–330.

Pendland, J. C., & Boucias, D. G. (1998). Characterization of monoclonal antibodies against cell wall epitopes of the insect pathogenic fungus, *Nomuraea rileyi*: differential binding to fungal surfaces and cross-reactivity with host hemocytes and basement membrane components. *European Journal of Cell Biology, 75*(2), 118–127.

Pendland, J. C., Hung, S. Y., & Boucias, D. G. (1993). Evasion of host defense by in vivo-produced protoplast-like cells of the insect mycopathogen *Beauveria bassiana*. *Journal of Bacteriology, 175*(18), 5962–5969.

Peng, G., & Xia, Y. (2015). Integration of an insecticidal scorpion toxin (BjalphaIT) gene into *Metarhizium acridum* enhances fungal virulence towards *Locusta migratoria manilensis*. *Pest Managment Science, 71*(1), 58–64.

Pick, C., Schneuer, M., & Burmester, T. (2010). Ontogeny of hemocyanin in the ovoviviparous cockroach *Blaptica dubia* suggests an embryo specific role in oxygen supply. *Journal of Insect Physiology, 56*, 455–460.

Pöppel, A.-K., Vogel, H., Wiesner, J., & Vilcinskas, A. (2015). Antimicrobial peptides expressed in medicinal maggots of the blow fly *Lucilia sericata* show combinatorial activity against bacteria. *Antimicrobial Agents and Chemotherapy, 59*, 2508–2514. http://dx.doi.org/10.1128/AAC.05180-14.

Price, C., & Ratcliffe, N. A. (1974). A reappraisal of insect hemocyte classification by examination of blood from fifteen insect orders. *Zeitschrift fur Zellforschung und Mikroskopische Anatomie, 147*, 537–549.

Qazi, S. S., & Khachatourians, G. G. (2007). Hydrated conidia of *Metarhizium anisopliae* release a family of metalloproteases. *Journal of Invertebrate Pathology, 95*(1), 48–59.

Rahnamaeian, R., Cytrynska, M., Zdybicka-Barabas, A., Dobslaff, K., Wiesner, J., Twyman, R. M. … Vilcinskas, A. (2015). Insect antimicrobial peptides show potentiating functional interactions against Gram-negative bacteria. *Proceedings of the Royal Society B, 282*, 2015.0293.

Rangel, D. E., Alder-Rangel, A., Dadachova, E., Finlay, R. D., Kupiec, M., Dijksterhuis, J. … Hallsworth, J. E. (August 2015). Fungal stress biology: a preface to the fungal stress responses special edition. *Current Genetics, 61*(3), 231–238. http://dx.doi.org/10.1007/s00294-015-0500-3. Epub June 27, 2015.

Rangel, D. E., Braga, G. U., Fernandes, É. K., Keyser, C. A., Hallsworth, J. E., & Roberts, D. W. (August 2015). Stress tolerance and virulence of insect-pathogenic fungi are determined by environmental conditions during conidial formation. *Current Genetics, 61*(3), 383–404. http://dx.doi.org/10.1007/s00294-015-0477-y. Epub March 20, 2015 1–22.

Rännbäck, L.-M., Cotes, B., Anderson, P., Rämert, B., & Meyling, N. V. (2015). Mortality risk from entomopathogenic fungi affects oviposition behavior in the parasitoid wasp *Trybliographa rapae*. *Journal of Invertebrate Pathology, 124*, 78–86.

Ratcliffe, N. A., & Gagen, S. J. (1977). Studies on the in vivo cellular reactions of insects: an ultrastructural analysis of nodule formation in *Galleria mellonella*. *Tissue & Cell, 9*(1), 73–85.

Ratcliffe, N. A., Rowley, A. F., Fitzgerald, S. W., & Rhodes, C. P. (1985). Invertebrate immunity, basic concepts and recent advances. *International Review of Cytology, 97*, 183–349.

Ravindran, K., Chitra, S., Wilson, A., & Sivaramakrishnan, S. (2014). Evaluation of antifungal activity of *Metarhizium anisopliae* against plant phytopathogenic fungi. In *Microbial diversity and biotechnology in food security* (pp. 251–255). India: Springer.

Reichhart, J. M., Gubb, D., & Leclerc, V. (2011). The *Drosophila* serpins: multiple functions in immunity and morphogenesis. *Methods in Enzymology, 499*, 205–225.

Riddell, C. E., Sumner, S., Adams, S., & Mallon, E. B. (2011). Pathways to immunity: temporal dynamics of the bumblebee *(Bombus terrestris)* immune response against a trypanosome gut parasite. *Insect Molecular Biology*, 529–540.

Robb, C. T., Dyrynda, E. A., Gray, R. D., Rossi, A. G., & Smith, V. J. (2014). Invertebrate extracellular phagocyte traps show that chromatin is an ancient defence weapon. *Nature Communications, 5*, 4627. http://dx.doi.org/10.1038/ncomms5627.

Rocha, L. F. N., Sousa, N. A., Rodrigues, J., Catão, A. M. L., Marques, C. S., Fernandes, É. K. K., & Luz, C. (2015). Efficacy of *Tolypocladium cylindrosporum* against *Aedes aegypti* eggs, larvae and adults. *Journal of Applied Microbiology, 119*, 1412–1419.

Roman, E., Arana, D. M., Nombela, C., Alonso-Monge, R., & Pla, J. (2007). MAP kinase pathways as regulators of fungal virulence. *Trends in Microbiology, 15*, 181–190.

Roseland, C. R., Grodowitz, M. J., Kramer, K. J., Hopkins, T. L., & Broce, A. B. (1985). Stabilization of mineralized and sclerotized puparial cuticle of muscid flies. *Insect Biochemistry, 15*(4), 521–528.

Rosengaus, R. R., Lefebvre, M. L., & Traniello, J. F. A. (2000). Inhibition of fungal spore germination by *Nasutitermes*: evidence for a possible antiseptic role of soldier defensive secretions. *Journal of Chemical Ecology, 26*, 21–39.

Rowley, A. F., & Powell, P. (2007). Invertebrate immune systems—specific, quasi-specific, or nonspecific? *The Journal of Immunology, 179*, 7209–7214.

Roy, H. E., Brown, P. M. J., Rothery, P., Ware, R. L., & Majerus, M. E. N. (2008). Interactions between the fungal pathogen *Beauveria bassiana* and three species of coccinellid: *Harmonia axyridis, Coccinella septempunctata* and *Adalia bipunctata*. *Biological Control, 53*, 265–276.

Sadd, B. M., & Siva-Jothy, M. T. (2006). Self-harm caused by an insect's innate immunity. *Proceedings of the Royal Society B: Biological Sciences, 273*, 2571–2574.

Saito, T., & Aoki, J. (1983). Toxicity of free fatty acids on the larval surfaces of 2 Lepidopterous insects towards *Beauveria bassiana* (Bals) Vuill and *Paecilomyces fumosoroseus* (Wize) Brown Et Smith (Deuteromycetes, Moniliales). *Applied Entomology and Zoology, 18*, 225–233.

Santi, L., da Silva, W. O. B., Berger, M., Guimarães, J. A., Schrank, A., & Vainstein, M. H. (2010). Conidial surface proteins of *Metarhizium anisopliae*: source of activities related with toxic effects, host penetration and pathogenesis. *Toxicon, 55*(4), 874–880.

Scheepmaker, J. W. A., & Butt, T. M. (2010). Natural and released inoculum levels of entomopathogenic fungal biocontrol agents in soil in relation to risk assessment and in accordance with EU regulations. *Biocontrol Science and Technology, 20*(5), 503–552.

Schrank, A., & Vainstein, M. H. (2010). *Metarhizium anisopliae* enzymes and toxins. *Toxicon, 56*(7), 1267–1274.

Schreiter, G., Butt, T. M., Beckett, A., Moritz, G., & Vestergaard, S. (1994). Invasion and development of *Verticillium lecanii* in the Western Flower Thrips, *Frankliniella occidentalis*. *Mycological Research, 98*, 1025–1034.

Schriner, S. E., Linford, N. J., Martin, G. M., Treuting, P., Ogburn, C. E., Emond, M. … Rabinovitch, P. S. (2005). Extension of murine life span by overexpression of catalase targeted to mitochondria. *Science, 308*(5730), 1909–1911.

Schumann, B., Seitz, V., Vilcinskas, A., & Podsiadlowski, L. (2003). Cloning and expression of gallerimycin, an antifungal peptide expressed in immune response of greater wax moth larvae, *Galleria mellonella*. *Archives of Insect Biochemistry and Physiology, 5*(3), 125–133.

Screen, S., Bailey, A., Charnley, K., Cooper, R., & Clarkson, J. (1997). Carbon regulation of the cuticle-degrading enzyme PR1 from *Metarhizium anisopliae* may involve a trans-acting DNA-binding protein CRR1, a functional equivalent of the *Aspergillus nidulans* CREA protein. *Current Genetics, 31*(6), 511–518.

Screen, S., Bailey, A. M., Charnley, K., Cooper, R., & Clarkson, J. (1998). Isolation of a nitrogen response regulator gene (nrr1) from *Metarhizium anisopliae*. *Gene, 221*(1), 17–24.

Seiedy, M., Saboori, A., Allahyari, H., Talaei-Hassanloui, R., & Tork, M. (2012). Functional response of *Phytoseiulus persimilis* (Acari: Phytoseiidae) on untreated and *Beauveria bassiana*-treated adults of *Tetranychus urticae* (Acari: Tetranychidae). *Journal of Insect Behavior, 25*(6), 543–553.

Seiedy, M., Saboori, A., & Zahedi-Golpayegani, A. (2013). Olfactory response of *Phytoseiulus persimilis* (Acari: Phytoseiidae) to untreated and *Beauveria bassiana*-treated adults of *Tetranychus urticae* (Acari: Tetranychidae) on cucumber plants. *Experimental and Applied Acarology, 60*, 219–227.

Sevim, A., Donzelli, B. G., Wu, D., Demirbag, Z., Gibson, D. M., & Turgeon, B. G. (2012). Hydrophobin genes of the entomopathogenic fungus, *Metarhizium brunneum*, are differentially expressed and corresponding mutants are decreased in virulence. *Current Genetics, 58*(2), 79–92.

Shah, F. A., Ansari, M. A., Prasad, M., & Butt, T. M. (2007). Evaluation of black vine weevil (*Otiorhynchus sulcatus*) control strategies using *Metarhizium anisopliae* with sublethal doses of insecticides in disparate horticultural growing media. *Biological Control, 40*(2), 246–252.

Shah, F. A., Gaffney, M., Ansari, M. A., Prasad, M., & Butt, T. M. (2008). Neem seed cake enhances the efficacy of the insect pathogenic fungus *Metarhizium anisopliae* for the control of black vine weevil, *Otiorhynuchs sulcatus* (Coleoptera: Curculionidae). *Biological Control, 44*(1), 111–115.

Shah, F. A., Wang, C. S., & Butt, T. M. (2005). Nutrition influences growth and virulence of the insect-pathogenic fungus *Metarhizium anisopliae*. *FEMS Microbiology Letters, 251*(2), 259–266.

Shang, Y., Chen, P., Chen, Y., Lu, Y., & Wang, C. (2015). MrSkn7 controls sporulation, cell wall integrity, autolysis, and virulence in *Metarhizium robertsii*. *Eukaryotic Cell, 14*(4), 396–405.

Shapiro-Ilan, D. I., Jackson, M., Reilly, C. C., & Hotchkiss, M. W. (2004). Effects of combining an entomopathogenic fungi or bacterium with entomopathogenic nematodes on mortality of *Curculio caryae* (Coleoptera: Curculionidae). *Biological Control, 30*(1), 119–126.

Singh, D., & Kaur, G. (2014). Production, HPLC analysis, and in situ apoptotic activities of swainsonine toward lepidopteran, Sf-21 cell line. *Biotechnology Progress, 30*(5), 1196–1205.

Slepneva, I. A., Komarov, D. A., Glupov, V. V., Serebrov, V. V., & Khramtsov, V. V. (2003). Influence of fungal infection on the DOPA-semiquinone and DOPA-quinone production in haemolymph of *Galleria mellonella* larvae. *Biochemical and Biophysical Research Communications, 300*, 188–191.

Small, C. L. N., & Bidochka, M. J. (2005). Up-regulation of Pr1, a subtilisin-like protease, during conidiation in the insect pathogen *Metarhizium anisopliae*. *Mycological Research, 109*(03), 307–313.

Smith, R. J., & Grula, E. A. (1982). Toxic components on the larval surface of the Corn-Earworm (*Heliothis zea*) and their effects on germination and growth of *Beauveria bassiana*. *Journal of Invertebrate Pathology, 39*, 15–22.

Smith, V. J. (2010). Immunology of invertebrates: cellular. In *Encyclopedia of life sciences*. Chichester: John Wiley & Sons, Ltd. http://dx.doi.org/10.1002/9780470015902.a0002344.pub2.

Söderhäll, K., & Ajaxon, R. (1982). Effect of quinones and melanin on mycelial growth of *Aphanomyces* spp. and extracellular protease of *Aphanomyces astaci*, a parasite on crayfish. *Journal of Invertebrate Pathology, 39*(1), 105–109.

Souza-Neto, J. A., Sim, S., & Dimopoulos, G. (2009). An evolutionary conserved function of the JAK-STAT pathway in anti-dengue defense. *Proceedings of the National Academy of Sciences of the United States of America, 106*, 17841–17846.

Sowa-Jasiłek, A., Zdybicka-Barabas, A., Staczek, S., Wydrych, J., Mak, P., Jakubowicz, T., & Cytrynska, M. (2014). Studies on the role of insect hemolymph polypeptides: *Galleria mellonella* anionic peptide 2 and lysozyme. *Peptides, 53*, 194–201.

Sree, K. S., & Padmaja, V. (2008a). Destruxin from *Metarhizium anisopliae* induces oxidative stress effecting larval mortality of the polyphagous pest *Spodoptera litura*. *Journal of Applied Entomology, 132*, 68–78.

Sree, K. S., & Padmaja, V. (2008b). Oxidative stress induced by destruxin from *Metarhizium anisopliae* (Metch.) involves changes in glutathione and ascorbate metabolism and instigates ultrastructural changes in the salivary glands of *Spodoptera litura* (Fab.) larvae. *Toxicon, 51*(7), 1140–1150.

Srisukchayakul, P., Wiwat, C., & Pantuwatana, S. (2005). Studies on the pathogenesis of the local isolates of *Nomuraea rileyi* against *Spodoptera litura*. *Science Asia, 31*, 273–276.

St. Leger, R. J., Bidochka, M. J., & Roberts, D. W. (1994). Isoforms of the cuticle-degrading Pr1 proteinase and production of a metalloproteinase by *Metarhizium-anisopliae*. *Archives of Biochemistry and Biophysics, 313*(1), 1–7.

St. Leger, R. J., Butt, T. M., Goettel, M. S., Staples, R., & Roberts, D. W. (1989). Production in vitro of appressoria by the entomopathogenic fungus *Metarhizium anisopliae*. *Experimental Mycology, 13*, 274–288.

St. Leger, R. J., Butt, T. M., Staples, R. C., & Roberts, D. W. (1989). Synthesis of proteins including a cuticle-degrading protease during differentiation of the entomopathogenic fungus *Metarhizium anisopliae*. *Experimental Mycology, 13*, 253–262.

St. Leger, R. J., Joshi, L., Bidochka, M. J., & Roberts, D. W. (1995). Protein-Synthesis in *Metarhizium anisopliae* growing on host cuticle. *Mycological Research, 99*, 1034–1040.

St. Leger, R. J., Joshi, L., Bidochka, M. J., & Roberts, D. W. (1996). Construction of an improved mycoinsecticide overexpressing a toxic protease. *Proceedings of the National Academy of Sciences of the United States of America, 93*(13), 6349–6354.

St. Leger, R. J., Joshi, L., & Roberts, D. (1998). Ambient pH is a major determinant in the expression of cuticle-degrading enzymes and hydrophobin by *Metarhizium anisopliae*. *Applied and Environmental Microbiology, 64*(2), 709–713.

St. Leger, R. J., Staples, R. C., & Roberts, D. W. (1992). Cloning and regulatory analysis of starvation-stress gene, ssgA, encoding a hydrophobin-like protein from the entomopathogenic fungus, *Metarhizium anisopliae*. *Gene, 120*, 119–124.

St. Leger, R. J., & Wang, C. (2010). Genetic engineering of fungal biocontrol agents to achieve greater efficacy against insect pests. *Applied Microbiology and Biotechnology, 85*, 901–907.

Stanley-Samuelson, D. W., Jensen, E., Nickerson, K. W., Tiebel, K., Ogg, C. L., & Howard, R. W. (1991). Insect immune response to bacterial infection is mediated by eicosanoids. *Proceedings of the National Academy of Sciences of the United States of America, 88*, 1064–1068.

Steenberg, T., & Kilpinen, O. (2014). Synergistic interaction between the fungus *Beauveria bassiana* and desiccant dusts applied against poultry red mites (*Dermanyssus gallinae*). *Exeriment and Applied Acarology, 62*, 511–524.

Stokes, B. A., Yadav, S., Shokal, U., Smith, L. C., & Eleftherianos, I. (2015). Bacterial and fungal pattern recognition receptors in homologous innate signalling pathways of insects and mammals. *Frontiers in Microbiology, 6*, 19.

Storey, G. K., Vandermeer, R. K., Boucias, D. G., & McCoy, C. W. (1991). Effect of fire ant (*Solenopsis invicta*) venom alkaloids on the in vitro germination and development of selected entomogenous fungi. *Journal of Invertebrate Pathology, 58*, 88–95.

Strand, M. R. (2008). The insect cellular immune response. *Insect Science, 15*(1), 1–14.

Strand, M. R., & Clark, K. D. (1999). Plasmatocyte spreading peptide induces spreading of plasmatocytes but represses spreading of granulocytes. *Archives of Insect Biochemistry and Physiology, 42*, 213–223.

Szafranek, B., Maliñski, E., Nawrot, J., Sosnowska, D., Ruszkowska, M., Pihlaja, K. … Szafranek, J. (2001). In vitro effects of cuticular lipids of the aphids *Sitobion avenae, Hyalopterus pruni* and *Brevicoryne brassicae* on growth and sporulation of the *Paecilomyces fumosoroseus* and *Beauveria bassiana*. *ARKIVOC, 2001*(iii), 81–94. ISSN 1424–6376.

Theopold, U., Li, D., Fabbri, M., Scherfer, C., & Schmidt, O. (2002). The coagulation of insect hemolymph. *Cellular and Molecular Life Sciences, 59*, 363–372.

Thompson, G. J., Crozier, Y. C., & Crozier, R. H. (2003). Isolation and characterisation of a termite transferrin gene up-regulated on infection. *Insect Molecular Biology, 12*, 1–7.

Thompson, S. R., & Brandenburg, R. L. (2005). Tunneling responses of mole crickets (Orthoptera: Gryllotalpidae) to the entomopathogenic fungus, *Beauveria bassiana*. *Environmental Entomology, 34*(1), 140–147.

Toledo, A. V., Alippi, A. M., & de Remes Lenicov, A. M. M. (2011). Growth inhibition of *Beauveria bassiana* by bacteria isolated from the cuticular surface of the corn leafhopper, *Dalbulus maidis* and the planthopper, *Delphacodes kuscheli*, two important vectors of maize pathogens. *Journal of Insect Science, 11*(1), 29.

Tragust, S., Mitteregger, B., Barone, V., Konrad, M., Line, V., & Ugelvig, S. C. (2013). Ants disinfect fungus-exposed brood by oral uptake and spread of their poison. *Current Biology, 23*, 76–82.

Uchida, R., Imasato, R., Yamaguchi, Y., Masuma, R., Shiomi, K., Tomoda, H., & Omura, S. (2005). New insecticidal antibiotics, hydroxyfungerins A and B, produced by *Metarhizium* sp. FKI-1079. *Journal of Antibiotics, 58*(12), 804–809.

Ugelvig, L. V., & Cremer, S. (2007). Social prophylaxis: group interaction promotes collective immunity in ant colonies. *Current Biology, 17*, 1967–1971.

Ulrich, K. R., Feldlaufer, M. F., Kramer, M., & Leger, R. J. S. (2015). Inhibition of the entomopathogenic fungus *Metarhizium anisopliae* sensu lato in vitro by the bed bug defensive secretions (E)-2-hexenal and (E)-2-octenal. *BioControl*, 1–10.

Urbanek, A., Szadziewski, R., Stepnowski, P., Boros-Majewska, J., Gabriel, I., Dawgul, M. … Golebiowski, M. (2012). Composition and antimicrobial activity of fatty acids detected in the hygroscopic secretion collected from the secretory setae of larvae of the biting midge *Forcipomyia nigra* (Diptera: Ceratopogonidae). *Journal of Insect Physiology, 58*, 1265–1276.

Uvell, H., & Engstrom, Y. (2007). A multi-layered defence against infection; combinatorial control of insect immune genes. *Trends in Genetics, 23*, 342–349.

Valles, S. M., & Pereira, R. M. (2005). *Solenopsis invicta* transferrin: cDNA cloning, gene architecture, and up-regulation in response to *Beauveria bassiana* infection. *Gene, 358*, 60–66.

Verma, M., Sharma, S., & Prasad, R. (2009). Biological alternatives for termite control: a review. *International Biodeterioration & Biodegradation, 63*(8), 959–972.

Vestergaard, S., Gillespie, A. T., Butt, T. M., Schreiter, G., & Eilenberg, J. (1995). Pathogenicity of the hyphomycete fungi *Verticillium lecanii* and *Metarhizium anisopliae* to the western flower thrips, *Frankliniella occidentalis*. *Biocontrol Science and Technology, 5*(2), 185–192.

Vey, A., Matha, V., & Dumas, C. (2002). Effects of the peptide mycotoxin destruxin E on insect haemocytes and on dynamics and efficiency of the multicellular immune reaction. *Journal of Invertebrate Pathology, 80*(3), 177–187.

Vilcinskas, A. (2010). Coevolution between pathogen-derived proteinases and proteinase inhibitors of host insects. *Virulence, 1*(3), 206–214.

Villani, M. G., Krueger, S. R., Schroeder, P. C., Consolie, F., Consolie, N. H., Preston-Wilsey, L. M., & Roberts, D. W. (1994). Soil application effects of *Metarhizium anisopliae* on Japanese beetle (Coleoptera: Scarabaeidae) behavior and survival in turfgrass microcosms. *Environmental Entomology, 23*(2), 502–513.

Vilmos, P., & Kurucz, E. (1998). Insect immunity, evolutionary roots of the mammalian immune system. *Immunology Letters, 62*, 59–66.

Walters, J. B., & Ratcliffe, N. A. (1983). Studies on the in vivo cellular reactions of insects: fate of pathogenic and non-pathogenic bacteria in *Galleria mellonella* nodules. *Journal of Insect Physiology, 29*, 417–424.

Wanchoo, A., Lewis, M. W., & Keyhani, N. O. (2009). Lectin mapping reveals stage-specific display of surface carbohydrates in in vitro and haemolymph-derived cells of the entomopathogenic fungus *Beauveria bassiana*. *Microbiology, 155*(9), 3121–3133.

Wang, C., Butt, T. M., & St. Leger, R. J. (2005). Colony sectorization of *Metarhizium anisopliae* is a sign of ageing. *Microbiology, 151*(10), 3223–3236.

Wang, C., Coa, Y., Wang, Z., Yin, Y., Peng, G., Li, Z.... Xia, Y. (2007). Differentially-expressed gylcoproteins in *Locusta migratoria* hemolymph infected with *Metarhizium anisopliae*. *Journal of Invertebrate Pathology, 96*, 230–236.

Wang, C., Duan, Z., & St. Leger, R. J. (2008). MOS1 osmosensor of *Metarhizium anisopliae* is required for adaptation to insect host hemolymph. *Eukaryotic Cell, 7*(2), 302–309.

Wang, S., Fang, W., Wang, C. S., & St. Leger, R. J. (2011). Insertion of an esterase gene into a specific locust pathogen (*Metarhizium acridum*) enables it to infect caterpillars. *PLoS Pathogens, 7*, e1002097. http://dx.doi.org/10.1371/journal.ppat.1002097.

Wang, B., Kang, Q., Lu, Y., Bai, L., & Wang, C. (2012). Unveiling the biosynthetic puzzle of destruxins in *Metarhizium* species. *Proceedings of the National Academy of Sciences of the United States of America, 109*(4), 1287–1292.

Wang, D., Kim, B. Y., Lee, K. S., Yoon, H. J., Cui, Z., Lu, W.... Jin, B. R. (2009). Molecular characterization of iron binding proteins, transferrin and ferritin heavy chain subunit from the bumblebee *Bombus ignitus*. *Comparative Physiology and Biochemistry Part B, 152*, 20–27.

Wang, S., Leclerque, A., Pava-Ripoll, M., Fang, W., & St. Leger, R. J. (2009). Comparative genomics using microarrays reveals divergence and loss of virulence-associated genes in host-specific strains of the insect pathogen *Metarhizium anisopliae*. *Eukaryotic Cell, 8*, 888–898.

Wang, C., Skrobek, A., & Butt, T. M. (2003). Concurrence of losing a chromosome and the ability to produce destruxins in a mutant of *Metarhizium anisopliae*. *FEMS Microbiology Letters, 226*(2), 373–378.

Wang, C., & St. Leger, R. J. (2005). Developmental and transcriptional responses to host and nonhost cuticles by the specific locust pathogen *Metarhizium anisopliae* var. *acridum*. *Eukaryotic Cell, 4*, 937–947.

Wang, C., & St. Leger, R. J. (2006). A collagenous protective coat enables *Metarhizium anisopliae* to evade insect immune responses. *Proceedings of the National Academy of Sciences of the United States of America, 103*(17), 6647–6652.

Wang, C., & St. Leger, R. J. (2007a). The MAD1 adhesin of *Metarhizium anisopliae* links adhesion with blastospore production and virulence to insects, and the MAD2 adhesin enables attachment to plants. *Eukaryotic Cell, 6*(5), 808–816.

Wang, C., & St. Leger, R. J. (2007b). The *Metarhizium anisopliae* perilipin homolog MPL1 regulates lipid metabolism, appressorial turgor pressure, and virulence. *Journal of Biological Chemistry, 282*, 21110–21115.

Wang, J., Sykes, B. D., & Ryan, R. O. (2002). Structural basis for the conformational adaptability of apolipophorin III, a helix-bundle exchangeable apolipoprotein. *Proceedings of the National Academy of Sciences of the United States of America, 99*, 1188–1193.

Wang, C., Typas, M. A., & Butt, T. M. (2002). Detection and characterisation of pr1 virulent gene deficiencies in the insect pathogenic fungus *Metarhizium anisopliae*. *FEMS Microbiology Letters, 213*(2), 251–255.

Wang, Z., Wilhelmsson, C., Hyrsl, P., Loof, T. G., Dobes, P., Klupp, M. … Theopold, U. (2010). Pathogen entrapment by transglutaminase—a conserved early innate immune mechanism. *PLoS Pathogens, 6*(2), e1000763.

Wang, Y., Yang, P., Cui, F., & Kang, L. (2013). Altered immunity in crowded locust reduced fungal (*Metarhizium anisopliae*) pathogenesis. *PLoS Pathogens, 9*, e1003102.

Wang, J., Zhou, G., Ying, S. H., & Feng, M. G. (2013). P-type calcium ATPase functions as a core regulator of *Beauveria bassiana* growth, conidiation and responses to multiple stressful stimuli through cross-talk with signalling networks. *Environmental Microbiology, 15*(3), 967–979.

Wekesa, V. W., Knapp, M., Maniania, N. K., & Boga, H. J. (2006). Effects of *Beauveria bassiana* and *Metarhizium anisopliae* on mortality, fecundity and egg fertility of *Tetranychus evansi*. *Journal of Applied Entomology, 130*(30), 155–159.

Whalon, M. E., Mota-Sanchez, D., & Hollingworth, R. M. (2008). *Global pesticide resistance in arthropods*. Oxfordshire, UK: Centre Agric Biosci Intl.

Whitten, M. M. A., Tew, I. F., Lee, B. L., & Ratcliffe, N. A. (2004). A novel role for an insect apolipoprotein (Apolipophorin III) in β-1,3-glucan pattern recognition and cellular encapsulation. *Journal of Immunology, 172*, 2177–2185.

Wiesner, A., Losen, S., Kopacek, P., Weise, C., & Gotz, P. (1997). Isolated apolipophorin III from *Galleria mellonella* stimulates the immune reactions of this insect. *Journal of Insect Physiology, 43*, 383–391.

Wilson-Rich, N., Spivak, M., Fefferman, N. H., & Starks, P. T. (2009). Genetic, individual, and group facilitation of disease resistance in insect societies. *Annual Review of Entomology, 54*, 405–423.

Wojda, I., & Jakubowicz, T. (2007). Humoral immune response upon mild heat-shock conditions in *Galleria mellonella* larvae. *Journal of Insect Physiology, 53*, 1134–1144.

Wraight, S. P., & Ramos, M. E. (2005). Synergistic interaction between *Beauveria bassiana*- and *Bacillus thuringiensis tenebrionis*-based biopesticides applied against field populations of Colorado potato beetle larvae. *Journal of Invertebrate Pathology, 90*, 139–150.

Wyrebek, M., & Bidochka, M. J. (2013). Variability in the insect and plant adhesins, *mad1* and *mad2*, within the fungal genus *Metarhizium* suggest plant adaptation as an evolutionary force. *PLoS One, 8*(3), e59357. http://dx.doi.org/10.1371/journal.pone.0059357.

Xavier-Santos, S., Magalhães, B., & Lima, E. A. (1999). Differentiation of the entomopathogenic fungus *Metarhizium* flavoviride (Hyphomycetes). *Revista de Microbiologia, 30*(1), 47–51.

Xiao, G., Ying, S. H., Zheng, P., Wang, Z. L., Zhang, S., Xie, X. Q. … Feng, M. G. (2012). Genomic perspectives on the evolution of fungal entomopathogenicity in *Beauveria bassiana*. *Scientific Reports, 2*, 483. http://dx.doi.org/10.1038/srep00483.

Xie, M., Zhang, Y.-J., Zhai, X.-M., Zhao, J.-J., De-Liang Peng, D.-L., & Wu, G. (2015). Expression of a scorpion toxin gene BmKit enhances the virulence of *Lecanicillium lecanii* against aphids. *Journal of Pest Science, 88*(3), 637–644.

Xiong, C., Xia, Y., Zheng, P., & Wang, C. (2013). Increasing oxidative stress tolerance and subculturing stability of *Cordyceps militaris* by overexpression of a glutathione peroxidase gene. *Applied Microbiology and Biotechnology, 97*(5), 2009–2015.

Xu, Y., Orozco, R., Wijeratne, E. K., Espinosa-Artiles, P., Gunatilaka, A. L., Stock, S. P., & Molnár, I. (2009). Biosynthesis of the cyclooligomer depsipeptide bassianolide, an insecticidal virulence factor of *Beauveria bassiana*. *Fungal Genetics and Biology, 46*(5), 353–364.

Xu, Y., Orozco, R., Wijeratne, E. K., Gunatilaka, A. L., Stock, S. P., & Molnár, I. (2008). Biosynthesis of the cyclooligomer depsipeptide beauvericin, a virulence factor of the entomopathogenic fungus *Beauveria bassiana*. *Chemistry & Biology, 15*(9), 898–907.

Yanagawa, A., & Shimzu, S. (2007). Resistance of the termite *Coptotermes formosans* Shiraki to *Metarhizium anisopliae* due to grooming. *BioControl, 52*, 75–85.

Ying, S. H., & Feng, M. G. (2011). A conidial protein (CP15) of *Beauveria bassiana* contributes to the conidial tolerance of the entomopathogenic fungus to thermal and oxidative stresses. *Applied Microbiology and Biotechnology, 90*(5), 1711–1720.

Ying, S. H., Ji, X. P., Wang, X. X., Feng, M. G., & Keyhani, N. O. (2014). The transcriptional co-activator multiprotein bridging factor 1 from the fungal insect pathogen, *Beauveria bassiana*, mediates regulation of hyphal morphogenesis, stress tolerance and virulence. *Environmental Microbiology, 16*(6), 1879–1897.

Yoshiga, T., Georgieva, T., Dunkov, B. C., Harizanova, N., Ralchev, K., & Law, J. H. (1999). *Drosophila melanogaster* transferrin. Cloning, deduced protein sequence, expression during the life-cycle, gene localisation and up-regulation on bacterial infection. *European Journal of Biochemistry, 260*, 414–420.

Yu, K. H., Kim, K., Lee, J., Heui Sam Lee, H. S., Kim, S. H., Cho, K. Y. … Lee, I. H. (2002). Comparative study on characteristics of lysozymes from the hemolymph of three lepidopteron larvae, *Galleria mellonella, Bombyx mori, Agrius convolvuli*. *Developmental and Comparative Immunology, 26*, 707–713.

Zdybicka-Barabas, A., & Cytrynska, M. (2011). Involvement of apolipophorin II in antibacterial defense of *Galleria mellonella* larvae. *Comparative Biochemistry and Physiology Part B, 158*, 90–98.

Zdybicka-Barabas, A., & Cytrynska, M. (2013). Apolipophorins and insect immune response. *Invertebrate Survival Journal, 10*, 58–68.

Zdybicka-Barabas, A., Januszanis, B., Mak, P., & Cytrynska, M. (2011). An atomic force microscopy study of *Galleria mellonella* apolipophorin III effect on bacteria. *Biochimica et Biophysica Acta, 1808*, 1896–1906.

Zdybicka-Barabas, A., Staczek, S., Mak, P., Piersiak, T., Skrzypiec, K., & Cytrynska, M. (2012). The effect of *Galleria mellonella* apolipophorin III on yeasts and filamentous fungi. *Journal of Insect Physiology, 58*, 164–177.

Zdybicka-Barabas, A., Staczek, S., Mak, P., Skrzypiec, K., Mendyk, E., & Cytrynska, M. (2013). Synergistic action of *Galleria mellonella* apolipophorin III and lysozyme against Gram-negative bacteria. *Biochimica et Biophysica Acta, 1828*, 1449–1456.

Zhang, Y. J., Feng, M. G., Fan, Y. H., Luo, Z. B., Yang, X. Y., Wu, D., & Pei, Y. (2008). A cuticle-degrading protease (CDEP-1) of *Beauveria bassiana* enhances virulence. *Biocontrol Science and Technology, 18*, 551–563.

Zhang, S., Widemann, E., Bernard, G., Lesot, A., Pinot, F., Pedrini, N., & Keyhani, N. O. (2012). CYP52X1, representing new cytochrome P450 subfamily, displays fatty acid hydroxylase activity and contributes to virulence and growth on insect cuticular substrates in entomopathogenic fungus *Beauveria bassiana*. *Journal of Biological Chemistry, 287*, 13477–13486.

Zhang, S. Z., Xia, Y. X., Kim, B., & Keyhani, N. O. (2011). Two hydrophobins are involved in fungal spore coat rodlet layer assembly and each play distinct roles in surface interactions, development and pathogenesis in the entomopathogenic fungus, *Beauveria bassiana*. *Molecular Microbiology, 80*, 811–826.

Zhang, Y., Zhang, J., Jiang, X., Wang, G., Luo, Z., Fan, Y. … Pei, Y. (2010). Requirement of a mitogen-activated protein kinase for appressorium formation and penetration of insect cuticle by the entomopathogenic fungus *Beauveria bassiana*. *Applied and Environmental Microbiology, 76*(7), 2262–2270.

Zhang, Y., Zhao, J., Fang, W., Zhang, J., Luo, Z., Zhang, M. … Pei, Y. (2009). Mitogen-activated protein kinase hog1 in the entomopathogenic fungus *Beauveria bassiana* regulates environmental stress responses and virulence to insects. *Applied and Environmental Microbiology, 75*(11), 3787–3795.

Zhang, Z.-T., & Zhu, S.-Y. (2009). Drosomycin, an essential component of antifungal defence in *Drosophila*. *Insect Molecular Biology, 18*(5), 549–556. http://dx.doi.org/10.1111/j.1365-2583.2009.00907.x.

Zhao, H., Charnley, A. K., Wang, Z. K., Yin, Y. P., Li, Z. L., Li, Y. L. … Xia, Y. X. (2006). Identification of an extracellular acid trehalase and its gene involved in fungal pathogenesis of *Metarizium anisopliae*. *Journal of Biochemistry, 140*(3), 319–327.

Zhao, H., Xu, C., Lu, H. L., Chen, X., St. Leger, R. J., & Fang, W. (2014). Host-to-pathogen gene transfer facilitated infection of insects by a pathogenic fungus. *PLoS Pathogens, 10*(4), e1004009.

Zhao, H., Lovett, B., & Fang, W. (2016). Genetically Engineering Entomopathogenic Fungi. *Advances in Genetics, 94*, 137–163.

Zheng, P., Xia, Y. L., Zhang, S. W., & Wang, C. S. (2013). Genetics of *Cordyceps* and related fungi. *Applied Microbiology and Biotechnology, 97*, 2797–2804.

Zhou, G., Wang, J., Qiu, L., & Feng, M. G. (2012). A Group III histidine kinase (mhk1) upstream of high-osmolarity glycerol pathway regulates sporulation, multi-stress tolerance and virulence of *Metarhizium robertsii*, a fungal entomopathogen. *Environmental Microbiology, 14*(3), 817–829.

Zhuang, J., Coates, C. J., Zhu, H., Zhu, P., Wu, Z., & Xie, L. (2015). Identification of candidate antimicrobial peptides derived from abalone hemocyanin. *Developmental and Comparative Immunology, 49*, 96–102.

Zimmermann, G. (2007). Review on safety of the entomopathogenic fungi *Beauveria bassiana* and *Beauveria brongniartii*. *Biocontrol Science and Technology, 17*(6), 553–596.

CHAPTER TEN

Molecular Genetics of Secondary Chemistry in *Metarhizium* Fungi

B.G.G. Donzelli*[1] and S.B. Krasnoff§[,1]
*Cornell University, Ithaca, NY, United States
§USDA-ARS, Ithaca, NY, United States
[1]Corresponding authors: E-mail: bdd1@cornell.edu; Stuart.Krasnoff@ars.usda.gov

Contents

Advances in Genetics, Volume 94
ISSN 0065-2660
http://dx.doi.org/10.1016/bs.adgen.2016.01.005

Abstract

As with many microbes, entomopathogenic fungi from the genus *Metarhizium* produce a plethora of small molecule metabolites, often referred to as secondary metabolites. Although these intriguing compounds are a conspicuous feature of the biology of the producing fungi, their roles in pathogenicity and other interactions with their hosts and competing microbes are still not well understood. In this review, secondary metabolites that have been isolated from *Metarhizium* are cataloged along with the history of their discovery and structural elucidation and the salient biological activities attributed to them. Newly available genome sequences revealed an abundance of biosynthetic pathways and a capacity for producing SMs by *Metarhizium* species that far exceeds the known chemistry. Secondary metabolism genes identified in nine sequenced *Metarhizium* species are analyzed in detail and classified into distinct families based on orthology, phylogenetic analysis, and conservation of the gene organization around them. This analysis led to the identification of seven hybrid polyketide/nonribosomal peptide synthetases (M-HPNs), two inverted hybrid nonribosomal peptide/polyketide synthetases (M-IHs), 27 nonribosomal peptide synthetases (M-NRPSs), 14 nonribosomal peptide synthetase—like (M-NPL) pathways, 32 polyketide synthases, and 44 terpene biosynthetic genes having a nonuniform distribution and largely following established phylogenetic relationships within the genus *Metarhzium*. This systematization also identified candidate pathways for known *Metarhizium* chemistries and predicted the presence of unknown natural products for this genus by drawing connections between these pathways and natural products known to be produced by other fungi.

1. INTRODUCTION

The bewildering profusion and variety of small molecules produced by bacteria and fungi creates the impression that what are commonly called "secondary metabolites" (SMs) are critical to the biology of these organisms. SMs from microbes have traditionally attracted attention as chemical platforms for building pharmaceuticals and agrochemicals. Virtually all free-living hyphomycetous fungi that have been studied at the genomic level reveal a capacity for producing small molecules that far exceeds the number of compounds actually isolated and identified from them. Thus, there appears to be a wealth of novel chemistry yet to be discovered that may afford huge pragmatic payoffs, as well as insights into how the producing organisms function. Entomopathogenic fungi in the genus *Metarhizium* typify this untapped resource and, because of the energy expended to synthesize these compounds they can be expected to influence the fitness of the producers and, thus, may play critical roles in the ability of these pathogens to successfully parasitize their hosts and ultimately contribute to the success or failure of these fungi as biological control agents of pestiferous insects and mites.

This review is divided into two sections. In the first section we describe the chemistry and biological activity of SMs that have been recorded in the literature as products of *Metarhizium* fungi; in the second section, we survey secondary metabolism (SMM) pathways identified in the genomes of nine *Metarhizium* strains. The chemistry and genomics surveys are separated because the biogenesis of most of the known compound types is unknown and the SMM end-products of most of the identifiable SMM genes are largely unknown. The goals in the coverage of chemistry are twofold: (1) to point out chemical relationships to compounds from other fungi so that extant genomic data relevant to the biosynthesis of these compounds, as well as data that become available in the future, can be cross-referenced for use in identifying biosynthetic pathways in *Metarhizium* for functional analysis and (2) to clarify the provenance of compounds, wherever possible with reference to current taxonomic information (Bischoff, Rehner, & Humber, 2009; Kepler, Humber, Bischoff, & Rehner, 2014). Accordingly, strains referred to as *Metarhizium anisopliae* in chemical characterization papers usually cannot be ascribed with certainty to the species now known as *Metarhizium anisopliae* (sensu stricto) and may belong to one of the other *Metarhizium* species established by Bischoff et al. (2009).

In the second part of the review we describe SMM biosynthetic genes and pathways identified in nine sequenced *Metarhizium* strains that can be definitively disposed to species, draw connections between genes and natural products identified both within and outside the genus, and classify these genes into groups based on orthology, phylogenetic approaches, and the organization of gene clusters.

2. THE SMALL MOLECULE METABOLITES OF *METARHIZIUM*

2.1 Peptides

2.1.1 Destruxins

Discussions of *Metarhizium* metabolites typically begin and end with the destruxins (DTXs). Since the first report of DTXs as insect toxins (Kodaira, 1961), hundreds of papers including several comprehensive reviews have described their chemistry and biological activities ranging from acute toxicity and immunosuppression in insects to a variety of biomedically relevant effects (Hu & Dong, 2015; Liu & Tzeng, 2012; Pedras, Zaharia, & Ward, 2002).

DTXs are six-residue depsipeptides comprising five amino acid units and a hydroxy acid that together form a 19-membered macrocyclic ring (Supplementary Fig. 1).

There are now >40 DTX analogues reported—most from *Metarhizium*, but some from other fungi including both insect and plant pathogens. In *Metarhizium*, DTXs occur in a complex microheterogeneous mixture that can be organized structurally in several ways into multiple series of congeners. In one such scheme, the series consists of analogues differing from one another in the structure of the moiety attached to the α-carbon of an α-hydroxy acid residue. Accordingly, in what can be considered the reference series (DTX A—F) the hydroxy acid unit is amidated by a proline (Pro) at position 2, and so on proceeding from N- to C-terminal with isoleucine (Ile) at position 3,N-methylvaline (N-MeVal) at position 4,N-methylalanine (N-MeAla) at position 5, which is amidated by the 3-amino group of the β-alanine at position 6, which, in turn, esterifies the α-hydroxy group of the hydroxy acid to complete the macrocycle.

The series differ from one another by amino acid substitutions at various positions in the ring. In the A1...E1 series the proline at position 2 is replaced by pipecolic acid. In the A2...E2 series the Ile is replaced by Val at position 3. In DTX A3, which has no B, C, D, or E counterparts, the Pro is replaced by N-MeAla. DTX F which has 2,4-dihydroxypentanoic acid has been reported only once (Wahlman & Davidson, 1993). For DTXs proper, the "desmethyl" prefix indicates the absence of the N-methylation of Ile or Val at position 4. N-MeAla occurs at position 5 in all DTX analogues, except for proto-DTX, which lacks both N-methylations.

DTX analogues from other fungi include DTX A4, A5, and homodestruxin B from an entomopathogenic *Aschersonia* sp. (Krasnoff,

Gibson, Belofsky, Gloer, & Gloer, 1996); roseotoxin B and roseocardin from *Trichothecium roseum* (Springer et al., 1984; Tsunoo, Kamijo, Taketomo, Sato, & Ajisaka, 1997); bursaphelocides from an unidentified fungus (Kawazu et al., 1993); pseudodestruxins from *Nigrosalbulum globosum* (Che, Swenson, Gloer, Koster, & Malloch, 2001) and *Beauveria felina* (Lira et al., 2006); and isaridins from an *Isaria* isolate (Ravindra et al., 2004; Sabareesh et al., 2007) and from *Beauveria felina* (Chung, El-Shazly, et al., 2013; Du et al., 2014; Langenfeld et al., 2011). The DTX biosynthetic pathway has been partially characterized (see under Section 3.1, M-NRPS19).

2.1.2 Serinocyclins

Serinocyclins were first identified from conidia of *Metarhizium robertsii* ARSEF 2575 (see Section 3.1, M-NRPS20) and structurally elucidated from *Metarhizium acridum* isolates, which yielded greater quantities of material (Krasnoff et al., 2007). Serinocyclin A, is a cyclic heptapeptide composed of a unit of 1-aminocyclopropane-1-carboxylic acid (ACC) acylating 4-hydroxyproline, followed by L-serine, D-4-hydroxylysine, β-alanine, D-serine, and L-serine which amidates ACC to complete a 22-membered macrocyle (Supplementary Fig. 2). Serinocyclin B is a minor analogue that lacks the 4-hydroxy group on the D-lysine.

Mosquito larvae immersed in ~60 ppm serinocyclin A were unable to control the position of their heads so that their typical backwards zig-zag pattern of locomotion was perturbed (Krasnoff et al., 2007). This suggested a neurophysiological effect on control of the hair tufts used as rudders or paddles to adjust head position in larval swimming (Brackenbury, 1999, Brackenbury, 2001). The relevance of the observation to the adaptive significance of serinocylins remains unknown. A 2014 virtual docking study suggested that the serinocyclin binds glutathione S-transferase (Sanivada & Challa, 2014).

Targeted gene knockout identified NPS1 as the core gene responsible for serinocyclin biosynthesis (see under Section 3.1, M-NRPS20) (Moon et al., 2008).

2.1.3 Metachelins

Siderophores are small molecules produced by plants and microbes that increase the bioavailability of highly insoluble Fe^{3+} and prevent the production of harmful free radicals within the cell (Haas, Eisendle, & Turgeon, 2008). A group of secreted, coprogen-type hydroxamate siderophores was isolated from fermentation broth of *M. robertsii* ARSEF 2575 grown in iron-depleted medium (Supplementary Fig. 3) (Krasnoff, Keresztes, Donzelli, & Gibson, 2014). The mixture included N^{α}-dimethyl coprogen (NADC) first isolated from *Alternaria longipes* and *Fusarium dimerum* (Jalal, Love, & van der Helm, 1988) and later from *Alternaria brassiciola* (Oide et al., 2006) and another known compound, dimerumic acid (DA), first reported as a degradation product of coprogen (Keller-Schierlein & Diekmann, 1970) and subsequently as a natural product from *Verticillium dahliae* (Harrington & Neilands, 1982), *Gliocladium virens* (Jalal, Love, & Van Der Helm, 1986), *Monascus anka* (Aniya et al., 2000), and *Penicillium chrysogenum* (Hördt, Römheld, & Winkelmann, 2000). DA is formed by the head-to-head condensation of two 5-anhydromevalonyl-*N*-5-hydroxyornithine (AHMO) units into a diketopiperazine ring. NADC is formed by the head-to-tail esterification of a third AHMO unit with one of the terminal hydroxyl groups of DA. In addition to these known compounds, four novel siderophores were reported from *M. robertsii*. Metachelin A, the major component of the siderophore mixture can be rationalized as a derivative of NADC in which both terminal hydroxyl groups are glycosylated by D-mannose and the dimethyl nitrogen is N-oxidized. Metachelin B lacks the N-oxide. The N-oxide analogue of NADC and a monomannoside analogue of DA were also characterized (Krasnoff et al., 2014).

As with other hydroxamate siderophores the metachelins form hexadentate chelation complexes with Fe^{+3} and other trivalent metal cations (Al^{+3} and Ga^{+3}). In a CAS-plate assay, the metachelins and related compounds from *M. robertsii* showed activity approximately equal to that of the bacterial siderophore ferrioxamine (Krasnoff et al., 2014). The NRPS responsible for metachelin biosynthesis has been identified (Donzelli, Gibson, & Krasnoff, 2015) (see Section 3.1 of this review M-NRPS18).

2.1.4 Ferricrocin

M. robertsii also synthesizes ferricrocin, a hexapeptide of the ferrichrome type (Supplementary Fig. 3) (Jalal, Hossain, van der Helm, & Barnes, 1988). Ferricrocin is produced in its ferrated form by mycelia and conidia of *M. robertsii* 2575, but is not secreted into fermentation broths. It is thus considered an intracellular siderophore, which is presumed to receive environmental iron scavenged by extracellular siderophores and to transport it to its target sites in the cell (Wallner et al., 2009).

Ferricrocin is a siderophore of the ferrichrome superfamily of compounds (which includes the asperchromes) with the sequence -Ser-Ser-Gly-Orn1-Orn2-Orn3- where the Orn units are all N^{δ}-acetyl-N^{δ}-hydroxyornithines.

Among the ferrichromes proper the N^{δ}-acyl groups in the Orn units are homogeneous, whereas in the asperchromes they, vary within a compound (Jalal, Hossain, et al., 1988).

Like the linear coprogens the cyclic ferrichromes all feature hydroxamate groups in modified ornithine units. Ferricrocin was first reported in 1963 from *Aspergillus* spp. (Zähner, Keller-Schierlein, Hütter, Hess-Leisinger, & Deer, 1963) and has since been identified in scores of fungi. The near ubiquity of ferricrocin in fungi as an intracellular siderophore stands in contrast to the combinatorial variation observed among extracellular siderophores. It is tempting to consider this variation as the result of interspecific competition for iron in a scenario in which some microbes can hijack the siderophore–iron complex of another species thereby driving selection for species-specific siderophores less likely to be "recognized" by a hijacker (van der Helm & Winkelmann, 1994; Lee, van Baalen, & Jansen, 2012; Renshaw et al., 2002).

The genetic base for ferricrocin biosynthesis has been identified (Donzelli, Gibson, & Krasnoff, 2015) (see Section 3.1 of this review, M-NRPS17).

2.2 Dipeptides and Didepsipeptides

2.2.1 Tyrosine Betaine

Tyrosine betaine, a dipeptide consisting of a unit of tyrosine acylated by an *N*-trimethyl glycine was isolated and characterized from *M. anisopliae* var. *anisopliae* strain ESALQ 1037 and then identified in an HPLC screen of conidial extracts of *M. acridum* (ARSEF 324, 3391, and 7486) and *Metarhizium brunneum* (ARSEF 1095, 5626, and 5749) (Carollo et al., 2010) (Supplementary Fig. 4). Based on mass spectrometric analyses, we observed that this compound co-occurs with serinocyclins and ferricrocin in extracts of conidia of *Metarhizium guizhouense* (ARSEF 683), *Metarhizium pingshaense* (ASEF 2106 and 5197), and *M. robertsii* (ARSEF 2575 and 4123) (Krasnoff, unpublished data). No biological activity has been reported to date for this compound. It should be noted that as a quaternary amine it bears a formal positive charge.

2.2.2 Metacytofilin

Metacytofilin is a two-residue depsipeptide produced by *Metarhizium* sp. TA2759 in which an α-hydroxyphenylalanine amidates an α-hydroxy *N*-MeLeu which also esterifies the α-hydroxyPhe to complete a six-membered 2,5-morpholine dione ring (Iijima et al., 1992) (Supplementary Fig. 4). Bassiatin from *Beauveria bassiana* (Kagamizono et al., 1995), and a diastereomer of bassiatin from *Isaria japonica* (Oh et al., 2002) similarly combine 2-hydroxy-3-methylbutanoic acid (deaminated Val) with N-MePhe to form an oxazine ring system, albeit with a methylated nitrogen. These compounds are similar and possibly related biogenetically, to the monomeric two-residue units of large cyclic despsipeptides like enniatin C, in which *N*-MeLeu amidates 2-hydroxy-3-methylbutanoic acid and beauvericin (trimer) and bassianolide (tetramer) in which the monomer comprises N-MePhe combined with 2-hydroxy-3-methylbutanoic acid (Hamill, Higgens, Boaz, & Gorman, 1969; Shemyakin et al., 1963; Suzuki et al., 1977).

2.3 Amino Acid Derivatives

2.3.1 Swainsonine

Observations of neurological symptoms as well as weight loss in livestock feeding on *Swainsona* spp. (Fabaceae) led to the discovery of a factor in *Swainsona canescens* that inhibits lysosomal α-mannosidase (Dorling, Huxtable, & Vogel, 1978). The structure of this factor was elucidated as indolizidine 1,2,8 triol and named swainsonine (Colegate, Dorling, & Huxtable, 1979) (Supplementary Fig. 4). A compound was then isolated from the fungus *Rhizoctonia leguminicola* and after complete structural elucidation was determined to be identical to swainsonine. This study corrected a previous structural assignment for the compound from *R. leguminicola* and revealed that swainsonine was, in actuality, first isolated from a fungus, and not a plant (Guengerich, DiMari, & Broquist, 1973). The contribution of swainsonine and the related alkaloid slaframine to the mycotoxicosis of cattle known as "slobbers syndrome" was reviewed by Croom, Hagler, Froetschel, and Johnson (1995). Swainsonine was subsequently isolated from *M. anisopliae* F-3622, a soil isolate, by Hino et al. (1985). The question of whether detection of swainsonine in plants is simply due to contamination by a fungus was addressed by Harris, Campbell, Molyneux, and Harris (1988) who concluded that it was indeed synthesized by both plants and fungi and that the biosynthetic pathway was likely transferred horizontally from plant to fungus or vice versa. Subsequently, the debate swung toward the side of exclusive fungal origins for swainsonine with the discovery of swainsonine-producing fungal endophytes where they have been looked for in swainsonine-producing plants (Cook, Gardner, & Pfister, 2014). The production of an economically important mycotoxin by *Metarhizium* species has obvious regulatory implications for the use of fungi in the genus as biocontrol agents.

2.3.2 Fungerins

Fungerin, previously identified from a *Fusarium* sp. (Singh et al., 2001), along with two novel analogues, hydroxyfungerin A and its regioisomer hydroxyfungerin B, were isolated from *Metarhizium* sp. FKI-1079 (Uchida et al., 2005) (Supplementary Fig. 4). These compounds feature an imidazole core, suggesting the involvement of histidine in their biosynthesis. The new compounds, unique to the *Metarhizium* strain, were only 1/12 as potent as fungerin in an acute toxicity assay against brine shrimp (*Artemia*), and were not active at 10 μg/disk against the nematode *Caenorhabditis elegans* or against a panel of microbes including nine bacteria and five fungi. Related compounds from *Fusarium* spp. include visoltricin and the recently isolated fusagerins (Visconti & Solfrizzo, 1994; Wen et al., 2015).

2.4 Polyketides

2.4.1 Aurovertins

New aurovertin analogues (F-H), as well as the previously described aurovertin D, were reported from *M. ansiopliae* HF260, an isolate "baited" from soil using dead chestnut weevil larvae (Supplementary Fig. 5) (Azumi et al., 2008). Aurovertins were first isolated from *Calcarisporium arbuscula* with the first structural elucidation presented for aurovertin B (Mulheirn, Beechey, Leworthy, & Osselton, 1974). Later, aurovertin F (as well as a novel analogue aurovertin I, and the known analogues D and E) was reported from *Pochonia chlamydisporia* (Niu et al., 2009).

Aurovertins inhibit the mitochondrial, bacterial, and chloroplast ATPases (F1) and thus, are used as probes for these critical enzymes. The aurovertin-binding site differs from that of two other well-characterized F1 inhibitors, efrapeptin and leucinostatin (Lardy, Reed, & Chiu, 1975), both of which, intriguingly, are also products of entomopathogenic fungi, from the genera *Tolypocladium* (Gupta et al., 1991) and *Paecilomyces* spp., respectively (Fukushima, Arai, Mori, Tsuboi, & Suzuki, 1983).

Aurovertins can be rationalized as nonaketides, most likely synthesized by an iterative type I polyketide synthase (PKS) (Fujii, Yoshida, Shimomaki, Oikawa, & Ebizuka, 2005), which feature a substituted 2,6-dioxabicyclo [3,2,1]octane moiety bridged by a conjugated triene to a substituted α-pyrone. They share structural similarity with citreoviridin, a mycotoxin from *Penicillium* spp. and related genera that presents a hazard from consumption of contaminated rice (Pitt, 2002), and with asteltoxins from *Pochonia bulbillosa* (Adachi et al., 2015).

2.5 Polyketide/Peptide Hybrids

2.5.1 *Cytochalasins*

Cytochalasins (CYT), a subset of the "cytochalasans" were thoroughly reviewed by Scherlach, Boettger, Remme, and Hertweck (2010) (Supplementary Fig. 6). The earliest identifications of CYTs (also known as phomins and zygosporins) came from three fungal sources, a *Phoma* sp. (Rothweiler & Tamm, 1966), *Helminthosporium dematioideum* (Aldridge, Armstrong, Speake & Turner, 1967), *Zygosporium mansonii* (Hayakawa, Matsushima, Kimura, Minato, & Katagiri, 1968), and *M. anisopliae* (sic), which afforded CYT D and C.

Subsequently many subclasses of compounds have been gathered under the cytochalasan rubric including the scoparisins, chaetoglobosins, penochalasins, aspochalsins, phomacins, and alachalasins (Scherlach et al., 2010). Two new CYT analogues, deacetyl-CYT C and an unnamed isomer were isolated from *M. anisopliae* in a screen for plant growth retardants (Supplementary Fig. 6) (Fujii, Tani, Ichinoe, & Nakajima, 2000).

Cellular effects attributed to cytochalasins suggest the possibility that they may function as hormones mediating changes in growth forms governed by endogenous timetables or in response to environmental conditions. It should be noted that inhibitory effects on growth have been observed not only in fungal germlings but also in plant apical shoots and pollen tubes. Considering the basic biological activities of the cytochalasins, it is somewhat surprising that they have been overshadowed by the DTXs in the collective effort to demonstrate a role for small molecules in *Metarhizium* as virulence factors, and in fact very little has been published on cytochalasins from *Metarhizium* since the early identification work. St. Leger, Roberts, and Staples (1991) used cytochalasins for their known depressive activity on filamentous actin formation to investigate the role of actin in differentiation in germlings of *M. anisopliae* (ME-1 parent strain of ARSEF 2575 = *M. robertsii*). CYT A completely halted appressorium formation at the lowest dose tested (2 μM) and also completely inhibited germination at higher levels (>8 μM). In addition, CYT A and E were more potent than CYT B or C, and CYT A had about 10 times the potency of CYT E as a germination inhibitor. Production of cytochalasins by the fungus was not reported. The potency comparisons contrast with Carter's early study of cytochalasins (Carter, 1967) in which he found that CYT C and D are 10-fold more potent than CYT A and B as inhibitors of cell motility and cytoplasmic cleavage. Vilcinskas, Mathia, and Götz (1997a, 1997b) investigated cytochalasins (along with DTXs) and reported that CYT B (not a known product of *Metarhizium* spp.) and CYT D perfused into blastospores of *M. anisopliae* (and *B. bassiana*) reduced phagocytosis of these "loaded" blastospores by *Galleria melonella* plasmatocytes in vitro suggesting that the compounds could have immunosuppressive activity in the pathosystem.

2.5.2 NG-391 and NG-393

Cultures of a serinocyclin-knockout mutant strain of *M. robertsii* ARSEF 2575 that contained an unknown second insertion attained a deep-yellow color that was not evident in other serinocyclin-knockout mutants or the wild type. Assuming overproduction of the colored material by this mutant, two compounds, NG-391 and NG-393 (NG39x), were isolated as yellow oils (Krasnoff et al., 2006). These compounds, also produced by the wild-type strain but in much lower quantity, had previously been isolated as neural growth stimulants from an undescribed *Fusarium* sp. (Sugawara, Shinonaga, Yoji, Yoshikawa, & Yamamoto, 1996). NG39x are 7-desmethyl analogues of fusarins, which are well-known mycotoxins from *Fusarium moniliforme* (Gelderblom et al., 1984) and other *Fusaria* (Cantalejo, Torondel, Amate, Carrasco, & Hernández, 1999; Song, Cox, Lazarus, & Simpson, 2004). As with fusarin C and its isomer (8*Z*)-fusarin C, NG-393 is produced nonenzymatically by light-activated isomerization of the eight to nine double bond in the polyene side chain of NG-391 (Supplementary Fig. 7).

Like the fusarins, NG-39x were found to be mutagenic in the Ames assay, but only in the presence of the S-9 liver enzyme fraction indicating that mutagenicity in mammals results from a metabolite of the fungal product that is produced in the liver (Krasnoff et al., 2006). In 2013 Subsequently, NG-391 was later found to depress nucleic acid synthesis in K-562 leukemia cells; however, it did not induce apoptosis (Bohnert, Dahse, Gibson, Krasnoff, & Hoffmeister, 2013).

Assuming the same biogenesis as fusarins, NG39x can be rationalized as heptaketides with three C-methyl groups (Niehaus et al., 2013). The methylated ketide is condensed with a homoserine by an NRPS module and the hybrid molecule is then tailored by enzymes that add a methyl ester at the first *S*-adenosylmethionine (SAM)-supplied methyl carbon, and catalyze the epoxidation and hydroxylation of the pyrrolidone on what would be the α-carbon of the homoserine unit (Niehaus et al., 2013; Song et al., 2004). Related fungal tetramate compounds with putatively similar biosynthetic pathways include tenellin (pentaketide + Tyr) and bassianin (hexaketide + Tyr) from *Beauveria tenella* and *B. bassiana* (McInnes, Smith, Wat, Vinning, & Wright, 1974), militarinone from *Cordyceps militaris* (heptaketide + Tyr) (Schmidt, Riese, Li, & Hamburger, 2003), and pramanicin (octaketide + Ser) from an unidentified fungus (mf5868 bw283) (Schwartz et al., 1994).

2.5.3 Metacridamides

Metacridamides A and B are 19-membered macrocylic lactones isolated from spores of *M. acridum* ARSEF 3341 that can be rationalized as nonaketides condensed with a phenylalanine unit (Supplementary Fig. 8) (Krasnoff et al., 2012).

Metacridamides showed neither insecticidal activity nor antimicrobial activity. However, metacridamide A (but not B) exhibited cytotoxicity against several cancer cell lines with EC_{50}s in the 10 μM range. Similar compounds include the torrubiellutins A–C, hexaketide + N-MePhe hybrids reported from the entomopathogenic *Torrubiella luteorostrata* (Pittayakhajonwut, Usuwan, Intaraudom, Khoyaiklang, & Supothina, 2009). Torrubiellutin C showed cytotoxicity against cancer cell lines in the 1–10 μM range. A heptaketide + Phe hybrid reported from an unidentified fungus (Jacquot, Poeschke, & Burger, 2009) showed activity at 0.4 μM in a TRG5 activation assay which indicated potential as a lead chemistry for an obesity control drug (Vallim & Edwards, 2009).

Metacridamides are presumed products of a hybrid PKS-NRPS that has yet to be identified. The PKS-NRPS hybrid thermolides, isolated as nematocides from *Talaromyces thermophilus*, combine nona- and decaketides with D-Ala and D-Val, but lack the olefin groups of the metacridamides and related compounds and thus exhibit a more reduced character (Guo et al., 2012; Niu et al., 2014).

2.6 Other Polyketide Hybrids

2.6.1 JBIR-19 and -20

A *Metarhizium* strain identified as *M. anisopliae* var. *anisopliae* (= *M. anisopliae* sensu stricto) based on comparisons of its ribosomal DNA and morphology of its conidiophores and conidia with *M. anisopliae* var. *anisopliae* strain EU307926, yielded JBIR-19 and JBIR-20, two 24-membered macrolides that differ from each other by one hydroxyl substitution (Supplementary Fig. 9) (Kozone et al., 2009). These compounds appear to be highly reduced dodecaketides that feature a rare 2-aminoethyl phosphate ester side chain.

They induced elongation of cells in *Saccharomyces cerevisiae* at 3 μM and antibiosis against *S. cerevisiae* at 200 μM. A *Eupencillium shearii* isolate yielded eushearilide (Hosoe et al., 2006), a 24-member macrolide with a choline phosphate ester positioned β to the carbonyl carbon as opposed to δ as in JBIR-19 and -20.

2.7 Terpenoids

2.7.1 Helvolic Acid and Related Compounds

Helvolic acid is a fusidane similar to fusidic acid and is built on the cyclopentanoperhydrophenanthrene skeleton common to steroids (Supplementary Fig. 10). It was originally isolated as "fumigacin," but not structurally elucidated, from *Aspergillus fumigatus* and *Aspergillus clavatus* by Waksman et al. (1943) as an antibactierial. Substructures were elucidated in the 1950s resulting in tentative structural hypotheses (Allinger, 1956; Allinger & Coke, 1961; Cram & Allinger, 1956; Okuda et al., 1964), but the full structure was not finally solved until 1970 (Iwasaki, Sair, Igarashi, & Okuda, 1970; Okuda et al., 1967, 1964). Turner and Aldridge (1983) reported helvolic acid from *M. anisopliae*, as well as from eight other fungal species, and helvolinic acid (Supplementary Fig. 10), a 6-deacetyl analogue of helvolic acid, from *M. anisopliae* and *M. brunneum* citing unpublished data. Helvolic acid was also reported from *M. anisopliae* (Espada & Dreyfuss, 1997) in a study testing the effect of addition of a 10 mg/L medium of a novel 10-residue bicyclic peptolide on production of DTXs versus helvolic acid in a panel of *M. anisopliae* isolates (Dreyfuss, Emmer, Grassberger, Rüedi, & Tscherter, 1989). The peptolide additive increased DTX production one- to fourfold at the expense of helvolic acid production, which decreased two- to tenfold. Helvolic acid as well as a 1,2-hydro analogue (Supplementary Fig. 10) were isolated from *M. anisopliae* strain HF293 and both compounds were shown to have antibacterial activity against *Staphylococcus aureus* (Lee, Kinoshita, Ihara, Igarashi, & Nihira, 2008). A helvolic acid biosynthetic cluster is present in several sequenced *Metarhizium* species (see Section 3.5).

2.7.2 Viridoxins

Viridoxins A and B, isolated from *Metarhizium flavoviride* (ARSEF 2133) have a diterpenoid core with a 6-methoxy-2,3-dimethyl-γ-pyrone moeity attached at C19 and with (2*R*,3*S*)-2-hydroxy-3-methylpentanoate and (*R*)-2-hydroxy-4-methylpentanoate, respectively, at C3 (Supplementary Fig. 11) (Gupta et al., 1993). Interestingly, these are the same α-hydroxy acids (deaminated Ile and Leu, respectively) found in bursaphelocide A and DTX B, respectively (vide supra). Viridoxins showed insecticidal activity against Colorado potato beetle (*Leptinotarsa decemlineata*) in a leaf contamination assay with $IC_{50}s$ of 40 and 51 ppm for A and B, respectively (Gupta et al., 1993).

2.7.3 Metarhizins

M. flavoviride also produces metarhizins A and B which are functionalized diterpenes similar to the viridoxins. They differ by the 4-hydroxy-5,6-dimethyl-α-pyrone moiety at C19. Metarhizin A has (2*R*,3*S*)-2-hydroxy-3-methylpentanoate at C3 as in viridoxin A, but metarhizin B has (*R*)-2-hydroxy-3-methylbutanoic acid (deaminated Val) (Supplementary Fig. 11) (Kikuchi et al., 2009). A compound named BR-050, which lacks the hydroxy acid unit at C3 (Singh et al., 2001), was also isolated along with the metarhizins. It was weaker by several orders of magnitude in cytotoxicity assays than the acylated analogues. This compound was also reported from *T. luteorostrata* and exhibited cytotoxicity in the 2–12 nM range, antimalarial activity at ~6 nM and anti-inflammatory activity inhibiting COX-2 (but not COX-1) in the low nanomolar range (Pittayakhajonwut et al., 2009).

Other related diterpenoid pyrones include the subglutinols (Lee, Lobkovsky, Pliam, Strobel, & Clardy, 1995), sesquicillin (Engel, Erkel, Anke, & Sterner, 1998), the candelalides (Singh et al., 2001), and nalanthalide from *Nalanthamala* sp. (Goetz et al., 2001) and *Chaunopycnis alba* MF6799 (Bills, Polishook, Goetz, Sullivan, & White, 2002), which is a natural product that is structurally identical to compound 6 of Gupta et al. (1993), the 3-acetyl derivative of colletochin from *Colletotrichum nicotianae* (García-Pajón & Collado, 2003).

2.7.4 Ovalicins

Mer-f3, or 12-hydroxy-ovalicin, is a monocyclic sesquiterpene extracted from fermentation broth of a *Metarhizium* sp. soil isolate (FERM P-15860—National Institute of Bioscience and Human Technology, Ibaraki Prefecture, Japan) grown in culture on potato starch/soybean meal broth (Supplementary Fig. 12) (Kuboki et al., 1999). This compound is a hydroxylated analogue of ovalicin, which was originally isolated from *Pseudeurotium ovalis* (Sigg & Weber, 1968). The ovalicins feature a highly oxygenated cyclohexane ring and two epoxide groups. Related compounds also isolated from fungi, include clovalicin from *Sporothrix* sp. (Takamatsu et al., 1996), in which the spiro epoxide of ovalicin is opened to the chlorohydrin and 5-demethyl ovalicin from *Chrysosporium lucknowense* (Son et al., 2002), which differs from ovalicin by the substitution of a hydroxyl group for the methoxy group at the C4 position. The most important related compound, fumagillin, has a similar sesquiterpenoid skeleton with a fully conjugated decatetraenedioic acid substituted at the C4 position. It was isolated from *A. fumigatus* in the late 1940s as an inhibitor of *S. aureus* 209 bacteriophage (Hanson & Eble, 1949). Fumagillin was subsequently shown to be a potent amebicide (McCowen, Callender, & Lawlis, 1951) and also showed activity against *Nosema* microsporidial diseases in honeybees and, against microsporidial infections in immune-compromised patients (Molina et al., 2002; van den Heever, Thompson, Curtis, Ibrahim, & Pernal, 2014). Reports of inhibitory activity against angiogenesis have brought it back into the spotlight as a potential chemotherapeutic for hard tumors (Ingber et al., 1990). Ovalicin exhibited immunosuppressive and antimitotic effects (Lazary & Stähelin, 1968). 12-Hydroxy-ovalicin showed potent cytotoxicity against four human cancer cell lines as well as human umbilical vein endothelial cells where its activity was very close to that of fumagillin. It also showed immunosuppressive activity in a mixed lymphocyte culture reaction assay and against L-1210 mouse leukemia cells (Kuboki et al., 1999). A candidate ovalicin biosynthetic gene is found in a pseurotin-like cluster in some sequenced *Metarhizium* strains (see Section 3.3, M-PKS14 and Section 3.5 M-TER44).

2.7.5 Taxol

Taxol® (paclitaxel) (Supplementary Fig. 13) the blockbuster cancer chemotherapeutic and other related taxanes were originally isolated from the bark of various yew tree species. The history of the structural chemistry is summarized by Kusari, Singh, and Jayabaskaran (2014). Taxol was subsequently reported from *Taxomyces andreanae*, an endophyte of *Taxus brevifolia*, Pacific yew (Stierle, Strobel, & Stierle, 1993), and since then it has been reported as a product of more than 200 species of endophytic fungi, including a plant endophyte isolated from Giant Panda dandruff (Gu et al., 2015). A controversy has arisen about whether taxol is indeed biosynthesized by fungi at all (Heinig, Scholz, & Jennewein, 2013), and if so, whether is it produced by a fungal version of the accepted plant pathway (Croteau, Ketchum, Long, Kaspera, & Wildung, 2006). Among the highest yields of taxol so far reported from an endophyte, 0.85 mg/L of fermentation broth, happens to be from *M. anisopliae* (H-27 Accession #FJ375161). Considering unresolved questions, the ultimate test of whether any *Metarhizium* species or any of a myriad of endophytes does indeed make taxanes will be the documentation of their absence in a mutant in which a fungal biosynthetic gene has been targeted and disrupted.

2.8 Miscellaneous Compound Types

N-(methyl-3-oxodec-6-enoyl)-2 pyrroline and *N*-(methyl-3-oxodecanoyl)-2 pyrroline, two substituted pyrrolines, were reported from *M. flavoviride* HF698 as weak inhibitors of plant pathogenic oomycetes *Phytophthora sojae* and *Aphanomyces cochlioides* (Supplementary Fig. 14) (Putri et al., 2014). These compounds were previously reported from *Penicillium brevicompactum* as juvenile hormone inhibitors and showed insecticidal activity against *Oncopeltus fasciatus* (Cantín et al., 1999; Moya et al., 1998).

3. MOLECULAR BASES OF SECONDARY METABOLISM IN THE GENUS *METARHIZIUM*

Biosynthesis of fungal natural products is based on three large, heterogeneous classes of core biosynthetic genes: PKSs, nonribosomal peptide synthetases (NRPSs), and terpene biosynthetic genes (Keller & Hohn, 1997). In the majority of the cases, biosynthesis of a single natural product requires the coordinated activity of multiple genes starting from production of the main backbone, followed by the tailoring of intermediates, and in many instances complemented by secretion and self-protection activities (Keller, Turner, & Bennett, 2005). To achieve functional coordination, these genes must be co-expressed, their product must be colocalized in the appropriate cellular compartment, and they must be transmitted from one generation to the other as a discrete group. It is not surprising that genes belonging to SMM pathways are physically grouped in the same genomic region to form so-called SMM gene clusters, as this organization makes simultaneous gene transmission more likely, and allows the control of gene expression by chromatin-remodeling regulators as well as pathway-specific transcription factors (Bayram et al., 2008; Bok & Keller, 2004; Brakhage, 2013; Campbell, Rokas, & Slot, 2012; Keller & Hohn, 1997; Proctor, McCormick, Alexander, & Desjardins, 2009; Slot & Rokas, 2011). The availability of sequenced fungal genomes has provided the unprecedented opportunity to identify virtually all the SMM core genes present in a particular strain. The invariable outcome of such surveys is that for a given fungal species, the number of known metabolites is much smaller than the number of core genes. Identification of the metabolites produced by silent pathways can be pursued through experimental work using a variety of approaches ranging, from the manipulation of growth conditions (Gerea et al., 2012; Marmann, Aly, Lin, Wang, & Proksch, 2014; Rateb et al., 2013; Tudzynski, 2014) to the chemical and genetic manipulation of both global and pathway-specific regulators (Bok & Keller, 2004; Chung, Wei, et al., 2013; Henrikson, Hoover, Joyner, & Cichewicz, 2009; Kosalkova et al., 2009; Wu, Oide, Zhang, Choi, & Turgeon, 2012). Comparative genomics, on the other hand, can be applied as an initial step to systematize the SMM pathways found in given strain or in a group of fungi, connect them to SMM pathways studied in other systems, and tentatively attribute genetic bases to known chemistries. Comparative genomics can also be used to place SMM genes and pathways in a taxonomic context in an attempt to better understand relationships among SM, lifestyle, and lineage evolution. For this section of the review we set out to survey and compare all the NRPS, PKS, and terpene genes in nine sequenced *Metarhizium* strains which include *M. acridum* CQMa 102, *Metarhizium album* ARSEF 1941, *M. anisopliae* ARSEF 549, *M. anisopliae* E6, *M. brunneum* ARSEF 3297, *M. robertsii* ARSEF 2575, *M. robertsii* ARSEF 23, *M. guizhouense* ARSEF 977, and *Metarhizium majus* ARSEF 297 (Gao et al., 2011; Hu et al., 2014; Staats et al., 2014). These strains encompass a timescale of about 120 million years, with species displaying different host ranges and geographic origin (Hu et al., 2014). Comparisons presented here are mainly aimed at categorizing SMM core genes in homogeneous sets or families likely to produce similar metabolites and to track their conservation within the genus. Identification of core genes was based on PFAM HMM models for adenylation (A)-domain (for NRPSs, hybrid NRPS-PKSs, and NRPS like), ketoacyl synthase (KS)-domain (for PKSs and hybrid NRPS-PKSs), terpene cyclase, and terpene synthetase domains as well as BLAST-based comparative searches. Grouping of genes in homogeneous sets was achieved by simultaneous use of Maximum Likelihood phylogenetic analysis (Price, Dehal, & Arkin, 2010), whole-protein ortholog search using Reciprocal Best Blast (RBB) (Wall, Fraser, & Hirsh, 2003) and comparison of the regions around biosynthetic core genes to those found in other fungi with MultiGeneBlast (Medema, Takano, & Breitling, 2013). Phylogenetic analyses conducted for PKSs and NRPSs was based on a panel of sequences extracted from 58 Ascomycete genomes, plus a panel of core genes experimentally linked to a natural product (84 KS-domain—containing proteins and 83 A-domain—containing proteins). Given the size of the final trees (the

A-domain tree had more than 4000 taxa), we are showing more manageable "pruned" versions of the original (Supplementary Figs. 18 and 19). The database used for MultiGeneBlast comparisons contained 64 Ascomycete genomes and 106 characterized fungal SMM biosynthetic gene clusters.

3.1 NRPS Pathways

NRPSs are enzymes capable of building peptidal natural products using amino acids and, less often, hydroxy acids as building blocks. These component parts are selected and activated by an A-domain while tethered to the synthase through a phosphopantetheinyl arm provided the thiolation (T)-domain. The condensation (C)-domain forms the amide or the ester bond between the growing chain and the new building unit. The C—A—T set constitutes a module, and in the vast majority of fungal NRPSs, one module corresponds to a single building block incorporated into a peptide. However, there are exceptions, and some fungal NRPSs (eg, enniatin, ferrichrome, and coprogen synthetases) can operate iteratively. Besides an astonishing variety of amino acids (including many not found in proteins), NRPSs can incorporate hydroxy acids, amines, and lipidic side chains derived from a PKS or a fatty acid synthase (Chen et al., 2013; Chiang et al., 2008; Jin et al., 2010; Marahiel, Stachelhaus, & Mootz, 1997). NRPSs can also carry modifying domains (epimerization, N-methyltransferase) that further increase the diversity of their end-products. A- and C-domains are selective toward the kind of building blocks destined to be incorporated into the elongating chain (Ehmann, Trauger, Stachelhaus, & Walsh, 2000; Gao et al., 2012; Rausch, Hoof, Weber, Wohlleben, & Huson, 2007; Stachelhaus, Mootz, Bergendahl, & Marahiel, 1998; Stachelhaus, Mootz, & Marahiel, 1999). Using phylogenetic analysis based on A-domains, NRPSs can be classified into groups, which can, at least in part, resolve protein families and help predict the nature of the final product (Bushley & Turgeon, 2010). NRPS-like enzymes form a distinct group of proteins that deviate from the canonical NRPS structure in that they resemble monomodular NRPSs, do not possess a C-domain, and usually carry one or more additional domains at the protein carboxy terminus not normally found in NRPSs. Their phylogenetic association with canonical NRPSs and other adenylating enzymes appears complex. For many NRPS-like enzymes, A-domain—based phylogenetic analysis supports a common origin with A-domains found in NRPSs, but they do not necessarily participate in SMM. Examples of NRPS-like enzymes nested within the larger NRPS clade are α-amino adipate reductase (AAR/LYS2), *Chnps10*, and *Chnps12/tmpL* (Bushley & Turgeon, 2010).

In other cases NRPS-like enzymes are more closely related to acyl-CoA synthetases (which are mostly involved in primary metabolic pathways) than canonical NRPSs. Yet, these NRPS-like enzymes participate in SMM, as do proteins linked to ochratoxin production and *tdiA*-like synthetases (Balibar, Howard, & Walsh, 2007; Karolewiez & Geisen, 2005). Using A-domain–based phylogenetic analysis (Supplementary Fig. 18), orthology, and composition of the genomic regions, we were able to group NRPS and NRPS-like genes from the analyzed *Metarhizium* strains into 27 and 14 families/pathways, respectively (Supplementary Tables 1 and 2). What follows is a description of these families.

M-NRPS1 is a three-module NRPS (Fig. 1-A) not found in *M. acridum* (Supplementary Table 1). M-NRPS1–like genes are found in *Ophiocordyceps sinensis*, *Tolypocladium inflatum*, and *Hirsutella thompsonii*. With the exception of an ABC transporter, genes found around M-NRPS1 are not conserved beyond the genus *Metarhizium*. No orthologous NRPS and associated clusters were identified in other Ascomycota.

M-NRPS2 is a monomodular NRPS (Fig. 1-B) found in all examined *Metarhizium* strains and in several Hypocreales. Phylogenetic analysis groups the A-domain with M-NRPS20 A2 (Supplementary Fig. 18). The presence of an N-terminal C-domain indicates that this synthetase may accept a specialized starter moiety (Rausch et al., 2007).

M-NRPS3 is a monomodular NRPS linked to the HR-PKS M-PKS9A presumably by a divergent promoter. We identified three variants of this synthetase (Supplementary Table 1). *M. robertsii*, *M. anisopliae*, *M. guizhouense*, and *M. majus* share a similar genic context around the variant dubbed M-NRPS3A (Fig. 1-C). *M. acridum* and *M. guizhouense* have a similar gene (M-NRPS3B) linked to the HR-PKS M-PKS9B and placed in a slightly different genic context. *M. majus* may possess pseudogenes derived from M-NRPS3B (models MAJ_10914 and MAJ_10915) colocated with M-PKS9B. Both genes are found on a small contig. In M-NRPS3A and -B the first C-domain may be used to accept the reduced polyketide moiety produced by their respective companion PKSs (Rausch et al., 2007). The single A-domain present in both M-NRPS3A and -B, group with A1/A3 of the *Cochliobolus carbonum* HC toxin synthetase, a cyclic peptide incorporating 2-amino-9,10-epoxi-8-oxodecanoic acid (AEO) produced by the fatty acid synthase TOXC (Ahn & Walton, 1997). The described gene clusters do not have significant matches in examined Ascomycetes. A third variant of M-NRPS3 (M-NRPS3C) is found only in *M. acridum* where it is located close to the M-PKS9C (Fig. 2-F).

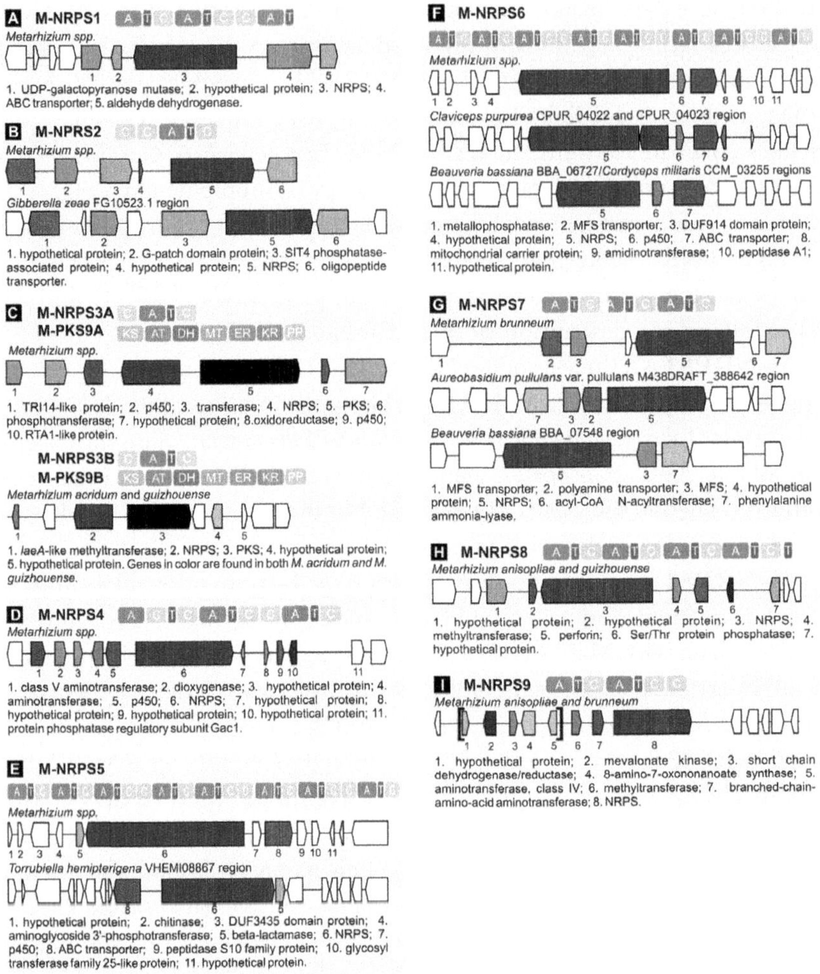

Figure 1 Structure and neighboring genes associated with NRPS genes identified in nine *Metarhizium* species. Domain name abbreviations: *C*, condensation; *A*, adenylation; *T*, thiolation; and *nMT*, N-methyltransferase. (See color plate)

M-NRPS4 is a three-module NRPS, found in all *Metarhizium* strains compared in this study. No similar gene or genomic regions were identified outside the genus *Metarhizium* (Fig. 1-D). Each A-domain has a different phylogenetic affiliation: A1 and A3 are nested within the peramine/DTX clade; A2 is part of the CCS clade (vide infra); A3 clusters within the *Cochliobolus heterostrophus NPS9* A3 clade (Supplementary Fig. 18).

M-NRPS5 and M-NRPS6 are eight-module NRPSs that share structure, phylogenetic relationship and limited presence beyond the

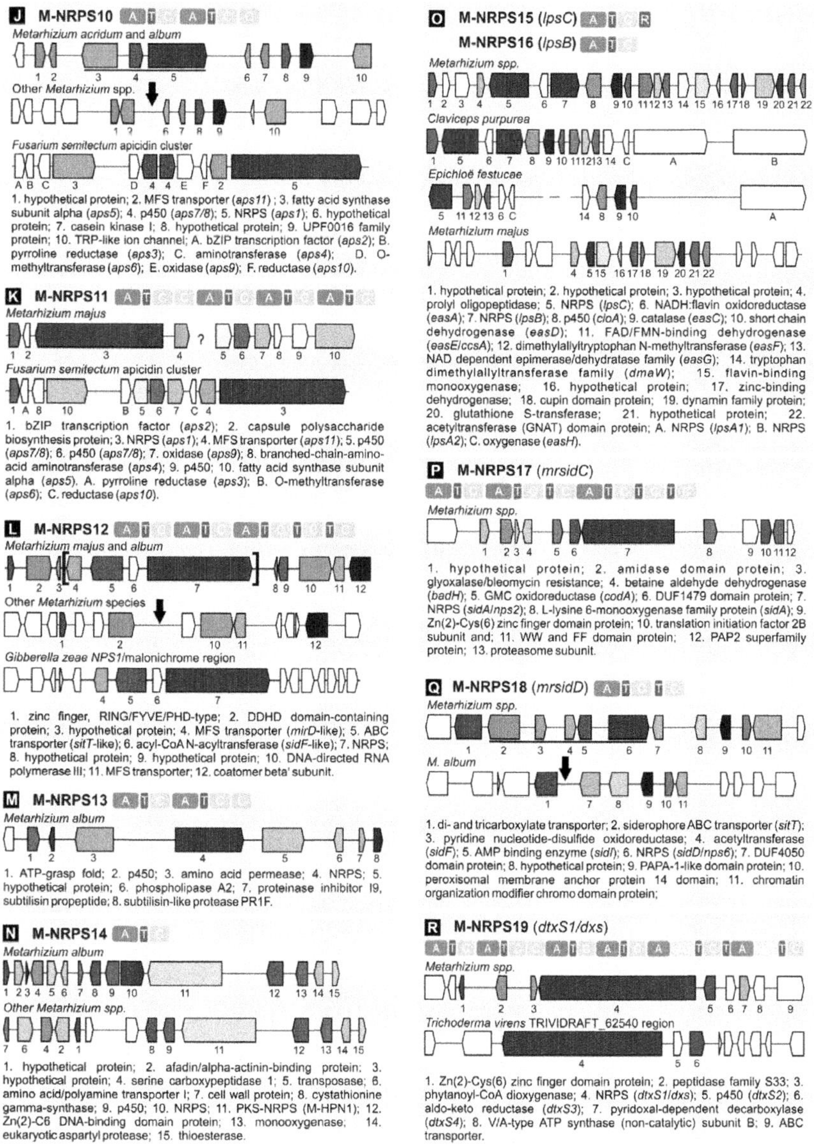

Figure 1 (*continued*).

Clavicipitaceae. For brevity we refer to these and related NRPSs as Clavicipitaceae- and Cordycipitaceae-specific (CCS) NRPSs. With the exception of A8 form M-NRPS5 (which is basal to the peramine clade), all A-domains from M-NRPS5 and M-NRPS6 form a well-supported monophyletic

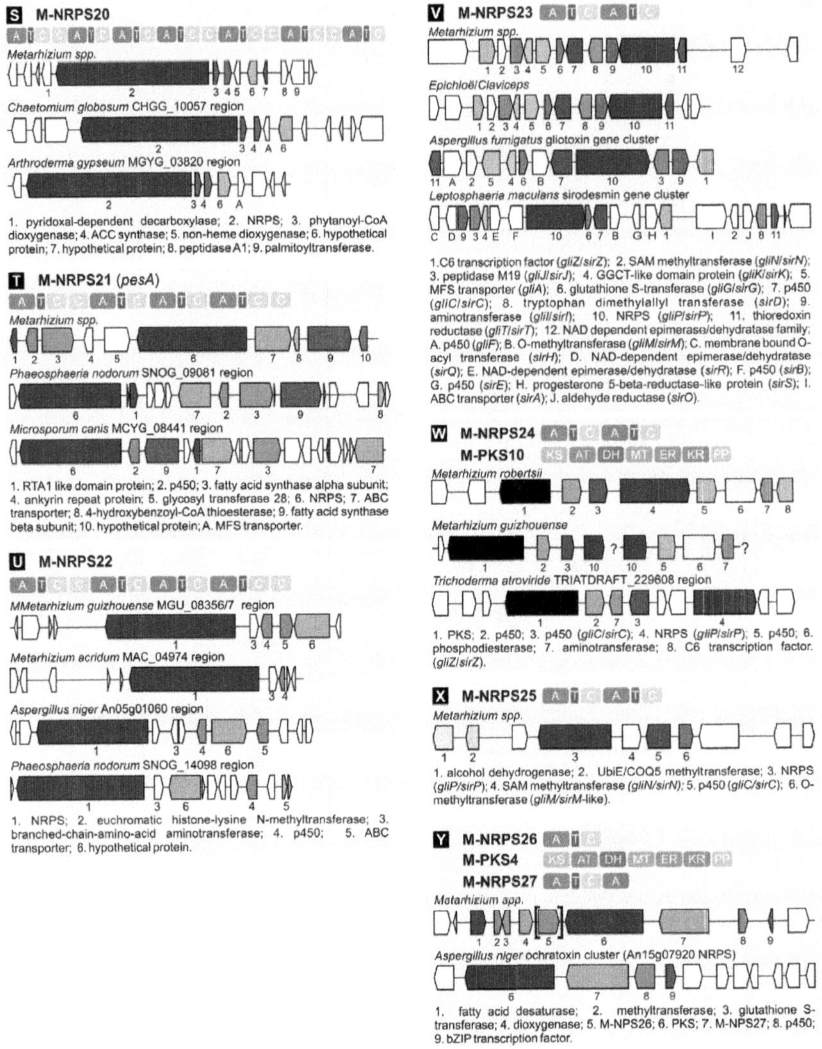

Figure 1 ***(continued)*****.**

group. This group likely arose by a process of duplication and recombination from an ancestor of the M-NRPS6 A2-domain as its subclade is basal to the CCS clade (Supplementary Table 1). The emergence of lineage-specific multimodular synthetases has well-known examples in the 11-module *T. inflatum* cyclosporine synthetase SIM1 and the 18-module *Trichoderma* peptaibol synthetase TEX1 (Bushley & Turgeon, 2010). Remarkably,

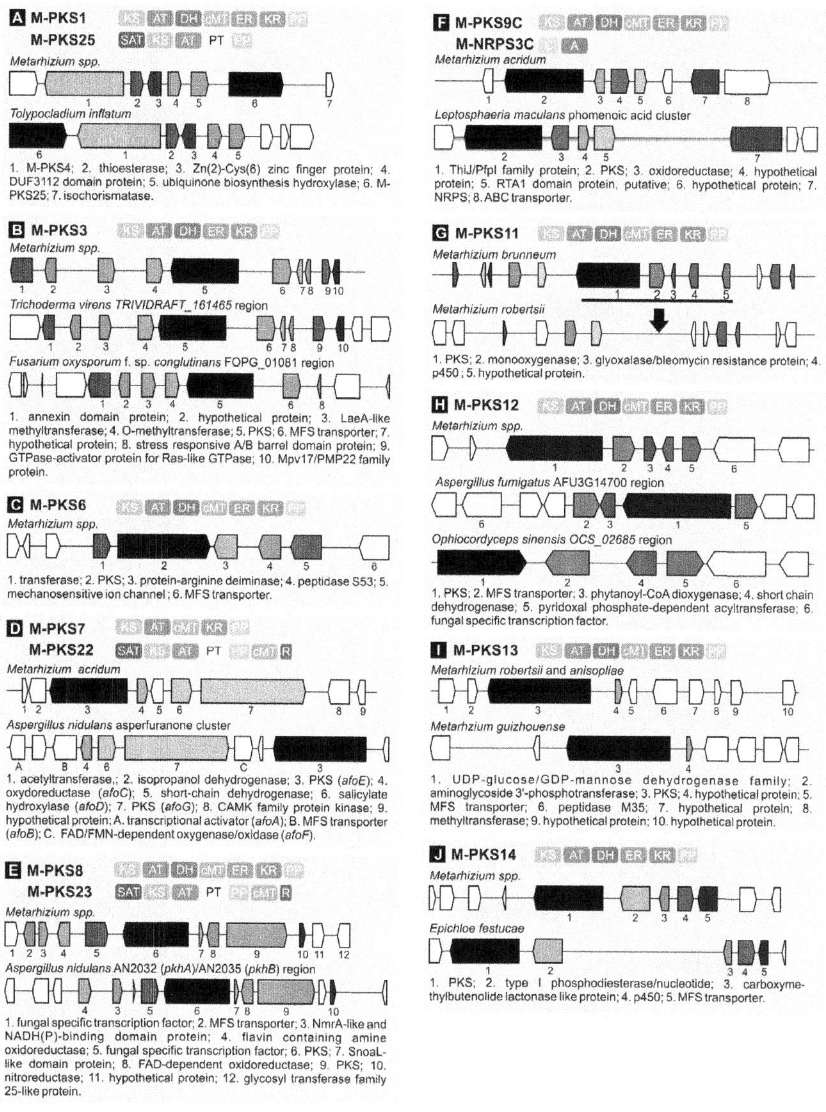

Figure 2 Structure and neighboring genes associated with PKS families identified in nine *Metarhizium* species. Domain name abbreviations: *KS*, ketoacyl synthetase; *SAT*, starter unit:ACP transacylase; *AT*, acyltransferase; *DH*, dehydratase; *cMT*, C-methyltransferase; *ER*, enoyl-reductase; *KR*, ketoreductase; *PT*, product template; *PP*, 4′-phosphopantetheine attachment; *R*, thioester reductase; *CARN*, choline/carnitine O-acyltransferase; *C*, condensation. (See color plate)

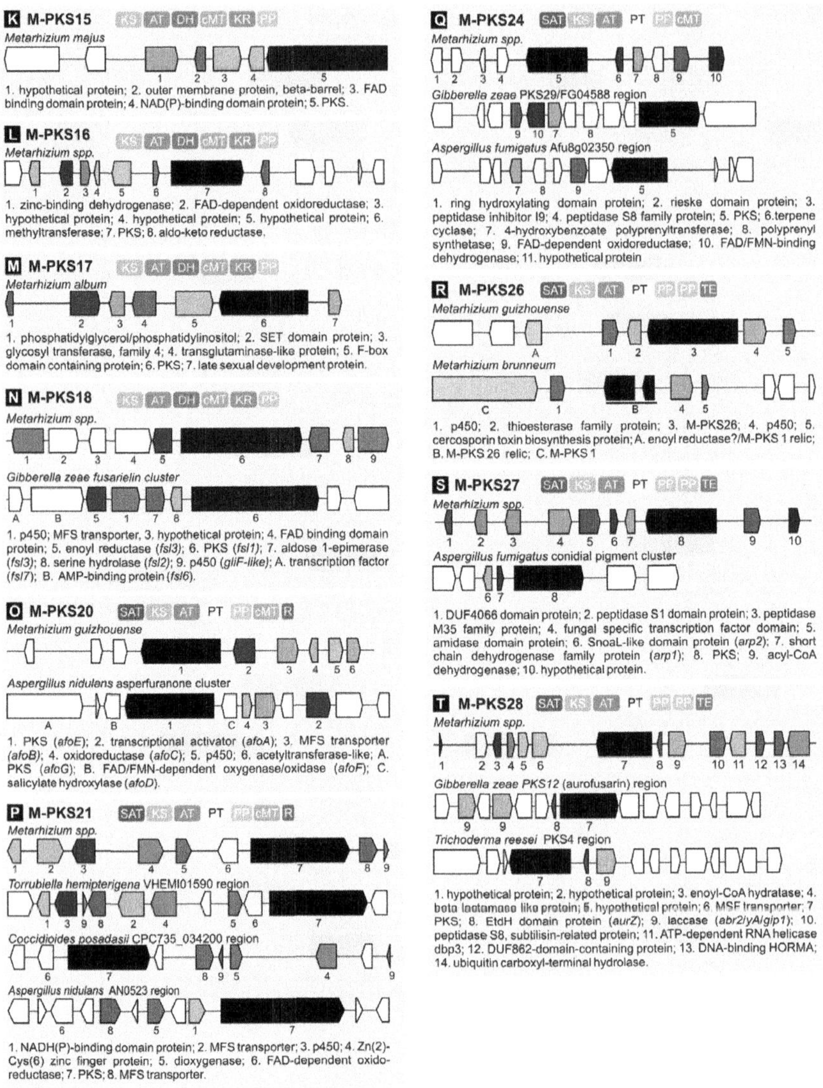

Figure 2 (*continued*).

with the exception of *M. album*, all the *Metarhizium* strains harbor both M-NRPS5 and M-NRPS6, with each gene showing the identical module arrangement in all analyzed species. The stability of these genes over the course of *Metarhizium* evolutionary history may indicate a strong selective pressure favoring the current module configuration, or the comparatively

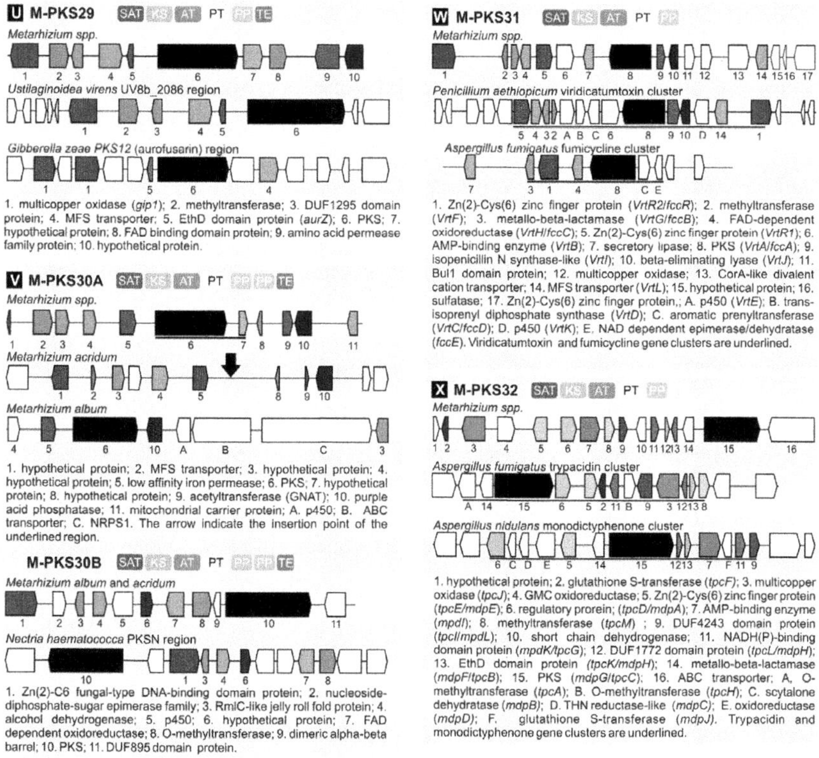

Figure 2 (*continued*).

recent emergence of this group of fungi (Hu et al., 2014). Genes surrounding M-NRPS5 and M-NRPS6 are also remarkably syntenic in all the examined *Metarhizium* strains. Comparing M-NRPS5 with M-NRPS6 genomic regions indicates that the ancestral gene cluster may have included an ABC transporter and a cytochrome p450 (Fig. 1-E and 1-F). The CCS clade also includes A-domains from *B. bassiana* (BBA_10105, BBA_06727), *C. militaris* (CCM_03255), *Claviceps purpurea* (CPUR_04022, CPUR_04023), *Epichloë festucae* (EfO2.006280.1, EfO2.006290.1, EfO2.043710.1, EfO2.043720.1), and *Torrubiella hemipterigena* (VHEMI08867) NRPSs (Fig. 1-E). A2 from the trimodular M-NRPS4 can be considered part of the CCS clade, and it is likely to derive from the M-NRPS6 A2 ancestor.

M-NRPS7 is a three-module NRPS found only in *M. brunneum* (Fig. 1-G). Genes encoding proteins with nearly identical domain structure (including a partial A2) and phylogenetic. Similar putative clusters are found in *B. bassiana* and *Aureobasidium pullulans* (Fig. 1-G).

M-NRPS8 is a four-module NRPS present only in *M. anisopliae* and *M. guizhouense* (Supplementary Table 1; Fig. 1-H). In other *Metarhizium* species the location contains an identical gene organization as in Fig. 1-H with an NRPS pseudogene. No similar cluster was located in other Ascomycota. The four A-domains have polyphyletic associations with various other NRPSs (Supplementary Fig. 18).

M-NRPS9 is a bimodular NRPS found only in *M. anisopliae* and *M. brunneum*. Both A-domains cluster within the DXS/peramine clade (Supplementary Fig. 18). Genes enclosed in square brackets in Fig. 1-I are missing in *M. anisopliae*. In other *Metarhizium* species the surrounding genes are conserved, which indicates a likely insertion of this cluster in the *M. anisopliae/M. brunneum* ancestor. The cluster does not have significant matches in other Ascomycota. It may produce a cyclic product.

M-NRPS10 is a bimodular NRPS found only in *M. acridum* and *M. album*, although the region surrounding the putative cluster is conserved in other *Metarhizium* species (Fig. 1-J). A1 groups with enniatin synthase A2 clade; A2 lays within the *C. heterostrophus* ChNPS1/ChNPS3 A1/A3 clade. The cluster associated with M-NRPS10 also has some similarities with that of histone deacetylase inhibitor apicidin found in *Fusarium seminfectum* (Jin et al., 2010) (Fig. 1-J). This metabolite is synthesized by a four-module NRPS, and the O-methyltransferase *aps6* is the only gene absolutely required for apicidin biosynthesis missing in the *Metarhizium* cluster (Jin et al., 2010) (Fig. 1-J).

M-NRPS11 is a four-module NRPS found only in *M. majus*. Its A domains display one-to-one phylogenetic relationship with both *F. seminfectum* apicidin and *C. carbonum* HC-toxin synthetases (Fig. 1-K). The small contig harboring M-NRPS11 (AZNE01000191, 29,839 nucleotides), plus the contigs AZNE01000236 (19,938 nucleotides) harbor eight of 11 genes found in the apicidin biosynthetic cluster (Jin et al., 2010) (Fig. 1-K). Both HC toxin and apicidin are cyclic tetrapeptides with a long aliphatic R-group (AEO in HC toxin and the related 2-amino-8-oxo-decanoic acid in apicidin) (Darkin-Rattray et al., 1996; Walton, 2006). Both are inhibitors of histone deacetylases (HDACs) and HC toxin is considered to be a host-selective toxin as it confers high virulence toward maize cultivars lacking functional carbonyl reductase HM1 and/or HM2 (Sindhu et al., 2008; Walton, 2006). Apicidin displays broad-spectrum antiprotozoan and anticancer activities (Darkin-Rattray et al., 1996; Kim et al., 2004).

M-NRPS12 is found in *M. album* and *M. majus* (Supplementary Table 1). Phylogenetic analysis places M-NRPS12 within the ferrichrome siderophore clade with a one-to-one module correspondence with *Gibberella zeae* NPS1/FGSG_11026. This NRPS was linked to the production of malonichrome, a ferrichrome family siderophore (see Section 2.1.4) identified in *F. roseum* (Emery, 1980) and found to assist the main extracellular siderophore in iron uptake in *G. zeae* (Oide, Berthiller, Wiesenberger, Adam, & Turgeon, 2015). The putative M-NRPS12 gene cluster (Fig. 1-L) includes MFS and ABC transporters as well as an acyl-CoA N-acyltransferase that is similar to *Aspergillus nidulans mirD*, *sitT*, and *sidF*, respectively, which are genes involved in either siderophore mobilization or biosynthesis (Schrettl et al., 2008). This group of genes is inserted in a region conserved in other *Metarhizium* strains (Fig. 1-L).

M-NRPS13 is a two-module NRPS found only in *M. album* (Supplementary Table 1) probably producing a cyclic peptide. No orthologs or similar putative gene clusters (Fig. 1-M) were identified in other fungi in our dataset. The closest match was an uncharacterized NRPS from *Glarea lozoyensis* ATCC 20868 displaying the same domain structure.

M-NRPS14 is a monomodular synthetase present only in *M. album* (Supplementary Table 1) where it is located next to the conserved hybrid PKS–NRPSs M-HPN1 (Fig. 1-N). In other *Metarhizium* species the same location has experienced several rearrangements (Fig. 1-N). Unexpectedly, the M-NRPS14 A-domain groups with sidL-type acyl CoA ligases, outside the large NRPS clade (Supplementary Fig. 18).

M-NRPS15 and M-NRPS16 are two monomodular NRPSs that form part of an ergot alkaloid–like pathway identified in *M. acridum* and *M. robertsii* (Gao et al., 2011). Expanded sequencing within *Metarhizium* clearly shows that the same ergot-alkaloid cluster is present in all the analyzed species except for *M. majus* (Supplementary Table 1; Fig. 1-O). Ergot alkaloids have been identified in several, often taxonomically unrelated fungi including *Claviceps*, *Aspergillus*, *Penicillium*, and *Epichloë* (Wallwey & Li, 2011). Archetypical pathways include one from *C. purpurea* leading to the production of ergot D-lysergic acid peptides alongside D-lysergic acid alkanolamides, and one from *A. fumigatus* which produces variants (eg, fumigaclavine C) lacking in peptide moieties. Remarkably, all the genes required for the biosynthesis of ergometrine have orthologs in *Metarhizium* (Fig. 1-O) (Ortel & Keller, 2009; Tsai, Wang, Gebler, Poulter, & Schardl, 1995; Wallwey & Li, 2011). *Metarhizium*, on the other hand, does not possess the trimodular NRPSs (*lpsA1* and *lpsA2*) responsible for the formation of tripeptide derivatives of D-lysergic acid (eg, ergotamine). It is possible that many *Metarhizium* species can produce lysergic acid α-hydroxyethylamide or ergonovine or a similar product (Young et al., 2015), although no compound consistent with this biogenetic origin has been reported from the genus. *M. majus* retains an *lps*C-like pseudogene indicating that the cluster was once present and subsequently lost.

M-NRPS17 and M-NRPS18 are siderophore biosynthetic NRPSs responsible for the biosynthesis of ferricrocin and coprogen-type siderophores (metachelins, in *M. robertsii*), respectively (see Sections 2.1.3 and 2.1.4) (Donzelli et al., 2015; Krasnoff et al., 2014). M-NRPS17 and -18 and their genomic regions are conserved in all *Metarhizium* strains (Fig. 1-P and -Q), except for *M. album* which lacks several genes involved in the biosynthesis of coprogen-type siderophores (Haas et al., 2008), including M-NRPS18 (underlined genes in Fig. 1-Q) (Donzelli et al., 2015). In place of a coprogen-type siderophore *M. album* may produce a malonichrome-like siderophore synthesized by M-NRPS12 (Supplementary Table 1 and Fig. 1-L). In pathogenic fungi impairment of either siderophore iron acquisition or siderophore iron-retention systems, lead to sensitivity to stressors, diminished virulence, and reduced fitness (Hof et al., 2007; Lopez-Berges et al., 2012; Oide et al., 2015, 2006; Schrettl et al., 2004). Targeted gene knockouts of the ferricrocin synthetase M-NRPS17/*mrsidC* in *M. robertsii* ARSEF 2575 resulted in a number of phenotypic changes including increased sensitivity to oxidative stress, sensitivity to iron depletion, and reduced germination speed (Donzelli et al., 2015). Compared to the wild type, ΔM-NRPS17/Δ*mrsidC* strains were about threefold less virulent against *Spodoptera exigua* larvae (Donzelli et al., 2015). Loss of the metachelin synthetase M-NRPS18/*mrsidD* was associated with increased sensitivity to iron deprivation mediated by Fe^{2+} chelators and increased mycelial sensitivity to hydrogen peroxide (Donzelli et al., 2015). As expected M-NRPS18/*mrsidD* was up-regulated in low iron conditions *in vitro* (Donzelli et al., 2015). However, during *S. exigua* infection, M-NRPS18/*mrsidD* was up-regulated only transiently, and its loss had no significant effect on *M. robertsii* virulence (Donzelli et al., 2015).

M-NRPS19 corresponds to the six-module DTX synthetase *dtxS1*/*dxs* (see Section 2.1.1). The genetic basis of DTX production was discovered independently by two research groups (Donzelli, Krasnoff, Sun-Moon, Churchill, & Gibson, 2012; Wang, Kang, Lu, Bai, & Wang, 2012). The core gene responsible for initiating DTX production was deleted in *M. robertsii* ARSEF 23 and *M. robertsii* ARSEF 2575 (MAA_10042, *dtxS1* and X797_010963, *dxs*, respectively). Wang et al. (2012) also determined the role of three additional genes (the p450 *dtxS2*, the aldo-keto reductase dtxS3, and the decarboxylase *dtxS4*) found in the DTX biosynthetic clusters by generating their respective deletion mutants. *ΔdtxS2* strains were able to produce only DTX B, B2, and desmethyl-B. When DTX B was fed to either *ΔdtxS1* or *ΔdtxS1/ΔdtxS2* mutants, production of DTX A, C, D, and E was restored only in the single mutant. These experiments elegantly established DTX B and B2 as end-products of the NRPS proper and *dtxS2* as an enzyme required for the appearance of DTXs C, D, A, and E. In the proposed biosynthetic scheme *dtxS2* alone catalyzes hydroxylation of one of the δ-carbons of the 2-hydroxy-4-methylpentanoic acid (HMPA) unit in DTX B to yield DTX C. A second hydroxylation and oxidation add the carbonyl group on the same carbon to afford DTX D. Then, the decarboxylation of DTX D combined with a reduction of the four to five bond of the 2-hydroxy-4-methylpentane dioic acid unit yields the olefin group in the 2-hydroxypent-4 enoic acid unit in DTX A. A final epoxidation of the olefin yields DTX E. Although catalysis of multiple successive oxidations by a single p450 has been reported from fungi (Lin et al., 2014), confirmation of this extreme multifunctionality in a p450 would require showing that *ΔdtxS2* cannot perform the predicted conversions of all intermediates in the proposed path from DTX B > C > D > E. The possibility remains that additional DTX tailoring enzymes are located elsewhere in the genome.

Δ*dtxS3* and *ΔdtxS4* knockouts were unable to produce DTX, and feeding experiments indicated that the probable function of the *dtxS3* product is to convert α-ketoisocaproic acid into HMPA, which is the hydroxy acid constituent of DTX B (Supplementary Fig. 18), while *dtxS4* decarboxylates aspartic acid to furnish β-alanine, another DTX constituent (Wang et al., 2012).

The survey of *Metarhizium* genomes included in this study shows that the biosynthetic capacity for DTXs is present in all the species surveyed except for *M. acridum* and *M. album* (Supplementary Table 1). Phylogenetic analysis based on A-domains assign the first four A-domains of the *dtxS1/dxs* to a clade also harboring both A-domains of the peramine synthetase. While *dtxS1/dxs* A5 and A6 form a sister clade to a larger group mostly harboring A-domains involved in the activation of N-methylated amino acids, which includes *T. inflatum* cyclosporine SIM1 and *F. equiseti* enniatin synthetase ESYN1 A2 (Supplementary Fig. 18). Using phylogenetic analysis we also determined that the uncharacterized six-module NRPS from *Trichoderma virens* TRIVIDRAFT_62540 has a one-to-one module homology with *dtxS1/dxs*, and a putative gene cluster contains orthologs of *dtxS2* (TRIVIDRAFT_212112) and *dtxS3* (TRIVIDRAFT_219641) (Fig. 1-R).

The earliest findings that DTXs were acutely toxic to insects (Kodaira, 1961, 1962; Roberts, 1966, 1969; Tamura, Kuyama, Kodaira, & Higashikawa, 1964) sparked interest in establishing the role that these toxins might play in pathogenicity and virulence of DTX-producing *Metarhizia*. Thus, the detection of DTXs in mycosed insects, the congruence of the symptomology of DTX intoxication (physiological tetanus followed by flaccid paralysis) with the symptomology in mycosed insects, and the positive correlation between DTX production levels and virulence in naturally occurring strains supported the notion of DTXs as virulence factors (Kershaw, Moorhouse, Bateman, Reynolds, & Charnley, 1999; Samuels, Charnley, & Reynolds, 1988). Intriguing reports of inhibitory effects of DTX on components of the insect immune response added further support to this hypothesis (Huxham, Lackie, & McCorkindale, 1989; Pal, St. Leger, & Wu, 2007; Vilcinskas et al., 1997b). However, *M. robertsii* DTX-minus strains showed little or insignificant changes in virulence whether insects were inoculated by injection (Wang et al., 2012) or by the topical route (Donzelli et al., 2012). On the other hand Δ*dtxS2* (still producing DTX B and B_2) showed a slight but significant reduction in virulence compared to the Δ*dtxS1* (where DTX production is completely abolished) (Wang et al., 2012).

While DTXs may contribute to a redundant system producing toxicity to an insect host, the lack of a dramatic reduction in virulence observed in independently produced DTX-minus strains supports the view that *Metarhizium* virulence and host range are much more complex phenomena than the effect of a single toxic metabolite can account for.

M-NRPS20 corresponds to the gene responsible for the biosynthesis of serinocylins (see Section 2.1.2) (Moon et al., 2008). Genomes of all *Metarhizium* species sequenced to date contain a copy of the serinocyclin synthetase and a highly conserved gene cluster (Fig. 1-S). The ACC synthase and a dioxygenase present in the putative cluster are likely to be involved in the conversion of methionine into AAC and proline into hydroxyproline, respectively (Houwaart, Youssar, & Huttel, 2014; Kakuta et al., 2001). A third gene, encoding for a pyridoxal-dependent decarboxylase (67% amino acid identity to DTX biosynthesis *dtxS4*) may be responsible for the conversion of glutamine to β-alanine (Wang et al., 2012). A gene encoding for a phytanoyl-CoA dioxygenase may be responsible for lysine hydroxylation. A-domain—phylogenetic analysis groups six of seven serinocyclin synthetase A-domains in a monophyletic clade sister to that harboring ergot-alkaloid synthetases *lpsA*1 and *lpsA*2 from *C. purpurea* (Supplementary Fig. 18).

Serinocyclin synthetase A2 forms a distinct group which includes A2 from *A. fumigatus* brevianamide F synthase *ftm*A as well as M-NRPS1 A3 and the single M-NRPS2 A-domain (Supplementary Fig. 18). Serinocyclin synthetase structure and genomic region are nearly identical to those found around the *Chaetomium globosum* seven-module NPS CHGG_10057 and dermatophyte *Arthroderma gypseum* seven-module MGYG_03820 (Fig. 1-S). Because of their similarity to the *Metarhizium* pathway these clusters are expected to produce a serinocyclin-like molecule.

M-NRPS21 corresponds to *pesA*, which encodes a four-module protein with unknown function and unknown product sequenced by Bailey et al. (1996). The *pesA* gene and putative cluster are not found in *M. acridum* (Supplementary Table 1). In other strains the putative gene cluster is almost completely conserved and the version found in *M. robertsii* is depicted in Fig. 1-T. Gene composition of the *pesA* region is similar to those surrounding the NRPSs AN7884 (*A. nidulans*, six modules), SNOG_09081 (*Parastagonospora nodorum*, four modules); PTRG_09101 (*Pyrenophora tritici-repentis*, five modules), and MCYG_08441 (*Microsporum canis*, five modules) (Fig. 1-T). The M-NRPS21/*pesA* genomic region is also similar to the apicidin biosynthetic gene cluster as it contains homologues of the p450s *aps7*/*aps8*, the fatty acid synthase alpha subunit *aps5*, and the ankyrin repeat protein *aps2* (Jin et al., 2010). The M-NRPS21 cluster includes two additional genes: an ankyrin repeat/bZIP protein (X797_008665/MAA_01641) and a glycosyltransferase 28 (X797_008666/MAA_01640). The bZIP protein may have a regulatory function; the GT28 belongs to a family of proteins involved in the biosynthesis of lipo-glyco peptide antibiotics in bacteria (eg, vancomycin), and no orthologs were identified outside the *Metarhizium* genus, although proteins belonging to this family are found in other Ascomycota.

M-NRPS22 is found only in *M. acridum* and *M. guizhouense* (Supplementary Table 1), and shows similarity to M-NRPS21/pesA in several features including module number, phylogenetic grouping of the A-domains and presence of two homologous genes (p450 and ABC transporter) within the putative biosynthetic cluster (Fig. 1-U). All A-domains in M-NRPS22 appear to be homologues to those in M-NRPS21/*pesA*, but their position is shuffled (Supplementary Fig. 18). M-NRPS22-like synthetases and gene clusters are found in other Ascomycota, but none of these NRPSs has yet to be associated with an SM (Fig. 1-U).

M-NRPS23, M-NRPS24, M-NRPS25, and **M-NRPS26** were identified by phylogenetic analysis as four distinct *Metarhizium* NRPSs grouping with the two-module NRPSs responsible for the production of the epipolythiodiketopiperazine (ETP) metabolites sirodermin (*sirP*) and gliotoxin (*gliP*) (Gardiner, Cozijnsen, Wilson, Pedras, & Howlett, 2004; Gardiner & Howlett, 2005) (Supplementary Fig. 18). Diketopiperazines are found in fungi and prokaryotes. They are dipeptides in which two amino acids both amidate and acylate each other in head-to-tail fashion to form a six-membered ring (Prakash et al., 2002). Many fungi produce ETP derivatives and some have been particularly prominent because of their association with virulence in *A. fumigatus* (Spikes et al., 2008; Sutton, Newcombe, Waring, & Mullbacher, 1994), *T. virens* (Atanasova, Le Crom, et al., 2013; Vargas et al., 2014), and *Leptosphaeria maculans* (Elliott et al., 2007). Several *Metarhizium* strains carry pathways likely to produce diketopiperazines. The bimodular *Metarhizium* M-NRPS23 and M-NRPS24 display a one-to-one A-domain phylogenetic relationship with the bimodular sirodesmin (*sirP*) and gliotoxin (*gliP*) synthetases (Supplementary Fig. 18). M-NRPS23 and M-NRPS24 differ in their putative biosynthetic cluster organization and distribution. M-NRPS23 is found in all *Metarhizium* species except *M. acridum* and *M. album*, while M-NRPS24 is restricted to *M. robertsii* and possibly to *M. guizhouense* (Supplementary Table 1). Both M-NRPS23 and M-NRPS24 regions share significant similarities to the gliotoxin and sirodesmin biosynthetic clusters (Fig. 1-V). The region surrounding M-NRPS23 is more similar to the sirodesmin cluster as it carries a prenyltransferase, which is not found in the gliotoxin pathway and it is virtually identical to a gene cluster found in both *E. festucae* and *C. purpurea* (Fig. 1-V). The M-NRPS24 genomic context is less similar to characterized ETP pathways, but similar to the region surrounding the NRPS TRIATDRAFT_229608 from *Trichoderma atroviride*. In *M. brunneum* the M-NRPS24 region is largely conserved with the exception of M-NRPS24 itself. In *M. guizhouense* M-NRPS24 appears to be split between scaffolds 77 and 219 (Fig. 1-W). It is unclear whether M-PKS10 is a functional part of the M-NRPS24 cluster. M-NRPS25 is an ETP-like monomodular NRPS found in some *Metarhizium* species (Supplementary Table 1). It carries T- and C-domains followed by a full NRPS module. M-NRPS25 and three additional genes (a p450 and two methyltransferases) are found consistently associated, and all display high similarity to homologues in known ETP pathways (Fig. 1-X). In *M. acridum* the reducing M-PKS2 is located in close proximity to M-NRPS25, while in the other *Metarhizium* species there is no such association.

The bimodular M-NRPS26 groups with the ETP2 clade, and it is found in *M. brunneum*, *M. anisopliae*, and *M. robertsii* (Supplementary Table 1). This

NRPS is nested in a more conserved gene cluster found around M-NRPS27/M-PKS4 couple (Fig. 1-Y).

Examples of fungal diketopiperazines among entomopathogenic species include SCH 54794, SCH 54796, and SCH 56396 from a *Tolypocladium* sp. (Chu et al., 1993; Chu, Truumees, Mierzwa, Patel, & Puar, 1997), bisdethiodi (methylthio)-1-demethylhyalodendrin and 1-demethylhyalodendrin tetrasulfide and the dimer vertihemiptellide from *T. hemipterigena* (Isaka, Palasarn, Rachtawee, Vimuttipong, & Kongsaeree, 2005; Nilanonta et al., 2003), and terezine D from *Paecilomyces cinnamomeus* and its teleomorph, *T. luteorostrata* (Isaka, Palasarn, Kocharin, & Hywel-Jones, 2007). Inasmuch as depsipeptides incorporating hydroxy acids by ester linkages can be biosynthesized by canonical NRPSs, it seems possible that the combination of an α-hydroxy acid and an amino acid to form a diketomorpholine ring as in metacytofilin (see Section 2.2.2) could be produced by genes that are predicted to produce diketopiperazines.

M-NRPS27 is a bimodular synthetase phylogenetically associated with ergot-type NRPSs and conserved in some of the *Metarhizium* species (Supplementary Table 1), where it is colocated with M-PKS4. Gene organization around M-NRPS27 is very similar to that found in ochratoxin clusters identified in several *Aspergilli* (Ferracin et al., 2012; Gallo et al., 2012; O'Callaghan, Caddick, & Dobson, 2003), and to that of An15g07910 (NRPS) and the An15g07920 (PKS) from *Aspergillus niger* strain CBS 513.88 (Fig. 1-Y) (Ferracin et al., 2012).The *A. niger* and *Metarhizium* homologues have identical module/domain structure. Moreover, both putative clusters contain a p450 and a bZIP transcription factor (Fig. 1-Y).

3.2 Nonribosomal Peptide Synthetase–Like Pathways

M-NPL1 is an NRPS-like gene restricted to some *Metarhizium* species (Supplementary Table 2). The single A-domain forms a sister clade to the enniatin synthetases ESYN1 A1 (Supplementary Fig. 18). In *M. majus* M-NPL1 includes a second A-domain at the N-terminus (Supplementary Fig. 15-A).

M-NPL2 is found in all the *Metarhizium* strains surveyed except for *M. acridum* and *M. album* (Supplementary Table 2), with no orthologs found outside the genus *Metarhizium*. In *M. robertsii*, *M. anisopliae*, and *M. brunneum*, the region containing M-NPL2 also harbors M-NPL6, which phylogenetic analysis tightly groups with M-NPL2 (Supplementary Fig. 18). A similar genomic region is present in *T. virens* (Supplementary Fig. 15-B).

M-NPL3A, **M-NPL3B**, and **M-NPL7** are NRPS-like genes phylogenetically associated with the ochratoxin clade (Supplementary Fig. 18). M-NPL3A and M-NPL3B are paralogs and the gene organization around each one is conserved across all *Metarhizium* species, but it differs between the two (Supplementary Fig. 15-C and 15-D). Both clusters lack a PKS as found in the ochratoxin biosynthetic pathways (Ferracin et al., 2012; Gallo et al., 2012). The gene organization around M-NPL3B is also conserved in *E. festucae* (Supplementary Fig. 15-D). M-NPL7 is colocalized with M-PKS5, a gene that carries a carnitine O-acyltransferase domain at its C-terminus (Supplementary Fig. 15-G). The M-NPL7/M-PKS5 cluster is not present in *M. acridum* and *M. album* (Supplementary Table 2), and organization around these core genes is partially conserved in different *Metarhizium* species (Supplementary Fig. 15-G). M-NPL7 may lead to a mixed PKS-NRPS product, but it is unclear how similar it is to ochratoxin.

M-NPL4 and **M-NPL5** are NRPS-like enzymes belonging to the same phylogenetic group as the *A. nidulans* terrequinone (*tdiA*, AN8513) and microperfuranone ligases (*micA*, AN3396) (Balibar et al., 2007; Bouhired, Weber, Kempf-Sontag, Keller, & Hoffmeister, 2007; Yeh et al., 2012) (Supplementary Fig. 18). Microperfuranone is produced from two phenylpyruvic acid molecules linked by the stand-alone NRPS-like enzyme *micA* (Yeh et al., 2012). In the terrequinone pathway, two deaminated L-tryptophan molecules (indole pyruvic acid) produced by *tdiD* are converted into didemethylasterriquinone by the NRPS-like *tdiA*. The end-product is then synthesized by the activities encoded in *tdiB*, *tdiC*, and *tdiE* (Balibar et al., 2007; Bouhired et al., 2007). In both instances these NRPS-like enzymes catalyze the formation of C—C bonds (Balibar et al., 2007; Yeh et al., 2012). In *Metarhzium* M-NPL5 and M-NPL4 are predicted to form a methylated product as in both cases a methyltransferase is localized in the same position as the prenyltransferase *tdiB* (Supplementary Fig. 15-E and 15-F). Neither M-NPL5 nor M-NPL4 is found in *M. acridum* and *M. album* (Supplementary Table 2). M-NPL5 is located within a putative gene cluster conserved in other *Metarhizium* strains and other Ascomycota including dermatophytes (Supplementary Fig. 15-F), *C. globosum* (around the *tdiA*-like gene CHGG_02302), and *L. maculans* (around the *tdiA*-like gene Lema_P126000).

M-NPL8 corresponds to the conserved α-amino adipate reductase (AAR/LYS2). LYS2 is the best studied NRPS-like enzyme and catalyzes the reduction of L-α-aminoadipic acid to L-α-amino adipate-6-semialdehyde as part of the lysine biosynthetic route (Zabriskie & Jackson, 2000). All

Metarhizium species analyzed here have a single copy of the *AAR/LYS2* ortholog except *M. guizhouense*, which has two paralogs (Supplementary Table 2 and Supplementary Fig. 18). It is worth noting that most of the *AAR/LYS2* genes from *Metarhizium* form a monophyletic group. However the *AAR/LYS2* MAC_09675 from *M. acridum* and the second *AAR/LYS2* from *M. guizhouense* (MGU_10504) are more similar to the homologues from *T. virens* and *T. atroviride* (Genbank accession # EHK16657.1 and EHK44312 respectively). MAM_06175 from *M. album* is more similar to two additional *LYS2* paralogs found in *T. virens* and *T. atroviride* (EHK27480 and EHK48382, respectively). These differences, plus the non-conserved genomic location indicate a complex evolutionary origin of M-NPL8 in *Metarhizium* (Supplementary Fig. 15-H).

M-NPL9 and M-NPL14 are NRPS-like genes that share a similar structure, where A- and T-domains are followed by a transmembrane, C-terminal region (Supplementary Figs. 15-I and 15-M). M-NPL9 is well conserved across *Metarhizium*, while M-NPL14 is found only in some species (Supplementary Table 2). Phylogenetic analysis indicates that the M-NPL9 A-domain is more similar to acyl CoA-ligases, while the M-NPL14 A-domain belongs to the NRPS clade (Supplementary Fig. 18). M-NPL9 is in a region conserved across all examined *Metarhizium* species and orthologs were identifiable in several Ascomycota. Remarkably, the same region is conserved in *E. festucae* except for the presence of M-NPL9 itself (Supplementary Fig. 15-I). M-NPL14 orthologs are discontinuously distributed in Ascomycetes and are present in *Mycosphaerella graminicola* (MYCGRDRAFT_77312), *C. purpurea* (CPUR_02583) and *Botryotinia fuckeliana* (BofuT4_P157880).

M-NPL10 is the *C. heterostrophus ChNPS10* ortholog in *Metarhizium*. This gene has no known function, but is conserved in Ascomycota which hints at a role in primary metabolism. *C. heterostrophus ChNPS10* mutants are sensitive to oxidative stress (Ohm et al., 2012). The region around M-NPL10 is conserved in *Metarhizium*.

M-NPL11 is an NRPS-like gene found in all *Metarhizium* except for *M. acridum* and *M. album* (Supplementary Table 2). Orthologs are found in other Ascomycota (*Verticillium* and *Cochliobolus*). This enzyme couples a pyoverdine/dityrosine biosynthesis domain (DIT1) with a single NRPS module (Supplementary Fig. 15-J). When present, M-NPL11 is consistently associated with a lipase, a putative acetyltransferase and one or two MFS transporters. Despite resembling a monomodular NRPS, M-NPL11 should be considered an NRPS-like enzyme because phylogenetic analysis groups the single A-domain with acyl CoA ligases and not with NRPSs (Supplementary Fig. 18). The presence of the DIT1-domain is puzzling as in *S. cerevisiae* the DIT1 protein (which does not contain an NRPS module) is involved in the modification of tyrosine into N,N'-bisformyl dityrosine which is then oxidized and dimerized by the p450 DIT2. The product of this process is involved in ascospore wall strengthening (Briza, Eckerstorfer, & Breitenbach, 1994). Moreover, the *S. cerevisiae* DIT1 ortholog is present in *Metarhizium* spp. (eg, *M. robertsii* X797_011338).

M-NPL12 is the *Metarhizium* ortholog of *C. heterostrophus ChNPS12* and *A. fumigatus tmpL* (Bushley & Turgeon, 2010; Kim et al., 2009). It is characterized by the presence of A- and T-domains followed by a ferredoxin reductase transmembrane domain. This protein has been characterized in *A. fumigatus* and *Alternaria brassicicola*. Δ*tmpL* strains show abnormal conidiogenesis, accelerated aging, enhanced oxidative burst during conidiation, hypersensitivity to oxidative stress, and reduced virulence (Kim et al., 2009). The *tmpL* product is associated with Woronin bodies and expressed during conidiation and plant invasion (Kim et al., 2009). Later, the *tmpL* ortholog from the white-rot Basidiomycete *Ceriporiopsis subvermispora* (GenBank accession EMD40260.1) was characterized biochemically and found to act as an L-serine reductase (Kalb, Lackner, & Hoffmeister, 2014). The M-NPL12 genomic region is conserved in *Metarhizium* and other *Hypocrealians* (Supplementary Fig. 15-K). Comparison with more distant fungi indicated that M-NPL12/*ChNPS12*/*tmpL* orthologs are consistently associated with an enoyl-CoA hydratase.

M-NPL13 is found only in *M. acridum* and *M. album* (Supplementary Table 2). Phylogenetic analysis associates its single A-domain with D-lysergic acid peptide synthetase *lpsA1* and *lpsA2* A1 (Supplementary Fig. 18). In the other *Metarhizium* species, M-NPL13 with several others genes are replaced by a four-gene cluster, which includes M-TER29, an ortholog of the dimethylallyl tryptophan synthase *xptA* from *A. fumigatus* and *nscD* from *A. nidulans* (see Section 3.5). In some *Metarhizium* species a pseudogene of the MFS transporter linked to M-NPL13 is still present (Supplementary Fig. 15-L). In *M. brunneum* both M-NPL13 and the four-gene terpenoid cluster are absent.

3.3 PKS Pathways

Polyketides are a large and highly diverse family of natural products assembled from relatively simple building blocks by a PKS. In fungi polyketides are produced by type I PKSs, multidomain enzymes that iteratively add ketide units (usually from acetate) to a variety of starter units. All type I PKSs have in common three basic domains: ketoacyl synthase (KS), acyl-transferase (AT), and acyl carrier protein (ACP or PP) domains. The presence of additional domains and the nature of the final product can be used to classify type I PKSs into partially reducing (PR), highly reducing (HR), and nonreducing (NR) PKSs (Kroken, Glass, Taylor, Yoder, & Turgeon, 2003). PR-PKSs possess two distinctive β-carbon processing domains: (1) the ketoreductase (KR)-domain, which uses NADPH to reduce the β-keto group to a hydroxyl group, and (2) the dehydratase (DH)-domain, which removes the hydroxyl group and introduces an α-β double bond. In addition to DH and KR-domains, HR PKSs also possess the enoyl reductase (ER)-domain that uses NADPH to hydrate the double bond to yield a fully reduced ketide unit (Kroken et al., 2003), and a C-methyltransferase (cMT) which adds a methyl group to the α-carbon of the ketide and uses SAM as a substrate. NR-PKSs have a different set of tailoring domains not found in reducing PKSs (Kroken et al., 2003): the N-terminal starter unit ACP transacylase (SAT)-domain, which selects a specific starter/primer unit (Crawford, Dancy, Hill, Udwary, & Townsend, 2006); the product template (PT)-domain, which mediates the cyclization of the completed polyketide backbone and establishes cyclization boundaries within the polyketide moiety (regioselectivity) (Crawford et al., 2009, Crawford et al., 2008; Li, Xu, & Tang, 2010); and the thioesterase (TE/CLC)-domain, which releases the finished product and also heavily influences its nature as well as the turnover level (Newman, Vagstad, Storm, & Townsend, 2014; Xu, Zhou, et al., 2013). Based on both PT-domain—dictated regioselectivity and domain structure, NR-PKSs can be further divided in eight groups, each one producing a well-defined range of cyclic products (Li et al., 2010; Liu et al., 2015). Phylogenetic analysis based on the KS-domain can readily resolve all these PKS groups and can help define distinct PKS families (Kroken et al., 2003) (Supplementary Fig. 19). The resolving power of such a small, but crucial, section of a large multidomain enzyme is probably because HR-, PR-, and NR-PKSs had independent evolutionary paths in fungi (Kroken et al., 2003). The following survey covers 32 distinct PKSs families in the genus *Metarhizium* (Supplementary Table 2), which we are going to describe briefly.

M-PKS1 and M-PKS25 are clustered genes that phylogenetic analysis groups with PKSs involved in the biosynthesis of resorcylic acid lactones (RAL), which are mycotoxins produced by a wide array of fungal species

(Supplementary Fig. 19). Examples of RALs are the estrogen agonist zearalenone (Kim et al., 2005), dehydrocurvularins (Xu, Espinosa-Artiles, et al., 2013), the MAP kinase inhibitors hypothemycins (Fukazawa et al., 2010), and the chlorinated HSP90 inhibitors pochonin/radicicol (Reeves, Hu, Reid, & Kealey, 2008) identified from the nematophagous endophytic fungus *Pochonia chlamydosporia* (Kepler et al., 2014). RAL biosynthesis requires the combined activity of two PKSs. In the prototypical pathway leading to zearalenone, *G. zeae* PKS4 (reducing PKS) produces a reduced hexaketide moiety that is used as a starter unit by the NR-PKS PKS13. *G. zeae* PKS4 and PKS13 are M-PKS1 and M-PKS25 orthologs. Not surprisingly these two PKSs are physically associated and likely part of a conserved biosynthetic cluster (Fig. 2-A), which have close counterparts in *T. inflatum* (Fig. 2-A), *Trichoderma*, and some *Penicillium* species, but which diverge significantly from characterized RAL pathways. For instance, the pathway leading to pochonin in the closely related *P. chlamydosporia* requires a heme halogenase, which is missing in the *Metarhizium* cluster (Reeves et al., 2008). The gene organization around M-PKS1 and M-PKS25 is maintained almost unaltered across *Metarhizium* except for *M. acridum*, where both genes are missing (Fig. 2-A).

M-PKS2 and **M-PKS3** are similar HR PKSs grouped together by both Maximum Likelihood —based phylogenetic analysis (Supplementary Fig. 19) and RBB. M-PKS2 is found only in *M. acridum* (MAC_08219), and it is located near the ETP-like M-NRPS25. M-PKS2 may operate as a stand-alone gene as the gene organization in the region is preserved in other species in the absence of the synthase. M-PKS3 members are found in all examined *Metarhizium* species as well as in many *Hypocreales* including *B. bassiana*, *Trichoderma*, *Nectria*, and *Fusarium*. The gene organization around M-PKS3 is also conserved in several *Hypocreales*. In Fig. 2-B we provide the M-PKS3-like cluster found in *T. virens* and *Fusarium oxysporum* f. sp. *conglutinans* as examples.

M-PKS4 is an HR-PKS located in a putative cluster together with M-NRPS27. The cluster is similar to that in *A. niger* strain CBS 513.88 where the NRPS An15g07910 and the PKS An15g07920 are found (Fig. 1-Y). This gene is not found in *M. album*, *M. acridum*, and *M. majus* (Supplementary Table 3).

M-PKS5 is an HR-PKS and carries a carnitine O-acyltransferase domain at its C-terminus. This structure is similar to that of *A. alternata* PKS Aft9-1, a gene that may be involved in AF-toxin biosynthesis (Kaneko, Katsuya, & Tsuge, 1997), and homologues with the same domain structure are found in other Ascomycota but no similar cluster was identified in our panel. M-PKS5 is linked to NRPS-like gene M-NPL7 (Supplementary Fig. 15-G) in a cluster conserved in all analyzed *Metarhizium* genomes but *M. acridum* and *M. album* (Supplementary Fig. 19).

M-PKS6 and its associated genes are found in all examined *Metarhizium* species except *M. acridum* and *M. album* (Supplementary Table 3). Phylogenetic analysis groups this PKS in a sister clade of *A. nidulans* emericellamide synthase *easB*, which is part of a mixed PKS-NRPS cluster (Chiang et al., 2008) (Supplementary Fig. 19). However, unlike *easA*, the M-PKS6 cluster does not include an NRPS (Fig. 2-C).

M-PKS7 and **M-PKS22** are PKSs found only in *M. acridum* and are located in the same putative cluster (Supplementary Table 3). Phylogenetic analysis groups M-PKS7 (HR-PKS) and M-PKS22 (NR-PKS) with synthases known to cooperate in the same pathway. M-PKS7 is found in the same clade as the *Aspergillus terreus* lovastatin diketide synthase *lovF*, *A. nidulans* asperfuranone *afoG*, and *A. nidulans pkhA* (Supplementary Fig. 19). M-PKS22 is related to *A. nidulans afoE* and *pkhB* (Ahuja et al., 2012) (Supplementary Fig. 19). M-PKS7, M-PKS22, and the remainder of the nearby genomic region have similarity to the pro-apoptotic *A. nidulans* asperfuranone biosynthetic cluster (Fig. 2-D) (Wang et al., 2010). In this pathway the HR-PKS AN1036/*afoG*, which has 43% amino acid identity with *M. acridum* M-PKS7/MAC_09707, produces the 3,5-dimethyloctadienone moiety used by the NR-PKS AN1034/*afoE* (52% amino acid identity with *M. acridum* M-PKS22/MAC_09711) as a starter unit, which is further extended by four malonyl-CoA units and one methyl group (Chiang et al., 2009). Two genes essential for asperfuranone formation, *afoC* and *afoD*, have orthologs in the asperfuranone-like pathway found in *M. acridum*; a third essential gene, *afoF* does not have a significant match in the M-PKS7/M-PKS22 cluster (Fig. 2-D) (Chiang et al., 2009).

M-PKS8 and **M-PKS23** are two clustered PKSs found in all surveyed *Metarhizium* species except for *M. acridum* and *M. album* (Supplementary Table 3). Phylogenetic analysis groups M-PKS8 with M-PKS7 and M-PKS23 with M-PKS22 (Supplementary Fig. 19). Both M-PKS8 and M-PKS23 have orthologs in a partially characterized *A. nidulans* cluster including the PKSs *pkhB* (AN2035) and *pkhA* (AN2032). *pkhA* is an HR-PKS providing an octatrienoyl starter for the NR-PKS AN2035 (*pkhB*), which in turn drives the biosynthesis of an orsellinaldehyde derivative (Ahuja et al., 2012). Amino acid identity between the *A. nidulans* and *Metarhizium* homologues is high: thus M-PKS8 and *pkhA* are 75% identical at the amino acid level, while M-PKS23 and *pkhB* are 72% identical. Genes located in the vicinity of the M-PKS8/M-PKS23 are virtually identical in all the analyzed *Metarhizium* strains (Fig. 2-E), with six displaying significant homology to those found in the *pkhA*/*pkhB* putative cluster.

M-PKS9A and M-PKS9B are probably part of a mixed pathway with **M-NRPS3** (Fig. 1-C, see Section 3.1).

M-PKS9C is a HR-PKS found only in *M. acridum* (Supplementary Table 3). Very similar gene clusters occur in the rice pathogens *Ustilaginoidea virens* around PKS UV8b_6705; in the canola pathogen *L. maculans*, around PKS LEMA_P002660.1; in *A. pullulans*, around the PKS M438DRAFT_324760; and in the wheat pathogen *P. nodorum*, around SNOG_04868. LEMA_P002660 (PKS2) was linked to the biosynthesis of the antifungal phomenoic acid (Elliott et al., 2013) (Fig. 2-F). Given the degree of homology to PKS2 and the conservation of the respective regions, M-PKS9C is likely to produce a phomenoic acid—like metabolite. *M. acridum*, *U. virens*, and *L. maculans* gene clusters are associated with a monomodular NRPS (M-NRPS3C/MAC_05898, UV8b_6701, and NRPS LEMA_P002700.1, respectively) (Fig. 2-F), all having identical structure and unknown function. The A-domains of these NRPSs group together in a sister cluster of the A1 from the HC-toxin synthetase (Supplementary Fig. 19), an NRPS that incorporates decanoic acid in its final product (Ahn & Walton, 1997). M-NRPS3C, if active, may incorporate a highly reduced product from M-PKS9C.

M-PKS10 is found in complete form only in *M. guizhouense* (MGU_09992). Those found in *M. brunneum* and *M. robertsii* may be pseudogenes as they lack the AT-domain. This HR-PKS is found in a similar genic context in *M. guizhouense*, *M. robertsii*, and *T. atroviride* which includes a diketopiperazine-like NRPS M-NRPS24 (Fig. 1-W).

M-PKS11 is a HR-PKS grouping within the fumonisin clade (Supplementary Fig. 19) found only in *M. brunneum* (Supplementary Table 3). In *M. majus*, M-PKS11 is interrupted by a transposable element. The putative cluster associated with this PKS is readily identifiable as adjacent regions are conserved in other *Metarhizium* species (Fig. 2-G). No similar clusters were identified in other fungal genomes.

M-PKS12 is a HR-PKS not found in *M. acridum* and *M. album* (Supplementary Table 3) that groups within the fumonisin clade (Supplementary Fig. 19). Both PKS and local gene composition are strongly conserved in *A. fumigatus* (AFU3G14700) and *O. sinensis* (OCS_02685) (Fig. 2-H). Similar clusters are also found around *Pseudocercospora fijiensis*

MYCFIDRAFT_166988, and *M. graminicola* MYCGRDRAFT_101493. All these pathways are uncharacterized.

M-PKS13 is an HR-PKS with orthologs identified only in entomopathogens, including some *Metarhizium* spp., *B. bassiana*, and *T. inflatum*. Phylogenetic analysis placed this PKS in a sister group to the fumonisin PKS clade (Supplementary Fig. 19). This HR-PKS is not found in *M. album*, *M. majus* (which may have a pseudogene), and *M. acridum* (Supplementary Table 3). It is not clear weather M-PKS13 is associated with other genes in a biosynthetic cluster (Fig. 2-I). In *M. guizhouense* gene organization around M-PKS13 differs from other *Metarhizium* species (Fig. 2-I).

M-PKS14 is a HR-PKS found in *M. robertsii*, *M. anisopliae*, *M. brunneum*, and *M. guizhouense* (Supplementary Table 3). It groups closely with *A. fumigatus* fumagillin PKS (*fma-PKS*) (Supplementary Fig. 19). Comparison of the M-PKS14 and *fma-PKS*-associated genes, reveal little similarity beyond the PKS and a DltD N-terminal domain—containing protein (*fma-AT* in *A. fumigatus*). Fumagillin is derived by the esterification of the terpenoid fumagillol with the *fma-PKS* -produced decatetraenedioic acid (see Section 2.7.4 and Supplementary Fig. 12). Fumagillol is structurally related to ovalicin, and it is synthesized by the terpene cyclase *fma-TC* (Lin et al., 2013) located within the interspersed fumagillin/pseurotin gene clusters (Wiemann et al., 2013). Interestingly, the homologous pseurotin cluster found in *Metarhizium* (M-HPN7) contains the *fma-TC* ortholog (M-TER44, see Section 3.5) but, in contrast to *A. fumigatus*, the gene is physically separated from M-PKS14. An M-PKS14-like cluster is present in *E. festucae* (Fig. 2-J).

M-PKS15 is an HR-PKS found in only *M. album* (Supplementary Table 3). Only low similarity clusters were identified in other Ascomycota (Fig. 2-K).

M-PKS16 is an HR-PKS found only in some *Metarhizium* species (Supplementary Table 3) that also displays a conserved genic context. Phylogenetic analysis places this PKS within the hybrid PKS-NRPS clade so it likely derived from a member of this clade following the loss of the NRPS section. (Supplementary Fig. 19). As with the vast majority of PKS-NRPS, M-PKS16 lacks an ER-domain. Based on ortholog distribution and gene conservation we hypothesize a gene cluster, as depicted in Fig. 2-L.

M-PKS17 is found only in *M. album* (Supplementary Table 3). It is phylogenetically close to equisetin and fusaridione synthetases (hybrid PKS-NRPSs, Supplementary Fig. 19), to which M-PKS17 is 49% and 47% identical at the amino acid level, respectively. The PKS is surrounded by genes that are not usually associated with PKS biosynthetic pathways: on one flank there are four genes consistently found clustered in many Hypocreales including *Metarhizium*, *Gibberella*, *Trichoderma*, and *Epichloë*, but never associated with a PKS (Fig. 2-M). On the other flank, the late sexual development protein is likewise never associated with a PKS in the examined *Metarhizium* species (Fig. 2-M). This PKS, if functional might operate as a stand-alone gene.

M-PKS18 and its putative gene cluster are found in all species except *M. acridum* (Supplementary Table 3). In *M. album*, M-PKS18 may be present as a pseudogene. M-PKS18 is phylogenetically close to both *A. fumigatus* Afu1g17740 (uncharacterized PKS) and *G. zeae* FGSG_10464 (PKS9/FLS1), which also share the same domain structure. *G. zeae* PKS9 is responsible for the biosynthesis of the weak antifungals/mycoestrogens fusarielins (Kobayashi, Sunaga, Furihata, Morisaki, & Iwasaki, 1995; Sorensen et al., 2012). Orthologs of the fusarielin tailoring genes FSL2, FSL3, and FSL5 are found in the immediate vicinity of M-PKS17 (Fig. 2-N).

M-PKS19 is found only in *M. robertsii* (Supplementary Table 3). KS-based phylogenetic analysis associates this protein with the hybrid PKS-NRPS M-HPN3 and *A. clavatus* CcsA (Supplementary Fig. 19) (Qiao, Chooi, & Tang, 2011). M-PKS19 is likely to derive from a hybrid PKS-NRPSs as it retains the C-domain but lacks the terminal A-, T-, and R-domains. This structure resembles that of the *A. terreus* lovastatin nonketide synthase *lovB* (Boettger, Bergmann, Kuehn, Shelest, & Hertweck, 2012). The gene organization around M-PKS19 is also similar to that found in the hybrid PKS-NRPS M-HPN3 (Supplementary Fig. 16-C).

M-PKS20 is a group VII-type NR-PKS, and it is found only in *M. guizhouense* (Supplementary Table 3). M-PKS20 shows 64% amino acid identity to *pkeA* (AN7903), and 56% amino acid identity to *afo*E (AN1034), which are PKSs from *A. nidulans* involved in the biosynthesis of the antibacterial DHMBA and asperfuranone, respectively (Chiang et al., 2009; Gerke et al., 2012). The region surrounding M-PKS20 however is more similar to the asperfuranone biosynthetic cluster, where an HR-PKS (*afoG*) produces a starter unit for the adjacent NR-PKS (*afo*E) (Chiang et al., 2009). Except for a pseudogene, no HR-PKS is found in the M-PKS20 vicinity (Fig. 2-O). It is noteworthy that an asperfuranone-like cluster is found in *M. acridum* (see M-PKS7 and M-PKS22).

M-PKS21 is found in all examined *Metarhizium* species (Supplementary Table 3). M-PKS21, a group VII NR-PKS (Supplementary Fig. 19), is an ortholog of *A. nidulans* AN7903 (*pkeA*), which is part of the 2,4-dihydroxy-3-methyl-6-(2-oxopropyl)benzaldehyde biosynthetic gene cluster

(Sanchez et al., 2010) (Supplementary Fig. 19). However, the gene organization around M-PKS21 is more similar to that associated with the nonorthologous PKS AN0523 (Fig. 2-P) which produces 2-ethyl-4,6-dhydroxy-3,5 dimethylbenzaldehyde. (Ahuja et al., 2012). The M-PKS21 putative gene cluster is highly conserved in *Metarhizium*, *T. hemipterigena*, and in *Coccidioides posadasii* (PKS CPC735_034200, Fig. 2-P).

M-PKS24 is a group VI NR-PKS found in all the species except *M. album* (Supplementary Table 3). It is an ortholog of *A. nidulans pkbA* (AN6448) and *ausA* (AN8383). These are part of the meroterpenoids cichorine (a phytotoxin) and austinol/dehydroaustinol biosynthetic pathways, respectively (Nielsen et al., 2011; Sanchez, Entwistle, Corcoran, Oakley, & Wang, 2012). Accordingly, phylogenetic analysis groups this PKS with *P. brevicompactum* mycophenolic acid (*mpaC*) and *A. nidulans* austinol (*aus*A) synthases (Lo et al., 2012; Nielsen et al., 2011; Regueira et al., 2011) (Supplementary Fig. 19). The M-PKS24 region is conserved within *Metarhizium* and is similar to that around *G. zeae* PKS29/FGSG_04588 (Sieber et al., 2014) and *A. fumigatus* Afu8g02350 (Fig. 2-Q). Phylogenetic affiliation and presence of both a polyprenyl tranferase (M-TER43) and a polyprenyl synthetase (M-TER33) in close proximity support the hypothesis that M-PKS24 is part of a pathway producing a mixed PKS—terpene natural product.

M-PKS26 is phylogenetically associated with the cercosporin biosynthesis PKS (group IV NR-PKS, Supplementary Fig. 19) and is found only in *M. guizhouense* (Supplementary Table 3). Conserved genes surrounding M-PKS26 define a hypothetical gene cluster that includes an O-methyltransferase similar to CbCTB2 that is required for cercosporin biosynthesis (Staerkel et al., 2013) (Fig. 2-R). An M-PKS26 pseudogene is found in *M. brunneum* (MBR_10206 and MBR_10205) together with genes matching the predicted M-PKS26 cluster (Fig. 2-R).

M-PKS27 and **M-PKS28** are group III NR-PKSs present in all *Metarhizium* species except *M. album* which lacks M-PKS27 (Supplementary Table 3). M-PKS28 orthologs are found in *C. militaris*, *T. hemipterigena*, *H. thompsonii*, *T. inflatum*, and *O. sinensis*. M-PKS28 is phylogenetically close to *G. zeae* aurofusarin PKS12 and *Trichoderma reesei* PKS4 (Atanasova, Knox, Kubicek, Druzhinina, & Baker, 2013; Malz et al., 2005). Both M-PKS28 and M-PKS27 have been characterized by Chen, Feng, Shang, Xu, and Wang (2015) in two *M. robertsii* strains (ARSEF 23 and ARSEF 2575) and were identified in their work as *MrPKS1* and *MrPKS2*, respectively. *MrPKS1*/M-PKS28 deletion resulted in reduced green pigmentation and significant alterations in the conidial wall structure. However, only ARSEF 23 mutants displayed increased sensitivity to UV irradiation (Chen et al., 2015). Disruption of *MrPKS2*/M-PKS27 did not result in significant changes in either pigmentation or tolerance to heat stress or UV irradiation. Neither *MrPKS1* nor *MrPKS2* contributed significantly to virulence (Chen et al., 2015). In *T. reesei*, loss of the *MrPKS1*/M-PKS28 ortholog PKS4 resulted in a slight reduction in mycoparasitism and a significant decrease in pigmentation and UV tolerance (Atanasova, Knox, Kubicek, Druzhinina, & Baker, 2013). Deletion of PKS12 in *G. zeae* did not affect either virulence or UV tolerance (Malz et al., 2005). Gene organization around MrPKS2/M-PKS27 is conserved in all *Metarhizium* species, but displays low similarity to clusters characterized in other fungi. One of the best matches is the *A. fumigatus* melanin biosynthetic cluster where only the 1,3,6,8-tetrahydroxynaphthalene reductase *arp2* and the scytalone dehydratase *arp1* display similarity to homologues in the M-PKS27 putative biosynthetic cluster (Chen et al., 2015) (Fig. 2-S). The MrPKS1/M-PKS28 genomic region also displays low similarity to characterized SMM clusters: M-PKS28 and *Trichoderma* PKS4 or *Gibberella* PKS12 genomic regions have homology limited to the PKSs, an EthD-domain protein (*aurZ* ortholog), and a laccase (*Gip1* ortholog) (Fig. 2-T) (Atanasova et al., 2013; Chen et al., 2015; Frandsen et al., 2011).

M-PKS29 is a group III NR-PKS found in all *Metarhizium*, including *M. guizhouense* and *M. majus* where they are not included in the current gene count (Supplementary Table 3). Phylogenetic analysis places M-PKS29 in a sister clade of that including the PKS12 from *G. zeae* (aurofusarin biosynthesis) (Malz et al., 2005), M-PKS28 and M-PKS27 (Supplementary Fig. 19). Gene organization around M-PKS29 is conserved in all *Metarhizium*, and a similar cluster is found in the rice pathogen *U. virens* (UV8b_2086) (Fig. 2-U).

M-PKS30 is a group III NR-PKS present in all analyzed *Metarhizium* species (Supplementary Fig. 19). Both amino acid similarity and gene context comparison support the presence of two M-PKS30 variants. M-PKS30A is not present in *M. acridum*, which has an otherwise nearly identical genomic region (Fig. 2-V). In *M. album* M-PKS30A is present, but the gene organization around it differs from other *Metarhizium* species (Fig. 2-V). The M-PKS30A cluster was not found in analyzed Ascomycota. The variant M-PKS30B is found only in *M. acridum* and *M. album*; its genomic region is similar to that surrounding the *Nectria haematococca* perithecial red pigment PKSN (Graziani, Vasnier, & Daboussi, 2004) (Fig. 2-V).

M-PKS31 is a group III NR-PKS conserved in all analyzed *Metarhizium* species (Supplementary Table 3) and is phylogenetically close to *A. nidulans* asperthecin (*AptA*/AN6000) (Szewczyk et al., 2008), *Penicillium aethiopicum* viridicatumtoxin (*VrtA*) (Chooi, Cacho, & Tang, 2010), and *A. fumigatus* fumicycline (*fccA*/Afu7g00160) (Konig et al., 2013) PKSs (Supplementary Fig. 19). Particularly significant is the resemblance of the M-PKS31 region to that of the viridicatumtoxin biosynthetic gene cluster (Fig. 2-W): these clusters share the same polyketide biosynthetic genes (starter unit producing acetoacetyl-CoA synthetase *vrtB*; the NR PKS *vrtA*; β-lactamase-type thioesterase operating the final ring closure, *vrtG*) and some of the tailoring enzymes (*vrtH, vrtI*) (Fig. 2-W). The cluster found in *Metarhizium*, however, does not include the prenyl synthetase (*vrtD*), the prenyltransferase (*vrtC*), and two p450s (*vrtE* and *vrtK*) (Fig. 2-W) (Chooi et al., 2010).

M-PKS32 belongs to group V NR-PKSs and is not found in *M. acridum* and *M. album* (Supplementary Table 3). M-PKS32 is an ortholog of the *A. fumigatus* trypacidin (*tpcC*/Afu4g14560) and the *A. nidulans* monodictyphenone/emodin (*mdpG*/AN0150) synthases. The gene organization around M-PKS32 is conserved and has similarity to the *Aspergillus* pathways (Chiang et al., 2010; Mattern et al., 2015) (Fig. 2-X). Trypacidin is an anti-phagocytic toxin active against protozoans and macrophages (Ebringer, Balan, Catar, Horakova, & Ebringerova, 1965; Gauthier et al., 2012; Mattern et al., 2015). Emodin and derived anthraquinones exhibit antibacterial, antifungal, and cytotoxic activities as well as therapeutic properties (Chien, Wu, Chen, & Yang, 2015; He et al., 2012; Izhaki, 2002; Srinivas, Babykutty, Sathiadevan, & Srinivas, 2007; Wei, Lin, Liu, & Wang, 2013).

3.4 Hybrid PKS-NRPS Pathways

In hybrid synthetases the PKS and NRPS machineries are fused into a single polypeptide capable of combining a polyketide chain to (usually) a single amino acid (Fisch, 2013). The PKS section of the protein bear similarities to HR-PKSs, although in most hybrids the ER activity is provided by a stand-alone protein (Halo et al., 2008; Kennedy et al., 1999; Xu, Cai, Jung, & Tang, 2010). Phylogenetic analysis based on either the adenylation or ketosynthase domain sequences supports a monophyletic origin for these genes (Bushley & Turgeon, 2010; Kroken et al., 2003; Lawrence, Kroken, Pryor, & Arnold, 2011). Analysis of the genome of nine *Metarhizium* species identified seven canonical PKS-NRPSs (M-HPNs) and two "inverted hybrids" (NRPS-domains precede PKS-domains, M-IHs).

M-HPN1 is found in all analyzed *Metarhizium* species (Supplementary Table 4), which also have a nearly identical putative gene cluster. In *M. album* ARSEF 1941 the cluster carries a monomodular NRPS (MAM_07839, M-NRPS14) (Supplementary Fig. 16-A). A similar gene organization is found in dermatophytes and in *H. thompsonii*. The *A. niger* PKS-NRPS An11g00250 (*PynA*) is an ortholog of M-HPN1 and was linked to the biosynthesis of pyranonigrin E (Awakawa, Yang, Wakimoto, & Abe, 2013). Pyranonigrins A—D are oxygen radical scavengers (Miyake, Ito, Itoigawa, & Osawa, 2007). In contrast to other PKS-NRPSs, both M-HPN1 and *PynA* carry an ER-domain that is likely to be active and lacks the cMT found in other PKS-NRPS hybrids. *Metarhizium* and *A. niger* clusters share six genes with amino acid identities ranging from 35% to 52% (Supplementary Fig. 16-A).

M-HPN2 corresponds to NGS1 synthetase responsible for the biosynthesis of the NG-391/393 in *M. robertsii* (Donzelli, Krasnoff, Churchill, Vandenberg, & Gibson, 2010). It is not present in *M. album* and *M. acridum* (Supplementary Table 4). This gene and the gene organization around it are highly similar to that of the fusarin C synthetase (FUSS/FusA) found in *Gibberella/Fusarium* (Niehaus et al., 2013; Song et al., 2004) except for the absence of a serine hydrolase (*fus5*) in *Metarhizium* (Supplementary Fig. 16-B). Very similar clusters, including the absence of a stand-alone ER, are found in *T. inflatum* and several *Trichoderma* species (Supplementary Fig. 16-B). NGS1/M-HPN2 knockout mutants in *M. robertsii* did not display changes in virulence against *S. exigua* larvae or significant phenotypic alterations (Donzelli et al., 2010). GFP expression driven by the NGS1 promoter showed that NGS1 is strongly up regulated during log phase, but expression timing and intensity were influenced by cell density (Donzelli et al., 2010).

M-HPN3 is a PKS-NRPS hybrid found only in *M. brunneum* and *M. guizhouense* (Supplementary Table 4). Phylogenetic analysis groups M-HPN3 with synthetases involved in the synthesis of cytochalasin E in *A. clavatus* (ACLA_078660/ccsA), chaetoglobosin A in *C. globosum* (CHGG_01239) and magnaporthepyrone in *Magnaporthe oryzae* ACE1 (Hu et al., 2012; Qiao et al., 2011; Song et al., 2015). The corresponding putative gene clusters also share significant similarities (Supplementary Fig. 16-C). M-HPN3 may be a good candidate for the cytochalasin biosynthesis gene in *Metarhizium*. M-HPN3 is more similar to ccsA (58% amino acid identity) than to CHGG_01239 (42% amino acid identity), and the region around it contains orthologs to genes involved in both chaetoglobosin A and cytochalasin E biosynthesis (Supplementary Fig. 16-C).

M-HPN4 is present in all analyzed *Metarhizum* species except for *M. brunneum* and *M. album* which have pseudogenes of this PKS-NRPS (Supplementary Table 4). In all these species, gene composition around the synthetase is conserved. In *M. anisopliae* strains E6 and ARSEF549 the gene may be present but incorrectly called. These strains are also missing three genes (including a stand-alone ER) identified in brackets in Supplementary Fig. 16-D. An M-HPN4-like cluster is found in the human pathogen *C. posadasii* (Supplementary Fig. 16-D).

M-HPN5 is found only in *M. album*. The cluster includes an ER-like/TRAM1-like chimeric gene model (MAM_011359). Similar clusters and core genes are found in *Glomerella graminicola* and *E. festucae* (Supplementary Fig. 16-E).

M-HPN6 is found in all *Metarhizium* species except for *M. album* and *M. acridum* (Supplementary Table 4). In *M. majus*, genes surrounding M-HPN6 are only partially conserved (PKS-NRPS, a thioesterase and a transferase) which may define the biosynthetic cluster (Supplementary Fig. 16-F). This putative cluster does not appear to be conserved outside the genus *Metarhizium*.

M-HPN7 is phylogenetically related to the pseurotin synthetase from *A. fumigatus* (Maiya, Grundmann, Li, Li, & Turner, 2007) (Supplementary Fig. 19) and is found in *M. anisopliae*, *M. brunneum*, and *M. robertsii* (Supplementary Table 4). Early studies defining the pseurotin biosynthetic pathway in *A. fumigatus* have been recently expanded to show that this gene cluster is intertwined and coregulated with those supporting the production of fumagillin and fumitremorgin (Maiya et al., 2007; Wiemann et al., 2013). Comparison of the gene surrounding PKS-NRPS7 and the pseurotin/fumitremorgin clusters revealed orthology for the vast majority of them, and thus the possible presence of a discontinuous/intertwined gene cluster in *Metarhizium* (Wiemann et al., 2013) (Supplementary Fig. 16-G). A candidate natural product for this cluster is not found in the literature.

M-IH1 contains single NRPS-like A- and T-domains, followed by a minimal PKS module and ending with a TE-domain (Supplementary Fig. 16-H). The gene composition is conserved across all analyzed *Metarhizium* species except for *M. album* where M-IH1 is missing (Supplementary Table 5). Phylogenetic analysis based on both KS and A-domains groups this protein within the previously described *C. heterostropus pks24/nps7* clade (Bushley & Turgeon, 2010) (Supplementary Fig. 19). Genes found around M-IH1 closely match those found in other fungi including several dermatophytes. M-IH1-like genes were likely transferred horizontally from a bacterial donor to a Pezizomycotina ancestor (Lawrence et al., 2011).

M-IH2 is present in all *Metrarhizium* except for *M. album* and *M. robertsii* (Supplementary Table 5). Orthologs can be found in *B. bassiana, C. militaris, T. hemipterigena, H. thompsonii*, and *T. inflatum*. NRPS-PKS2 is consistently found associated with a divergently transcribed transporter (Supplementary Fig. 16-I). A-domain—based phylogenetic analysis places M-IH2 in a clade basal to that containing *C. heterostrophus Chnps12/tmplL* (Supplementary Fig. 18). KS-domain phylogenetic analysis confines this gene to the fatty acids synthase clade (Supplementary Fig. 19).

3.5 Terpenoid Pathways

Terpenoids originate as linear chains that use five-carbon building blocks (prenyl units) dimethyl—allyl pyrophosphate (DMAPP, starter unit) and isopentenyl diphosphate (IPP elongation unit). Multiple IPP head-to-tail condensations and cyclization lead to the biosynthesis of mono- (C10), sesqui- (C15), and di- (C20) terpenoids by isoprenyl diphosphate synthases. Tri- (C30) and tetra- (C40) terpenoids are synthesized by the condensation of two C15 or C20 units by squalene and phytoene synthetases, respectively. Terpenoid chains can be further modified by tailoring enzymes (monooxygenases, oxidoreductases, methyltransferases, and cyclases) and/or linked to molecules of different origin to generate mixed natural products as in meroterpenoids and indole diterpenoids (Geris & Simpson, 2009; Quin, Flynn, & Schmidt-Dannert, 2014; Young, McMillan, Telfer, & Scott, 2001). Analysis of nine *Metarhizium* genomes identified the terpene biosynthetic genes listed in Supplementary Table 6. Some of these genes are involved in primary metabolism including sterol biosynthesis, protein prenylation, and other basic cellular functions, as they are orthologs of well-studied genes in model organisms (Supplementary Table 6). Others have no clear function, as they do not match genes involved in primary metabolism, but have some similarity to genes known to be involved in SMM in other fungi (Supplementary Table 6). M-TER32, is a *pyr4*/paxB-like terpene cyclase with homologs that are part of several terpene pathways including pyripyropene and paxilline (Itoh et al., 2010; Scott et al., 2013). This particular gene is conserved in the genus *Metarhizium*, but has no characterized orthologs in other fungi and apparently does not belong to a biosynthetic gene cluster. Another such gene M-TER13, which is an ortholog of the uncharacterized geranylgeranyl diphosphate synthase Afu6g09770, located just outside the boundaries of the gliotoxin gene cluster in *A. fumigatus* (Inglis et al., 2013). M-TER15 is restricted to *M. guizhouense* and contains a hydroxyneurosporene synthase (CrtC) fused to geranylgeranyl pyrophosphate synthase domain. BlastP search identified only one fungal match in *Pseudogymnoascus pannorum* VKM F-4246, while highly significant matches (e-value < 1e-20) were found in prokaryotes and Euglenozoa. Hydroxyneurosporene synthase is involved in carotenoid metabolism in prokaryotes (Van Dien, Marx, O'Brien, & Lidstrom, 2003). M-TER1, M-TER2A, M-TER2B, M-TER6, M-TER16, M-TER21, and M-TER25 and their genomic regions tend to be conserved within the genus. These genes do not appear to be part of biosynthetic clusters and have orthologs sometimes found in other Ascomycetes. For instance the region harboring M-TER16 is conserved in several *Trichoderma* species, with the exception of M-TER16 itself.

M-TER1 is found in *M. guizhouense* only as a likely stand-alone gene, displaying similarities to both fungal and bacterial squalene-hopene cyclases. Its ortholog in *A. fumigatus* (Afu7g00260) is a hopene cyclase indirectly linked to hopanoid biosynthesis (http://www.aspergillusgenome.org/) (Kannenberg & Poralla, 1999; Wang et al., 2009). M-TER2A and M-TER2B are uncharacterized genes restricted to *Metarhizium* and *N. haematococca.*

Some other terpenoid biosynthetic genes identified in *Metarhizium* are parts of gene clusters likely to lead to the biosynthesis of mixed origin products. M-TER14 (geranylgeranyl pyrophosphate synthetase), M-TER33 (terpene cyclase) and M-TER43 (UbiA prenyltransferase) are all located in the M-PKS24 region (Fig. 2-Q) which is similar to meroterpenoid-producing gene clusters identified in other fungi (Lo et al., 2012; Nielsen et al., 2011; Regueira et al., 2011). M-TER33 and M-TER43 are homologues *ausL* and *ausN* from *A. nidulans*. These catalyze terpene cyclization and C-alkylation of the 3,5-dimethylorsellinic acid in the austinol biosynthetic pathway (Lo et al., 2012).

M-TER27 is a dimethylallyl tryptophan synthase associated with the ETP—like cluster M-NRPS23 and an ortholog of *sirD*. This gene catalyzes the first step in the biosynthesis of sirodesmin in *L. maculans* (Kremer & Li, 2010) (Fig. 1-V).

M-TER29 does not belong to a biosynthetic cluster but may participate in a mixed pathway. M-TER29 is an ortholog of the prenyltransferases *nscD* (Afu7g00170) from *A. fumigatus* and *xptA*, *xptB* (AN6784, AN12402) from *A. nidulans* that use PKS-derived polycyclic substrates to produce C-prenylated xanthones (Chooi et al., 2013; Konig et al., 2013; Nielsen et al., 2011; Sanchez et al., 2011). In *A. nidulans* these polycyclic backbones derive from the monodictyphenone pathway, located in different regions of the genome from *xptA* and *xptB* (Chiang et al., 2010; Sanchez et al., 2011). Thus, in *Metarhizium* species the pathways associated with M-PKS31 (viridicatumtoxin-like) and M-PKS32 (monodictyphenone/trypacidin-like) may produce polycyclic polyketide substrates for M-TER29.

M-TER30 is the ortholog of *dmaW* a dimethylallyl tryptophan synthase that catalyzes the first step in the biosynthesis ergot alkaloids in *C. purpurea* and fumigaclavine C in *A. fumigatus* (Coyle & Panaccione, 2005; Tsai et al., 1995). M-TER30 is found within the ergot alkaloid—like cluster together with M-NRPS15 and M-NRPS16.

M-TER37 is a trichodiene synthase, a family of sesquiterpene biosynthetic genes widely distributed in fungi (Hohn & Plattner, 1989). Trichodiene synthases are known to convert farnesyl pyrophosphate to trichodiene, the precursor of trichothecenes and other derivatives (Gledhill, Hesketh, Bycroft, Dewick, & Gilbert, 1991; Hohn & Vanmiddlesworth, 1986; Pitel, Arsenault, & Vining, 1971). M-TER37 is an ortholog of FG10397/CLM1 from *Fusarium graminearum* that is responsible for the biosynthesis of the sesquiterpene alcohol longiborneol, an intermediate leading to the antifungal and insecticidal compound culmorin (Dowd, Miller, & Greenhalgh, 1989; McCormick, Alexander, & Harris, 2010; Strongman, Miller, Calhoun, Findlay, & Whitney, 1987). M-TER37 is found in all examined *Metarhizium* species except *M. album* (Supplementary Table 6). As for FG10397/CLM1 from *F. graminearum*, it is not clear if M-TER37 is part of a gene cluster.

Metarhizium has two additional sesquiterpene synthases, M-TER38 and M-TER39 (Supplementary Table 6). M-TER38 is present in all the strains except for *M. album* and is part of a conserved region; M-TER39 is restricted to *M. robertsii*, *M. majus*, and *M. brunneum*, and associated with a different cluster which includes, among others, an ERG9-like squalene cyclase (M-TER7) and an ERG13-like gene (3-hydroxy-3-methylglutaryl-CoA-synthase). In *M. robertsii* this three-gene group is inserted between a region containing an NRPS pseudogene plus the putative helvolic acid biosynthetic cluster and the M-PKS32 cluster (Supplementary Fig. 17-A). In *M. brunneum* the cluster associated with M-TER7 is partially conserved and found in a different location. In *M. majus* M-TER39 is in a small contig also harboring an ERG13-like gene. Neither M-TER38 nor M-TER39 sesquiterpene synthases are linked to a specific SM.

M-TER44 is a prenyltransferase located within the M-HPN7 cluster (Supplementary Fig. 16-G). This gene is an ortholog of *fma-TC*/Afu8g00520, which produces β-trans-bergamotene and the ovalicin-related fumagillol as part of the fumagillin biosynthetic route (see Section 2.7.4 and Supplementary Fig. 12) (Lin et al., 2013). In contrast to *A. fumigatus*, where pseurotin and fumagillin synthases are intertwined and form a supercluster (Niehaus et al., 2013), in *Metarhizium* the pseurotin-like and fumagillin-like clusters (M-HPN7 and M-PKS14, respectively) are physically separated, with the putative ovalicin synthetase M-TER44 residing in the "wrong" (M-HPN7) cluster. A genomic region closely matching that of the *A. fumigatus* helvolic acid biosynthetic cluster (Mitsuguchi et al., 2009) (see Section 2.7.1) is conserved in all the strains except *M. acridum* and *M. album* (Supplementary Fig. 17-B). This region also corresponds to a sequence deposited (accession HQ129929) in GenBank linked to an unpublished report.

M-TER10, M-TER11, M-TER24, M-TER26, and M-TER31 are part of a conserved cluster closely resembling those involved in indole diterpene production in other Ascomycetes including *Neotyphodium lolii* (lolitrem), *T. inflatum* (terpendole E-like cluster), and *Penicillium paxilli* (paxilline) (Nicholson et al., 2015; Scott et al., 2013; Young et al., 2001) (Supplementary Fig. 17-C). Six of seven genes required for paxilline biosynthesis are found in *Metarhizium* with the exception of the terpene cylase *paxA* (Saikia, Parker, Koulman, & Scott, 2006; Scott et al., 2013; Tagami et al., 2013). Similarly, nine of 10 genes found in the *N. lolii* lolitrem cluster have homologues in *Metarhizium* (Supplementary Fig. 17-C) (Young et al., 2006). It is thus conceivable that *Metarhizium* has the ability to produce a novel indole diterpene. Many indole diterpenes are tremorgenic mycotoxins, and some are insecticidal (Furutani et al., 2014; Ostlind et al., 1997; Shiono, Akiyama, & Hayashi, 2000). Although indole diterpenes have been isolated from several Clavicipiataceae (Gallagher et al., 1980; Gatenby, Munday-Finch, Wilkins, & Miles, 1999; Motoyama, Hayashi, Hirota, Ueki, & Osada, 2012; Rowan, 1993; Schroeder et al., 2007), there are no reports of such chemistries in *Metarhizium*.

M-TER34, M-TER35, and M-TER36 are similar to ent-kaurene and taxadiene synthases identified in other fungi that are involved in the biosynthesis of gibberellins and aphidicolin (Kawaide, Imai, Sassa, & Kamiya, 1997; Toyomasu et al., 2004). These genes do not appear to be part of clusters and have uncharacterized orthologs in a wide range of Ascomycota. Ent-kaurene is a precursor in gibberellin biosynthesis (Kawaide et al., 1997), and taxadiene synthase catalyzes the first step of taxol biosynthesis (Wildung & Croteau, 1996).

4. CONCLUSIONS

In this review we surveyed the known chemistry of SMs from *Metarhizium* and cataloged identifiable SMM pathways in nine sequenced *Metarhizium* strains (Gao et al., 2011; Hu et al., 2014; Staats et al., 2014). This analysis confirms the view that the known secondary chemistry in the genus *Metarhizium* represents the tip of a very large iceberg. The vast majority of the pathways recognized herein are unexplored chemically and our understanding of the services that their SMM end-products perform for the producing fungi especially in interactions with other microbes, insect hosts, and plants is rudimentary at best. To date functional analysis of SMM pathways have failed to show that any *Metarhizium* SM is a sine qua non for pathogenicity. The only exception is ferricrocin, which many do not consider an SM at all, since its primary function, intracellular iron sequestration, is provided by primary metabolites in other organisms (Silva & Faustino, 2015). Some of the pathways characterized in this survey correspond to known products (serinocyclins, DTXs, and NG39X), some are likely to be responsible for known chemistries (eg, cytochalasins and ovalicin), others are tantalizingly similar to pathways identified in other fungi (ergot, diketopipearzine, and resorcylic acid lactones), but their putative SM products are still unknown from *Metarhizium*, and still others defy attempts to predict the type of molecule they code for.

Although comparing SMs with other entomopathogens was not our primary focus, it is quite clear that, with few exceptions, SMM pathways of *Metarhizium* (Clavicipitaceae) have little in common with those of *B. bassiana* and *C. militaris* (Cordycipitaceae). More often, in our comparisons we encountered similar pathways in sequenced genomes from species in the Ophiocordicipitaceae (eg, *Tolypocladium* spp.). The survey also reveals considerable variability in the number of SMM genes among species in the genus. The lowest number of SMM genes is found in *M. album* (48 in total) and the highest is in *M. guizhouense* (93 in total), with the proportion of each SMM gene class showing some variation (Fig. 3A and B). Interestingly, in *M. album* and *M. acridum* NRPSs represent a higher percentage of SMM biosynthetic genes compared to PKSs, while in all the other strains

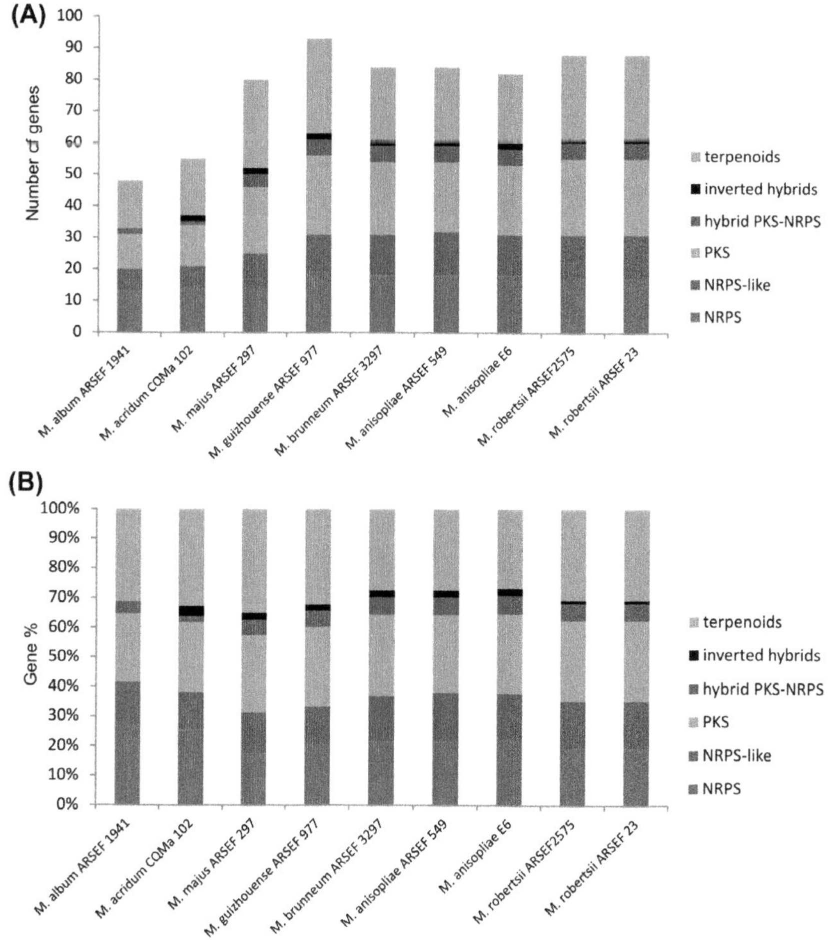

Figure 3 Absolute (A) and relative (B) abundance of six classes of core secondary metabolite biosynthetic genes in nine *Metarhizium* species. (See color plate)

the situation is reversed (B). Also, the number of hybrid PKS-NRPSs was proportionally lower in *M. album* and *M. acridum* (2–4%) than in the rest of the strains (5–6%) (B). A subset of the core SMM genes (M-NRPS2, M-NRPS4, M-NRPS5, M-NRPS6, M-NRPS17, M-PKS3, M-PKS21, M-PKS29, M-PKS31, M-NPL3b, M-NPL9, M-HPN1, and an indole diterpene-like pathway) is found in all *Metarhizium* strains. These genes are likely to be present in the common ancestors of the nine *Metarhizium* species and their conservation suggests that they might provide activities relevant for *Metarhizium* survival in their common niche. Some of SMM genes (eg, M-NRPS4, M-NRPS5, and M-NRPS6) appear to have emerged within the genus.

The long history of chemical studies of *Metarhizium* SMs has left us with an abundance of compounds from *M. anisopliae* and thus with a provenance that is uncertain in light of current taxonomic thinking. One challenge that looms ahead in this field is establishing which of the currently recognized species of *Metarhizium* produce which compounds. Hopes of applying modern techniques of molecular taxonomy to correctly dispose the original source strains of these compounds to species may be dim. Accessing strains used in the original chemistry reports, many from >30 years ago, may not be possible. A more likely approach would be looking for these compounds in strains whose identity has already been established. Confirmation of provenance for these diverse chemistries will set the stage for pursuing functional analysis of likely biosynthetic genes. Unlocking the vast treasure trove represented by the biosynthetic capacity in *Metarhizium* may produce new lead chemistries with agroeconomic and biomedical importance and will indubitably increase our understanding of the adaptive significance of these pathways and the intriguing compounds they produce.

SUPPLEMENTARY DATA

Supplementary data related to this article can be found online at http://dx.doi.org/10.1016/bs.adgen.2016.01.005.

REFERENCES

Adachi, H., Doi, H., Kasahara, Y., Sawa, R., Nakajima, K., Kubota, Y.... Nomoto, A. (2015). Asteltoxins from the entomopathogenic fungus *Pochonia bulbillosa* 8-H-28. *Journal of Natural Products, 78*, 1730–1734.

Ahn, J. H., & Walton, J. D. (1997). A fatty acid synthase gene in *Cochliobolus carbonum* required for production of HC-toxin, cyclo(D-prolyl-L-alanyl-D-alanyl-L-2-amino-9, 10-epoxi-8-oxodecanoyl). *Molecular Plant-Microbe Interactions, 10*, 207–214.

Ahuja, M., Chiang, Y. M., Chang, S. L., Praseuth, M. B., Entwistle, R., Sanchez, J. F.... Wang, C. C. (2012). Illuminating the diversity of aromatic polyketide synthases in *Aspergillus nidulans*. *Journal of the American Chemical Society, 134*, 8212–8221.

Aldridge, D., Armstrong, J., Speake, R., & Turner, W. (1967). The cytochalasins, a new class of biologically active mould metabolites. *Chemical Communications (London)*, 26–27.

Allinger, N. (1956). The nuclear magnetic resonance spectrum of helvolic acid. *The Journal of Organic Chemistry, 21*, 1180–1182.

Allinger, N. L., & Coke, J. L. (1961). The structure of helvolic acid. III. *The Journal of Organic Chemistry, 26*, 4522–4529.

Aniya, Y., Ohtani, I. I., Higa, T., Miyagi, C., Gibo, H., Shimabukuro, M.... Taira, J. (2000). Dimerumic acid as an antioxidant of the mold, *Monascus anka*. *Free Radical Biology and Medicine, 28*, 999–1004.

Atanasova, L., Knox, B. P., Kubicek, C. P., Druzhinina, I. S., & Baker, S. E. (2013). The polyketide synthase gene pks4 of *Trichoderma reesei* provides pigmentation and stress resistance. *Eukaryotic Cell, 12*, 1499–1508.

Atanasova, L., Le Crom, S., Gruber, S., Coulpier, F., Seidl-Seiboth, V., Kubicek, C. P., & Druzhinina, I. (2013). Comparative transcriptomics reveals different strategies of *Trichoderma* mycoparasitism. *BMC Genomics, 14*, 121.

Awakawa, T., Yang, X. L., Wakimoto, T., & Abe, I. (2013). Pyranonigrin E: a PKS-NRPS hybrid metabolite from *Aspergillus niger* identified by genome mining. *ChemBioChem, 14*, 2095–2099.

Azumi, M., Ishidoh, K., Kinoshita, H., Nihira, T., Ihara, F., Fujita, T., & Igarashi, Y. (2008). Aurovertins F-H from the entomopathogenic fungus *Metarhizium anisopliae*. *Journal of Natural Products, 71*, 278–280.

Bailey, A. M., Kershaw, M. J., Hunt, B. A., Paterson, I. C., Charnley, A. K., Reynolds, S. E. … Clarkson, J. M. (1996). Cloning and sequence analysis of an intron-containing domain from a peptide synthetase-encoding gene of the entomopathogenic fungus *Metarhizium anisopliae*. *Gene, 173*, 195–197.

Balibar, C. J., Howard, A. R., & Walsh, C. T. (2007). Terrequinone A biosynthesis through L-tryptophan oxidation, dimerization and bisprenylation. *Nature Chemical Biology, 3*, 584–592.

Bayram, O., Krappmann, S., Ni, M., Bok, J. W., Helmstaedt, K., Valerius, O. … Braus, G. H. (2008). VelB/VeA/LaeA complex coordinates light signal with fungal development and secondary metabolism. *Science, 320*, 1504–1506.

Bills, G. F., Polishook, J. D., Goetz, M. A., Sullivan, R. F., & White, J. F., Jr. (2002). *Chaunopycnis pustulata* sp. nov., a new clavicipitalean anamorph producing metabolites that modulate potassium ion channels. *Mycological Progress, 1*, 3–17.

Bischoff, J. F., Rehner, S. A., & Humber, R. A. (2009). A multilocus phylogeny of the *Metarhizium anisopliae* lineage. *Mycologia, 101*, 512–530.

Boettger, D., Bergmann, H., Kuehn, B., Shelest, E., & Hertweck, C. (2012). Evolutionary imprint of catalytic domains in fungal PKS–NRPS hybrids. *ChemBioChem, 13*, 2363–2373.

Bohnert, M., Dahse, H.-M., Gibson, D. M., Krasnoff, S. B., & Hoffmeister, D. (2013). The fusarin analog NG-391 impairs nucleic acid formation in K-562 leukemia cells. *Phytochemistry Letters, 6*(2), 189–192.

Bok, J. W., & Keller, N. P. (2004). LaeA, a regulator of secondary metabolism in *Aspergillus* spp. *Eukaryotic Cell, 3*, 527–535.

Bouhired, S., Weber, M., Kempf-Sontag, A., Keller, N. P., & Hoffmeister, D. (2007). Accurate prediction of the *Aspergillus nidulans* terrequinone gene cluster boundaries using the transcriptional regulator *LaeA*. *Fungal Genetics and Biology, 44*, 1134–1145.

Brackenbury, J. (1999). Regulation of swimming in the *Culex pipiens* (Diptera, Culicidae) pupa: kinematics and locomotory trajectories. *The Journal of Experimental Biology, 202*, 2521.

Brackenbury, J. (2001). The vortex wake of the free-swimming larva and pupa of *Culex pipiens* (Diptera). *Journal of Experimental Biology, 204*, 1855–1867.

Brakhage, A. A. (2013). Regulation of fungal secondary metabolism. *Nature Reviews Microbiology, 11*, 21–32.

Briza, P., Eckerstorfer, M., & Breitenbach, M. (1994). The sporulation-specific enzymes encoded by the DIT1 and DIT2 genes catalyze a two-step reaction leading to a soluble LL-dityrosine-containing precursor of the yeast spore wall. *Proceedings of the National Academy of Sciences of the United States of America, 91*, 4524–4528.

Bushley, K. E., & Turgeon, B. G. (2010). Phylogenomics reveals subfamilies of fungal nonribosomal peptide synthetases and their evolutionary relationships. *BMC Evolutionary Biology, 10*.

Campbell, M. A., Rokas, A., & Slot, J. C. (2012). Horizontal transfer and death of a fungal secondary metabolic gene cluster. *Genome Biology and Evolution, 4*, 289–293.
Cantalejo, M. J., Torondel, P., Amate, L., Carrasco, J. M., & Hernández, E. (1999). Detection of fusarin C and trichothecenes in *Fusarium* strains from Spain. *Journal of Basic Microbiology, 39*, 143–153.
Cantín, Á., Moya, P., Castillo, M. A., Primo, J., Miranda, M. A., & Primo-Yúfera, E. (1999). Isolation and synthesis of n-(2-methyl-3-oxodec-8-enoyl)-2-pyrroline and 2-(hept-5-enyl)-3-methyl-4-oxo-6, 7, 8, 8a-tetrahydro-4h-pyrrolo [2, 1-b] 1, 3-oxazine—two new fungal metabolites with in vivo anti-juvenile-hormone and insecticidal activity. *European Journal of Organic Chemistry, 1999*, 221–226.
Carollo, C., Calil, A. L. A., Schiave, L. A., Guaratini, T., Roberts, D. W., & Lopes, N. P. (2010). Tyrosine betaine: novel secondary metabolite isolated form conidia of the entompathogenic fungus *Metarhizium anisopliae* var. *anisopliae*. *Mycological Research, 114*.
Carter, S. (1967). Effects of cytochalasins on mammalian cells. *Nature, 213*, 261.
Che, Y., Swenson, D. C., Gloer, J. B., Koster, B., & Malloch, D. (2001). Pseudodestruxins A and B: new cyclic depsipeptides from the coprophilous fungus *Nigrosabulum globosum*. *Journal of Natural Products, 64*, 555–558.
Chen, L., Yue, Q., Zhang, X., Xiang, M., Wang, C., Li, S. … An, Z. (2013). Genomics-driven discovery of the pneumocandin biosynthetic gene cluster in the fungus *Glarea lozoyensis*. *BMC Genomics, 14*, 339.
Chen, Y. X., Feng, P., Shang, Y. F., Xu, Y. J., & Wang, C. S. (2015). Biosynthesis of non-melanin pigment by a divergent polyketide synthase in *Metarhizium robertsii*. *Fungal Genetics and Biology, 81*, 142–149.
Chiang, Y. M., Szewczyk, E., Davidson, A. D., Entwistle, R., Keller, N. P., Wang, C. C. C., & Oakley, B. R. (2010). Characterization of the *Aspergillus nidulans* monodictyphenone gene cluster. *Applied and Environmental Microbiology, 76*, 2067–2074.
Chiang, Y.-M., Szewczyk, E., Davidson, A. D., Keller, N., Oakley, B. R., & Wang, C. C. (2009). A gene cluster containing two fungal polyketide synthases encodes the biosynthetic pathway for a polyketide, asperfuranone, in *Aspergillus nidulans*. *Journal of the American Chemical Society, 131*, 2965–2970.
Chiang, Y. M., Szewczyk, E., Nayak, T., Davidson, A. D., Sanchez, J. F., Lo, H. C. … Wang, C. C. (2008). Molecular genetic mining of the *Aspergillus* secondary metabolome: discovery of the emericellamide biosynthetic pathway. *Chemistry & Biology, 15*, 527–532.
Chien, S. C., Wu, Y. C., Chen, Z. W., & Yang, W. C. (2015). Naturally occurring anthraquinones: chemistry and therapeutic potential in autoimmune diabetes. *Evidence-Based Complementary and Alternative Medicine, 2015*.
Chooi, Y. H., Cacho, R., & Tang, Y. (2010). Identification of the viridicatumtoxin and griseofulvin gene clusters from *Penicillium aethiopicum*. *Chemistry & Biology, 17*, 483–494.
Chooi, Y. H., Fang, J., Liu, H., Filler, S. G., Wang, P., & Tang, Y. (2013). Genome mining of a prenylated and immunosuppressive polyketide from pathogenic fungi. *Organic Letters, 15*, 780–783.
Chu, M., Mierzwa, R., Truumees, I., Gentile, F., Patel, M., Gullo, V. … Puar, M. S. (1993). Two novel diketopiperazines isolated from the fungus *Tolypocladium* sp. *Tetrahedron Letters, 34*, 7537–7540.
Chu, M., Truumees, I., Mierzwa, R., Patel, M., & Puar, M. S. (1997). Sch 56396: a new c-fos proto-oncogene inhibitor produced by the fungus *Tolypocladium* sp. *Journal of Antibiotics (Tokyo), 50*, 1061–1063.
Chung, Y.-M., El-Shazly, M., Chuang, D.-W., Hwang, T.-L., Asai, T., Oshima, Y. … Chang, F. R. (2013). Suberoylanilide hydroxamic acid, a histone deacetylase inhibitor, induces the production of anti-inflammatory cyclodepsipeptides from *Beauveria felina*. *Journal of Natural Products, 76*, 1260–1266.

Chung, Y. M., Wei, C. K., Chuang, D. W., El-Shazly, M., Hsieh, C. T., Asai, T.... Chang, F. R. (2013). An epigenetic modifier enhances the production of anti-diabetic and anti-inflammatory sesquiterpenoids from *Aspergillus sydowii*. *Bioorganic & Medicinal Chemistry, 21*, 3866–3872.

Colegate, S., Dorling, P., & Huxtable, C. (1979). A spectroscopic investigation of swainsonine: an α-mannosidase inhibitor isolated from *Swainsona canescens*. *Australian Journal of Chemistry, 32*, 2257–2264.

Cook, D., Gardner, D. R., & Pfister, J. A. (2014). Swainsonine-containing plants and their relationship to endophytic fungi. *Journal of Agricultural and Food Chemistry, 62*, 7326–7334.

Coyle, C. M., & Panaccione, D. G. (2005). An ergot alkaloid biosynthesis gene and clustered hypothetical genes from *Aspergillus fumigatus*. *Applied and Environmental Microbiology, 71*, 3112–3118.

Cram, D. J., & Allinger, N. L. (1956). Mold Metabolites. VIII. contribution to the elucidation of the structure of helvolic acid. *Journal of the American Chemical Society, 78*, 5275–5284.

Crawford, J. M., Dancy, B. C. R., Hill, E. A., Udwary, D. W., & Townsend, C. A. (2006). Identification of a starter unit acyl-carrier protein transacylase domain in an iterative type I polyketide synthase. *Proceedings of the National Academy of Sciences of the United States of America, 103*, 16728–16733.

Crawford, J. M., Korman, T. P., Labonte, J. W., Vagstad, A. L., Hill, E. A., Kamari-Bidkorpeh, O.... Townsend, C. A. (2009). Structural basis for biosynthetic programming of fungal aromatic polyketide cyclization. *Nature, 461*, 1139–U1243.

Crawford, J. M., Thomas, P. M., Scheerer, J. R., Vagstad, A. L., Kelleher, N. L., & Townsend, C. A. (2008). Deconstruction of iterative multidomain polyketide synthase function. *Science, 320*, 243–246.

Croom, W., Hagler, W., Froetschel, M., & Johnson, A. (1995). The involvement of slaframine and swainsonine in slobbers syndrome: a review. *Journal of Animal Science, 73*, 1499–1508.

Croteau, R., Ketchum, R. E., Long, R. M., Kaspera, R., & Wildung, M. R. (2006). Taxol biosynthesis and molecular genetics. *Phytochemistry Reviews, 5*, 75–97.

Darkin-Rattray, S. J., Gurnett, A. M., Myers, R. W., Dulski, P. M., Crumley, T. M., Allocco, J. J.... Schmatz, D. M. (1996). Apicidin: a novel antiprotozoal agent that inhibits parasite histone deacetylase. *Proceedings of the National Academy of Sciences of the United States of America, 93*, 13143–13147.

Donzelli, B. G., Krasnoff, S. B., Churchill, A. C., Vandenberg, J. D., & Gibson, D. M. (2010). Identification of a hybrid PKS-NRPS required for the biosynthesis of NG-391 in *Metarhizium robertsii*. *Current Genetics, 56*, 151–162.

Donzelli, B. G. G., Gibson, D. M., & Krasnoff, S. B. (2015). Intracellular siderophore but not extracellular siderophore is required for full virulence in *Metarhizium robertsii*. *Fungal Genetics and Biology, 82*, 56–68.

Donzelli, B. G. G., Krasnoff, S. B., Sun-Moon, Y., Churchill, A. C. L., & Gibson, D. M. (2012). Genetic basis of destruxin production in the entomopathogen *Metarhizium robertsii*. *Current Genetics, 58*, 105–116.

Dorling, P., Huxtable, C., & Vogel, P. (1978). Lysosomal storage in *Swainsona* spp. toxicosis: an induced mannosidosis. *Neuropathology and Applied Neurobiology, 4*, 285–295.

Dowd, P. F., Miller, J. D., & Greenhalgh, R. (1989). Toxicity and interactions of some *Fusarium graminearum* metabolites to caterpillars. *Mycologia, 81*, 646–650.

Dreyfuss, M. M., Emmer, G., Grassberger, M., Rüedi, K., & Tscherter, H. (1989). *Pipecolic acid containing peptolides, their preparation and pharmaceutical compositions containing them.* Europe Patent No. 0 360 760 A2 and A3. European Patent Office.

Du, F.-Y., Zhang, P., Li, X.-M., Li, C.-S., Cui, C.-M., & Wang, B.-G. (2014). Cyclohexadepsipeptides of the isaridin class from the marine-derived fungus *Beauveria felina* EN-135. *Journal of Natural Products, 77*, 1164–1169.

Ebringer, L., Balan, J., Catar, G., Horakova, K., & Ebringerova, J. (1965). Effect of trypacidin on *Toxoplasma gondii* in tissue culture and in mice. *Experimental Parasitology, 16*, 182–189.

Ehmann, D. E., Trauger, J. W., Stachelhaus, T., & Walsh, C. T. (2000). Aminoacyl-SNACs as small-molecule substrates for the condensation domains of nonribosomal peptide synthetases. *Chemistry & Biology, 7*, 765–772.

Elliott, C. E., Callahan, D. L., Schwenk, D., Nett, M., Hoffmeister, D., & Howlett, B. J. (2013). A gene cluster responsible for biosynthesis of phomenoic acid in the plant pathogenic fungus, *Leptosphaeria maculans*. *Fungal Genetics and Biology, 53*, 50–58.

Elliott, C. E., Gardiner, D. M., Thomas, G., Cozijnsen, A., Van De Wouw, A., & Howlett, B. J. (2007). Production of the toxin sirodesmin PL by *Leptosphaeria maculans* during infection of *Brassica napus*. *Molecular Plant Pathology, 8*, 791–802.

Emery, T. (1980). Malonichrome, a new iron chelate from *Fusarium roseum*. *Biochimica et Biophysica Acta, 629*, 382–390.

Engel, B., Erkel, G., Anke, T., & Sterner, O. (1998). Sesquicillin, an inhibitor of glucocorticoid mediated signal transduction. *The Journal of Antibiotics, 51*, 518–521.

Espada, A., & Dreyfuss, M. M. (1997). Effect of the cyclopeptolide 90-215 on the production of destruxins and helvolic acid by *Metarhizium anisopliae*. *Journal of Industrial Microbiology & Technology, 19*, 7–11.

Ferracin, L. M., Fier, C. B., Vieira, M. L. C., Monteiro-Vitorello, C. B., de Mello Varani, A., Rossi, M. M. … Fungaro, M. H. (2012). Strain-specific polyketide synthase genes of *Aspergillus niger*. *International Journal of Food Microbiology, 155*, 137–145.

Fisch, K. M. (2013). Biosynthesis of natural products by microbial iterative hybrid PKS–NRPS. *RSC Advances, 3*, 18228–18247.

Frandsen, R. J. N., Schutt, C., Lund, B. W., Staerk, D., Nielsen, J., Olsson, S., & Giese, H. (2011). Two novel classes of enzymes are required for the biosynthesis of aurofusarin in *Fusarium graminearum*. *Journal of Biological Chemistry, 286*, 10419–10428.

Fujii, Y., Tani, H., Ichinoe, M., & Nakajima, H. (2000). Zygosporin D and two new cytochalasins produced by the fungus *Metarrhizium anisopliae*. *Journal of Natural Products, 63*, 132–135.

Fujii, I., Yoshida, N., Shimomaki, S., Oikawa, H., & Ebizuka, Y. (2005). An iterative type I polyketide synthase PKSN catalyzes synthesis of the decaketide alternapyrone with regio-specific octa-methylation. *Chemistry & Biology, 12*, 1301–1309.

Fukazawa, H., Ikeda, Y., Fukuyama, M., Suzuki, T., Hori, H., Okuda, T., & Uehara, Y. (2010). The resorcylic acid lactone hypothemycin selectively inhibits the mitogen-activated protein kinase kinase-extracellular signal-regulated kinase pathway in cells. *Biological & Pharmaceutical Bulletin, 33*, 168–173.

Fukushima, K., Arai, T., Mori, Y., Tsuboi, M., & Suzuki, M. (1983). Studies on peptide antibiotics leucinostatins I. separation physicochemical properties and biological activities of leucinostatins a and b. *Journal of Antibiotics, 36*, 1606–1612.

Furutani, S., Nakatani, Y., Miura, Y., Ihara, M., Kai, K., Hayashi, H., & Matsuda, K. (2014). GluCl a target of indole alkaloid okaramines: a 25 year enigma solved. *Scientific Reports, 4*, 6190.

Gallagher, R. T., Finer, J., Clardy, J., Leutwiler, A., Weibel, F., Acklin, W., & Arigoni, D. (1980). Paspalinine, a tremorgenic metabolite from *Claviceps paspali* Stevens et Hall. *Tetrahedron Letters, 21*, 235–238.

Gallo, A., Bruno, K. S., Solfrizzo, M., Perrone, G., Mule, G., Visconti, A., & Baker, S. E. (2012). New insight into the ochratoxin A biosynthetic pathway through deletion of a nonribosomal peptide synthetase gene in *Aspergillus carbonarius*. *Applied and Environmental Microbiology, 78*, 8208–8218.

Gao, Q. A., Jin, K., Ying, S. H., Zhang, Y. J., Xiao, G. H., Shang, Y. F. … Wang, C. (2011). Genome sequencing and comparative transcriptomics of the model entomopathogenic fungi *Metarhizium anisopliae* and *M. acridum*. *PLoS Genetics, 7*.

Gao, X., Haynes, S. W., Ames, B. D., Wang, P., Vien, L. P., Walsh, C. T., & Tang, Y. (2012). Cyclization of fungal nonribosomal peptides by a terminal condensation-like domain. *Nature Chemical Biology, 8*, 823–830.

García-Pajón, C., & Collado, I. G. (2003). Secondary metabolites isolated from *Colletotrichum* species. *Natural Product Reports, 20*, 426–431.

Gardiner, D. M., Cozijnsen, A. J., Wilson, L. M., Pedras, M. S., & Howlett, B. J. (2004). The sirodesmin biosynthetic gene cluster of the plant pathogenic fungus *Leptosphaeria maculans*. *Molecular Microbiology, 53*, 1307–1318.

Gardiner, D. M., & Howlett, B. J. (2005). Bioinformatic and expression analysis of the putative gliotoxin biosynthetic gene cluster of *Aspergillus fumigatus*. *FEMS Microbiology Letters, 248*, 241–248.

Gatenby, W. A., Munday-Finch, S. C., Wilkins, A. L., & Miles, C. O. (1999). Terpendole M, a novel indole–diterpenoid isolated from *Lolium perenne* infected with the endophytic fungus *Neotyphodium lolii*. *Journal of Agricultural and Food Chemistry, 47*, 1092–1097.

Gauthier, T., Wang, X., Sifuentes Dos Santos, J., Fysikopoulos, A., Tadrist, S., Canlet, C. ... Puel, O. (2012). Trypacidin, a spore-borne toxin from *Aspergillus fumigatus*, is cytotoxic to lung cells. *PLoS One, 7*, e29906.

Gelderblom, W. C. A., Marasas, W. F., Steyn, P. S., Thiel, P. G., van der Merwe, K. J., van Rooyen, P. H. ... Wessels, P. L. (1984). Structure elucidation of fusarin C, a mutagen produced by *Fusarium moniliforme*. *Journal of the Chemical Society, Chemical Communications*, 122–124.

Gerea, A. L., Branscum, K. M., King, J. B., You, J., Powell, D. R., Miller, A. N. ... Cichewicz, R. H. (2012). Secondary metabolites produced by fungi derived from a microbial mat encountered in an iron-rich natural spring. *Tetrahedron Letters, 53*, 4202–4205.

Geris, R., & Simpson, T. J. (2009). Meroterpenoids produced by fungi. *Natural Product Reports, 26*, 1063–1094.

Gerke, J., Bayram, Ö., Feussner, K., Landesfeind, M., Shelest, E., Feussner, I., & Braus, G. H. (2012). Breaking the silence: protein stabilization uncovers silenced biosynthetic gene clusters in the fungus *Aspergillus nidulans*. *Applied and Environmental Microbiology, 78*, 8234–8244.

Gledhill, L., Hesketh, A. R., Bycroft, B. W., Dewick, P. M., & Gilbert, J. (1991). Biosynthesis of trichothecene mycotoxins: cell-free epoxidation of a trichodiene derivative. *FEMS Microbiology Letters, 65*, 241–245.

Goetz, M. A., Zink, D. L., Dezeny, G., Dombrowski, A., Polishook, J. D., Felix, J. P. ... Singh, S. B. (2001). Diterpenoid pyrones, novel blockers of the voltage-gated potassium channel Kv1. 3 from fungal fermentations. *Tetrahedron Letters, 42*, 1255–1257.

Graziani, S., Vasnier, C., & Daboussi, M. J. (2004). Novel polyketide synthase from *Nectria haematococca*. *Applied and Environmental Microbiology, 70*, 2984–2988.

Gu, Y., Wang, Y., Ma, X., Wang, C., Yue, G., Zhang, Y. ... Wu, R. (2015). Greater taxol yield of fungus *Pestalotiopsis hainanensis* from dermatitic scurf of the giant panda (*Ailuropoda melanoleuca*). *Applied Biochemistry and Biotechnology, 175*, 155–165.

Guengerich, F. P., DiMari, S. J., & Broquist, H. P. (1973). Isolation and characterization of a l-pyrindine fungal alkaloid. *Journal of the American Chemical Society, 95*, 2055–2056.

Guo, J.-P., Zhu, C.-Y., Zhang, C.-P., Chu, Y.-S., Wang, Y.-L., Zhang, J.-X. ... Niu, X.-M. (2012). Thermolides, potent nematocidal PKS-NRPS hybrid metabolites from thermophilic fungus *Talaromyces thermophilus*. *Journal of the American Chemical Society, 134*, 20306–20309.

Gupta, S., Krasnoff, S. B., Renwick, J. A. A., Roberts, D. W., Steiner, J. R., & Clardy, J. (1993). Viridoxins A and B: novel toxins from the fungus *Metarhizium flavoviride*. *The Journal of Organic Chemistry, 58*, 1062–1067.

Gupta, S., Krasnoff, S. B., Roberts, D. W., Renwick, J. A. A., Brinen, L. S., & Clardy, J. (1991). Structures of the efrapeptins-potent inhibitors of mitochondrial ATPase from the fungus *Tolypocladium niveum*. *Journal of the American Chemical Society, 113*, 707–709.
Haas, H., Eisendle, M., & Turgeon, B. G. (2008). Siderophores in fungal physiology and virulence. *Annual Review of Phytopathology, 46*, 149–187.
Halo, L. M., Marshall, J. W., Yakasai, A. A., Song, Z., Butts, C. P., Crump, M. P. … Cox, R. J. (2008). Authentic heterologous expression of the tenellin iterative polyketide synthase nonribosomal peptide synthetase requires coexpression with an enoyl reductase. *ChemBioChem, 9*, 585–594.
Hamill, R. L., Higgens, C. E., Boaz, J. E., & Gorman, M. (1969). The structure of beauvericin, a new depsipeptide antibiotic toxic to *Artemia salina*. *Tetrahedron Letters, 49*, 4255–4258.
Hanson, F. R., & Eble, T. E. (1949). An antiphage agent isolated from *Aspergillus* sp. *Journal of Bacteriology, 58*, 527.
Harrington, G., & Neilands, J. (1982). Isolation and characterization of dimerum acid from *Verticillium dahliae*. *Journal of Plant Nutrition, 5*, 675–682.
Harris, C. M., Campbell, B. C., Molyneux, R. J., & Harris, T. M. (1988). Biosynthesis of swainsonine in the diablo locoweed (*Astragalus oxyphyrus*). *Tetrahedron Letters, 29*, 4815–4818.
Hayakawa, S., Matsushima, T., Kimura, T., Minato, H., & Katagiri, K. (1968). Zygosporin A, a new antibiotic from *Zygosporium masonii*. *The Journal of Antibiotics, 21*, 523–524.
He, Q. X., Liu, K. C., Wang, S. F., Hou, H. R., Yuan, Y. Q., & Wang, X. M. (2012). Toxicity induced by emodin on zebrafish embryos. *Drug and Chemical Toxicology, 35*, 149–154.
van den Heever, J. P., Thompson, T. S., Curtis, J. M., Ibrahim, A., & Pernal, S. F. (2014). Fumagillin: an overview of recent scientific advances and their significance for apiculture. *Journal of Agricultural and Food Chemistry, 62*, 2728–2737.
Heinig, U., Scholz, S., & Jennewein, S. (2013). Getting to the bottom of Taxol biosynthesis by fungi. *Fungal Diversity, 60*, 161–170.
van der Helm, D., & Winkelmann, G. (1994). Hydroxamates and polycarboxylates as iron transport agents (siderophores) in fungi. In G. Winkelmann, & D. R. Winge (Eds.), *Metal ions in fungi* (Vol. 11, pp. 39–98). New York: Marcel Dekker, Inc.
Henrikson, J. C., Hoover, A. R., Joyner, P. M., & Cichewicz, R. H. (2009). A chemical epigenetics approach for engineering the in situ biosynthesis of a cryptic natural product from *Aspergillus niger*. *Organic & Biomolecular Chemistry, 7*, 435–438.
Hino, M., Nakayama, O., Tsurumi, Y., Adachi, K., Shibata, T., Terano, H. … Imanaka, H. (1985). Studies of an immunomodulator, swainsonine. I. Enhancement of immune response by swainsonine in vitro. *The Journal of Antibiotics, 38*, 926–935.
Hof, C., Eisfeld, K., Welzel, K., Antelo, L., Foster, A. J., & Anke, H. (2007). Ferricrocin synthesis in *Magnaporthe grisea* and its role in pathogenicity in rice. *Molecular Plant Pathology, 8*, 163–172.
Hohn, T. M., & Plattner, R. D. (1989). Expression of the trichodiene synthase gene of *Fusarium sporotrichioides* in *Escherichia coli* results in sesquiterpene production. *Archives of Biochemistry and Physics, 275*, 92–97.
Hohn, T. M., & Vanmiddlesworth, F. (1986). Purification and characterization of the sesquiterpene cyclase trichodiene synthetase from *Fusarium sporotrichioides*. *Archives of Biochemistry and Physics, 251*, 756–761.
Hördt, W., Römheld, V., & Winkelmann, G. (2000). Fusarinines and dimerum acid, mono- and dihydroxamate siderophores from *Penicillium chrysogenum*, improve iron utilization by strategy I and strategy II plants. *Biometals, 13*, 37–46.
Hosoe, T., Fukushima, K., Takizawa, K., Itahashi, T., Kawahara, N., Vidotto, V., & Kawai, K. (2006). A new antifungal macrolide, eushearilide, isolated from *Eupenicillium shearii*. *The Journal of Antibiotics, 59*, 597–600.

Houwaart, S., Youssar, L., & Huttel, W. (2014). Pneumocandin biosynthesis: involvement of a trans-selective proline hydroxylase. *ChemBioChem, 15*, 2365–2369.

Hu, Q., & Dong, T. (2015). Non-ribosomal peptides from entomogenous fungi. In K. S. Sree, & A. Varma (Eds.), *Biocontrol of lepidopteran pests* (pp. 169–206). Cham, Switzerland: Springer.

Hu, X., Xiao, G. H., Zheng, P., Shang, Y. F., Su, Y., Zhang, X. Y.... Wang, C. (2014). Trajectory and genomic determinants of fungal-pathogen speciation and host adaptation. *Proceedings of the National Academy of Sciences of the United States of America, 111*, 16796–16801.

Hu, Y., Hao, X. R., Lou, J., Zhang, P., Pan, J., & Zhu, X. D. (2012). A PKS gene, pks-1, is involved in chaetoglobosin biosynthesis, pigmentation and sporulation in *Chaetomium globosum*. *Science China-Life Sciences, 55*, 1100–1108.

Huxham, I. M., Lackie, A. M., & McCorkindale, N. J. (1989). Inhibitory effects of cyclodepsipeptides, destruxins, from the fungus *Metarhizium anisopliae* on cellular immunity in insects. *Journal of Insect Physiology, 35*, 97–106.

Iijima, M., Masuda, T., Nakamura, H., Naganawa, H., Kurasawa, S., Okami, Y.... Litake, Y. (1992). Metacytofilin, a novel immunomodulator produced by *Metarhizium* sp. TA2759. *The Journal of Antibiotics, 45*, 1553–1556.

Ingber, D., Fujita, T., Kishimoto, S., Sudo, K., Kanamaru, T., Brem, H., & Folkman, J. (1990). Synthetic analogues of fumagillin that inhibit angiogenesis and suppress tumour growth. *Nature, 348*, 555–557.

Inglis, D. O., Binkley, J., Skrzypek, M. S., Arnaud, M. B., Cerqueira, G. C., Shah, P.... Sherlock, G. (2013). Comprehensive annotation of secondary metabolite biosynthetic genes and gene clusters of *Aspergillus nidulans, A. fumigatus, A. niger* and *A. oryzae*. *BMC Microbiology, 13*, 91.

Isaka, M., Palasarn, S., Kocharin, K., & Hywel-Jones, N. L. (2007). Comparison of the bioactive secondary metabolites from the scale insect pathogens, anamorph paecilomyces cinnamomeus, and teleomorph torrubiella luteorostrata. *The Journal of Antibiotics, 60*, 577–581.

Isaka, M., Palasarn, S., Rachtawee, P., Vimuttipong, S., & Kongsaeree, P. (2005). Unique diketopiperazine dimers from the insect pathogenic fungus *Verticillium hemipterigenum* BCC 1449. *Organic Letters, 7*, 2257–2260.

Itoh, T., Tokunaga, K., Matsuda, Y., Fujii, I., Abe, I., Ebizuka, Y., & Kushiro, T. (2010). Reconstitution of a fungal meroterpenoid biosynthesis reveals the involvement of a novel family of terpene cyclases. *Nature Chemistry, 2*, 858–864.

Iwasaki, S., Sair, M. I., Igarashi, H., & Okuda, S. (1970). Revised structure of helvolic acid. *Chemical Communications, 500*, 1119–1120.

Izhaki, I. (2002). Emodin – a secondary metabolite with multiple ecological functions in higher plants. *New Phytologist, 155*, 205–217.

Jacquot, D., Poeschke, O., & Burger, C. (2009). *New terpenes and macrocycles*. International Patent WO/2009/146,772. W. I. P. O.-I. Bureau.

Jalal, M. A. F., Hossain, M. B., van der Helm, D., & Barnes, C. L. (1988). Structure of ferrichrome-type siderophores with dissimilar N^{δ}-acyl groups: asperchrome B1, B2, B3, D1, D2, and D3. *Biology of Metals, 1*, 77–89.

Jalal, M. A. F., Love, S. K., & Van Der Helm, D. (1986). Siderophore mediated iron-iii uptake in *Gliocladium virens*. 1. Properties of *cis*-fusarinine *trans*- fusarinine, dimerum acid, and their ferric complexes. *Journal of Inorganic Biochemistry, 28*, 417–430.

Jalal, M. A. F., Love, S. K., & Van Der Helm, D. (1988). N^{α}-dimethylcoprogens three novel trihydroxamate siderophores from pathogenic fungi. *Biology of Metals, 1*, 4–8.

Jin, J. M., Lee, S., Lee, J., Baek, S. R., Kim, J. C., Yun, S. H.... Lee, Y. W. (2010). Functional characterization and manipulation of the apicidin biosynthetic pathway in *Fusarium semitectum*. *Molecular Microbiology, 76*, 456–466.

Kagamizono, T., Nishino, E., Matsumoto, K., Kawashima, A., Kishimoto, M., Sakai, N. ... Morimoto, S. (1995). Bassiatin, a new platelet aggregation inhibitor produced by *Beauveria bassiana* K-717. *The Journal of Antibiotics, 48*, 1407–1413.

Kakuta, Y., Igarashi, T., Murakami, T., Ito, H., Matsui, H., & Honma, M. (2001). 1-aminocyclopropane-1-carboxylate synthase of *Penicillium citrinum*: primary structure and expression in *Escherichia coli* and *Saccharomyces cerevisiae*. *Bioscience Biotechnology and Biochemistry, 65*, 1511–1518.

Kalb, D., Lackner, G., & Hoffmeister, D. (2014). Functional and phylogenetic divergence of fungal adenylate-forming reductases. *Applied and Environmental Microbiology, 80*, 6175–6183.

Kaneko, I., Katsuya, S., & Tsuge, T. (1997). Structural analysis of the plasmid pAAT56 of the filamentous fungus *Alternaria alternata*. *Gene, 203*, 51–57.

Kannenberg, E. L., & Poralla, K. (1999). Hopanoid biosynthesis and function in Bacteria. *Naturwissenschaften, 86*, 168–176.

Karolewiez, A., & Geisen, R. (2005). Cloning a part of the ochratoxin A biosynthetic gene cluster of *Penicillium nordicum* and characterization of the ochratoxin polyketide synthase gene. *Systematic and Applied Microbiology, 28*, 588–595.

Kawaide, H., Imai, R., Sassa, T., & Kamiya, Y. (1997). Ent-kaurene synthase from the fungus *Phaeosphaeria* sp. L487. cDNA isolation, characterization, and bacterial expression of a bifunctional diterpene cyclase in fungal gibberellin biosynthesis. *Journal of Biological Chemistry, 272*, 21706–21712.

Kawazu, K., Murakami, T., Ono, Y., Kanzaki, H., Kobayashi, A., Mikawa, T., & Yoshikawa, N. (1993). Isolation and characterization of 2 novel nematicidal depsipeptides from an imperfect fungus, strain D1084. *Bioscience Biotechnology and Biochemistry, 57*, 98–101.

Keller, N. P., & Hohn, T. M. (1997). Metabolic pathway gene clusters in filamentous fungi. *Fungal Genetics and Biology, 21*, 17–29.

Keller, N. P., Turner, G., & Bennett, J. W. (2005). Fungal secondary metabolism—from biochemistry to genomics. *Nature Reviews Microbiology, 3*, 937–947.

Keller-Schierlein, W., & Diekmann, H. (1970). Stoffwechselprodukte von Mikroorganismen. 85. Mitteilung [1]. Zur Konstitution des Coprogens. *Helvetica Chimica Acta, 53*, 2035–2044.

Kennedy, J., Auclair, K., Kendrew, S. G., Park, C., Vederas, J. C., & Hutchinson, C. R. (1999). Modulation of polyketide synthase activity by accessory proteins during lovastatin biosynthesis. *Science, 284*, 1368–1372.

Kepler, R. M., Humber, R. A., Bischoff, J. F., & Rehner, S. A. (2014). Clarification of generic and species boundaries for *Metarhizium* and related fungi through multigene phylogenetics. *Mycologia, 106*, 811–829.

Kershaw, M. J., Moorhouse, E. R., Bateman, R., Reynolds, S. E., & Charnley, A. K. (1999). The role of destruxins in the pathogenicity of *Metarhizium anisopliae* for three species of insect. *Journal of Invertebrate Pathology, 74*, 213–223.

Kikuchi, H., Hoshi, T., Kitayama, M., Sekiya, M., Katou, Y., Ueda, K. ... Oshima, Y. (2009). New diterpene pyrone-type compounds, metarhizins A and B, isolated from entomopathogenic fungus, *Metarhizium flavoviride* and their inhibitory effects on cellular proliferation. *Tetrahedron, 65*, 469–477.

Kim, K. H., Willger, S. D., Park, S. W., Puttikamonkul, S., Grahl, N., Cho, Y. ... Lawrence, C. B. (2009). *TmpL*, a transmembrane protein required for intracellular redox homeostasis and virulence in a plant and an animal fungal pathogen. *PLoS Pathogens, 5*.

Kim, S. H., Ahn, S., Han, J.-W., Lee, H.-W., Lee, H. Y., Lee, Y.-W. ... Hong, S. (2004). Apicidin is a histone deacetylase inhibitor with anti-invasive and anti-angiogenic potentials. *Biochemical and Biophysical Research Communications, 315*, 964–970.

Kim, Y. T., Lee, Y. R., Jin, J. M., Han, K. H., Kim, H., Kim, J. C. ... Lee, Y. W. (2005). Two different polyketide synthase genes are required for synthesis of zearalenone in *Gibberella zeae*. *Molecular Microbiology, 58*, 1102–1113.

Kobayashi, H., Sunaga, R., Furihata, K., Morisaki, N., & Iwasaki, S. (1995). Isolation and structures of an antifungal antibiotic, fusarielin A, and related compounds produced by a *Fusarium* sp. *The Journal of Antibiotics, 48*, 42–52.

Kodaira, Y. (1961). Toxic substances to insects, produced by *Aspergillus ochraceus* and *Oospora destructor*. *Agricultural and Biological Chemistry, 25*, 261–262.

Kodaira, Y. (1962). Studies on the new toxic substances to insects, destruxin A and B, produced by *Oospora destructor*. Part I. Isolation and purification of destruxin A and B. *Agricultural and Biological Chemistry, 26*, 36–42.

Konig, C. C., Scherlach, K., Schroeckh, V., Horn, F., Nietzsche, S., Brakhage, A. A., & Hertweck, C. (2013). Bacterium induces cryptic meroterpenoid pathway in the pathogenic fungus *Aspergillus fumigatus*. *ChemBioChem, 14*, 938–942.

Kosalkova, K., Garcia-Estrada, C., Ullan, R. V., Godio, R. P., Feltrer, R., Teijeira, F.... Martin, J. F. (2009). The global regulator LaeA controls penicillin biosynthesis, pigmentation and sporulation, but not roquefortine C synthesis in *Penicillium chrysogenum*. *Biochimie, 91*, 214–225.

Kozone, I., Ueda, J., Watanabe, M., Nogami, S., Nagai, A., Inaba, S.... Shin-ya, K. (2009). Novel 24-membered macrolides, JBIR-19 and-20 isolated from *Metarhizium* sp. fE61. *The Journal of Antibiotics, 62*, 159–162.

Krasnoff, S. B., Englich, U., Miller, P. G., Shuler, M. L., Glahn, R. P., Donzelli, B. G. G., & Gibson, D. M. (2012). Metacridamides a and B, macrocycles from conidia of the entomopathogenic fungus *Metarhizium acridum*. *Journal of Natural Products, 75*, 175–180.

Krasnoff, S. B., Gibson, D. M., Belofsky, G. N., Gloer, K. B., & Gloer, J. B. (1996). New destruxins from the entomopathogenic fungus *Aschersonia* sp. *Journal of Natural Products, 59*, 485–489.

Krasnoff, S. B., Keresztes, I., Donzelli, B. G., & Gibson, D. M. (2014). Metachelins, mannosylated and N-oxidized coprogen-type siderophores from *Metarhizium robertsii*. *Journal of Natural Products, 77*, 1685–1692.

Krasnoff, S. B., Keresztes, I., Gillilan, R. E., Szebenyi, D. M. E., Donzelli, B. G. G., Churchill, A. C. L., & Gibson, D. M. (2007). Serinocyclins A and B, cyclic heptapeptides from *Metarhizium anisopliae*. *Journal of Natural Products, 70*, 1919–1924.

Krasnoff, S. B., Sommers, C. H., Moon, Y.-S., Donzelli, B. G., Vandenberg, J. D., Churchill, A. C., & Gibson, D. M. (2006). Production of mutagenic metabolites by *Metarhizium anisopliae*. *Journal of Agricultural and Food Chemistry, 54*, 7083–7088.

Kremer, A., & Li, S.-M. (2010). A tyrosine O-prenyltransferase catalyses the first pathway-specific step in the biosynthesis of sirodesmin PL. *Microbiology, 156*, 278–286.

Kroken, S., Glass, N. L., Taylor, J. W., Yoder, O. C., & Turgeon, B. G. (2003). Phylogenomic analysis of type I polyketide synthase genes in pathogenic and saprobic Ascomycetes. *Proceedings of the National Academy of Sciences of the United States of America, 100*, 15670–15675.

Kuboki, H., Tsuchida, T., Wakazono, K., Isshiki, K., Kumagai, H., & Yoshioka, T. (1999). Mer-f3, 12-hydroxy-ovalicin, produced by *Metarrhizium* sp. f3. *The Journal of Antibiotics, 52*, 590–593.

Kusari, S., Singh, S., & Jayabaskaran, C. (2014). Rethinking production of Taxol® (paclitaxel) using endophyte biotechnology. *Trends in Biotechnology, 32*, 304–311.

Langenfeld, A., Blond, A., Gueye, S., Herson, P., Nay, B., Dupont, J., & Prado, S. (2011). Insecticidal cyclodepsipeptides from *Beauveria felina*. *Journal of Natural Products, 74*, 825–830.

Lardy, H., Reed, P., & Chiu, L. C. H. (1975). Antibiotic inhibitors of mitochondrial ATP synthesis. *Federation Proceedings, 34*, 1707–1710.

Lawrence, D. P., Kroken, S., Pryor, B. M., & Arnold, A. E. (2011). Interkingdom gene transfer of a hybrid NPS/PKS from bacteria to filamentous Ascomycota. *PLoS One, 6*, e28231.

Lazary, S., & Stähelin, H. (1968). Immunosuppressive and specific antimitotic effects of ovalicin. *Experientia, 24*, 1171–1173.
Lee, J. C., Lobkovsky, E., Pliam, N. B., Strobel, G., & Clardy, J. (1995). Subglutinols A and B: immunosuppressive compounds from the endophytic fungus *Fusarium subglutinans*. *The Journal of Organic Chemistry, 60*, 7076–7077.
Lee, S., Kinoshita, H., Ihara, F., Igarashi, Y., & Nihira, T. (2008). Identification of novel derivative of helvolic acid from *Metarhizium anisopliae* grown in medium with insect component. *Journal of Bioscience and Bioengineering, 105*, 476–480.
Lee, W., van Baalen, M., & Jansen, V. A. A. (2012). An evolutionary mechanism for diversity in siderophore-producing bacteria. *Ecology Letters, 15*, 119–125.
Li, Y. R., Xu, W., & Tang, Y. (2010). Classification, prediction, and verification of the regioselectivity of fungal polyketide synthase product template domains. *Journal of Biological Chemistry, 285*, 22762–22771.
Lin, H. C., Chooi, Y. H., Dhingra, S., Xu, W., Calvo, A. M., & Tang, Y. (2013). Fumagillin biosynthesis in *Aspergillus fumigatus*: a cryptic terpene cyclase gene is involved in the formation of beta-trans-bergamotene. *Planta Medica, 79*, 825.
Lin, H.-C., Tsunematsu, Y., Dhingra, S., Xu, W., Fukutomi, M., Chooi, Y.-H. ... Tang, Y. (2014). Generation of complexity in fungal terpene biosynthesis: discovery of a multifunctional cytochrome P450 in the fumagillin pathway. *Journal of the American Chemical Society, 136*, 4426–4436.
Lira, S. P., Vita-Marques, A. M., Seleghim, M. H. R., Bugni, T. S., Labarbera, D. V., Sette, L. D. ... Berlinck, R. G. (2006). New destruxins from the marine-derived fungus *Beauveria felina*. *The Journal of Antibiotics, 59*, 553–563.
Liu, B.-L., & Tzeng, Y.-M. (2012). Development and applications of destruxins: a review. *Biotechnology Advances, 30*, 1242–1254.
Liu, L., Zhang, Z., Shao, C.-L., Wang, J.-L., Bai, H., & Wang, C.-Y. (2015). Bioinformatical analysis of the sequences, structures and functions of fungal polyketide synthase product template domains. *Scientific Reports, 5*.
Lo, H.-C., Entwistle, R., Guo, C.-J., Ahuja, M., Szewczyk, E., Hung, J.-H. ... Wang, C. C. C. (2012). Two separate gene clusters encode the biosynthetic pathway for the meroterpenoids austinol and dehydroaustinol in *Aspergillus nidulans*. *Journal of the American Chemical Society, 134*, 4709–4720.
Lopez-Berges, M. S., Capilla, J., Turra, D., Schafferer, L., Matthijs, S., Jochl, C. ... Di Pietro, A. (2012). HapX-mediated iron homeostasis is essential for rhizosphere competence and virulence of the soilborne pathogen *Fusarium oxysporum*. *The Plant Cell, 24*, 3805–3822.
Maiya, S., Grundmann, A., Li, X., Li, S. M., & Turner, G. (2007). Identification of a hybrid PKS/NRPS required for pseurotin A biosynthesis in the human pathogen *Aspergillus fumigatus*. *ChemBioChem, 8*, 1736–1743.
Malz, S., Grell, M. N., Thrane, C., Maier, F. J., Rosager, P., Felk, A. ... Giese, H. (2005). Identification of a gene cluster responsible for the biosynthesis of aurofusarin in the *Fusarium graminearum* species complex. *Fungal Genetics and Biology, 42*, 420–433.
Marahiel, M. A., Stachelhaus, T., & Mootz, H. D. (1997). Modular peptide synthetases involved in nonribosomal peptide synthesis. *Chemical Reviews, 97*, 2651–2674.
Marmann, A., Aly, A. H., Lin, W., Wang, B., & Proksch, P. (2014). Co-cultivation—a powerful emerging tool for enhancing the chemical diversity of microorganisms. *Marine Drugs, 12*, 1043–1065.
Mattern, D. J., Schoeler, H., Weber, J., Novohradská, S., Kraibooj, K., Dahse, H.-M. ... Brakhage, A. A. (2015). Identification of the antiphagocytic trypacidin gene cluster in the human-pathogenic fungus *Aspergillus fumigatus*. *Applied Microbiology and Biotechnology*, 1–11.
McCormick, S. P., Alexander, N. J., & Harris, L. J. (2010). CLM1 of *Fusarium graminearum* encodes a longiborneol synthase required for culmorin production. *Applied and Environmental Microbiology, 76*, 136–141.

McCowen, M. C., Callender, M. E., & Lawlis, J. F. (1951). Fumagillin (H-3), a new antibiotic with amebicidal properties. *Science, 113*, 202–203.

McInnes, G. A., Smith, D. G., Wat, C.-K., Vinning, L. C., & Wright, J. L. C. (1974). Tenellin and bassianin, metabolites of *Beauveria* species. Structure elucidation with ^{15}N- and doubly ^{13}C-enriched compounds using ^{13}C nuclear magnetic resonance spectroscopy. *Journal of the Chemical Society, Chemical Communications, 8*, 281–282.

Medema, M. H., Takano, E., & Breitling, R. (2013). Detecting sequence homology at the gene cluster level with MultiGeneBlast. *Molecular Biology and Evolution, 30*, 1218–1223.

Mitsuguchi, H., Seshime, Y., Fujii, I., Shibuya, M., Ebizuka, Y., & Kushiro, T. (2009). Biosynthesis of steroidal antibiotic fusidanes: functional analysis of oxidosqualene cyclase and subsequent tailoring enzymes from *Aspergillus fumigatus*. *Journal of the American Chemical Society, 131*, 6402–6411.

Miyake, Y., Ito, C., Itoigawa, M., & Osawa, T. (2007). Isolation of the antioxidant pyranonigrin-A from rice mold starters used in the manufacturing process of fermented foods. *Bioscience, Biotechnology, and Biochemistry, 71*, 2515–2521.

Molina, J.-M., Tourneur, M., Sarfati, C., Chevret, S., de Gouvello, A., Gobert, J.-G.… Agence Nationale de Recherches sur le SIDA 090 Study Group. (2002). Fumagillin treatment of intestinal microsporidiosis. *New England Journal of Medicine, 346*, 1963–1969.

Moon, Y.-S., Donzelli, B. D., Krasnoff, S., McLane, H., Griggs, M. H., Cooke, P.… Churchill, A. C. (2008). *Agrobacterium*-mediated disruption of a nonribosomal peptide synthetase gene in the invertebrate pathogen *Metarhizium anisopliae* reveals a peptide spore factor. *Applied and Environmental Microbiology, 74*, 4366–4380.

Motoyama, T., Hayashi, T., Hirota, H., Ueki, M., & Osada, H. (2012). Terpendole E, a kinesin Eg5 inhibitor, is a key biosynthetic intermediate of indole-diterpenes in the producing fungus *Chaunopycnis alba*. *Chemistry & Biology, 19*, 1611–1619.

Moya, P., Cantín, Á., Castillo, M.-A., Primo, J., Miranda, M. A., & Primo-Yúfera, E. (1998). Isolation, structural assignment, and synthesis of *N*-(2-Methyl-3-oxodecanoyl)-2-pyrroline, a new natural product from *Penicillium brevicompactum* with in vivo anti-juvenile hormone activity. *The Journal of Organic Chemistry, 63*, 8530–8535.

Mulheirn, L., Beechey, R., Leworthy, D., & Osselton, M. (1974). Aurovertin B, a metabolite of *Calcarisporium arbuscula*. *Journal of the Chemical Society, Chemical Communications, 1974*, 874–876.

Newman, A. G., Vagstad, A. L., Storm, P. A., & Townsend, C. A. (2014). Systematic domain swaps of iterative, nonreducing polyketide synthases provide a mechanistic understanding and rationale for catalytic reprogramming. *Journal of the American Chemical Society, 136*, 7348–7362.

Nicholson, M. J., Eaton, C. J., Starkel, C., Tapper, B. A., Cox, M. P., & Scott, B. (2015). Molecular cloning and functional analysis of gene clusters for the biosynthesis of indole-diterpenes in *Penicillium crustosum* and *P. janthinellum*. *Toxins, 7*, 2701–2722.

Niehaus, E. M., Kleigrewe, K., Wiemann, P., Studt, L., Sieber, C. M., Connolly, L. R.… Humpf, H. U. (2013). Genetic manipulation of the *Fusarium fujikuroi* fusarin gene cluster yields insight into the complex regulation and fusarin biosynthetic pathway. *Chemistry & Biology, 20*, 1055–1066.

Nielsen, M. L., Nielsen, J. B., Rank, C., Klejnstrup, M. L., Holm, D. K., Brogaard, K. H.… Mortensen, U. H. (2011). A genome-wide polyketide synthase deletion library uncovers novel genetic links to polyketides and meroterpenoids in *Aspergillus nidulans*. *FEMS Microbiology Letters, 321*, 157–166.

Nilanonta, C., Isaka, M., Kittakoop, P., Saenboonrueng, J., Rukachaisirikul, V., Kongsaeree, P., & Thebtaranonth, Y. (2003). New diketopiperazines from the entomopathogenic fungus *Verticillium hemipterigenum* BCC 1449. *The Journal of Antibiotics, 56*, 647–651.

Niu, X., Chen, L., Yue, Q., Wang, B., Zhang, J., Zhu, C.… An, Z. (2014). Characterization of thermolide biosynthetic genes and a new thermolide from sister thermophilic fungi. *Organic Letters, 16*, 3744–3747.

Niu, X.-M., Wang, Y.-L., Chu, Y.-S., Xue, H.-X., Li, N., Wei, L.-X.... Zhang, K. Q. (2009). Nematodetoxic aurovertin-type metabolites from a root-knot nematode parasitic fungus *Pochonia chlamydosporia*. *Journal of Agricultural and Food Chemistry, 58*, 828–834.

O'Callaghan, J., Caddick, M. X., & Dobson, A. D. (2003). A polyketide synthase gene required for ochratoxin A biosynthesis in *Aspergillus ochraceus*. *Microbiology, 149*, 3485–3491.

Oh, H., Kim, T., Oh, G., Pae, H., Hong, K., Chai, K.... Lee, H. S. (2002). (3*R*, 6*R*)-4-Methyl-6-(1-methylethyl)-3-phenyl-methylperhydro-1, 4-oxazine-2, 5-dione: an apoptosis-inducer from the fruiting bodies of *Isaria japonica*. *Planta Medica, 68*, 345–348.

Ohm, R. A., Feau, N., Henrissat, B., Schoch, C. L., Horwitz, B. A., Barry, K. W.... Grigoriev, I. V. (2012). Diverse lifestyles and strategies of plant pathogenesis encoded in the genomes of eighteen Dothideomycetes fungi. *PLoS Pathogens, 8*, e1003037.

Oide, S., Berthiller, F., Wiesenberger, G., Adam, G., & Turgeon, B. G. (2015). Individual and combined roles of malonichrome, ferricrocin, and TAFC siderophores in *Fusarium graminearum* pathogenic and sexual development. *Frontiers in Microbiology, 5*.

Oide, S., Moeder, W., Krasnoff, S., Gibson, D., Haas, H., Yoshioka, K.... Turgeon, B. G. (2006). NPS6, encoding a nonribosomal peptide synthetase involved in siderophore-mediated iron metabolism, is a conserved virulence determinant of plant pathogenic Ascomycetes. *The Plant Cell, 18*, 1–18.

Okuda, S., Iwasaki, S., Sair, M. I., Machida, Y., Inoue, A., & Tsuda, K. (1967). Stereochemistry of helvolic acid. *Tetrahedron Letters, 24*, 2295–2302.

Okuda, S., Iwasaki, S., Tsuda, K., Sano, Y., Hata, T., Udagawa, S.... Yamaguchi, H. (1964). The structure of helvolic acid. *Chemical & Pharmaceutical Bulletin, 12*, 121.

Ortel, I., & Keller, U. (2009). Combinatorial assembly of simple and complex D-lysergic acid alkaloid peptide classes in the ergot fungus *Claviceps purpurea*. *Journal of Biological Chemistry, 284*, 6650–6660.

Ostlind, D. A., Felcetto, T., Misura, A., Ondeyka, J., Smith, S., Goetz, M.... Mickle, W. (1997). Discovery of a novel indole diterpene insecticide using first instars of *Lucilia sericata*. *Medical and Veterinary Entomology, 11*, 407–408.

Pal, S., St. Leger, R., & Wu, L. (2007). Fungal peptide destruxin A plays a specific role in suppressing the innate immune response in *Drosophila melanogaster*. *Journal of Biological Chemistry, 282*, 8969.

Pedras, M. S. C., Zaharia, I. L., & Ward, D. E. (2002). The destruxins: synthesis, biosynthesis, biotransformation, and biological activity. *Phytochemistry, 59*, 579–596.

Pitel, D. W., Arsenault, G. P., & Vining, L. C. (1971). Cyclonerodiol, a sesquiterpene metabolite of *Gibberella fujikuroi*. *The Journal of Antibiotics, 24*, 483–484.

Pitt, J. I. (2002). *Biology and ecology of toxigenic* Penicillium *species* Mycotoxins and food safety (pp. 29–41). Springer.

Pittayakhajonwut, P., Usuwan, A., Intaraudom, C., Khoyaiklang, P., & Supothina, S. (2009). Torrubiellutins A and C, from insect pathogenic fungus *Torrubiella luteorostrata* BCC 12904. *Tetrahedron, 65*, 6069–6073.

Prakash, K. R., Tang, Y., Kozikowski, A. P., Flippen-Anderson, J. L., Knoblach, S. M., & Faden, A. I. (2002). Synthesis and biological activity of novel neuroprotective diketopiperazines. *Bioorganic & Medicinal Chemistry, 10*, 3043–3048.

Price, M. N., Dehal, P. S., & Arkin, A. P. (2010). FastTree 2—approximately maximum-likelihood trees for large alignments. *PLoS One, 5*, e9490.

Proctor, R. H., McCormick, S. P., Alexander, N. J., & Desjardins, A. E. (2009). Evidence that a secondary metabolic biosynthetic gene cluster has grown by gene relocation during evolution of the filamentous fungus *Fusarium*. *Molecular Microbiology, 74*, 1128–1142.

Putri, S. P., Ishido, K.-I., Kinoshita, H., Kitani, S., Ihara, F., Sakihama, Y. … Nihira, T. (2014). Production of antioomycete compounds active against the phytopathogens *Phytophthora sojae* and *Aphanomyces cochlioides* by clavicipitoid entomopathogenic fungi. *Journal of Bioscience and Bioengineering, 117*, 557–562.

Qiao, K., Chooi, Y. H., & Tang, Y. (2011). Identification and engineering of the cytochalasin gene cluster from *Aspergillus clavatus* NRRL 1. *Metabolic Engineering, 13*, 723–732.

Quin, M. B., Flynn, C. M., & Schmidt-Dannert, C. (2014). Traversing the fungal terpenome. *Natural Product Reports, 31*, 1449–1473.

Rateb, M. E., Hallyburton, I., Houssen, W. E., Bull, A. T., Goodfellow, M., Santhanam, R. … Ebel, R. (2013). Induction of diverse secondary metabolites in *Aspergillus fumigatus* by microbial co-culture. *RSC Advances, 3*, 14444–14450.

Rausch, C., Hoof, I., Weber, T., Wohlleben, W., & Huson, D. H. (2007). Phylogenetic analysis of condensation domains in NRPS sheds light on their functional evolution. *BMC Evolutionary Biology, 7*, 78.

Ravindra, G., Ranganayaki, R. S., Raghothama, S., Srinivasan, M. C., Gilardi, R. D., Karle, I. L., & Balaram, P. (2004). Two novel hexadepsipeptides with several modified amino acid residues isloated from the fungus *Isaria*. *Chemistry and Biodiversity, 1*, 489–504.

Reeves, C. D., Hu, Z. H., Reid, R., & Kealey, J. T. (2008). Genes for the biosynthesis of the fungal polyketides hypothemycin from hypomyces subiculosus and radicicol from *Pochonia chlamydosporia*. *Applied and Environmental Microbiology, 74*, 5121–5129.

Regueira, T. B., Kildegaard, K. R., Hansen, B. G., Mortensen, U. H., Hertweck, C., & Nielsen, J. (2011). Molecular basis for mycophenolic acid biosynthesis in *Penicillium brevicompactum*. *Applied and Environmental Microbiology, 77*, 3035–3043.

Renshaw, J. C., Robson, G. D., Trinci, A. P. J., Wiebe, M. G., Livens, F. R., Collison, D. … Taylor, R. J. (2002). Fungal siderophores: structures, functions and applications. *Mycological Research, 106*, 1123–1142.

Roberts, D. W. (1966). Toxins from the entomogenous fungus *Metarrhizium anisopliae*: I. Production in submerged and surface cultures, and in inorganic and organic nitrogen media. *Journal of Invertebrate Pathology, 8*, 212–221.

Roberts, D. W. (1969). Toxins from the entomogenous fungus *Metarrhizium anisopliae*: isolation of destruxins from submerged cultures. *Journal of Invertebrate Pathology, 14*, 82–88.

Rothweiler, W., & Tamm, C. (1966). Isolation and structure of phomin. *Cellular and Molecular Life Sciences, 22*, 750–752.

Rowan, D. D. (1993). Lolitrems, peramine and paxilline: mycotoxins of the ryegrass/endophyte interaction. *Agriculture, Ecosystems & Environment, 44*, 103–122.

Sabareesh, V., Ranganayaki, R. S., Raghothama, S., Bopanna, M. P., Balaram, H., Srinivasan, M. C., & Balaram, P. (2007). Identification and characterization of a library of microheterogeneous cyclohexadepsipeptides from the fungus *Isaria*. *Journal of Natural Products, 70*, 715.

Saikia, S., Parker, E. J., Koulman, A., & Scott, B. (2006). Four gene products are required for the fungal synthesis of the indole-diterpene, paspaline. *FEBS Letters, 580*, 1625–1630.

Samuels, R. I., Charnley, A. K., & Reynolds, S. E. (1988). The role of destruxins in the pathogenicity of 3 strains of *Metarhizium anisopliae* for the tobacco hornworm *Manduca sexta*. *Mycopathologia, 104*, 51–58.

Sanchez, J. F., Chiang, Y. M., Szewczyk, E., Davidson, A. D., Ahuja, M., Elizabeth Oakley, C. … Wang, C. C. (2010). Molecular genetic analysis of the orsellinic acid/F9775 gene cluster of *Aspergillus nidulans*. *Molecular Biosystematics, 6*, 587–593.

Sanchez, J. F., Entwistle, R., Corcoran, D., Oakley, B. R., & Wang, C. C. C. (2012). Identification and molecular genetic analysis of the cichorine gene cluster in *Aspergillus nidulans*. *MedChemComm, 3*, 997–1002.

Sanchez, J. F., Entwistle, R., Hung, J. H., Yaegashi, J., Jain, S., Chiang, Y. M.... Oakley, B. R. (2011). Genome-based deletion analysis reveals the prenyl xanthone biosynthesis pathway in *Aspergillus nidulans*. *Journal of the American Chemical Society, 133*, 4010–4017.

Sanivada, S. K., & Challa, M. M. (2014). Computational interaction of entomopathogenic fungal secondary metabolites with proteins involved in human xenobiotic detoxification. *International Journal of Pharmacy and Pharmaceutical Sciences, 6*.

Scherlach, K., Boettger, D., Remme, N., & Hertweck, C. (2010). The chemistry and biology of cytochalasans. *Natural Product Reports, 27*, 869–886.

Schmidt, K., Riese, U., Li, Z., & Hamburger, M. (2003). Novel tetramic acids and pyridone alkaloids, militarinones B, C, and D, from the insect pathogenic fungus *Paecilomyces militaris*. *Journal of Natural Products, 66*, 378–383.

Schrettl, M., Bignell, E., Kragl, C., Joechl, C., Rogers, T., Arst, H. N.... Haas, H. (2004). Siderophore biosynthesis but not reductive iron assimilation is essential for *Aspergillus fumigatus* virulence. *The Journal of Experimental Medicine, 200*, 1213–1219.

Schrettl, M., Kim, H. S., Eisendle, M., Kragl, C., Nierman, W. C., Heinekamp, T.... Haas, H. (2008). SreA-mediated iron regulation in *Aspergillus fumigatus*. *Molecular Microbiology, 70*, 27–43.

Schroeder, F. C., Gibson, D. M., Churchill, A. C. L., Wursthorn, E. J., Krasnoff, S. B., & Clardy, J. (2007). Differential analysis of 2D NMR spectra: new natural products from a pilot-scale fungal extract library. *Angewandte Chemie International Edition, 46*, 901–904.

Schwartz, R.-E., Helms, G.-L., Bolessa, E.-A., Wilson, K.-E., Giacobbe, R.-A., Tkacz, J.-S.... Onishi, J. C. (1994). Pramanicin, a novel antimicrobial agent from a fungal fermentation. *Tetrahedron, 50*, 1675–1686.

Scott, B., Young, C. A., Saikia, S., McMillan, L. K., Monahan, B. J., Koulman, A.... Jameson, G. B. (2013). Deletion and gene expression analyses define the paxilline biosynthetic gene cluster in *Penicillium paxilli*. *Toxins, 5*, 1422–1446.

Shemyakin, M. M., Ovchinnikov, Y. A., Ivanov, V. T., Kiryushkin, A. A., Zhdanov, G. L., & Ryabova, I. D. (1963). The structure-antimicrobial relation of depsipeptides. *Experientia, 19*, 566–568.

Shiono, Y., Akiyama, K., & Hayashi, H. (2000). Okaramines N, O, P, Q and R, new okaramine congeners, from *Penicillium simplicissimum* ATCC 90288. *Bioscience, Biotechnology, and Biochemistry, 64*, 103–110.

Sieber, C. M., Lee, W., Wong, P., Munsterkotter, M., Mewes, H. W., Schmeitzl, C.... Guldener, D. (2014). The *Fusarium graminearum* genome reveals more secondary metabolite gene clusters and hints of horizontal gene transfer. *PLoS One, 9*, e110311.

Sigg, H., & Weber, H. (1968). Isolierung und Strukturaufklärung von Ovalicin. *Helvetica Chimica Acta, 51*, 1395–1408.

Silva, B., & Faustino, P. (2015). An overview of molecular basis of iron metabolism regulation and the associated pathologies. *Biochimica et Biophysica Acta (BBA)-Molecular Basis of Disease, 1852*, 1347–1359.

Sindhu, A., Chintamanani, S., Brandt, A. S., Zanis, M., Scofield, S. R., & Johal, G. S. (2008). A guardian of grasses: specific origin and conservation of a unique disease-resistance gene in the grass lineage. *Proceedings of the National Academy of Sciences of the United States of America, 105*, 1762–1767.

Singh, S. B., Zink, D. L., Dombrowski, A. W., Dezeny, G., Bills, G. F., Felix, J. P.... Goetz, M. A. (2001). Candelalides AC: novel diterpenoid pyrones from fermentations of *Sesquicillium candelabrum* as blockers of the voltage-gated potassium channel Kv1. 3. *Organic Letters, 3*, 247–250.

Slot, J. C., & Rokas, A. (2011). Horizontal transfer of a large and highly toxic secondary metabolic gene cluster between fungi. *Current Biology, 21*, 134–139.

Son, K.-H., Kwon, J.-Y., Jeong, H.-W., Kim, H.-K., Kim, C.-J., Chang, Y.-H. … Kwon, B.-M. (2002). 5-Demethylovalicin, as a methionine aminopeptidase-2 inhibitor produced by *Chrysosporium*. *Bioorganic & Medicinal Chemistry, 10*, 185–188.

Song, Z., Cox, R. J., Lazarus, C. M., & Simpson, T. J. (2004). Fusarin C biosynthesis in *Fusarium moniliforme* and *Fusarium venenatum*. *ChemBioChem, 5*, 1196–1203.

Song, Z. S., Bakeer, W., Marshall, J. W., Yakasai, A. A., Khalid, R. M., Collemare, J. … Cox, R. J. (2015). Heterologous expression of the avirulence gene ACE1 from the fungal rice pathogen *Magnaporthe oryzae*. *Chemical Science, 6*, 4837–4845.

Sorensen, J. L., Hansen, F. T., Sondergaard, T. E., Staerk, D., Lee, T. V., Wimmer, R. … Frandsen, R. J. (2012). Production of novel fusarielins by ectopic activation of the polyketide synthase 9 cluster in *Fusarium graminearum*. *Environmental Microbiology, 14*, 1159–1170.

Spikes, S., Xu, R., Nguyen, C. K., Chamilos, G., Kontoyiannis, D. P., Jacobson, R. H. … May, G. S. (2008). Gliotoxin production in *Aspergillus fumigatus* contributes to host-specific differences in virulence. *Journal of Infectious Diseases, 197*, 479–486.

Springer, J. P., Cole, R. J., Dorner, J. W., Cox, R. H., Richard, J. L., Barnes, C. L., & Van der Helm, D. (1984). Structure and conformation of roseotoxin B. *Journal of the American Chemical Society, 106*, 2388–2392.

Srinivas, G., Babykutty, S., Sathiadevan, P. P., & Srinivas, P. (2007). Molecular mechanism of emodin action: transition from laxative ingredient to an antitumor agent. *Medicinal Research Reviews, 27*, 591–608.

St. Leger, R., Roberts, D., & Staples, R. (1991). A model to explain differentiation of appressoria by germlings of *Metarhizium anisopliae*. *Journal of Invertebrate Pathology, 57*, 299–310.

Staats, C. C., Junges, A., Guedes, R. L., Thompson, C. E., de Morais, G. L., Boldo, J. T. … Schrank, A. (2014). Comparative genome analysis of entomopathogenic fungi reveals a complex set of secreted proteins. *BMC Genomics, 15*, 822.

Stachelhaus, T., Mootz, H. D., Bergendahl, V., & Marahiel, M. A. (1998). Peptide bond formation in nonribosomal peptide biosynthesis. Catalytic role of the condensation domain. *Journal of Biological Chemistry, 273*, 22773–22781.

Stachelhaus, T., Mootz, H. D., & Marahiel, M. A. (1999). The specificity-conferring code of adenylation domains in nonribosomal peptide synthetases. *Chemistry & Biology, 6*, 493–505.

Staerkel, C., Boenisch, M. J., Kroger, C., Bormann, J., Schafer, W., & Stahl, D. (2013). CbCTB2, an O-methyltransferase is essential for biosynthesis of the phytotoxin cercosporin and infection of sugar beet by *Cercospora beticola*. *BMC Plant Biology, 13*.

Stierle, A., Strobel, G., & Stierle, D. (1993). Taxol and taxane production by *Taxomyces andreanae*, an endophytic fungus of pacific yew. *Science, 260*, 214–216.

Strongman, D. B., Miller, J. D., Calhoun, L., Findlay, J. A., & Whitney, N. J. (1987). The biochemical basis for interference competition among some lignicolous marine fungi. *Botanica Marina, 30*, 21–26.

Sugawara, T., Shinonaga, H., Yoji, S., Yoshikawa, R., & Yamamoto, K. (1996). *Polyene-based compounds*. Japan Patent No. 319289. Japan Kokai Tokkyo Koho.

Sutton, P., Newcombe, N. R., Waring, P., & Mullbacher, A. (1994). In vivo immunosuppressive activity of gliotoxin, a metabolite produced by human pathogenic fungi. *Infection and Immunity, 62*, 1192–1198.

Suzuki, A., Kanaoka, M., Isogai, A., Murakoshi, S., Ichinoe, M., & Tamura, S. (1977). Bassianolide, a new insecticidal cyclodepsipeptide from *Beauveria bassiana*, and *Verticillium lecanii*. *Tetrahedron Letters, 1977*, 2167–2170.

Szewczyk, E., Chiang, Y. M., Oakley, C. E., Davidson, A. D., Wang, C. C. C., & Oakley, B. R. (2008). Identification and characterization of the asperthecin gene cluster of *Aspergillus nidulans*. *Applied and Environmental Microbiology, 74*, 7607–7612.

Tagami, K., Liu, C., Minami, A., Noike, M., Isaka, T., Fueki, S.... Oikawa, H. (2013). Reconstitution of biosynthetic machinery for indole-diterpene paxilline in *Aspergillus oryzae*. *Journal of the American Chemical Society, 135*, 1260–1263.

Takamatsu, S., Kim, Y.-P., Komiya, T., Sunazuka, T., Hayashi, M., Tanaka, H.... Omura, S. (1996). Chlovalicin, a new cytocidal antibiotic produced by *Sporothrix* sp. FO-4649. II. Physicochemical properties and structural elucidation. *The Journal of Antibiotics, 49*, 635–638.

Tamura, S., Kuyama, S., Kodaira, Y., & Higashikawa, S. (1964). The structure of destruxin B, a toxic metabolite of *Oospora destructor*. *Agricultual and Biological Chemistry, 28*, 137–138.

Toyomasu, T., Nakaminami, K., Toshima, H., Mie, T., Watanabe, K., Ito, H.... Oikawa, H. (2004). Cloning of a gene cluster responsible for the biosynthesis of diterpene aphidicolin, a specific inhibitor of DNA polymerase alpha. *Bioscience, Biotechnology, and Biochemistry, 68*, 146–152.

Tsai, H. F., Wang, H., Gebler, J. C., Poulter, C. D., & Schardl, C. L. (1995). The *Claviceps purpurea* gene encoding dimethylallyltryptophan synthase, the committed step for ergot alkaloid biosynthesis. *Biochemical and Biophysical Research Communications, 216*, 119–125.

Tsunoo, A., Kamijo, M., Taketomo, N., Sato, Y., & Ajisaka, K. (1997). Roseocardin, a novel cardiotonic cyclodepsipeptide from *Trichothecium roseum* TT103. *Journal of Antibiotics, 50*, 1007–1013.

Tudzynski, B. (2014). Nitrogen regulation of fungal secondary metabolism in fungi. *Frontiers in Microbiology, 5*, 656.

Turner, W., & Aldridge, D. (1983). *Fungal metabolites II*. London: Academic Press.

Uchida, R., Imasato, R., Yamaguchi, Y., Masuma, R., Shiomi, K., Tomoda, H., & Omura, S. (2005). New insecticidal antibiotics, hydroxyfungerins A and B, produced by *Metarhizium* sp. FKI-1079. *Journal of Antibiotics, 58*, 804–809.

Vallim, T., & Edwards, P. (2009). Bile acids have the gall to function as hormones. *Cell Metabolism, 10*, 162–164.

Van Dien, S. J., Marx, C. J., O'Brien, B. N., & Lidstrom, M. E. (2003). Genetic characterization of the carotenoid biosynthetic pathway in *Methylobacterium extorquens* AM1 and isolation of a colorless mutant. *Applied and Environmental Microbiology, 69*, 7563–7566.

Vargas, W. A., Mukherjee, P. K., Laughlin, D., Wiest, A., Moran-Diez, M. E., & Kenerley, C. M. (2014). Role of gliotoxin in the symbiotic and pathogenic interactions of *Trichoderma virens*. *Microbiology, 160*, 2319–2330.

Vilcinskas, A., Mathia, V., & Götz, P. (1997a). Effects of the entomopathogenic fungus *Metarhizium anisopliae* and its secondary metabolites on morphology and cytoskeleton of plasmatocytes isolated from the greater wax moth, *Galleria mellonella*. *Journal of Insect Physiology, 43*, 1149–1159.

Vilcinskas, A., Mathia, V., & Götz, P. (1997b). Inhibition of phagocytic activity of plasmatocytes isolated from *Galleria mellonella* by entomogenous fungi and their secondary metabolites. *Journal of Insect Physiology, 43*, 475–483.

Visconti, A., & Solfrizzo, M. (1994). Isolation, characterization and biological activity of visoltricin, a novel metabolite of *Fusarium tricinctum*. *Journal of Agricultural and Food Chemistry, 42*, 195–199.

Wahlman, M., & Davidson, B. S. (1993). New destruxins from the entomopathogenic fungus *Metarhizium anisopliae*. *Journal of Natural Products, 56*, 643–647.

Waksman, S. A., Horning, E. S., & Spencer, E. L. (1943). Two antagonistic fungi, *Aspergillus fumigatus* and *Aspergillus clavatus*, and their antibiotic substances. *Journal of Bacteriology, 45*, 233.

Wall, D. P., Fraser, H. B., & Hirsh, A. E. (2003). Detecting putative orthologs. *Bioinformatics, 19*, 1710–1711.

Wallner, A., Blatzer, M., Schrettl, M., Sarg, B., Lindner, H., & Haas, H. (2009). Ferricrocin, a siderophore involved in intra- and transcellular iron distribution in *Aspergillus fumigatus*. *Applied and Environmental Microbiology, 75*, 4194–4196.

Wallwey, C., & Li, S. M. (2011). Ergot alkaloids: structure diversity, biosynthetic gene clusters and functional proof of biosynthetic genes. *Natural Product Reports, 28*, 496–510.

Walton, J. D. (2006). HC-toxin. *Phytochemistry, 67*, 1406–1413.

Wang, B., Kang, Q. J., Lu, Y. Z., Bai, L. Q., & Wang, C. S. (2012). Unveiling the biosynthetic puzzle of destruxins in *Metarhizium* species. *Proceedings of the National Academy of Sciences of the United States of America, 109*, 1287–1292.

Wang, C. C., Chiang, Y. M., Praseuth, M. B., Kuo, P. L., Liang, H. L., & Hsu, Y. L. (2010). Asperfuranone from *Aspergillus nidulans* inhibits proliferation of human non-small cell lung cancer A549 cells via blocking cell cycle progression and inducing apoptosis. *Basic & Clinical Pharmacology & Toxicology, 107*, 583–589.

Wang, W. L., Liu, P. P., Zhang, Y. P., Li, J., Tao, H. W., Gu, Q. Q., & Zhu, W. M. (2009). 2-Hydroxydiplopterol, a new cytotoxic pentacyclic triterpenoid from the halotolerant fungus *Aspergillus variecolor* B-17. *Archives of Pharmacal Research, 32*, 1211–1214.

Wei, W. T., Lin, S. Z., Liu, D. L., & Wang, Z. H. (2013). The distinct mechanisms of the antitumor activity of emodin in different types of cancer. *Oncology Reports, 30*, 2555–2562.

Wen, H., Li, Y., Liu, X., Ye, W., Yao, X., & Che, Y. (2015). Fusagerins A–F, new alkaloids from the fungus *Fusarium* sp. *Natural Products and Bioprospecting*, 1–9.

Wiemann, P., Guo, C. J., Palmer, J. M., Sekonyela, R., Wang, C. C. C., & Keller, N. P. (2013). Prototype of an intertwined secondary-metabolite supercluster. *Proceedings of the National Academy of Sciences of the United States of America, 110*, 17065–17070.

Wildung, M. R., & Croteau, R. (1996). A cDNA clone for taxadiene synthase, the diterpene cyclase that catalyzes the committed step of taxol biosynthesis. *Journal of Biological Chemistry, 271*, 9201–9204.

Wu, D., Oide, S., Zhang, N., Choi, M. Y., & Turgeon, B. G. (2012). ChLae1 and ChVel1 regulate T-toxin production, virulence, oxidative stress response, and development of the maize pathogen *Cochliobolus heterostrophus*. *PLoS Pathogens, 8*, e1002542.

Xu, W., Cai, X., Jung, M. E., & Tang, Y. (2010). Analysis of intact and dissected fungal polyketide synthase-nonribosomal peptide synthetase in vitro and in *Saccharomyces cerevisiae*. *Journal of the American Chemical Society, 132*, 13604–13607.

Xu, Y. Q., Espinosa-Artiles, P., Schubert, V., Xu, Y. M., Zhang, W., Lin, M. … Molnar, I. (2013). Characterization of the biosynthetic genes for 10,11-dehydrocurvularin, a heat shock response-modulating anticancer fungal polyketide from *Aspergillus terreus*. *Applied and Environmental Microbiology, 79*, 2038–2047.

Xu, Y. Q., Zhou, T., Zhang, S. W., Xuan, L. J., Zhan, J. X., & Molnar, I. (2013). Thioesterase domains of fungal nonreducing polyketide synthases act as decision gates during combinatorial biosynthesis. *Journal of the American Chemical Society, 135*, 10783–10791.

Yeh, H. H., Chiang, Y. M., Entwistle, R., Ahuja, M., Lee, K. H., Bruno, K. S. … Wang, C. C. (2012). Molecular genetic analysis reveals that a nonribosomal peptide synthetase-like (NRPS-like) gene in *Aspergillus nidulans* is responsible for microperfuranone biosynthesis. *Applied Microbiology and Biotechnology, 96*, 739–748.

Young, C., McMillan, L., Telfer, E., & Scott, B. (2001). Molecular cloning and genetic analysis of an indole-diterpene gene cluster from *Penicillium paxilli*. *Molecular Microbiology, 39*, 754–764.

Young, C. A., Felitti, S., Shields, K., Spangenberg, G., Johnson, R. D., Bryan, G. T. … Scott, B. (2006). A complex gene cluster for indole-diterpene biosynthesis in the grass endophyte *Neotyphodium lolii*. *Fungal Genetics and Biology, 43*, 679–693.

Young, C. A., Schardl, C. L., Panaccione, D. G., Florea, S., Takach, J. E., Charlton, N. D. … Jaromczyk, J. (2015). Genetics, genomics and evolution of ergot alkaloid diversity. *Toxins, 7*, 1273–1302.
Zabriskie, T. M., & Jackson, M. D. (2000). Lysine biosynthesis and metabolism in fungi. *Natural Product Reports, 17*, 85–97.
Zähner, H., Keller-Schierlein, W., Hütter, R., Hess-Leisinger, K., & Deer, A. (1963). Stoffwechselprodukte von Mikroorganismen 40. Mittleilung: Sideramine aus Aspergillaceen. *Archives of Microbiology, 45*, 119–135.

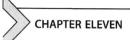

CHAPTER ELEVEN

From So Simple a Beginning: The Evolution of Behavioral Manipulation by Fungi

D.P. Hughes*[,1], J.P.M. Araújo*, R.G. Loreto*[,§], L. Quevillon*, C. de Bekker[¶] and H.C. Evans[†]

*Pennsylvania State University, University Park, PA, United States
[§]CAPES Foundation, Ministry of Education of Brazil, Brasília, DF, Brazil
[¶]Ludwig-Maximilians-University Munich, Munich, Germany
[†]CAB International, Surrey, United Kingdom
[1]Corresponding author: E-mail: dhughes@psu.edu

Contents

Abstract

Parasites can manipulate the behavior of their hosts in ways that increase either their direct fitness or transmission to new hosts. The Kingdom Fungi have evolved a diverse array of strategies to manipulate arthropod behavior resulting in some of the most complex and impressive examples of behavioral manipulation by parasites. Here we provide an overview of these different interactions and discuss them from an evolutionary perspective. We discuss parasite manipulation within the context of Niko

Advances in Genetics, Volume 94
ISSN 0065-2660
http://dx.doi.org/10.1016/bs.adgen.2016.01.004

Tinbergen's four questions (function, phylogeny, causation, and ontogeny) before detailing the proximate mechanisms by which fungi control arthropod behavior and the evolutionary pathways to such adaptations. We focus on some systems for which we have recently acquired new knowledge (such as the zombie ant fungus, *Ophiocordyceps unilateralis s.l.*), but a major goal is also to highlight how many interesting examples remain to be discovered and investigated. With this in mind, we also discuss likely examples of manipulated spiders that are largely unexplored ("zombie spiders"). Armed with advanced tools in evolutionary biology (from serial block face SEM to RNAseq) we can discover how the fungi, a group of microbes capable of coordinated activity, have evolved the ability to direct animal behavior. In short, we have the ability to understand how the organism without the brain controls the one with the brain. We hope such a goal, coupled with the knowledge that many diverse examples of control exist, will inspire other organismal biologists to study the complex adaptations that have arisen from "so simple a beginning."

1. INTRODUCTION

Fungi have an intimate association with animal life on planet Earth. Life first arose in the sea and subsequently colonized the land. The most diverse groups of animals to have evolved from "so simple a beginning" (Darwin, 1859) are the insects that today have almost one million described species, while spiders are also among the most successful of terrestrial organisms, with over 42,000 known species. As the arthropods were emerging to become the dominant animals in all terrestrial ecosystems, another dominant group of eukaryotes, the Fungi, were also colonizing the land. These two phylogenetically and ecologically diverse taxa (Phylum Arthropoda and Kingdom Fungi) have, over the last 400 million years, evolved a wide array of intimate interactions with one another (Vega & Blackwell, 2005). These interactions run the gamut and include mutualistic endosymbiosis (Suh, Noda, & Blackwell, 2001); fungi as obligate food sources, such as those found in fungus-gardening ants (Mueller, Gerardo, Aanen, Six, & Schultz, 2005); sexually- and behaviorally transmitted parasites, such as Laboulbeniales (DeKesel, 1996); and pathogens that have pronounced effects on host populations (Evans, 1974). Entomopathogens occur in all the major phyla of the Kingdom Fungi and the exploitation of the host body for food has evolved independently and repeatedly (Araújo & Hughes, 2016). Despite this knowledge, fungal–arthropod associations remain an understudied area of fungal biodiversity and likely harbor one of the largest reservoirs of undocumented taxonomic, functional, and genetic diversity within the Fungi (Vega & Blackwell, 2005).

Insects (Class Insecta) belong to the Phylum Athropoda, which includes the familiar spiders (Order Araneae) and mites and ticks (subclass Acari); both taxa are also known to be hosts of fungal parasites (Evans & Samson, 1987; Evans, 2013). The spiders are notable because there are records of very large die off events driven by fungal pathogens (Evans, 2013; Samson & Evans, 1973), which are similar to the graveyard events that occur when fungi infect ants (Pontoppidan, Himaman, Hywel-Jones, Boomsma, & Hughes, 2009). This high occurrence coupled with the observation that many spiders die on the underside of leaves, which is likewise observed in ants and other insects, suggests that as part of the life cycle, fungi that infect spiders may also manipulate host behavior to increase transmission to new hosts.

The observation that, as a Kingdom, Fungi have many parasitic taxa (at the specific, generic, or familial level), does not distinguish them from other major groups. Parasitism is a very common mode of life that has evolved repeatedly and probably more times than predation as a life history strategy (Poulin & Morand, 2000, 2005). What is notable is the apparently high frequency of parasitic fungi that have evolved not just to infect animals, but also to adaptively manipulate animal behavior in ways that increase the fitness of the fungus. In this chapter, we explore the diversity and origins of such behavioral manipulation in insects and spiders before considering the mechanisms by which fungal pathogens control arthropod nervous systems.

2. WHAT IS BEHAVIORAL MANIPULATION?

One of the most distinctive features of animals is their ability to express complex behaviors. Honeybees can use a dance language to signal the location of high quality flowers, wolf packs can act in concert to chase down large prey that an individual pack member could not handle alone, and in peafowl, the peacocks display their genetic quality by parading elaborate tails to potential mates (Alcock, 1993). In recent years, behavioral ecologists, researchers who study the evolution of behaviors within an ecological context, have begun searching for patterns of behavior peculiar to animals that are infected by parasites (Moore, 2002). In many cases, the behavior observed is a general sickness that is the consequence of pathogen growth and development within the animal, and the associated change in either the immune system of the host or its general physiology as it reacts to the stress of parasitism. But in some host–parasite associations, the parasites have gone further than sequestering resources and have evolved the ability to adaptively alter the behavior of the animal in which they live.

It turns out that parasites can be the reason for wholly novel behaviors in animals. Such behaviors can be as complex and novel as the waggle dance of the honeybee (Hughes, 2014). The purpose of such parasite-mediated change of animal behavior is to use the animal as vehicle for parasite genes that are transmitted to either a new host or a new habitat. In all cases, the behaviors are those that the animals would not normally express because such behaviors are costly, and oftentimes fatal, to the animal. Some prominent examples of such behavioral manipulation are *Dicrocoelium dendriticum* (brainworms) that induce ants to bite into leaves to reach the guts of ruminants (Moore, 2002); hairworms causing crickets to jump into water to achieve parasite mating (Thomas et al., 2002); or *Toxoplasma* changing the behavior of rats to induce a fatal feline attraction for the parasite so it reaches its definitive host where it reproduces (Berdoy, Webster, & Macdonald, 2000; Webster, 2001). Such examples of parasites affecting the behavior and morphology of hosts in ways that increase transmission have come to be known as parasite-extended phenotypes (Dawkins, 1982, 1990, 2004, 2012). Here, natural selection has shaped parasite genomes to control host phenotypes and multiple lines of evidence are emerging to illustrate the mechanisms by which parasites achieve this end (Adamo, 2012; Adamo & Webster, 2013; de Bekker, Merrow, & Hughes, 2014; Biron & Loxdale, 2013; Hughes, 2013; Lefevre et al., 2009; Van Houte, Ros, & Oers, 2013).

Although parasitism as a life history trait is common, it is not true that behavioral manipulation of animal behavior is also common. It is difficult to estimate what percent of all parasites (in any taxa) have evolved complex control of behavior in their life cycle, but a parsimonious position to take is that it is a small minority (Hughes, 2014). The reasons for this are probably related to the high costs involved in controlling the central nervous system of another organism (Poulin, 1994). This implies that there must have been a strong selective force leading to the evolution of behavioral manipulation. For whatever reasons, these selective forces operate frequently on entomopathogens because the number, range, and diversity of behavioral manipulation of animals by these fungi are high. In the next section, we examine this diversity.

3. DIVERSITY OF FUNGI CONTROLLING ANIMAL BEHAVIOR

There are estimated to be between 1.5 and 5 million species of fungi (Blackwell, 2011; Hawksworth & Rossman, 1997), but only around about 100,000 have been described so far (Kirk, Canon, Minter, & Staplers, 2008). Those are currently organized into seven phyla (Microsporidia, Neocallimastigomycota, Chytridiomycota, Glomeromycota, Entomophthoromycota, Basidiomycota, and Ascomycota), with some groups not assigned to any phylum due to lack of data (Hibbett et al., 2007). Entomopathogenic species are known for all phyla except Neocallimastigomycota, which are anaerobic, inhabiting the rumen of large herbivorous mammals and Glomeromycota, a group formed almost exclusively by arbuscular mycorrhizal fungi, with a single exception *Geosiphon pyriformis* that forms symbiosis with cyanobacteria (Kützing, 1849).

A recent review of these entomopathogenic associations found that approximately 65% of all insect orders (19 of the 30) are known to be infected by fungi (Araújo & Hughes, 2016). Microsporidia infect 14 orders of insects, Ascomycota (mostly species in the order Hypocreales) and Entomophthoromycota infect 13 and 10 orders, respectively, Chytridiomycota infect 3 and Basidiomycota infect 2 orders. Until recently, due to lack of host data (Evans, 2013), such calculations could not be made for spiders (Order Araneae). However, a pioneering study concentrating on one spider-specific fungal genus (*Gibellula*) offers a tantalizing glimpse into the potential diversity of hosts affected by these fungi (Costa, 2014; Evans, unpublished data). Thus far, spiders in 10 families have been recorded as hosts of pathogenic fungi, all in the Order Hypocreales of the Ascomycota, but sharing no common species with the insect pathogens (Evans, 2013). When other spider host records are included, this approximates to over 10% of the 110 known families of spiders (Nentwig, 2013).

Such a broad representation across the taxonomic levels has resulted in heterogeneous ecological groups in many aspects. One example of this heterogeneity is the variation on display in fungal morphology. Chytrids exhibit flagellated zoospores that are adapted to "seek," recognize, and penetrate the host cuticle (Barr & Désaulniers, 1988); whereas, the extremely small spores in some microsporidians (eg, 3 μm long) shoot a harpoon-like structure to inject the protoplasm into the host's cell (Araújo & Hughes, 2016); while complex ascospores (eg, some species of *Ophiocordyceps*) exhibit specific shapes to improve aerodynamics upon dispersion (hosts are often attached to plant material up to 2 m high) and germinate secondary structures once on the forest floor (eg, capilliconidia and capilliconidiophore) (Araújo, Geiser, Evans, & Hughes, 2015; Evans, Elliot, & Hughes, 2011a, 2011b), which are analogous to and demonstrate convergent evolution with the capillispores and capillisporophores of the Entomophthorales. In terms of habitats, we can find equally impressive diversity. Entomopathogenic fungi are found from African deserts (Evans & Shah, 2002) to aquatic environments like ponds, streams, or even leaf axils that collect water (Frances, Sweeney, & Humber, 1989). However, the greatest diversity is found in tropical forests worldwide. There, we find fungi infecting arthropods inhabiting soil (eg, trapdoor spiders) to leaf litter (eg, beetle larvae and caterpillars) to the understory (eg, ants, wasps, bees) to high canopy (eg, homopterans) (Figs. 1–5).

However, one of the most fascinating aspects of these fungal–arthropod associations is the host diversity (Figs. 2–3). The chytrids (the only aquatic group among the entomopathogenic fungi) are known to infect almost exclusively mosquito larvae, including important disease vectors (eg, *Aedes*, *Anopheles*, and *Culex*) with very rare exceptions (eg, *Myiophagus* sp. infecting the purple scale *Lepidosaphes beckii* (Muma & Clancy, 1961)). Although the majority of entomopathogenic microsporidians also infect Diptera, they are known to parasitize a broad range of hosts such as Zygentoma, Ephemeroptera, Odonata, Plecoptera, Orthoptera, Isoptera, Psocoptera, Hemiptera, Coleoptera, Hymenoptera, Siphonaptera, Trichoptera, and Lepidoptera. Although there are more than 32,000 species described for Basidiomycota (Kirk et al., 2008), less than 1% have evolved to live inside the insect body (Araújo & Hughes, 2016), with almost all entomopathogenic species belonging to a single genus, *Septobasidium*, infecting Diaspididae scale insects (Couch, 1938). Entomophthoromycota and Ascomycota are the phyla that exhibit the highest diversity among entomopathogenic fungi. In both phyla, we also see repeated origins of a complex strategy of infection and transmission with the manipulation of host behavior.

Within the Entomophthoromycota, host behavioral manipulation has evolved at least twice and in Ascomycota multiple origins happened over the long evolutionary history of the group. For entomophthoralean fungi, there are several classic examples, *Pandora* infecting the ant genus *Formica* ants (Fig. 2G), *Entomophaga* infecting acridid hosts (especially locusts, Fig. 4A), and *Entomophthoromycota* infecting flies (eg, *Musca domestica* and

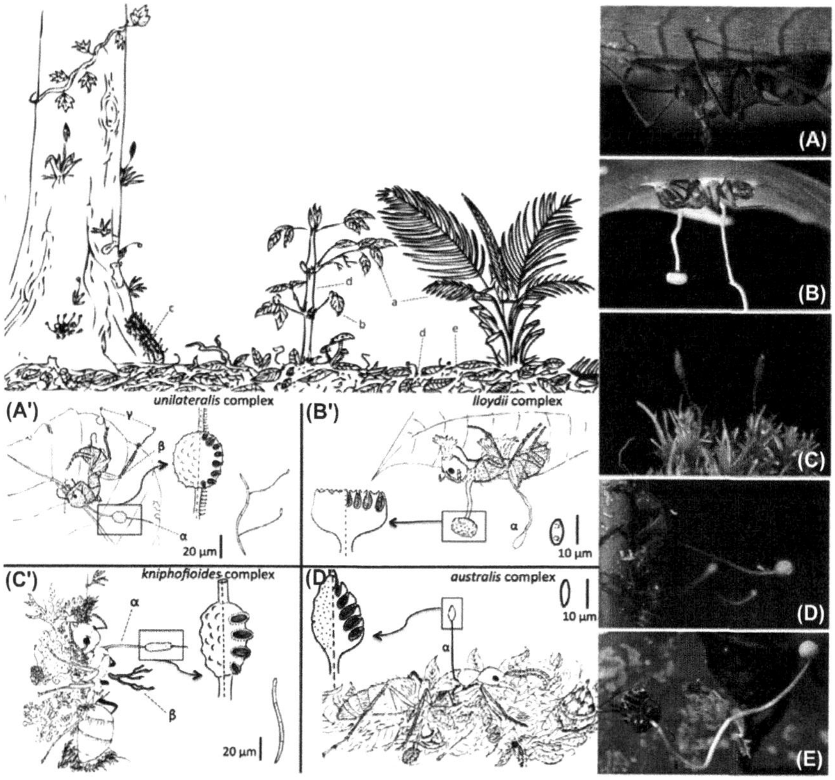

Figure 1 ***Diversity of behavioral manipulation of ants by fungi in a tropical forest setting.*** Represented is some of the extensive taxonomic and functional diversity of *Ophiocordyceps* in ants. Top left shows a composite forest scene with six locations where cadavers are found: under leaves (a = *Ophiocordyceps unilateralis*), (b = *Ophiocordyceps lloydii*), tree bole/bark (c = *Ophiocordyceps kniphofioides*), leaf litter (d = *Ophiocordyceps australis, Ophiocordyceps myrmecophila, Ophiocordyceps irangensis, O. kniphofioides*), and stem (e = *O. australis* in Ghana). Each specific name represents a complex with more than one species. Line diagrams show the functional morphology (represented as a composite, all morphologies do not occur on one species). In (A') *unilateralis* has one teleomorph (sexual stage) and three anamorphs (asexual, α–γ). The teleomorph (ascoma) shows the outside and inside where ascospores are produced and the ascospore is drawn with capilliconidia (secondary spores on hairs). In (B') *lloydii* and (D') *australis* only part spores are produced and they do not produce secondary structures and only one anamorph is found (asexual, α). In *lloydii* the ant is not biting but glued via hyphae from mouth (photo b). The complex (C') *kniphofioides* (here on ant species *Dolichoderus bispinosus,* which is hidden from view in moss) is related to *unilateralis* and we can see two anamorphs (one does not occur on *Dolichoderus,* but only on *Cephalotes atratus).* The photos A–E show host ants and position of death: A–E *Polyrhachis armata,* Thailand, *Camponotus atriceps*, Brazil, *Dolichoderus bispinosus*, Brazil, *Paltothyreus tarsatus*, Ghana, and *Polyrhachis robsoni*, Australia. Among these are two undescribed fungal species from the following complexes: *unilateralis* (A') *and australis* (D'). (See color plate)

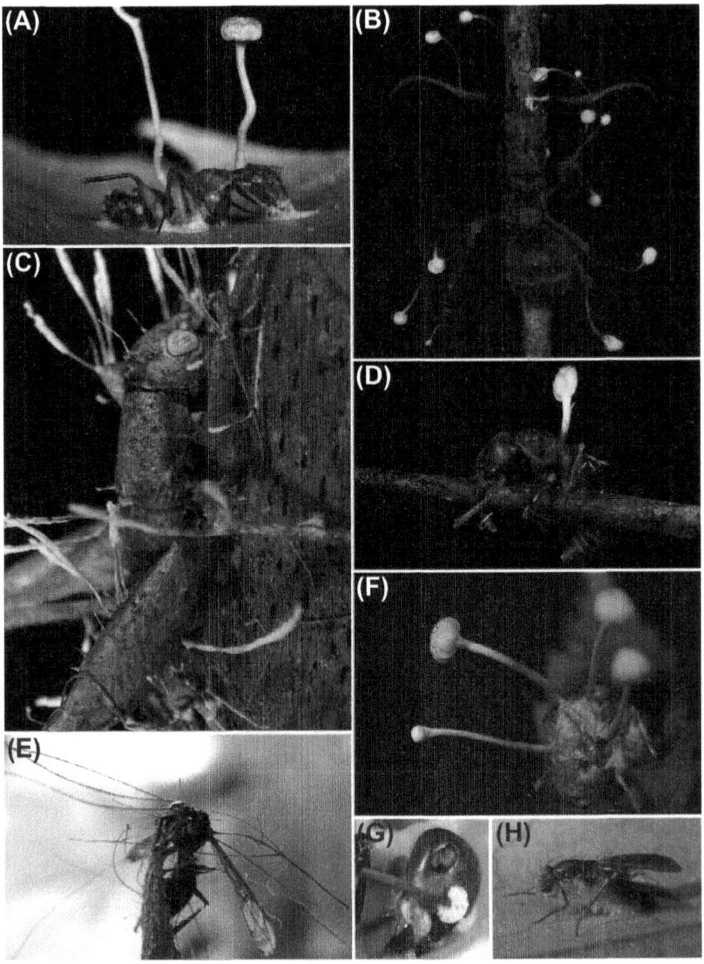

Figure 2 ***Diversity of behavioral manipulation where insects are controlled to die attached to arboreal surfaces*** (A) *Ophiocordyceps lloydii* on *Camponotus atriceps* (Brazilian Amazon); (B) *Stilbella burmensis* on *Polyrhachis cf. militaris* (Ghana); (C) anamorphic Hypocreales on Orthopera (Atlantic rainforest, Brazil); (D) *Ophiocordyceps* sp. on *Pachycondyla impressa* (Brazilian Amazon); (E) *Hirsutella saussurei* (anamorph of *Ophiocordyceps humberti*, Atlantic rainforest in Brazil) on Polistinae wasp; (F) *Ophiocordyceps dipterigena s.l.* (Brazilian Amazon) on unidentified fly; (G) *Pandora formicae* on *Formica* ant (Finland); (H) *Ophiocordyceps dipterigena s.l.* early developmental stage (Atlantic Rainforest in Brazil) on unidentified fly. (See color plate)

Scatophaga stercoraria) (Humber, 1989; Maitland, 1994; Małagocka, Grell, Lange, Eilenberg, & Jensen, 2015). After infection, the fungus proliferates within the host, manipulates the behavior (ie, controlling the host to reach an elevated position on plants, called "summit disease"), kills the host, and in

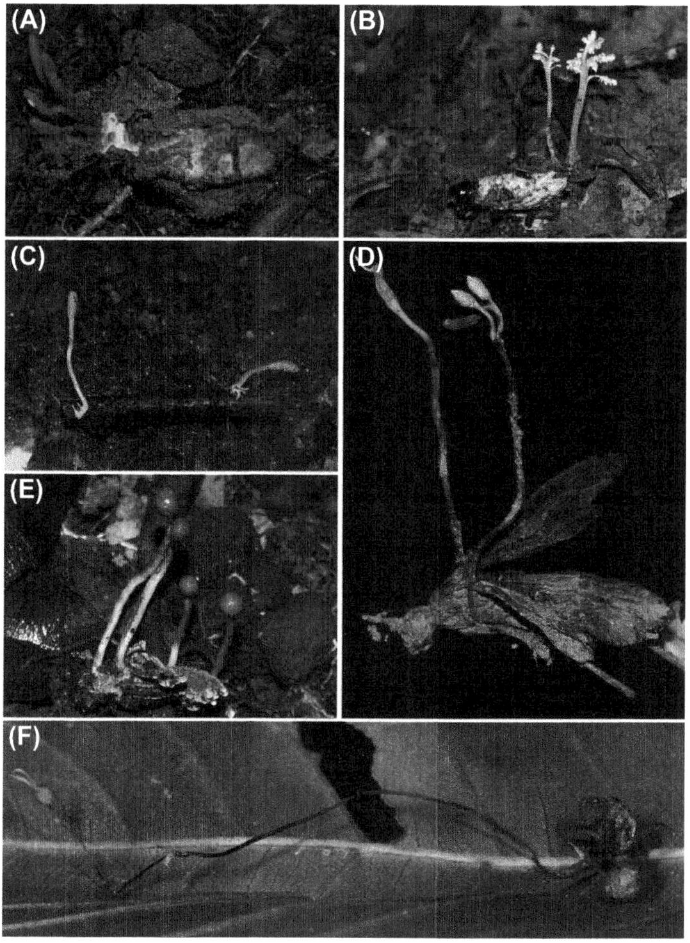

Figure 3 ***Diversity of interactions where insects die in the soil where behavioral manipulation is not assumed to occur*** (A) *Ophiocordyceps* sp. on trapdoor spider (Brazilian Amazon); (B) *Isaria* sp. on Lepidoptera pupa (Brazilian Amazon); (C) *Ophiocordyceps cf. cardinalis* on Coleoptera larva (Brazilian amazon); (D) *Ophiocordyceps* sp. (*Neocordyceps* group) on wasp. (E) *Ophiocordyceps amazonica s.l.* on Orthoptera (Colombia); (F) *Ophiocordyceps australis s.l.* on *Polyrhachis* sp. (Ghana). (See color plate)

certain genera (*Erynia*) creates fungal structures (rhizoids) to attach the host more securely to the substrate (Małagocka et al., 2015) (see Fig. 2G), but more typically this is a "death lock," particularly in grasshoppers and locusts which grasp vegetation with their legs before dying (see Fig. 4A, *Entomophaga grylli*). This precedes fungal growth from the interior to the exterior of the insect followed by sporulation from these exterior structures (Roy,

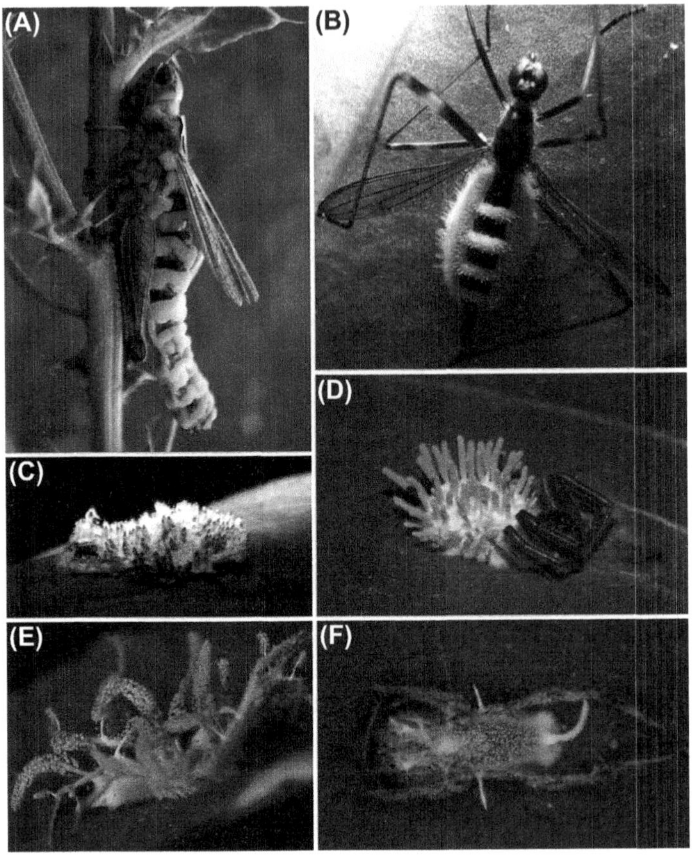

Figure 4 ***Further examples of death on arboreal surfaces*** (A) *Entomophaga grylli* on a locust host, near the summit of a thistle plant, in the grasslands of Outer Mongolia (China). Note the legs clasping the stem and the creamy-white fungal bands bursting from the intersegmental sutures. All the thistles in the immediate area had infected locusts in the same position, mostly with two to three corpses per plant. (B) *Erynia* on an unknown dipteran host, attached to the underside of a shrub leaf by rhizoids, Atlantic rainforest, Mata do Paraíso, Minas Gerais, Brazil. A white spore halo is beginning to form on the leaf surface. (C) *Gibellula* sp. nov. on a huntsman spider (*Caayguara* sp., Sparassidae), underside of shrub leaf, Mata do Paraíso. The asexual stage is nonsynnematal and reduced to sporing heads on the fore legs, while the abdomen bears an abundance of flask-shaped perithecia with prominent necks. The host genus was erected in 2010 and the spider is a fast-moving predator living in tree bark. (D) *Gibellula*, of the *leiopus* group on a spider, on the underside of shrub leaf, Reserva Ducke, near Manaus, Amazonas, Brazil. (E) *Gibellula* sp. nov. on an unknown ghost spider (Anyphaenidae), underside of shrub leaf, Mata do Paraíso. Note the abundance of lilac-colored synnemata arising from the mycelial-covered abdomen. (F) *Gibellula* sp. nov. on a ghost spider (*Iguarima sensoria*, Anyphaenidae), underside of shrub leaf, Mata do Paraíso; showing similar synnematal production to above. This is a fast-moving spider in the forest understory and rests in silken retreats (sacs) in the litter, tree bark, and vegetation. (See color plate)

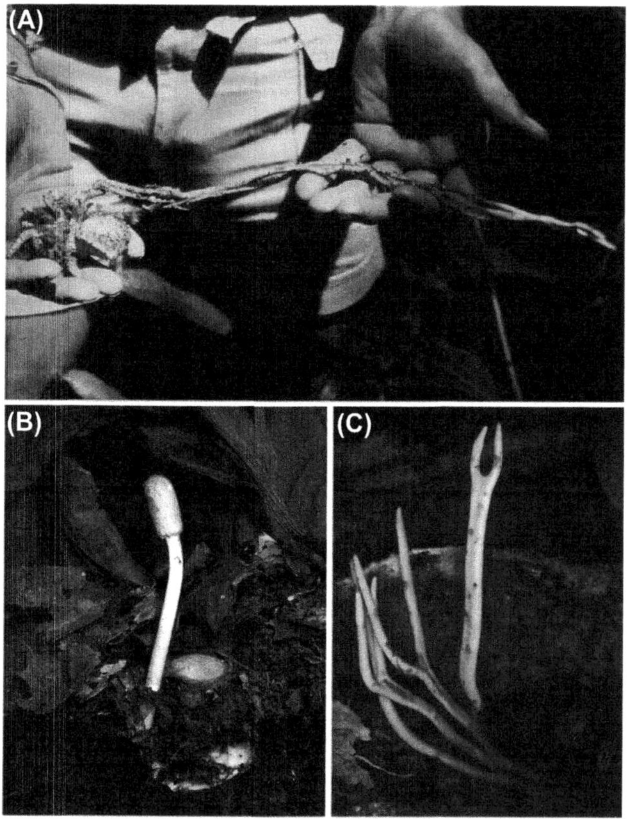

Figure 5 ***Infection of ground dwelling spiders.*** (A) The entire fungus (*Ophiocordyceps caloceroides*) extricated from the deep burrow, showing the mycelial-covered but easily recognizable bird-eating mygalomorph host. (B) *Cordyceps* sp. emerging from the burrow of a trapdoor spider (Ctenizidae), Rio Negro, Amazonian Brazil. The fungal clava has pushed open the trapdoor (right) and formed a yellow perithecial-bearing head (ascostroma), with the host enveloped in a white mycelial mat (C). Clavae or ascomata of *O. caloceroides* (Hypocreales) emerging from the burrow of a *Mygale* spider (Ctenizidae), forest litter, Rio Napo, Amazonian Ecuador. The paler-colored apex contains embedded perithecia and the structures resemble the *Calocera* mushroom genus. (See color plate)

Steinkraus, Eilenberg, Hajek, & Pell, 2006). In some genera, the forcibly discharged primary infective spores (ballistospores) have evolved the ability to form secondary sticky spores (capillispores), if they miss their aerial targets, thereby creating "minefields" around cadavers, potentially, to entrap crawling targets.

Summit disease optimizes both the formation of fungal spores and their subsequent dispersal. Earlier reports also indicate that host manipulation by these "primitive" fungi might be more widespread. For example, sugar-beet aphids that live and feed below ground, emerge when infected by *Erynia aphidis* and ascend the stem to die (Harper, 1958); while *Entomophthora*-infected carrot flies move away from the crop with the females completely altering their soil egg-laying habit so that they deposit their eggs on the foliage of hedgerow trees (Eilenberg, 1986). Such actions would reduce vertical transmission of the parasite within the population, but movement between populations (horizontal transmission) would be enhanced.

For Ascomycota species, the manipulation is well known for some ant—pathogenic species within the genus *Ophiocordyceps* infecting Camponotini ants (Araújo et al., 2015; Evans et al., 2011a, 2011b; Hughes, Anderson, et al., 2011; Hughes, Wappler, & Labandeira, 2011; Loreto, Elliot, Freitas, Pereira, & Hughes, 2014). In this case, about a week after infection, the fungus will induce the host to leave its nest and climb onto the vegetation. Once there, the ant will lock its jaws into the plant tissue and die. After the death of the host, the fungus starts to grow a fruiting body from the back of its head and, in a few weeks of subsequent growth, starts to produce spores that rain down on passing ants as they move on the forest floor or on branches below (Andersen et al., 2009; Hughes, Anderson, et al., 2011; Hughes, Wappler, et al., 2011). As such, the behavioral manipulation functions to provide a platform for the release of forcibly ejected spores (ascospores) that then infect susceptible individuals, thus continuing the cycle (Figs. 1 and 2E represents similar situation in a wasp host). These ascospores, if missing their targets, are capable of producing secondary sticky spores (condia) on long needle-like outgrowths (capilliconidiophore), and it is probably a common occurrence providing an insurance mechanism to ensure successful infection; the spores attaching to the target hosts as they crawl over the substrate. Thus, there is an analogous situation in the genus *Hypocrella* and the Phylum Entomophthoromycota.

The behavioral change that leads infected insects and spiders to die elevated on vegetation prior to host death (Figs. 2 and 4) is not the only complex behavioral manipulation observed. In some cases, the fungus keeps the host alive and controls its flight behavior so that the insect becomes a moving vehicle for spore release. One prominent example is the infection caused by *Strongwellsea castrans* in *Hylemya brassicae* and *H. platura* (Diptera) (Araújo & Hughes, 2016). The fungus causes a large circular hole to develop on the lateral side of the host's abdomen. This hole is filled with fungal tissue and conidiophores (spore-producing cells) that are released during a flight pattern that is described as being stereotypical and centered on a narrow area of fields (R. Humber, personal communication). This presumably aids in spore release to distinct areas of the environment where uninfected flies are present. Another similar case occurs with *Massospora cicadina*, which attack cicadas (Araújo & Hughes, 2016). This fungus also initiates sporulation when the host is still alive (Humber, 1982; Thaxter, 1888). Over time the abdomen falls apart until just the head and thorax of the living insect remain. The ability to fly is retained, increasing dispersion of spores in the environment, especially in the case of infected male cicadas which attempt to attract and copulate with females and even continue to feed (Evans, 1988; Soper, 1963; Soper, Delyzer, & Smith, 1976), which suggests that the central nervous system is functioning normally.

As mentioned previously, an interpretation of such pathogen—host relationships for spiders has not been even remotely possible due to incomplete host identification, especially for the araneomorphs, which tend to be completely overgrown by the fungus, compared to the much larger mygalomorphs (Figs. 4C—F and 5A). Thus, this is unchartered territory and needs to be explored with some urgency given the pivotal position that spiders occupy in ecological networks and the key role that they play in ecosystem functioning (Nentwig, 2013). Now, thanks to a spider taxonomist (Renner L.C. Baptista, Universidade Federal do Rio de Janeiro), new light is being shed onto these associations. In an on-going study, in a fragment of Atlantic rainforest in Minas Gerais (Brazil), almost 80 specimens of spiders infected by the entomopathogenic genus *Gibellula* have been examined: all covered by the fungal stroma and attached to the underside of understory shrubs or small trees. Examination involved excising the upper leaf surface to reveal the under body of the spider—specifically the genital area—to allow for accurate identification, without disturbing the fungal structures. Thus far, 14 genera in 10 families of spiders have been identified; comprising new species of the genus *Gibellula*, as well as new host taxa and only recently described spider genera. Over 50% belong to two families, the Anyphaenidae (ghost spiders) and the Pholcidae (cellar spiders), the former are nocturnal hunters, spending their days in silken retreats ("sleep sacs"), while the latter construct rudimentary webs around leaves and under bark. Other commoner hosts, in the Corinnidae, Salticidae, Sparassidae, Theridiidae, Thomisidae, and Zodariidae, are either fast-moving, free-living predators on vegetation and the forest floor or construct sticky webs or drag lines in the litter. The overall conclusion, therefore, is that, like many of the insect examples, infected spiders move away from their natural habitats to climb and die on understory plants, invariably on the underside of leaves (see Fig. 4C—F). In sharp contrast, the burrow-dwelling mygalomorphs of the family Ctenizidae always die in their underground nests; necessitating the production of complex phototrophic structures (stromata or clubs) by the fungus in order to ensure that the embedded sporulating organs (perithecia) are carried above ground to liberate their spores (see Fig. 5A—C). Conversely, the perithecia of the exposed *Gibellula*-infected spiders are formed directly on the spider cadaver (see Fig. 4C—F).

4. TINBERGEN'S FOUR QUESTIONS AS THEY APPLY TO BEHAVIORAL MANIPULATION OF ARTHROPODS BY FUNGI

One of the most important papers in the field of animal behavior is the classic by Niko Tinbergen, (Tinbergen, 1963) "On aims and methods of ethology." Tinbergen was a founding father of the field of animal behavior and, together with Karl Von Frisch and Konrad Lorenz, shared the Nobel Prize in Physiology or Medicine in 1973, awarded for their contributions to animal behavior (Burkhardt, 2005). Tinbergen suggested that animal behavior can be better understood when we ask four complementary questions. We can ask why a behavior exists by studying its (1) function, (2) phylogeny, (3) causation, and (4) ontogeny. In Fig. 6, we place one prominent example of a manipulated behavior, the death grip induced by species in the complex *Ophiocordyceps unilateralis s.l.*, within this four-question framework. This behavior (infected ants biting a leaf) can then be examined from four complimentary approaches. In the next section we examine each of these approaches in turn.

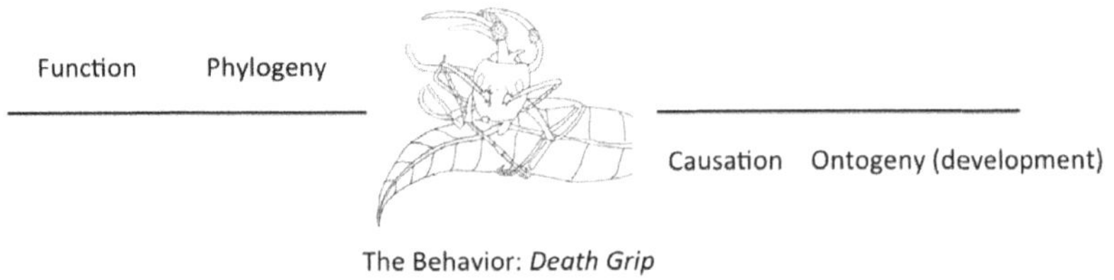

Figure 6 ***The death grip behavior with the framework of Tinbergen's four questions.*** A dead carpenter ant is seen attached to a leaf with the fungus emerging from just behind the head. The manipulated behavior (when the ant was alive) is the product of natural selection acting on fungal genes to control ant behavior (extended phenotype). We can examine this behavior in four complementary ways.

4.1 Function

The first question (function or adaptive value) examines the behavior from the perspective of an organism's fitness. In which way is the behavior adaptive? So, males in some species of birds sing to gain mates or wolves collectively hunt to feed the group. In the case of fungi controlling animal behavior, the framework of the extended phenotype laid out by Richard Dawkins (a student of Tinbergen) argues that the altered behavior in the host benefits parasite genes (Dawkins, 1982, 1990, 2004, 2012). For many examples of parasites controlling behavior, the adaptive value of the altered behavior is inferred. If the behavior is complex, highly stereotyped and not part of the host's repertoire, but possibly benefiting parasite transmission, then we might parsimoniously suggest the altered behavior is an adaptation for the parasite (Poulin, 2011). The death grip behavior in ants infected by *O. unilateralis s.l.* is unusual among examples of parasites controlling behavior because its adaptive value has been tested, twice. In the first test, carried out in a lowland tropical forest in Southern Thailand, the leaves that the manipulated ants bit were experimentally relocated either to the high canopy (15 m), or to the forest floor (0 cm) and, in both cases the fungus failed to develop and thus had zero fitness (Andersen et al., 2009). This experiment supported the claim that manipulating ants to bite into leaves in the narrow understory of the forest was adaptive for the fungus as it provided the parasite with a suitable microhabitat in which to develop. A second experiment, this time in the Atlantic rainforest of Brazil with a species within the *Ophiocordyceps unilateralis* complex (namely *Ophiocordyceps camponoti-rufipedis*) demonstrated that the fungus could not develop inside the ant colony (Loreto et al., 2014). This study offers supportive evidence that fungal manipulated ants leave the colony and bite into leaves which provide species in the complex *O. unilateralis s.l.* with a platform to develop a stalk and release spores that eventually infect other ants. Such stalk formation and spore release, it was shown, could not happen inside the ant nest. While the available evidence would support the conclusion that the complex behavioral changes observed in ants infected by species in the complex *O. unilateralis s.l.* is adaptive for the fungus, it is important to always consider such claims in the context of fungal transmission and specifically spores.

In most examples of fungi infecting insects and spiders, the transmission is direct (from infected animal to susceptible animal). There are examples where the life cycle of the fungus has two different hosts. For example within the Chytridiomycota the species *Coelomomyces psophorae* infects both copepods and mosquito larvae (Whisler, Zebold, & Shemanchuk, 1975). And, there is a recent example from the Ascomycota where an infected Cerambycid beetle larva in its tunnel in a tree produced an outgrowth (synnema), characteristic of the insect—pathogenic genus *Hirsutella,* with additional side branches forming structures typical of the genus *Harposporium*, a pathogen of nematodes with crescent-shaped spores that lodge in the buccal cavity (Evans & Whitehead, 2005). It is supposed that these spores infest the bark and inner wood and are ingested by free-living nematodes. However, most of the life cycles of fungi infecting arthropods are direct, as far as we can ascertain. What is also apparently the default mode is host death as a developmental necessity (but not always, as discussed above). This means that before they can produce transmissible spores the fungal pathogen kills the host insect or spider. Therefore, transmission by the parasite is postmortem for the host. For parasites generally (ie, all taxa and not just Kingdom Fungi), the production of the transmissible stage following host death is rare. It is in fact so rare that a special word, parasitoid, is used to describe the phenomenon (Kuris, 1974). The word parasitoid generally refers to parasitic insects that kill their host arthropod during the course of development. In those cases, the parasite does not transmit from the body of the host it kills, but rather emerges from it to engage upon a free-living stage that often involves feeding, mating, and diapause (Askew, 1971; Godfray, 1994). Technically, fungi that also kill their hosts as a developmental necessity are parasitoids, but the term is generally not used (Andersen et al., 2009). What direct life cycles and postmortem sporulation mean is that where the host animal dies is the point from where the next infection begins. There are two ways then for infection to occur: the susceptible host either touches the cadaver of the insect or spider (becoming infected via contact with spores), or spores are released and the susceptible host encounters them.

We cannot estimate how common transmission to a susceptible host is after that individual touches a sporulating cadaver. It surely occurs because some taxa grow sporodochia, which are infectious spores that are not released and are capped by sticky material that adheres to passing insects. In some cases, the fungus may imitate the smell or visual appearance of sexually receptive female insects to lure males that touch them and become infected, such as within the Entomophthoromycota when the species *Eryniopsis lampyridarum* infects chantarid beetles (Araújo & Hughes, 2016). Another likely unappreciated arena for contact transmission is the soil where burrowing insects encounter the sporulating cadavers and become infected.

For example, cadavers of reproductive ants that die before building a nest have infectious sclerotia that may infect future burrowing reproductive ants (Hughes, Evans, Hywel-Jones, Boomsma, & Armitage, 2009).

The majority of transmission is not via direct contact with a sporulating cadaver, but occurs from spores released from specialized structures such as the sexual ascomata in Phyla Ascomycota or from specialized asexual spore-producing cells (conidiophores). There is a very wide range of sizes, shapes, and masses of entomopathogen spores (Araújo & Hughes, 2016) such that some travel very short distances from the cadaver (millimeters) and others enter the airstream to travel longer distances (meters to presumably kilometers).

4.2 Phylogeny

The second question asks how can we understand a behavior by looking at the species displaying that behavior in a phylogenetic context? Perhaps the reason the organism behaves in such a way is because all members of the clade (genus, family etc.) have such a behavior. Previously, Hughes, Wappler, et al. (2011) and Hughes, Andersen, et al. (2011) argued that the death grip behavior observed in *Ophiocordyceps* and *Pandora*, for example, has evolved convergently because these fungi are separated by 500 million years of evolution (Hibbett et al., 2007). The parsimonious explanation is that both examples evolved independently and convergently, as opposed to the hypothesis that the common ancestor of both fungi manipulated insects to bite vegetation and it was subsequently lost in many other taxa.

With increasing resolution in fungal phylogenies, achieved either by better taxon sampling or more genes, we now are in a better position to make a more refined assessment of the role of phylogeny in fungal-extended phenotypes (ie, manipulation of host behavior). For example, the death grip of ants infected by *O. unilateralis s.l.* can be studied in the context of sister taxa. We know from recent studies (Quandt et al., 2014; Sung et al., 2007) that the genus *Ophiocordyceps* infects insects from nine different orders across the Class Insecta, as well as spiders, which are an order themselves. Those in the species complex *O. unilateralis s.l.* are sister to other complexes that infect beetles (eg, *Ophiocordyceps rhizoidea* on beetle larvae) and other ants (*Ophiocordyceps kniphofioides* on ants). In no case is the behavior manipulation as complex as in *O. unilateralis s.l.*, and while further phylogenetic reconstructions that have more taxon sampling will likely rearrange the sister taxa relationships, it is already clear that this approach allows us to better understand the evolutionary pathways to manipulation.

A strong complement to a molecular phylogeny is the use of fossils that can provide calibration points. For fungi infecting insects, an amber fossil of an infected plant feeding insect (Hemiptera) exists from the Cretaceous period *Paleoophiocordyceps coccophagus* (Sung, Poinar, & Spatafora, 2008). This fossil allowed researchers to suggest that the hypocrealean fungi were at least present in the Early Jurassic (193 million years ago with CI of 158–232 million years). Another fossil, this time of a leaf, was used to propose that the complex manipulation of ant behavior by fungi, which leaves telltale marks on leaves, has been occurring since the Eocene, 47 million years ago (Hughes, Andersen, et al., 2011; Hughes, Wappler, et al., 2011). In the future, the discovery of other fossils could provide important data on the evolution of key innovations as fungi both colonized insects and eventually evolved to control them.

4.3 Causation

The third complementary question is causation. How do behaviors occur? In this approach the focus shifts from the ultimate (or evolutionary) scale down to the proximate (or mechanistic) scale. The above two approaches (function and phylogeny) are ultimate in scope and causation and ontogeny (below) are proximate (Fig. 6). How parasites control host behavior is of considerable interest to many areas of biology (Adamo, 2012; Adamo & Webster, 2013). This is because parasites that have evolved the ability to manipulate the nervous systems of their hosts represent independent experiments in evolution. Traditionally, in trying to understand the mechanistic basis of animal behavior we examine genes, chemicals, or neuronal architecture to gain insights into what factors account for complex behaviors. Parasites that evolved to control behavior represent an independent outcome of natural selection acting on a genome to control host behavior. For this reason we can observe the designation of these parasites as neuroengineers (Adamo & Webster, 2013).

Fungi are particularly fascinating because they are microbial. Thus, in examples of fungi controlling animal behavior, it is the brainless organisms controlling the brain. In Section 5, we expand on recent advances that have been made in understanding the question of causation. Here, we would like to stress that the question of causation should be approached using multiple tools. There may be a temptation to rely upon gene expression studies, but the addition of small molecule surveys (metabolomics and proteomics) together with direct visualization will go a long way in helping us understand how fungi control host behavior.

4.4 Ontogeny

Behaviors are typically expressed at relevant points in the organism's life cycle. So, male birds who sing either to establish a territory or to attract a mate only do so when they achieve sexual maturity (Alcock, 1993). Additionally, the timing of singing is important (both during the breeding season and at the correct time of day). From this it is clear that timing is important to the expression of behavior, and it was for this reason Tinbergen included ontogeny in his four complementary approaches. Previous to this, only the triumvirate approach of function, phylogeny, and causation were considered (Tinbergen, 1963). The addition of ontogeny has been crucial to the study of animal behavior, as we can now examine the survival value of a behavior (function), its evolutionary history (phylogeny), or its mechanistic basis (causation) in addition to when in the organism's life cycle the behavior occurs.

The "when" of behavior is especially relevant to studies of parasites altering animal behavior because the novel behavior, typically, is not part of the organism's natural repertoire. The death grip of carpenter ants infected by either *Pandora* or *Ophiocordyceps* (Figs. 1–2) is not something uninfected ants do. So, once infected and before they are killed, the individual ant (or other insect) is taken over by the developing fungus and begins to execute the behavioral change that ultimately enables fungal transmission. At some point in the infection process, the ant stops being a normal member of its colony and becomes a fungus in ant's clothing. Through the process of kin selection, individual worker ants (and other social insects), act altruistically to benefit their colony and themselves via indirect fitness benefits, but at some point in the infection cycle, the behavior of the ant switches from increasing the fitness of the ant colony to increasing the fitness of the colony of fungi growing inside its body. A major frontier in future proximate levels studies will be to understand when ants (and indeed other arthropods) make this switch to being vehicles for fungal fitness.

5. MECHANISMS OF BEHAVIORAL MANIPULATION

Let us return to the major and important question of how do fungi control animal behavior? As we have emphasized, there are a wide diversity of parasites altering the behavior of their hosts ranging from viruses (eg, *Baculovirus*) to bacteria (*Wolbachia*) to fungi (this review) to worms (hairworms, cestodes and nematodes) to insects such as parasitic flies that infect other insects (eg, Phorids). In all cases, a major question is what are the mechanisms by which control occurs? In recent years, the broader field of parasite and host behavior has seen some major advances in our understanding of the proximate mechanisms of parasites controlling host behavior (Adamo, 2012; Adamo & Webster, 2013; Biron & Loxdale, 2013).

Fungi probably represent a special case study in this general field because of several unique factors peculiar to this Kingdom. The first and most prominent is the range and complexity of behavioral manipulation by fungi of arthropods. As reviewed here and by Araújo and Hughes (2016), this ranges from species in the genus *Ophiocordyceps* that control worker ants to seek out and bite leaves near ant trails to the altered flight behavior of living insects (cicadas and flies infected by *Massospora* and *Strongswellsea,* respectively) so that they act as spore dispersal factories. It is difficult and perhaps futile to rank manipulation across different kingdoms of life and argue that fungal manipulation is more complex than that observed when the manipulator is in the Kingdom Animalia (eg, trematodes). However, what is clear is that the diversity of strategies is greater than that observed in other groups. In addition, it is evident that behavioral manipulation has arisen multiple times independently (Hughes, Andersen, et al., 2011; Hughes, Wappler, et al., 2011). These two factors (diversity and multiple origins) offer the researcher interested in the proximate mechanisms by which one organism controls the behavior of another sufficient materials to perform a comparative analysis. For example, the observation that species in the genus *Pandora* and *Ophiocordyceps* both induce ants to bite into leaves and the fact that these two genera are in groups that diverged over 500 million years ago (Hibbett et al., 2007) offers the potential to study independent experiments in evolution. Further, within the genus *Ophiocordyceps* (which contains most of the entomopathogens) the availability of high-quality phylogenetic reconstructions (Quandt et al., 2014), together with detailed field work and analysis of past collections (Fawcett, 1886), allows us to determine that even within this genus, parasite manipulation of host behavior has occurred independently. Here, we discuss some recent advances in understanding the mechanisms of behavioral control and the corresponding changes in the host.

5.1 Molecular Basis of Fungal Control of Insect Behavior (Ants As a Case Study)

The regulation of something as plastic as behavior is rather complex, and, therefore, the mechanisms needed to change it so precisely as done by parasites are likely to be equally complex. Substantial progress on elucidating these mechanisms can be made through the use of RNA and DNA sequencing technology, proteomics and metabolomics to identify the compounds in play, and the development of controlled laboratory infections followed by forward genetics techniques. In addition, advances in microscopy allow us to better visualize the host–parasite interface. Controlled laboratory experiments allow for the elimination of fluctuating environmental factors in the field, which add another level of complexity to the process. Their development is therefore extremely beneficial to tease apart the many aspects that are likely to be involved in behavioral manipulation.

A substantial leap forward toward learning how certain fungal species can exploit their ant hosts has been the recent publication of the transcriptomes of two fungal entomopathogenic manipulators: *Pandora formicae* and *O. unilateralis s.l.* (de Bekker et al., 2015; Malagocka et al., 2015). For *O. unilateralis s.l.*, this work was paired with the assembly and annotation of a draft genome. These fungi reside within completely different phyla (Zygomycota and Ascomycota) and infect ants from different genera (*Formica* and *Camponotus*). Yet, their manipulated hosts display similar behavioral aspects. In both systems, infected ants are manipulated prior to death to leave the nest, climb up vegetation and fix themselves there with their mandibles (Boer, 2008; Marikovsky, 1962). This elevated position, away from the ant's nest, promotes disease transmission through spore dispersal. In the case of *Pandora,* conidia are actively discharged from soft parts of the ant exoskeleton, while *Ophiocordyceps* shoots sexual ascospores from a fruiting body (ascoma) that has sprouted from behind the ant's head (Fig. 1). Laboratory infections of several *Camponotus* species with *Ophiocordyceps* followed by behavioral observations demonstrated that this entomopathogen could in fact kill all species tested (de Bekker, Quevillon, et al., 2014). However, it was only able to manipulate those species that were under natural conditions. Moreover, *Ophiocordyceps* seemed additionally hampered in its development since the characteristic switch after host death from yeast-like to hyphal growth, and subsequent host mummification, appeared to not take place in those nonmanipulated individuals. This work was accompanied by a metabolomic analysis of ex vivo fungal–ant brain interactions of *Ophiocordyceps* with various ant species. The results suggested a molecular basis for the species-specific manipulation observed as the fungus displayed a significantly heterogeneous secretome upon interaction with the different ant species' brains it was presented with (de Bekker, Merrow, et al., 2014). Unfortunately, similar experiments have not been done with *Pandora* since laboratory conditions for this system have not yet been established.

The transcriptomics studies done on *Pandora* and *Ophiocordyceps* infecting ants had different approaches, making them not directly comparable to one another. Since controlled infections cannot be performed for *Pandora*, field samples were used for RNA extraction. This limited the sampling to ants that recently died and did not display any sporulation yet, and ants carrying infective conidia that died at least one day before sampling (Małagocka et al., 2015). Thus, the manipulation event had already taken place by the time sampling was performed. In the *Ophiocordyceps* system, controlled infection studies could be performed, which made it possible to sample ants one step earlier; during the manipulated biting event. In addition, *Ophiocordyceps* can be cultured in insect cell culture media, so fungal baseline expression levels irrespective of manipulation could be established (de Bekker et al., 2015). Moreover, this allowed for the construction of a good quality draft genome to aid in gene expression analysis and gene annotation. The entomophthoralean fungi unfortunately suffer from a lack of genome and transcriptome data, leaving a large part of the *Pandora* data unannotated at this time. The study into *P. formicae* revealed the various enzymes this entomopathogen employs to rapidly change from growth within the host's body (nonsporulating phase) to the production of infective conidia (sporulation phase). The fungal pathogen goes through an intensive morphological reorganization to establish this. Enzyme production appears to be carefully orchestrated with the upregulation of various pathogenic subtilisin- and trypsin-like serine proteases and catalases protecting against host oxidative defenses during the nonsporulating phase. These catalases have been suggested to be regulating the fungal stress responses. During the following sporulating phase, lipases, chitinases, and GTPases can be found among the highest upregulated protein functions. These enzymes are necessary for the switch to invasive growth with lipases facilitating the switch from yeast to hyphal type of growth (Małagocka et al., 2015). In line with this study, subtilisin- and trypsin-like serine proteases were found to be upregulated during manipulated biting behavior in the *O. unilateralis* study as well. Moreover, many enzymes involved in oxidation—reduction processes were also found during this parasite—host interaction. These processes, together with the secretion of lipocalins might be regulating stress responses in this particular fungus—ant interaction. In addition, clues for morphological reorganization from yeast-like to hyphal cells were found in the form of genes encoding for lectin-like flocculation proteins in this system (de Bekker et al., 2015).

The transcriptomics study on experimentally *Ophiocordyceps*-infected ants centers itself around the event of manipulated biting behavior prior to death and with that proposes mechanisms for fungal control of insect behavior (de Bekker et al., 2015). Generally, genes encoding for proteins involved in sugar metabolism are downregulated during manipulation, while, as expected, pathogenicity-related genes are upregulated. After manipulation has taken place, these pathogenicity features go down again, and sugar metabolism becomes a priority. Similarly, oxidation—reduction processes that are a hallmark for parasite—host interactions are overrepresented among the genes that are active during manipulated biting and inactivated again when death sets in. Differential expression analysis also revealed that the fungal entomopathogen dynamically changes the expression of the genes that encode for secreted proteins depending on the status of infection and manipulation. This is in line with earlier reported metabolomics studies on the secretomes of fungal entomopathogens (de Bekker, Merrow, et al., 2014; de Bekker, Quevillon, et al., 2014). Among the secreted enzymes that are upregulated during manipulated biting behavior are many pathogenicity-related genes such as the lectins and proteases mentioned above. In addition, an aegerolysin with homology to the highly toxic Asp-hemolysin of *Aspergillus fumigatus* was found, as well as 21 (out of a total of 34) genes encoding for enterotoxins. Fungal enterotoxins are not well-described at this point, and genome comparison shows that, while they seem to be present in the genomes of Hypocrealean entomopathogens, fungal plant pathogens of the Phylum Ascomycota do not necessarily have genes encoding for them. The transcriptomics data, however, show that *Ophiocordyceps* dynamically tailors the expression of its secreted enterotoxins. These could be impairing the host's chemosensory pathways by reducing the production of chemosignaling molecules. Enterotoxins could also be complementing the upregulated secretion of the acid sphingomyelinase and, as such, contribute to the extensive muscle atrophy that is observed as a hallmark of *O. unilateralis* infection (Hughes, Andersen, et al., 2011; Hughes, Wappler, et al., 2011). Acid sphingomyelinase is also an important enzyme in sphingolipid metabolism, which determines the composition of biological membranes. Altering the composition of these membranes alters cell signaling, which, in the case of neuron cells, can result in neurological disorders. Additionally, the reported upregulation of different types of alkaloid metabolism can result in signaling issues, since these alkaloids could function as (ant)agonists of various receptors. Such alterations in receptor signaling could result in an altered behavioral output by the brain. An additional possibility is that the fungus targets the peripheral nervous system and in particular the motor neurons (Hughes, Andersen, et al., 2011; Hughes, Wappler, et al., 2011).

Other aspects that were found in this study that are suggestive to changing behavioral outputs are secreted fungal enzymes that could be changing serotonin and dopamine levels, as well as bioactive small secreted proteins, polyketides and nonribosomal proteins that have unknown function at present (de Bekker et al., 2015). Last but not least, this study demonstrated that behavioral manipulators across kingdoms could have mechanisms in common. The secreted enzyme protein tyrosine phosphatase (PTP) was upregulated >110-fold by *Ophiocordyceps*. The gene-encoding PTP in baculoviruses was found to be responsible for the enhanced locomotion activity observed in the caterpillars they infected (van Houte et al., 2012; Kamita et al., 2005). Late in the infection of the so-called "treetop disease" this gene gets activated when caterpillars are moved to the upper plant foliage where they die. *Ophiocordyceps*-infected ants similarly move to elevated positions where they die while biting. This suggests that the induction of enhanced locomotion activity through PTP could also be incorporated in the suite of mechanisms employed by the fungus to control ant behavior. Of course, these mechanisms are all still suggestive, as functional studies are needed to confirm their involvement.

5.2 How Host Brains Are Controlled (Ants As a Case Study)

In parasite manipulation of host behavior there is interplay between host and parasite. Although the abnormal host behavior depends on the parasite genotype, it is expected that the successful manipulation will depend on the host physiology and genotype as well (Lefèvre et al., 2008). Where fungi control host behavior, we have begun to elucidate the parasite genome and transcriptome (as discussed above); however, given that all host-parasite interactions are an interplay of both organisms, it is not possible to completely understand the mechanisms of manipulation without elucidating both parties of the interactions. Manipulative parasites are restricted to their parasitic life and their molecular activity is specialized to infect and manipulate the host. For the host, studying the changes at the molecular level could be more complicated. As in any pathology, the host will display a generalized response to parasite invasion, sickness, and impending death, at the same moment it is being manipulated. This makes it difficult to untangle manipulation effects from pathology responses, especially because they could overlap.

An ideal system to explore mechanisms of behavioral manipulation would allow us to study the host under different parasite infections, such as nonmanipulative parasite, manipulative parasite, as well as the healthy condition. However, this approach has not yet been employed, but independent studies already indicate its importance. The infection of ants by the generalist parasitic fungus *Metarhizium brunneun* results in the host upregulating immune-related genes expression (Yek, Boomsma, & Schiøtt, 2013), while the infection of ants by *O. unilateralis s.l.* induced an overall downregulation of these genes (de Bekker et al., 2015). These opposite results could indicate a strategy of the manipulative parasite that requires a longer time to develop inside the host to successful achieve the manipulation. However, as these ants were sampled shortly after infection by *M. brunneun* and shortly prior to being killed by *O. unilateralis s.l.*, an alternative explanation would be the manipulated ant is about to die and there is no more investment in the immune system. The downregulation of the immune system has also been shown in other parasites that do not manipulate the behavior of their hosts (Barribeau, Sadd, du Plessis, & Schmid-Hempel, 2014). Thus, studies on host manipulation, at the molecular level, can be very inconclusive without the suggested approach of controlling for generalized responses to infection.

Molecular approaches can be combined with direct visualization enabled by advances in histology. When a manipulative fungus infects its host, the aberrant behavior is accompanied by other phenotypic alterations as the colony of fungi grow inside the animal (Hughes, Andersen, et al., 2011; Hughes, Wappler, et al., 2011). In the case of ants infected by the fungus *O. unilateralis s.l.*, the fungal cells cause the mandibles of the ant to penetrate the plant substrate. This is accompanied by atrophied mandibular muscles that causes the "lock-jaw" so typical of the death grip phenotype. The function of such behavior is to ensure the fixation of the host after death (Hughes, Andersen, et al., 2011; Hughes, Wappler, et al., 2011). In line with histological observations, a recent study on the transcriptome of the host showed the downregulation of muscle maintenance and integrity-related genes, such as genes encoding collagen, indicating a possible pathway alteration in the host relevant to the manipulation (de Bekker et al., 2015). While the mandibular muscles are atrophied, the brain morphology appears to be preserved, suggesting the importance of the central nervous system of the host for the manipulation. The act of biting is suggested to be related to changes in the dopamine pathways (de Bekker et al., 2015), which is known to mediate aggressive behavior in ants, resulting in opening mandibles and biting (Szczuka et al., 2013). Another characteristic phenotype of ants manipulated by *O. unilateralis s.l.* is the reduced response to external stimuli, which could result from the downregulation of odorant receptors (de Bekker et al., 2015).

This could also explain why the manipulated ants do not follow the foraging trails (Hughes, Andersen, et al., 2011; Hughes, Wappler, et al., 2011). Although very intuitive, the correlation between phenotype and gene expression is not conclusive, and more investigation is necessary to confirm the role the suggested pathways have on the manipulation.

The role of the host responses in the mechanisms of behavioral manipulation has been studied in other systems, such as hairworms manipulating crickets to jump into the water (Biron & Loxdale, 2013). However, the fungal development within the host is substantially different and parallels with other systems like hairworms (an animal) might not be very helpful. On the other hand, many species of *Ophiocordyceps* fungi manipulate other ants (Fig. 1) and wasps to bite (Fig. 2E) as well as fungi in the genus *Pandora,* which also manipulate ants to bite (Fig. 2G). A parallel among these systems would be valuable to identify the convergence on the host histology, physiology, and gene expression that results in the same biting behavior. Additionally, because the fungus takes between 15 and 24 days to manipulate the ant after the infection, another approach to exploit the mechanisms of manipulation is to investigate the changes as the infection progress, for both host and parasite, at different levels. This is the important aspect of causation discussed above. Finally, as mentioned before, the complexity of manipulation by fungal parasites is a spectrum, from precise positioning of the host prior to death to dispersal of spores from the live host (Loreto et al., 2014), and we have just begun to understand one aspect of the total complexity implied by such a spectrum.

6. CAN BEHAVIORAL MANIPULATION BE EVOLVED IN SILICO?

When we consider the complexity of behavioral manipulation of animal behavior by fungi it is often a challenge to understand how it occurs. In Section 5, we discussed how advances are being made in uncovering these proximate mechanisms by focusing on genomic features, transcription, or the production of small molecules that affect behavior, as well direct visualization. A complementary approach to this empirical work is to undertake theoretical experiments asking how complex behavioral changes could result from "so simple a beginning" as killing a host at the point where it was infected. As Fig. 1 shows, there is a diversity of locations where insects are killed prior to the postmortem development of the fungus. If we take the parsimonious position that most entomopathogens kill their insect hosts in the same location as where they were infected (soil, leaf litter, decaying wood), then the evolution of manipulation to the underside of leaves or on other parts of plants represents a derived condition. In evolutionary biology, one approach to understanding the evolutionary pathways to a derived trait is via genetic algorithms. In this section we introduce these and argue that they have utility for understanding the evolution of complex manipulation of animal behavior by fungi.

Genetic algorithms (GAs), a subset of the broader field of evolutionary computation, are an optimization technique that borrows principles from natural selection to adaptively search phenotypic or genotypic space for fitness maximums (Mitchell, 1998). The basic components of a GA consist of a population of solutions ("individuals") that are evaluated for their fitness in solving a particular problem. After their fitness is calculated, the individuals undergo selection and reproduction (ie, the top X% of the population survive and reproduce, or reproduction is fitness proportional). During reproduction, the solutions represented by the parent individuals undergo random mutation and/or recombination to produce offspring, and these offspring and parents merge to form a new population ("replacement"). This new population then repeats the same process of selection, reproduction, and replacement until either convergence to a particular solution is reached or after a target number of generations has occurred. GAs have been used in many different contexts, from finding solutions to the iterated prisoners' dilemma problem (Axelrod, 1987) and optimizing travel routes (Grefenstette, Gopal, Rosmaita, & Van Gucht, 1985) to predicting gene–gene interactions (Hahn, Ritchie, & Moore, 2003). In behavioral ecology, genetic algorithms have been used to model optimal tradeoff decisions between singing for mates and foraging for survival in birds (Sumida, Houston, McNamara, & Hamilton, 1990), and to explore the best antipredator vigilance strategies for animals foraging in groups (Ruxton & Beauchamp, 2008). In the preceding examples, the results from using GAs were similar to those obtained by analytical methods, but GAs have the flexibility to be used for problems that are otherwise analytically intractable.

One potential way to use GAs to explore the evolution of behavioral manipulation in silico would be to represent combinations of fungal fitness components (eg, cadaver placement, spore morphology, production, infectiousness, etc.) as individuals in a population of many different solutions. Fitness could be assessed by competing these individuals in an agent-based model of insect foraging and quantifying the number of insects that they each successfully infect. The most successful individuals of each round would reproduce with some amount of mutation and recombination, and the algorithm would continue until convergence on a particular best individual (combination of fungal traits) or until a given number of generations had been reached. Crucially, by competing these fungal strategies in an agent-based model of insect foraging, we allow for the particular details of host foraging ecology to be included, and, thus, it is likely that we could identify a diversity of "best" strategies for behavioral manipulation.

7. CONCLUSION

Fungi can control arthropod behavior in spectacular and complex ways. In this chapter we sought to present an overview of this complexity and discuss the multiple approaches we can take to study such complex adaptations (eg, Tinbergen questions or in silico genetic algorithms) as well as the advances that have been made so far (eg, mechanisms of host behavior). Despite these advances, we are just at the very tip of what is a considerable iceberg of complex interactions. In recent years, diverse efforts have revealed details of how *O. unilateralis s.l.* controls ant behaviors, but this is just one complex of at least 11 that infect and manipulate ants (Fig. 1). Within the Hymenoptera (the order to which ants belong), there are also wasps and bees that are manipulated. Among the insects, approximately 65% of all orders are infected (Araújo & Hughes, 2016), and in some cases complex manipulation occurs. Outside of the insects, we also suspect spiders are similarly controlled, as evidenced by recent studies (Costa, 2014; Evans, unpublished data). We are thus in a golden age of discovery. Armed with advanced tools in evolutionary biology (from SEM to RNAseq), we are in a position to discover how the fungi, a group of microbes capable of coordinated activity, have evolved the ability to direct animal behavior. In short, we have the ability to understand how the organism without the brain controls the one with the brain. We hope such a goal and the knowledge that many diverse examples of control exist inspires future organismal biologists to study the complex adaptations that have arisen from "so simple a beginning" (Darwin, 1859).

ACKNOWLEDGMENTS

We thank Ray St. Leger and Brian Lovett for inviting this review. We are very grateful to the many national park and research station staff around the world that have enabled our work by providing access to forest sites. Thanks to Dr. Renner L.C. Baptista, Universidade Federal do Rio de Janeiro for help in identifying spiders. This work is supported in part by the NSF foundation through a grant to DPH, No. 1414296 as part of the joint NSF-NIH-USDA Ecology and Evolution of Infectious Diseases program.

REFERENCES

Adamo, S. A. (2012). The strings of the puppet master: how parasites change host behavior. In D. P. Hughes, J. Brodeur, & F. Thomas (Eds.), *Host manipulation by parasites*. Oxford Oxford University Press.

Adamo, S. A., & Webster, J. P. (2013). Neural parasitology: how parasites manipulate host behaviour. *The Journal of Experimental Biology, 216*(1), 1–2.

Alcock, J. (1993). *Animal behavior: An evolutionary approach*. Sinauer Associates.

Andersen, S. B., Gerritsma, S., Yusah, K. M., Mayntz, D., Hywel-Jones, N. L., Billen, J. … Hughes, D. P. (2009). The life of a dead ant: the expression of an adaptive extended phenotype. *American Naturalist, 174*(3), 424–433. http://dx.doi.org/10.1086/603640.

Araújo, J., Geiser, D. M., Evans, H. C., & Hughes, D. P. (2015). Three new species of *Ophiocordyceps* fungi infecting Carpenter ants from the Amazon. *Phytotaxa, 220*(3), 224–238.

Araújo, J., & Hughes, D. P. (2016). Diversity of entomopathogenic Fungi: Which groups conquered the insect body? *Genetics and Molecular Biology of Entomopathogenic Fungi, 94*.

Askew, R. R. (1971). *Parasitic insects*. New York: American Elsevier Publishing Company. Inc.

Axelrod, R. (1987). The evolution of strategies in the iterated prisoner's dilemma. *The Dynamics of Norms*, 1–16.

Barr, D. J., & Désaulniers, N. L. (1988). Precise configuration of the chytrid zoospore. *Canadian Journal of Botany, 66*(5), 869–876.

Barribeau, S. M., Sadd, B. M., du Plessis, L., & Schmid-Hempel, P. (2014). Gene expression differences underlying genotype-by-genotype specificity in a host–parasite system. *Proceedings of the National Academy of Sciences of the United States of America, 111*(9), 3496–3501.

de Bekker, C., Merrow, M., & Hughes, D. P. (2014). From behavior to mechanisms: an integrative approach to the manipulation by a parasitic fungus (*Ophiocordyceps unilateralis* s.l.) of its host ants (*Camponotus* spp.). *Integrative and Comparative Biology*, icu063.

de Bekker, C., Quevillon, L., Smith, P. B., Fleming, K., Patterson, A. D., & Hughes, D. P. (2014). Species-specific ant brain manipulation by a specialized fungal parasites involves secondary metabolites. *BMC Evolutionary Biology, 14*(166). http://dx.doi.org/10.1186/s12862-014-0166-3.

de Bekker, C., Ohm, R. A., Loreto, R. G., Sebastian, A., Albert, I., Merrow, M. … Hughes, D. P. (2015). Gene expression during zombie ant biting behavior reflects the complexity underlying fungal parasitic behavioral manipulation. *BMC Genomics, 16*.

Berdoy, M., Webster, J. P., & Macdonald, D. W. (2000). Fatal attraction in rats infected with *Toxoplasma gondii*. *Proceedings of the Royal Society of London, B, 267*(1452), 1591–1594.

Biron, D. G., & Loxdale, H. D. (2013). Host–parasite molecular cross-talk during the manipulative process of a host by its parasite. *The Journal of Experimental Biology, 216*(1), 148–160.

Blackwell, M. (2011). The Fungi: 1, 2, 3... 5.1 million species? *American Journal of Botany, 98*(3), 426–438. http://dx.doi.org/10.3732/ajb.1000298.

Boer, P. (2008). Observations of summit disease in *Formica rufa* Linnaeus, 1761 (Hymenoptera: Formicidae). *Myrmecological News, 11*, 63–66.

Burkhardt, R. W. J. (2005). *Patterns of Behavior: Konrad Lorenz, Niko Tinbergen, and the Founding of Ethology*. University of Chicago Press.

Costa, P. P. (2014). *Gibellula spp. associadas a aranhas da Mata do Paraíso, Viçosa-MG* (M.Sc.). Minas Gerais, Brazil: Universidade Federal de Viçosa, Universidade Federal de Viçosa.

Couch, J. N. (1938). *The genus* Septobasidium.

Darwin, C. (1859). *On the origin of species by means of natural selection, or the preservation of favoured races in the struggle for life*. London: John Murray.

Dawkins, R. (1982). *The extended phenotype*. Oxford: W.H. Freeman.
Dawkins, R. (1990). Parasites, desiderata lists and the paradox of the organism. *Parasitology, 100*, S63–S73.
Dawkins, R. (2004). Extended phenotype — but not too extended. A reply to Laland, Turner and Jablonka. *Biology & Philosophy, 19*(3), 377–396.
Dawkins, R. (2012). Foreword to host manipulation by parasites. In D. P. Hughes, J. Brodeur, & F. Thomas (Eds.), *Host manipulation by parasites* (pp. xi–xiii). Oxford: Oxord Univeristy Press.
DeKesel, A. (1996). Host specificity and habitat preference of *Laboulbenia slackensis*. *Mycologia, 88*(4), 565–573.
Eilenberg, J. (1986). Effect of *Entomophthora muscae* (C.) Fres. on egg-laying behavior of female carrot flies (*Psila rosae* F.). In R. Samson, J. M. Vlak, & D. Peters (Eds.), *Fundamental and applied aspects of invertebrate pathology* (p. 235). Wageningen.
Evans, H., & Samson, R. (1987). Fungal pathogens of spiders. *Mycologist, 1*(4), 152–159.
Evans, H. C. (1974). Natural control of Arthropods with special reference to Ants (Formicidae) by Fungi in the tropical high forest of Ghana. *The Journal of Applied Ecology, 11*(1), 37–49.
Evans, H. C. (1988). Coevolution of entomogenous fungi and their insect hosts. In K. A. Pirozynski, & D. L. Hawksworth (Eds.), *Coevolution of fungi with plants and animals* (pp. 149–171). London Academic Press.
Evans, H. C. (2013). *Fungal pathogens of spiders Spider Ecophysiology* (pp. 107–121). Springer.
Evans, H. C., Elliot, S. L., & Hughes, D. P. (2011a). Hidden diversity behind the zombie-ant fungus *Ophiocordyceps unilateralis*: four new species described from carpenter ants in Minas Gerais, Brazil. *PLoS One, 6*, e17024. http://dx.doi.org/10.1371/journal.pone.0017024.
Evans, H. C., Elliot, S. L., & Hughes, D. P. (2011b). *Ophiocordyceps unilateralis*: a keystone species for unraveling ecosystem functioning and biodiversity of fungi in tropical forests? *Communicative & Integrative Biology, 4*(5), 598–602.
Evans, H. C., & Shah, P. A. (2002). Taxonomic status of the genera *Sorosporella* and *Syngliocladium* associated with grasshoppers and locusts (Orthoptera: Acridoidea) in Africa. *Mycological Research, 106*(06), 737–744.
Evans, H. C., & Whitehead, P. F. (2005). Entomogenous fungi of arboreal Coleoptera from Worcestershire, England, including the new species *Harposporium bredonense*. *Mycological Progress, 4*, 91–99.
Fawcett, W. (1886). Description of *Cordyceps lloydii* in ants. *Annals and Magazine of Natural History, 5*(XVIII), 317.
Frances, S., Sweeney, A., & Humber, R. (1989). *Crypticola clavulifera* gen. et sp. nov. and *Lagenidium giganteum*: oomycetes pathogenic for dipterans infesting leaf axils in an Australian rain forest. *Journal of Invertebrate Pathology, 54*(1), 103–111.
Godfray, H. C. J. (1994). *Parasitoids*. Princeton, NJ: Princeton University Press.
Grefenstette, J., Gopal, R., Rosmaita, B., & Van Gucht, D. (1985). Genetic algorithms for the traveling salesman problem. In *Paper presented at the proceedings of the first international conference on genetic algorithms and their applications*.
Hahn, L. W., Ritchie, M. D., & Moore, J. H. (2003). Multifactor dimensionality reduction software for detecting gene–gene and gene–environment interactions. *Bioinformatics, 19*(3), 376–382.
Harper, A. M. (1958). Notes on the behaviour of *Pemphigus betae* Doane (Homoptera: Aphididae) infected with *Entomophthora aphidis* Hoffm. *Canadian Journal of Entomology, 90*, 439–440.
Hawksworth, D. L., & Rossman, A. Y. (1997). Where are all the undescribed fungi? *Phytopathology, 87*(9), 888–891.
Hibbett, D. S., Binder, M., Bischoff, J. F., Blackwell, M., Cannon, P. F., Eriksson, O. E. … Zhang, N. (2007). A higher-level phylogenetic classification of the Fungi. *Mycological Research, 111*, 509–547. http://dx.doi.org/10.1016/j.mycres.2007.03.004.

van Houte, S., Ros, V. I., Mastenbroek, T. G., Vendrig, N. J., Hoover, K., Spitzen, J., & van Oers, M. M. (2012). Protein tyrosine phosphatase-induced hyperactivity is a conserved strategy of a subset of baculoviruses to manipulate lepidopteran host behavior. *PLoS One, 7*(10).

Hughes, D. P. (2013). Pathways to understanding the extended phenotype of parasites in their hosts. *The Journal of Experimental Biology, 216*(1), 142–147.

Hughes, D. P. (2014). On the origins of parasite extended phenotypes. *Integrative and Comparative Biology, 54*(2), 210–217.

Hughes, D. P., Andersen, S., Hywel-Jones, N. L., Himaman, W., Bilen, J., & Boomsma, J. J. (2011). Behavioral mechanisms and morphological symptoms of zombie ants dying from fungal infection. *BMC Ecology, 11*, 13. http://dx.doi.org/10.1186/1472-6785-1111-1113.

Hughes, D. P., Evans, H., Hywel-Jones, N., Boomsma, J., & Armitage, S. (2009). Novel fungal disease in complex leaf-cutting ant societies. *Ecological Entomology, 34*(2), 214–220.

Hughes, D. P., Wappler, T., & Labandeira, C. C. (2011). Ancient death-grip leaf scars reveal ant–fungal parasitism. *Biology Letters*. http://dx.doi.org/10.1098/rsbl.2010.0521 (August 18, 2010).

Humber, R. (1982). *Strongwellsea* vs. *Erynia*: the case for a phylogenetic classification of the Entomophthorales (Zygomycetes). *Mycotaxon, 15*, 167–184.

Humber, R. A. (1989). *Synopsis of a revised classification for the Entomophthorales (Zygomycotina)*. USA: Mycotaxon.

Kamita, S. G., Nagasaka, K., Chua, J. W., Shimada, T., Mita, K., Kobayashi, M.… Hammock, B. D. (2005). A baculovirus-encoded protein tyrosine phosphatase gene induces enhanced locomotory activity in a lepidopteran host. *Proceedings of the National Academy of Sciences of the United States of America, 102*(7), 2584–2589.

Kirk, P. M., Canon, P. F., Minter, D. W., & Staplers, J. A. (2008). *Dictionary of the Fungi* (10th ed.). CABI.

Kuris, A. M. (1974). Trophic interactions: similarity of parasitic castrators to parasitoids. *Quarterly Review of Biology, 49*, 129–148.

Kützing, F. T. (1849). *Species Algarum*. Lipsiae.

Lefevre, T., Adamo, S. A., Biron, D. G., Misse, D., Hughes, D., & Thomas, F. (2009). Invasion of the body snatchers: the diversity and evolution of manipulative strategies in host-parasite interactions. *Advances in Parasitology, 68*, 45–83. http://dx.doi.org/10.1016/S0065-308x(08)00603-9.

Lefèvre, T., Roche, B., Poulin, R., Hurd, H., Renaud, F., & Thomas, F. (2008). Exploiting host compensatory responses: the 'must' of manipulation? *Trends in Parasitology, 24*(10), 435–439.

Loreto, R. G., Elliot, S. L., Freitas, M. L., Pereira, T. M., & Hughes, D. P. (2014). Long-term disease dynamics for a specialized parasite of ant societies: a field study. *PLoS One, 9*(8), e103516.

Maitland, D. (1994). A parasitic fungus infecting yellow dungflies manipulates host perching behaviour. *Proceedings of the Royal Society of London B: Biological Sciences, 258*(1352), 187–193.

Małagocka, J., Grell, M. N., Lange, L., Eilenberg, J., & Jensen, A. B. (2015). Transcriptome of an entomophthoralean fungus (*Pandora formicae*) shows molecular machinery adjusted for successful host exploitation and transmission. *Journal of Invertebrate Pathology, 128*, 47–56.

Marikovsky, P. I. (1962). On some features of behaviour of the ants *Formica rufa* L. infected with fungous disease. *Insectes Sociaux, 9*, 173–179.

Mitchell, M. (1998). *An introduction to genetic algorithms*. MIT press.

Moore, J. (2002). *Parasites and the behavior of animals*. Oxford: Oxford University Press.

Mueller, U. G., Gerardo, N. M., Aanen, D. K., Six, D. L., & Schultz, T. R. (2005). The evolution of agriculture in insects. *Annual Review of Ecology, Evolution, and Systematics*, 563–595.

Muma, M. H., & Clancy, D. (1961). Parasitism of purple scale in Florida citrus groves. *Florida Entomologist*, 159–165.

Nentwig, W. (Ed.). (2013). *Spider Ecophysiology*. Berlin: Springer-Verlag.

Pontoppidan, M.-B., Himaman, W., Hywel-Jones, N. L., Boomsma, J. J., & Hughes, D. P. (2009). Graveyards on the move: the spatio-temporal distribution of dead *Ophiocordyceps*-infected ants. *PLoS One, 4*(3), e4835.

Poulin, R. (1994). The evolution of parasite manipulation of host behavior: a theoretical analysis. *Parasitology, 109*, S109–S118.

Poulin, R. (2011). Parasite manipulation of host behavior: an update and frequently asked questions. In H. J. Brockmann (Ed.), *Advances in the study of behavior*, (Vol. 41, pp. 151–186). Burlington: Elsevier.

Poulin, R., & Morand, S. (2000). The diversity of parasites. *Quarterly Review of Biology, 75*(3), 277–293.

Poulin, R., & Morand, S. (2005). *Parasite biodiversity*. Washington: Smithsonian Books.

Quandt, C. A., Kepler, R. M., Gams, W., Araújo, J. P., Ban, S., Evans, H. C. ... Li, Z. (2014). Phylogenetic-based nomenclatural proposals for *Ophiocordycipitaceae* (*Hypocreales*) with new combinations in *Tolypocladium*. *IMA Fungus, 5*(1), 121.

Roy, H. E. D., Steinkraus, C., Eilenberg, J., Hajek, A. E., & Pell, J. K. (2006). Bizarre interactions and endgames: entomopathogenic fungi and their arthropod hosts. *Annual Review of Entomology, 51*, 331–357.

Ruxton, G. D., & Beauchamp, G. (2008). The application of genetic algorithms in behavioural ecology, illustrated with a model of anti-predator vigilance. *Journal of Theoretical Biology, 250*(3), 435–448.

Samson, R. A., & Evans, H. (1973). Notes on entomogenous fungi from Ghana 1: the genera Gibellula and Pseudogibellula. *Acta Botanica Neerlandica, 22*(5), 522–528.

Soper, R. S. (1963). *Massospora laevispora*, a new species of fungus pathogenic to the cicada, *Okanagana rimosa*. *Canadian Journal of Botany, 41*, 875–878.

Soper, R. S., Delyzer, A. J., & Smith, F. L. R. (1976). The genus *Massospora*, entomopathogenic for cicadas. *Annals of the Entomolgical Society of America, 69*, 88–95.

Suh, S. O., Noda, H., & Blackwell, M. (2001). Insect symbiosis: derivation of yeast-like endosymbionts within an entomopathogenic filamentous lineage. *Molecular Biology and Evolution, 18*(6), 995–1000.

Sumida, B. H., Houston, A., McNamara, J., & Hamilton, W. (1990). Genetic algorithms and evolution. *Journal of Theoretical Biology, 147*(1), 59–84.

Sung, G. H., Hywel-Jones, N. L., Sung, J. M., Luangsa-Ard, J. J., Shrestha, B., & Spatafora, J. W. (2007). Phylogenetic classification of Cordyceps and the clavicipitaceous fungi. *Studies in Mycology, 57*, 5–59. http://dx.doi.org/10.3114/sim.2007.57.01.

Sung, G. H., Poinar, G. O., & Spatafora, J. W. (2008). The oldest fossil evidence of animal parasitism by fungi supports a Cretaceous diversification of fungal-arthropod symbioses. *Molecular Phylogenetics and Evolution, 49*(2), 495–502. http://dx.doi.org/10.1016/j.ympev.2008.08.028.

Szczuka, A., Korczyńska, J., Wnuk, A., Symonowicz, B., Szwacka, A. G., Mazurkiewicz, P. ... Godzińska, E. J. (2013). The effects of serotonin, dopamine, octopamine and tyramine on behavior of workers of the ant *Formica polyctena* during dyadic aggression tests. *Acta Neurobiologiae Experimentalis, 73*, 495–520.

Thaxter, R. (1888). *The Entomophthoreae of the United States*. Boston Society of Natural History.

Thomas, F., Schmidt-Rhaesa, A., Martin, G., Manu, C., Durand, P., & Renaud, F. (2002). Do hairworms (Nematomorpha) manipulate the water seeking behaviour of their terrestrial hosts? *Journal of Evolutionary Biology, 15*(3), 356–361.

Tinbergen, N. (1963). On aims and methods of ethology. *Zeitschrift für Tierpsychologie, 20*(4), 410–433.

Van Houte, S., Ros, V. I., & Oers, M. M. (2013). Walking with insects: molecular mechanisms behind parasitic manipulation of host behaviour. *Molecular Ecology, 22*(13), 3458–3475.

Vega, F. E., & Blackwell, M. (Eds.). (2005). *Insect-fungal associations: Ecology and evolution.* Oxford: Oxford University Press.

Webster, J. P. (2001). Rats, cats, people and parasites: the impact of latent toxoplasmosis on behaviour. *Microbes and Infection, 3*(12), 1037–1045.

Whisler, H. C., Zebold, S. L., & Shemanchuk, J. A. (1975). Life history of *Coelomomyces psorophorae. Proceedings of the National Academy of Sciences of the United States of America, 72*(2), 693–696.

Yek, S. H., Boomsma, J. J., & Schiøtt, M. (2013). Differential gene expression in Acromyrmex leaf-cutting ants after challenges with two fungal pathogens. *Molecular Ecology, 22*(8), 2173–2187.

INDEX

'*Note*: Page numbers followed by "f" indicate figures "t" indicate tables.'

A

C

F

G

H

T

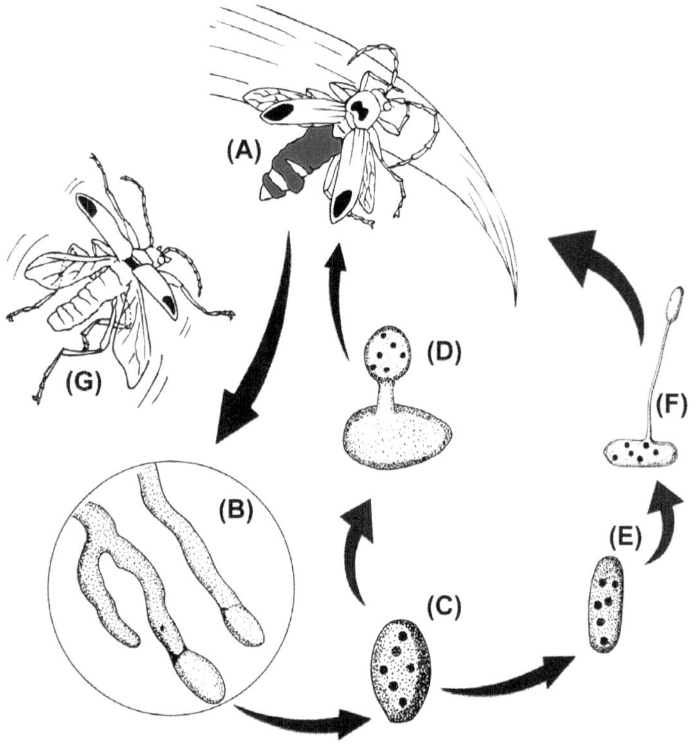

J.P.M. Araújo and D.P. Hughes, Figure 2 ***Entomophthoromycota*****—Eryniopsis lampyridarum.** (A) Cantharid beetle infected by *E. lampyridarum* died with its mandibles attached to flowering plants or grass. The elytra and wings gradually open as the fungus grows through the host's body; (B) Conidiophores emerges directly from the host's body; (C) Primary conidium; (D) Primary conidium bearing mature secondary conidium at the tip; (E) Secondary conidium; (F) Secondary conidia eventually will produce capilliconidia in absence of a suitable host; (G) Another cantharid beetle will get infected if it touches the exposed fungal hymenium. *References: (Humber, 1984; Roy et al., 2006; Thaxter, 1888).*

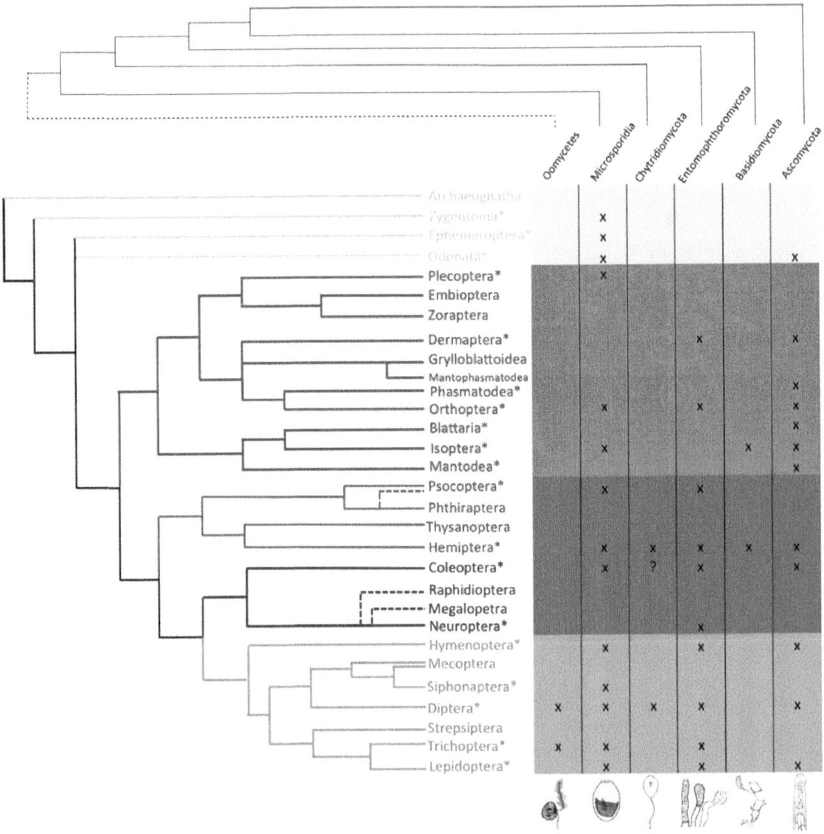

J.P.M. Araújo and D.P. Hughes, Figure 5 ***Insect orders × fungal phyla (and oomycetes): the parasitic relationship between entomopathogens and their hosts.*** On the left, the phylogeny of insect orders (*adapted from Grimaldi, D. & Engel, M. S. (2005). Evolution of the Insects. Cambridge University Press.*); on the top the phylogeny of fungal phyla and oomycetes (*adapted from James, T. Y., Kauff, F., Schoch, C. L., Matheny, P. B., Hofstetter, V., Cox, … Miadlikowska, J. (2006). Reconstructing the early evolution of fungi using a six-gene phylogeny. Nature, 443(7113), 818–822.*); the table shows which fungal group infects each insect order. The uncertainty of a record is denoted with a question mark.

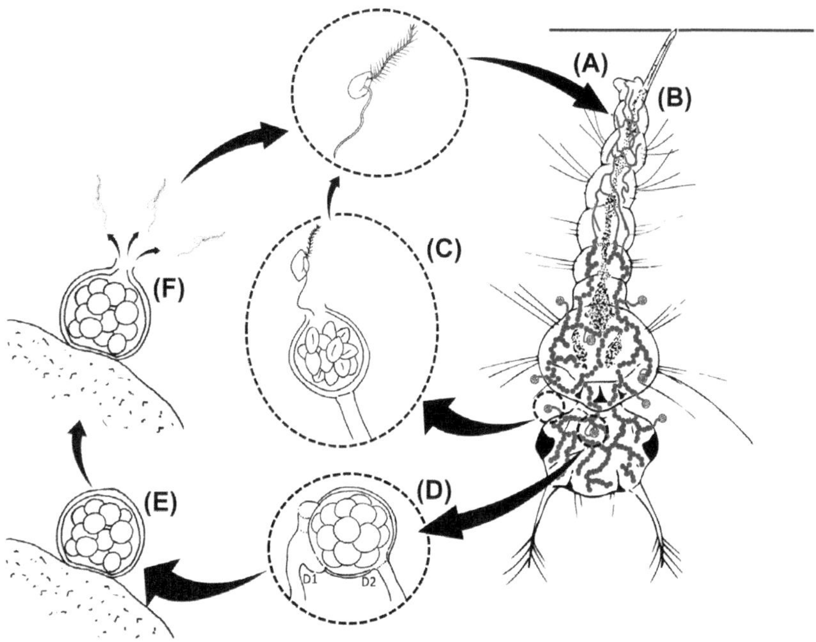

J.P.M. Araújo and D.P. Hughes, Figure 7 ***Oomycetes*****—Lagenidium giganteum.** (A) Zoospore (n) adhere and penetrate the cuticle, starting the infection; (B) Mycelium starts to grow and proliferates within the larva's body; (C) Zoosporangium releasing asexual zoospores; (D) Oospore (Sexual part of the cycle)—(D1) Antheridium fertilizing (D2) oogonium; (E) Oospore detaches from the hyphae and settles down in the environment; (F) Releasing of sexual zoospores (2n).

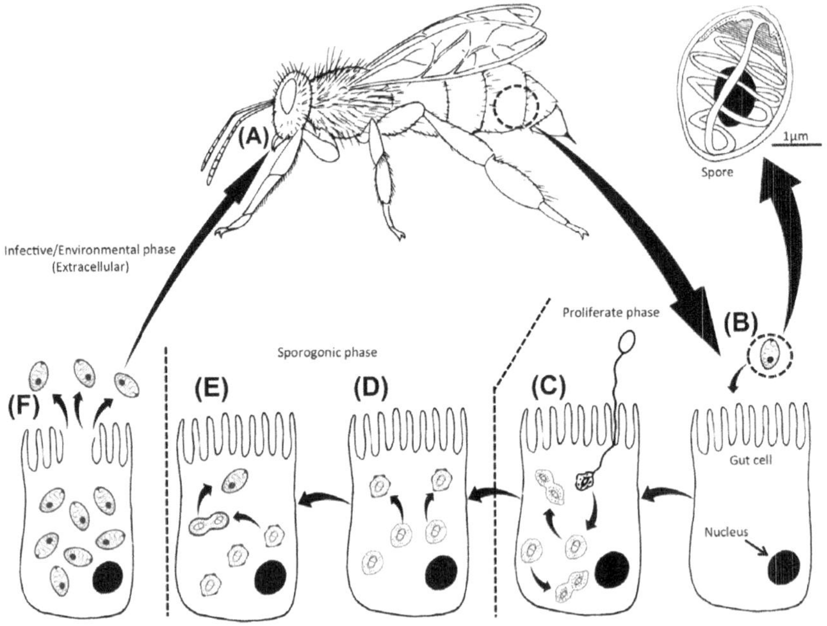

J.P.M. Araújo and D.P. Hughes, Figure 8 ***Microsporidia*****—Nosema *sp.*** (A) Ingestion of spores; (B) Spore reaches the gut of the bee and is activated by its environment, triggering the polar tube to inject the sporoplasm into the host's cell; (C) Cellular multiplication (proliferate phase); (D and E) Transition from sporoplasm to spore; (F) Spores are released into the gut again, and will be spread in the bee's feces or will reinfect the same individual.

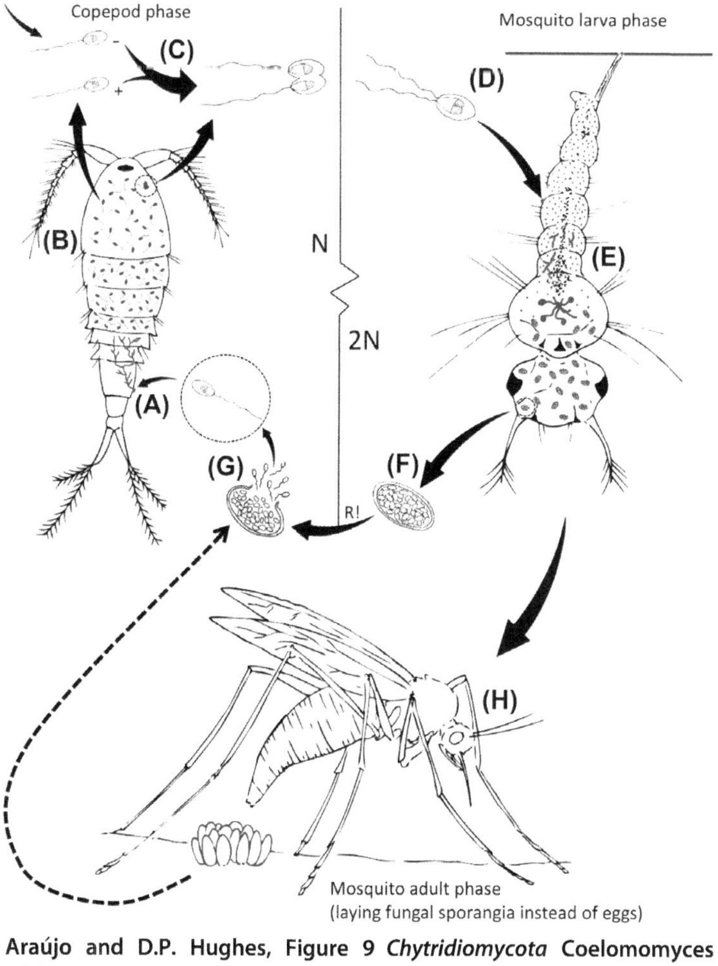

J.P.M. Araújo and D.P. Hughes, Figure 9 ***Chytridiomycota*** **Coelomomyces psophorae.** (A) Zoospores attach and penetrate the copepod cuticle; (B) Development of the gametophytic phase and dispersion of gametes into the environment; (C) Fusion of compatible gametes, inside the copepod or in the environment (plasmogamy); (D) Formation of zygote (kariogamy = 2n) and attachment to the cuticle of the mosquito larva; (E) Colonization and development of the sporophytic phase and formation of sporangium; (F) Resting sporangium released into the environment after the larva's death; (G) Meiosis and release of asexual zoospores; (H) If the larvae reach the adult stage, the fungus will migrate to the ovaries. Instead of laying eggs, the mosquito will lay fungal sporangia.

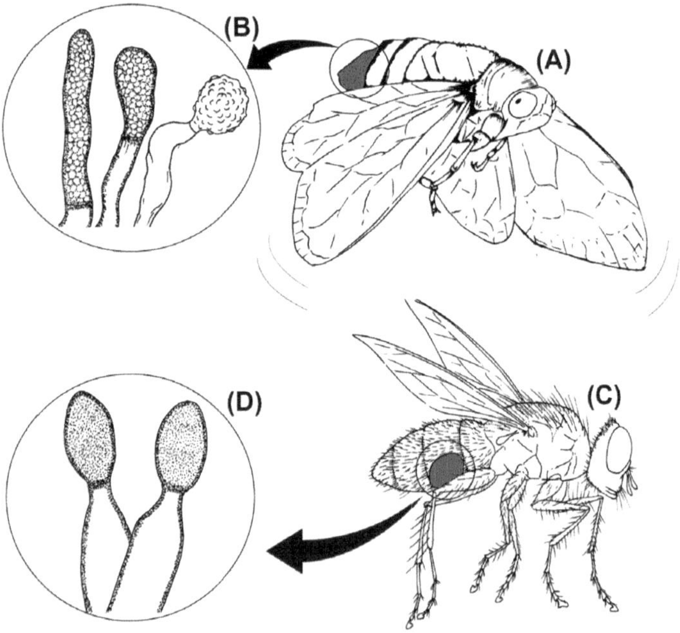

J.P.M. Araújo and D.P. Hughes, Figure 10 ***Entomophthoromycota*****—Massospora cicadina** ***(A and B),*** **Strongwellsea castrans** ***(C and D).*** (A) A living cicada flying and dispersing spores while its body disintegrates due to fungal activity. (B) Spore-producing cells (Conidiophores) in different stages of development. (C) Fly exhibiting a hole on the abdomen caused by the fungal infection. (D) Conidiophores exhibiting a terminal spore.

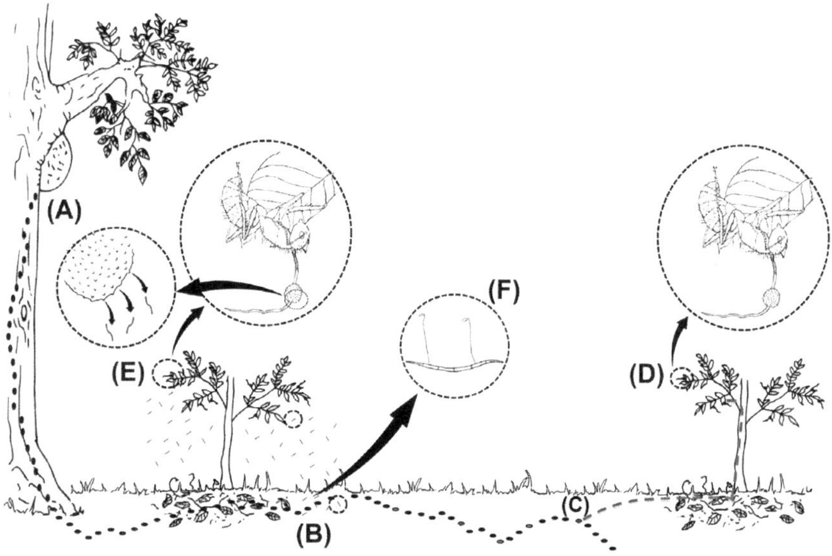

J.P.M. Araújo and D.P. Hughes, Figure 11 *Ascomycota*—**Ophiocordyceps unilateralis *s.l.*** (A) Ants leave the nest to forage on the forest floor; (B) Eventually they get infected with *Ophiocordyceps* ascospores that were previously shot on the forest floor; (C and D) About 10 days after infection (depending on the species and the geographical location) the infected ant leaves the nest to die on an elevated position, biting the edge or the main vein of a leaf. The fungus places the ant on a precise location, which is optimal for fungal development and further dispersion of the spores; (E) Two to eight weeks after the ant's death, depending on the weather conditions, the fungus starts to produce spores and shoot them into the environment; (F) From 24 to 72 h after being shot, the spores will germinate and form a secondary spore, the capilliconidiospore.

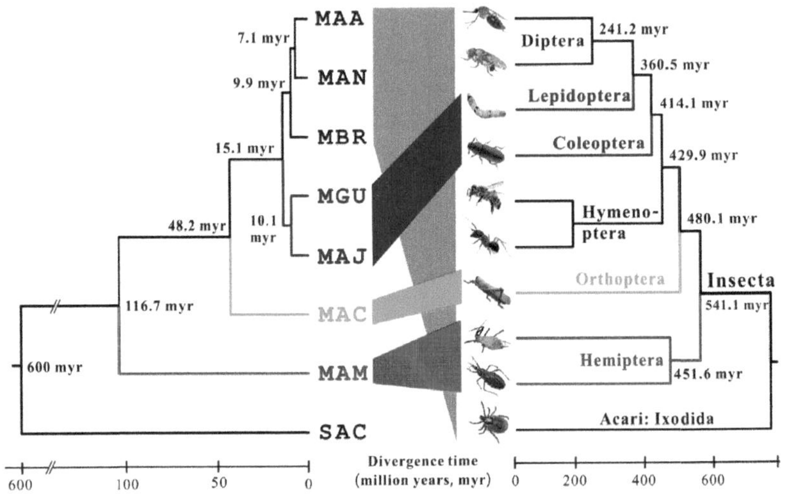

J.B. Wang, *et al.*, Figure 2 Reconstructed phylogeny of *Metarhizium* showing their insect host ranges and divergence time. *MAA, Metarhizium robertsii; MAN, Metarhizium anisopliae; MBR, Metarhizium brunneum; MGU, Metarhizium guizhouense; MAJ, Metarhizium majus; MAC, Metarhizium acridum; MAM, Metarhizium album; SAC, Saccharomyces cerevisiae. Rebuilt from Hu, X., Xiao, G., Zheng, P., Shang, Y., Su, Y., Zhang, X., … Wang, C. S. (2014). Trajectory and genomic determinants of fungal-pathogen speciation and host adaptation.* Proceedings of the National Academy of Sciences of the United States of America, 111, *16796–16801.*

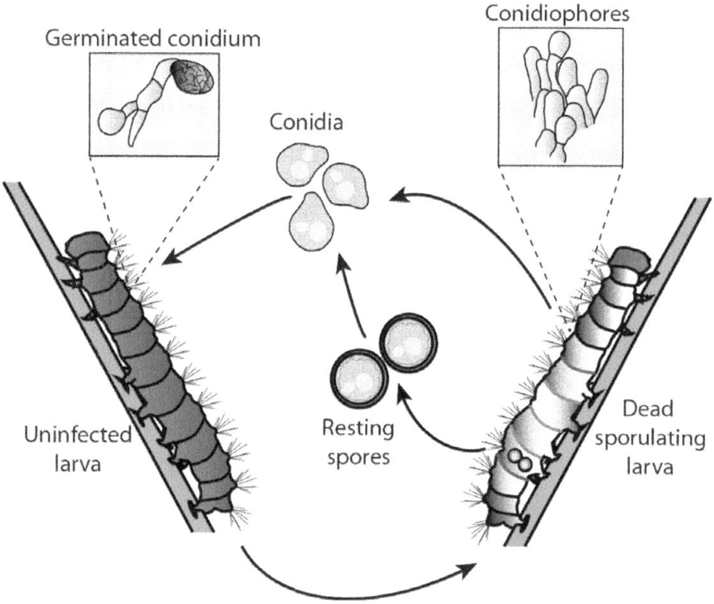

H.H. De Fine Licht, *et al.*, Figure 2 Drawing of the life cycle of *Entomophaga maimaiga*. Actively ejected conidia land on the uninfected host insect cuticle, germinate, and form an appressorium, which is a fungal cell enabling penetration of the cuticle. *Entomophaga maimaiga* grows inside the lepidopteran larva until it eventually kills the host, grows back outside of the body through the intersegmental membranes and the cuticle, and forms conidia-bearing conidiophores. Alternatively or in addition to conidial formation, *E. maimaiga* may develop thick-walled resting spores (azygospores) inside the dead larva, which often dies hanging onto a surface above ground level. The insect carcass containing resting spores eventually falls to the soil, the cuticle breaks, and resting spores are released onto and into the soil, where they may lie dormant from one to many years.

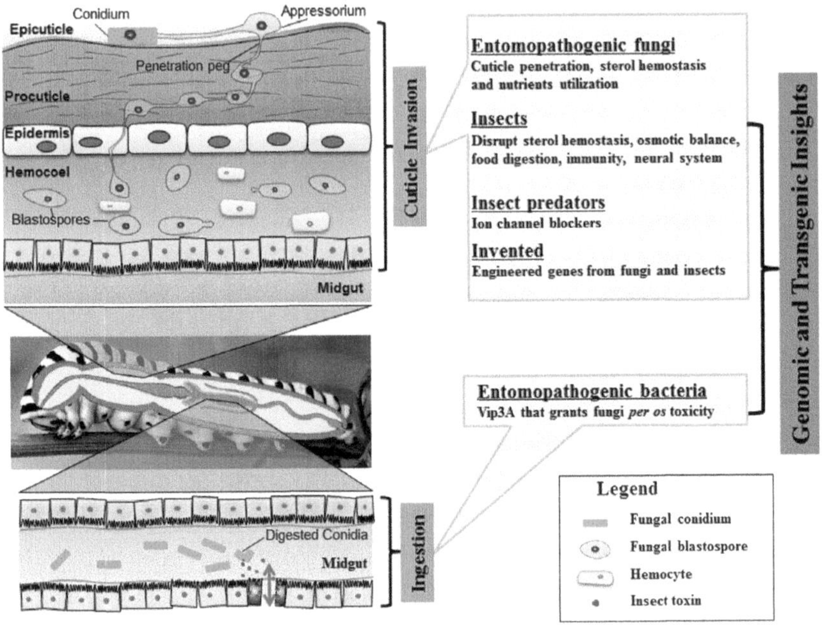

H. Zhao, *et al.*, Figure 1 Strategies for genetically engineering to improve the virulence of entomopathogenic fungi. Left panel: infection pathways of entomopathogenic fungi through cuticle contact (upper zoom) and per os toxicity to insects of ingested transgenic strains expressing Bt toxins such as Vip3A (lower zoom). Right panel: currently exploited sources of genes for improving virulence of entomopathogenic fungi.

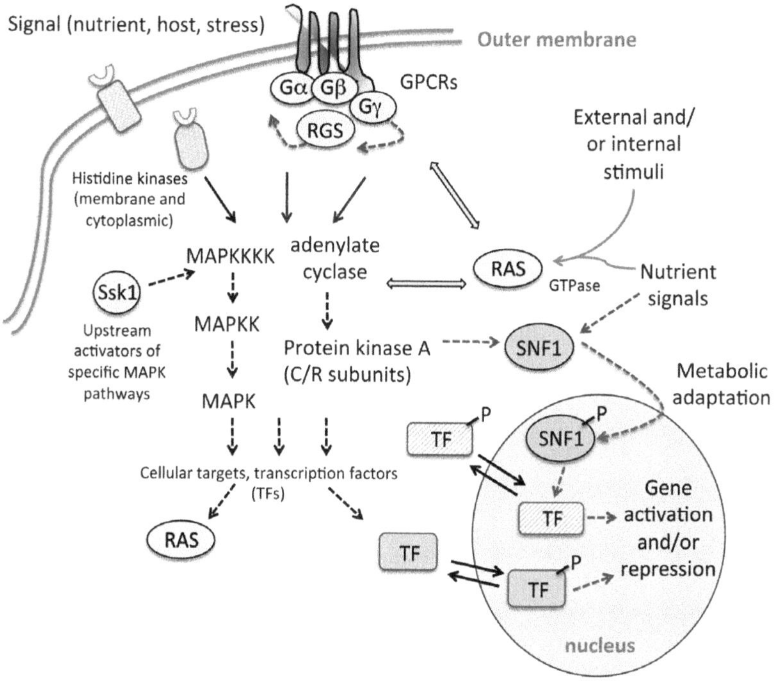

A. Ortiz-Urquiza and N.O. Keyhani, Figure 2 Overview of select *Beauveria bassiana* genes involved in signal transduction. *GPCR*, G-protein-coupled receptor; *MAPKs*, mitogen-activated protein kinase pathways; *RAS*, RAS family of GTPases; *SNF1*, sucrose non-fermenting factor (metabolic adaptor protein); *TF*, transcription factor.

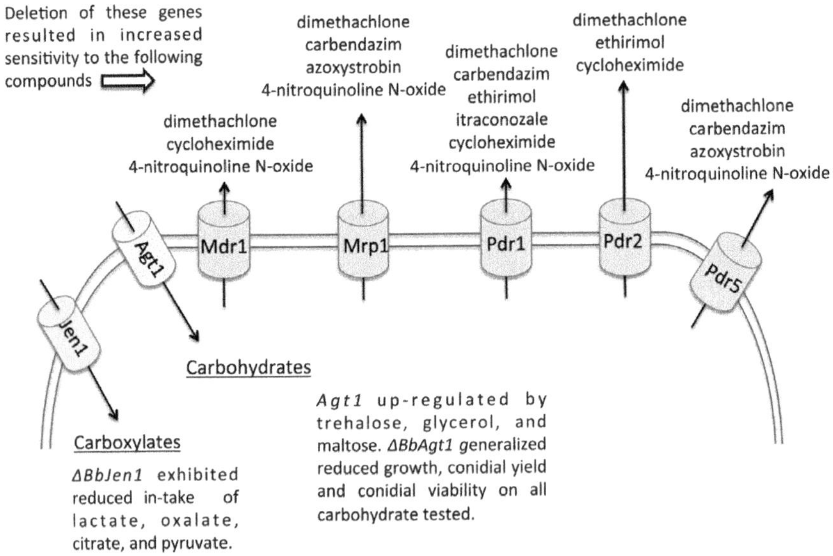

A. Ortiz-Urquiza and N.O. Keyhani, Figure 3 Overview of select *Beauveria bassiana* transporters. Jen1 (light blue), carboxylic acid transporter, Agt1 (light green), carbohydrate transporter, Mdr1, Mrp1, Pdr1, Pdr2, and Pdr5 (gray), ABC-type (multidrug) transporters.

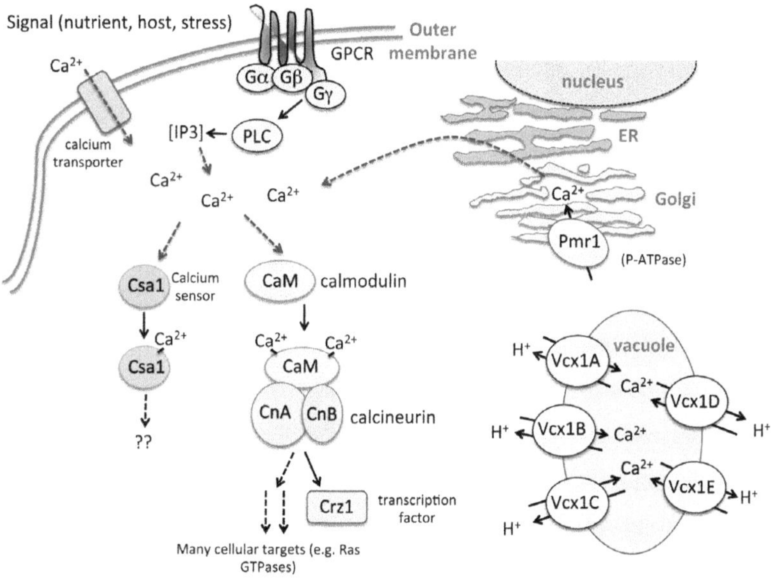

A. Ortiz-Urquiza and N.O. Keyhani, Figure 4 Overview of select *Beauveria bassiana* genes involved in calcium signaling and transport. *GPCR*, G-protein-coupled receptor; *PLC*, phospholipase C; *Csa1*, calcium sensor acidification (Ca^{2+} binding protein); *CaM*, calmodulin; *CnA/B*, calcineurin catalytic and regulatory subunits; *Crz1*, calcineurin responsive transcription factor; *Vcx*, vacuolar Ca^{2+} transporters; *Pmr1*, Golgi Ca^{2+} transporter.

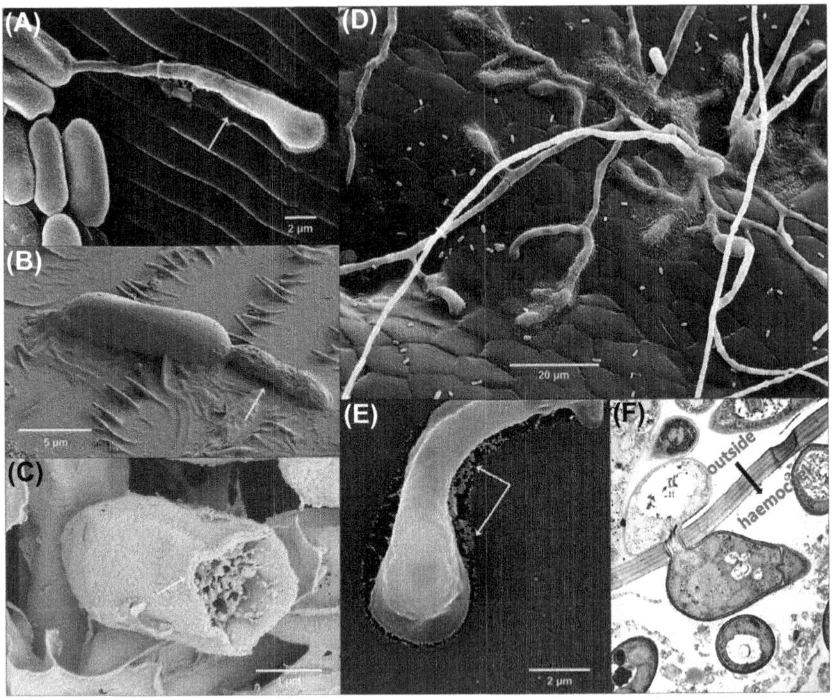

T.M. Butt, *et al.*, Figure 5 ***Key morphological features of entomopathogenic fungi interacting with the insect cuticle.*** (A) Blastospores develop infection structures covered in mucilage. (B) Appressorium formation. The different surface characteristics of the germinating blastospore can be distinguished. (C) The hydrophobin rodlet layer of an aerial conidium. (D) Differential hyphae colonize the insect cuticle. Some rod-shaped bacteria can also be seen in the image. (E) Removing the mucilage of an invading *Metarhizium* hypha revealed distinct enzymatic erosion of the integument. (F) EPF penetration of the insect cuticle layers. Key features are indicated using *arrows. Modified from Schreiter, G., Butt, T. M., Beckett, A., Moritz, G. & Vestergaard, S. (1994). Invasion and development of Verticillium lecanii in the Western Flower Thrips, Frankliniella occidentalis.* Mycological Research, 98, *1025–1034.*

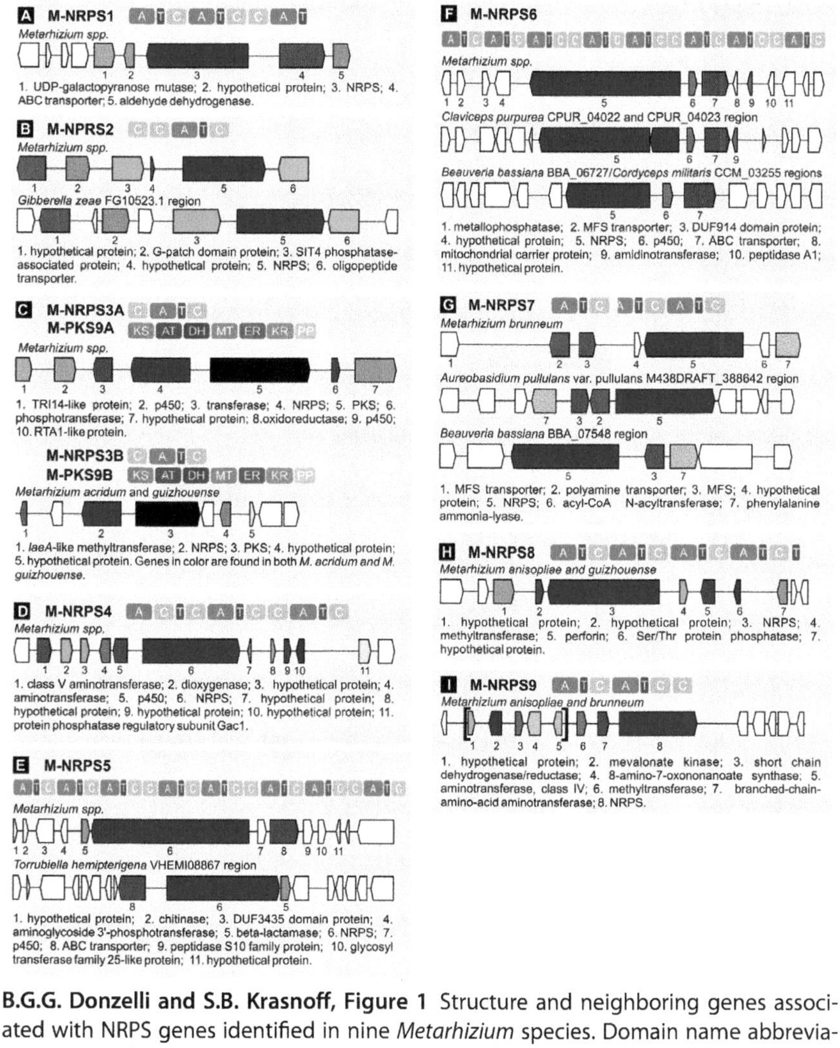

B.G.G. Donzelli and S.B. Krasnoff, Figure 1 Structure and neighboring genes associated with NRPS genes identified in nine *Metarhizium* species. Domain name abbreviations: *C*, condensation; *A*, adenylation; *T*, thiolation; and *nMT*, N-methyltransferase.

J M-NRPS10 A T C A T C C

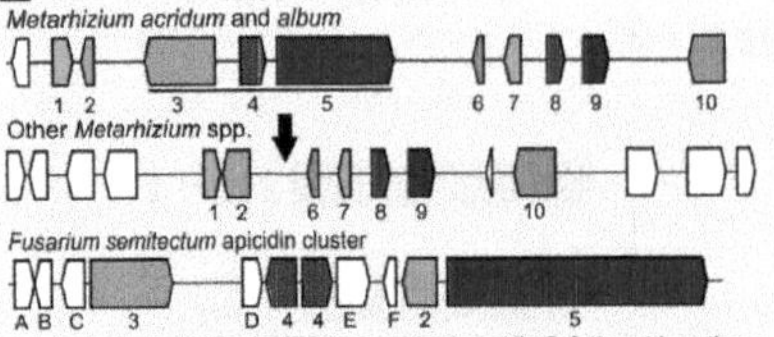

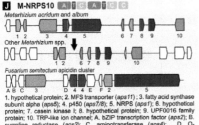

1. hypothetical protein; 2. MFS transporter (*aps11*); 3. fatty acid synthase subunit alpha (*aps5*); 4. p450 (*aps7/8*); 5. NRPS (*aps1*); 6. hypothetical protein; 7. casein kinase I; 8. hypothetical protein; 9. UPF0016 family protein; 10. TRP-like ion channel; A. bZIP transcription factor (*aps2*); B. pyrroline reductase (*aps3*); C. aminotransferase (*aps4*); D. O-methyltransferase (*aps6*); E. oxidase (*aps9*); F. reductase (*aps10*).

K M-NRPS11 A T C C A T C A T C A T C

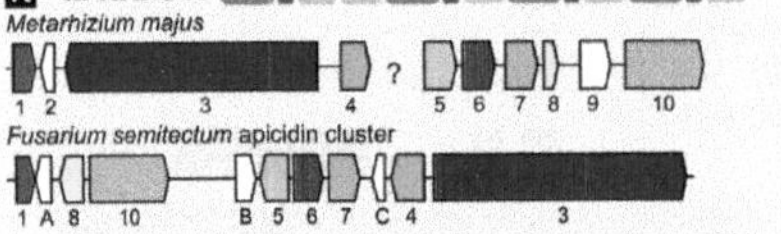

1. bZIP transcription factor (*aps2*); 2. capsule polysaccharide biosynthesis protein; 3. NRPS (*aps1*); 4. MFS transporter (*aps11*); 5. p450 (*aps7/8*); 6. p450 (*aps7/8*); 7. oxidase (*aps9*); 8. branched-chain-amino-acid aminotransferase (*aps4*); 9. p450; 10. fatty acid synthase subunit alpha (*aps5*). A. pyrroline reductase (*aps3*); B. O-methyltransferase (*aps6*); C. reductase (*aps10*).

L M-NRPS12 A T C A T C A T C T C T C

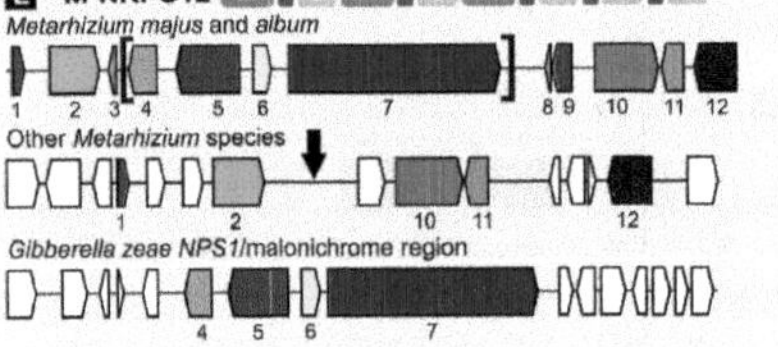

1. zinc finger, RING/FYVE/PHD-type; 2. DDHD domain-containing protein; 3. hypothetical protein; 4. MFS transporter (*mirD*-like); 5. ABC transporter (*sitT*-like); 6. acyl-CoA N-acyltransferase (*sidF*-like); 7. NRPS; 8. hypothetical protein; 9. hypothetical protein; 10. DNA-directed RNA polymerase III; 11. MFS transporter; 12. coatomer beta' subunit.

M M-NRPS13 A T C A T C C

1. ATP-grasp fold; 2. p450; 3. amino acid permease; 4. NRPS; 5. hypothetical protein; 6. phospholipase A2; 7. proteinase inhibitor I9, subtilisin propeptide; 8. subtilisin-like protease PR1F.

N M-NRPS14 A T C

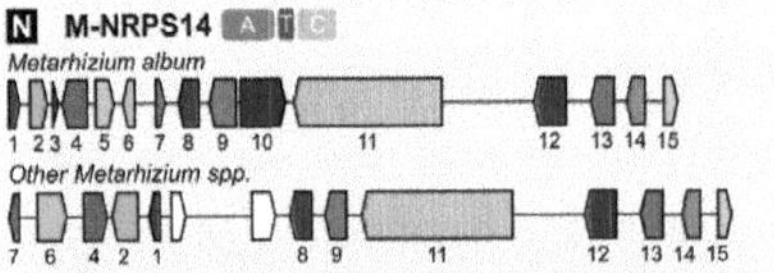

1. hypothetical protein; 2. afadin/alpha-actinin-binding protein; 3. hypothetical protein; 4. serine carboxypeptidase 1; 5. transposase; 6. amino acid/polyamine transporter I; 7. cell wall protein; 8. cystathionine gamma-synthase; 9. p450; 10. NRPS; 11. PKS-NRPS (M-HPN1); 12. Zn(2)-C6 DNA-binding domain protein; 13. monooxygenase; 14. eukaryotic aspartyl protease; 15. thioesterase.

O M-NRPS15 (*lpsC*) A T C R

M-NRPS16 (*lpsB*) A T

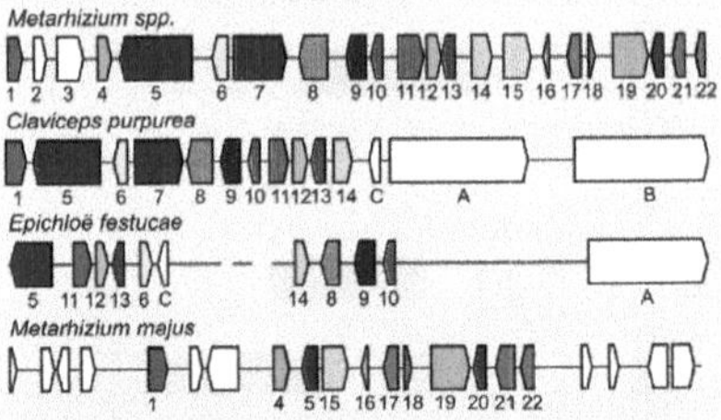

1. hypothetical protein; 2. hypothetical protein; 3. hypothetical protein; 4. prolyl oligopeptidase; 5. NRPS (*lpsC*); 6. NADH:flavin oxidoreductase (*easA*); 7. NRPS (*lpsB*); 8. p450 (*cloA*); 9. catalase (*easC*); 10. short chain dehydrogenase (*easD*); 11. FAD/FMN-binding dehydrogenase (*easE/ccsA*); 12. dimethylallyltryptophan N-methyltransferase (*easF*); 13. NAD dependent epimerase/dehydratase family (*easG*); 14. tryptophan dimethylallyltransferase family (*dmaW*); 15. flavin-binding monooxygenase; 16. hypothetical protein; 17. zinc-binding dehydrogenase; 18. cupin domain protein; 19. dynamin family protein; 20. glutathione S-transferase; 21. hypothetical protein; 22. acetyltransferase (GNAT) domain protein; A. NRPS (*lpsA1*); B. NRPS (*lpsA2*); C. oxygenase (*easH*).

P M-NRPS17 (*mrsidC*) A T C A T C T C A T C T C T C

1. hypothetical protein; 2. amidase domain protein; 3. glyoxalase/bleomycin resistance; 4. betaine aldehyde dehydrogenase (*badH*); 5. GMC oxidoreductase (*codA*); 6. DUF1479 domain protein; 7. NRPS (*sidA/nps2*); 8. L-lysine 6-monooxygenase family protein (*sidA*); 9. Zn(2)-Cys(6) zinc finger domain protein; 10. translation initiation factor 2B subunit and; 11. WW and FF domain protein; 12. PAP2 superfamily protein; 13. proteasome subunit.

Q M-NRPS18 (*mrsidD*) A T C T C

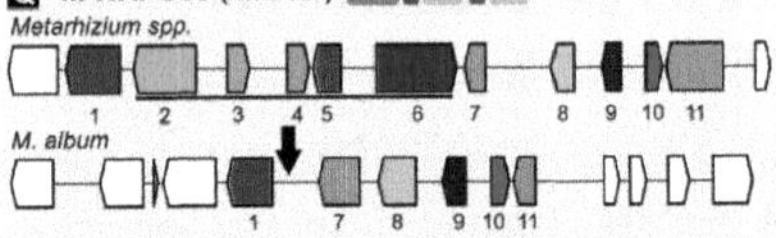

1. di- and tricarboxylate transporter; 2. siderophore ABC transporter (*sitT*); 3. pyridine nucleotide-disulfide oxidoreductase; 4. acetyltransferase (*sidF*); 5. AMP binding enzyme (*sidI*); 6. NRPS (*sidD/nps6*); 7. DUF4050 domain protein; 8. hypothetical protein; 9. PAPA-1-like domain protein; 10. peroxisomal membrane anchor protein 14 domain; 11. chromatin organization modifier chromo domain protein;

R M-NRPS19 (*dtxS1/dxs*) A T C A T C C A T C A T C A MT C T A MT T C

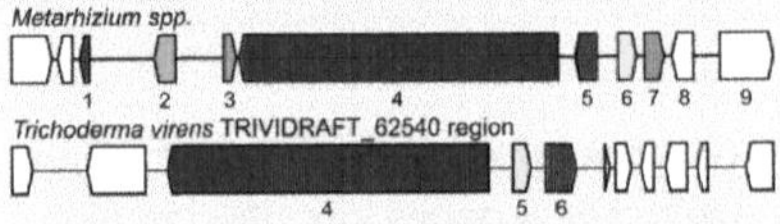

1. Zn(2)-Cys(6) zinc finger domain protein; 2. peptidase family S33; 3. phytanoyl-CoA dioxygenase; 4. NRPS (*dtxS1/dxs*); 5. p450 (*dtxS2*); 6. aldo-keto reductase (*dtxS3*); 7. pyridoxal-dependent decarboxylase (*dtxS4*); 8. V/A-type ATP synthase (non-catalytic) subunit B; 9. ABC transporter.

B.G.G. Donzelli and S.B. Krasnoff, Figure 1 (*continued*).

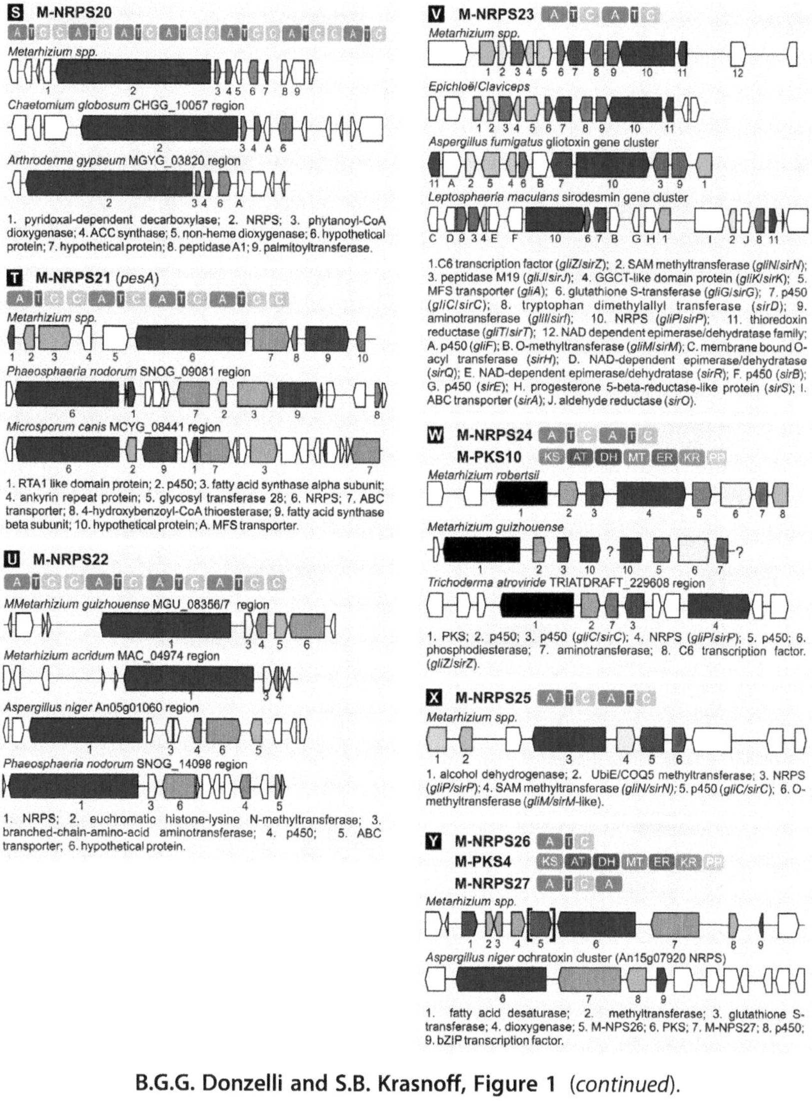

B.G.G. Donzelli and S.B. Krasnoff, Figure 1 *(continued)*.

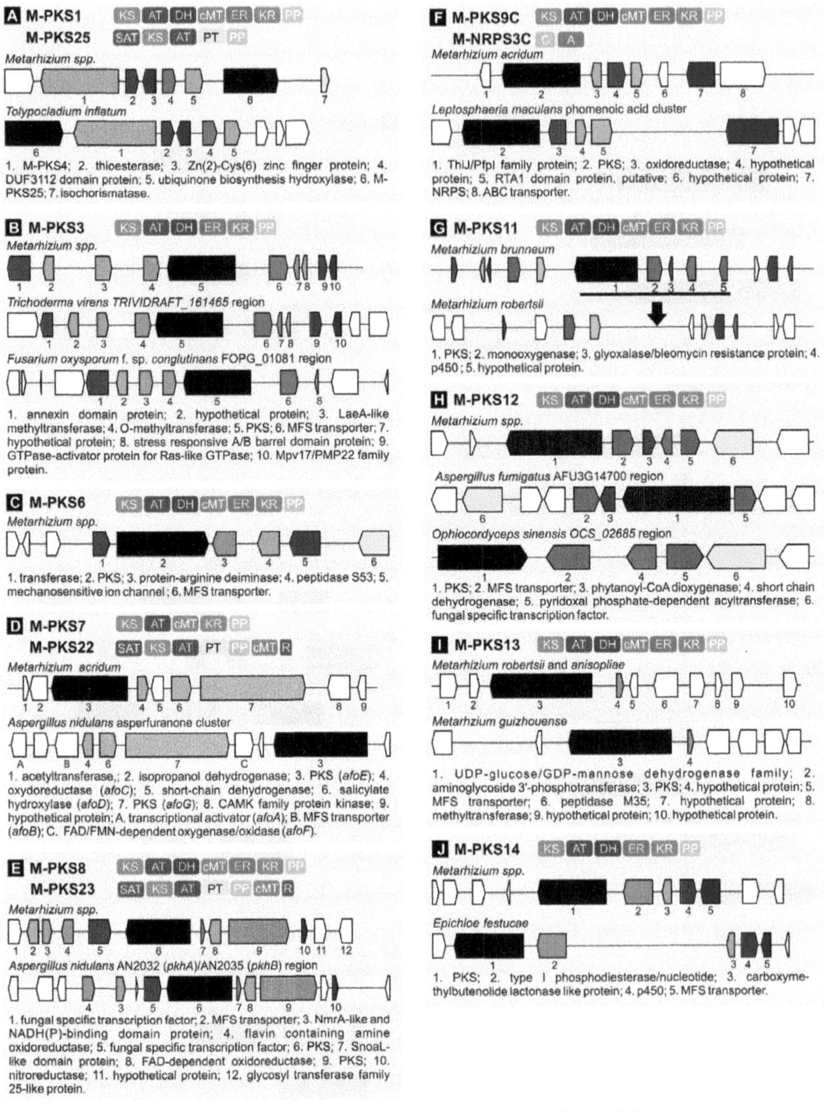

B.G.G. Donzelli and S.B. Krasnoff, Figure 2 Structure and neighboring genes associated with PKS families identified in nine *Metarhizium* species. Domain name abbreviations: *KS*, ketoacyl synthetase; *SAT*, starter unit:ACP transacylase; *AT*, acyltransferase; *DH*, dehydratase; *cMT*, C-methyltransferase; *ER*, enoyl-reductase; *KR*, ketoreductase; *PT*, product template; *PP*, 4′-phosphopantetheine attachment; *R*, thioester reductase; *CARN*, choline/carnitine O-acyltransferase; *C*, condensation.

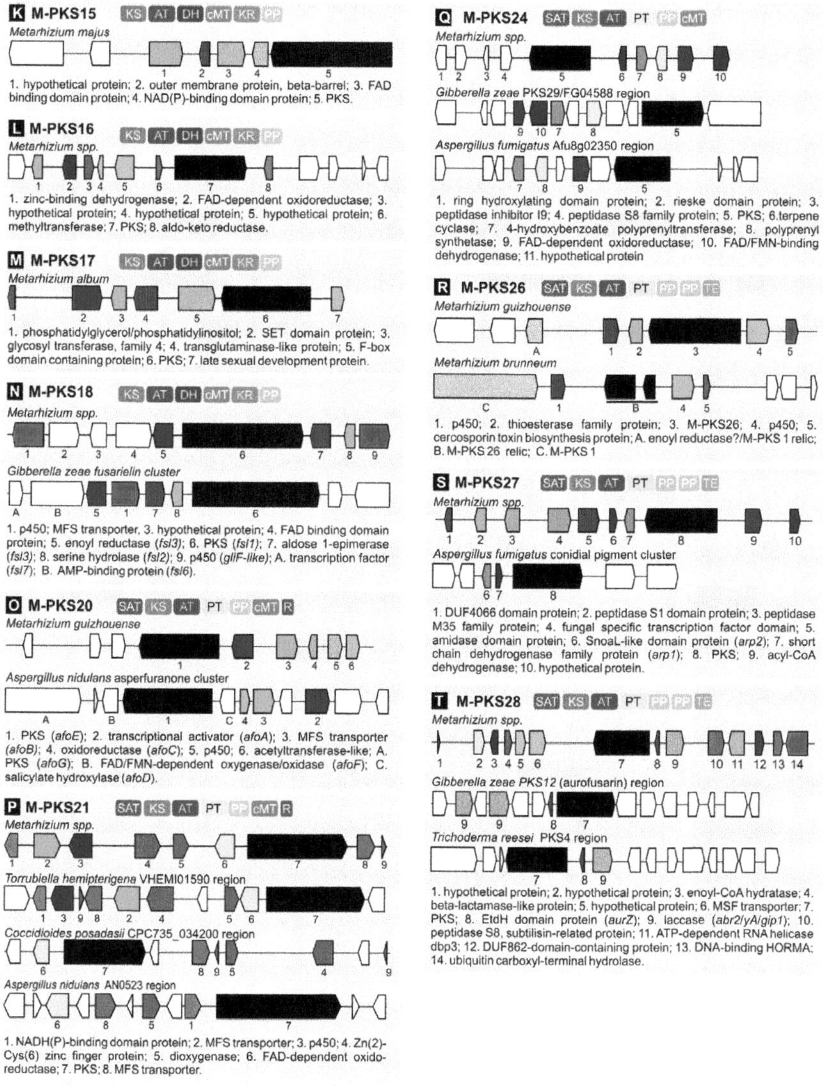

B.G.G. Donzelli and S.B. Krasnoff, Figure 2 (*continued*).

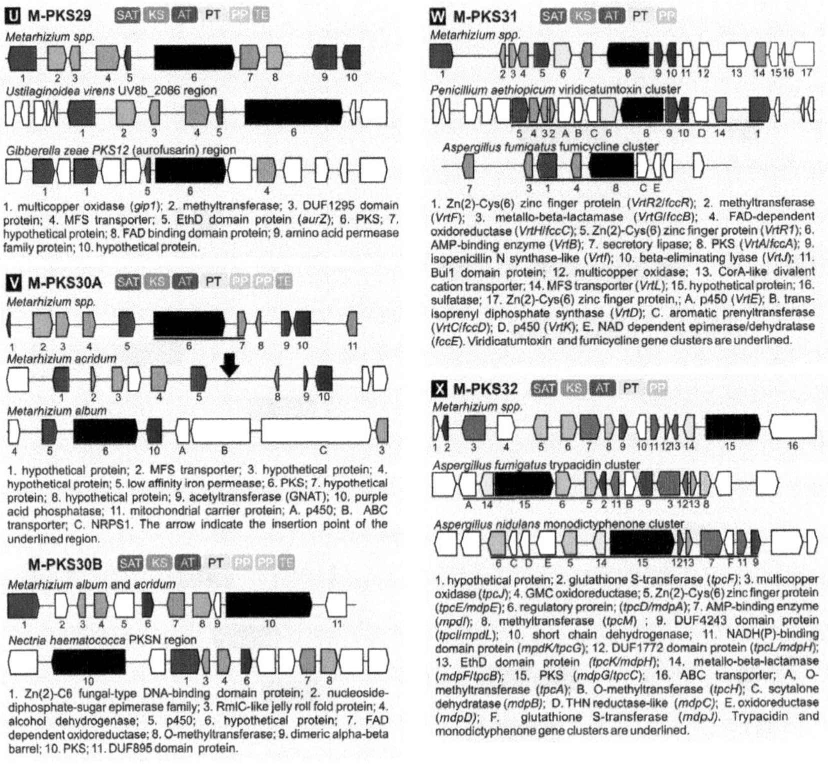

U M-PKS29 SAT KS AT PT PP TE
Metarhizium spp.
1 2 3 4 5 6 7 8 9 10
Ustilaginoidea virens UV8b_2086 region
1 2 3 4 5 6
Gibberella zeae PKS12 (aurofusarin) region
1 1 5 6 4
1. multicopper oxidase (gip1); 2. methyltransferase; 3. DUF1295 domain protein; 4. MFS transporter; 5. EthD domain protein (aurZ); 6. PKS; 7. hypothetical protein; 8. FAD binding domain protein; 9. amino acid permease family protein; 10. hypothetical protein.
V M-PKS30A SAT KS AT PT PP PP TE
Metarhizium spp.
1 2 3 4 5 6 7 8 9 10 11
Metarhizium acridum
1 2 3 4 5 8 9 10
Metarhizium album
4 5 6 10 A B C 3
1. hypothetical protein; 2. MFS transporter; 3. hypothetical protein; 4. hypothetical protein; 5. low affinity iron permease; 6. PKS; 7. hypothetical protein; 8. hypothetical protein; 9. acetyltransferase (GNAT); 10. purple acid phosphatase; 11. mitochondrial carrier protein; A. p450; B. ABC transporter; C. NRPS1. The arrow indicate the insertion point of the underlined region.
M-PKS30B SAT KS AT PT PP PP TE
Metarhizium album and acridum
1 2 3 4 5 6 7 8 9 10 11
Nectria haematococca PKSN region
10 1 3 4 6 7 8
1. Zn(2)-C6 fungal-type DNA-binding domain protein; 2. nucleoside-diphosphate-sugar epimerase family; 3. RmlC-like jelly roll fold protein; 4. alcohol dehydrogenase; 5. p450; 6. hypothetical protein; 7. FAD dependent oxidoreductase; 8. O-methyltransferase; 9. dimeric alpha-beta barrel; 10. PKS; 11. DUF895 domain protein.
W M-PKS31 SAT KS AT PT PP
Metarhizium spp.
1 2 3 4 5 6 7 8 9 10 11 12 13 14 15 16 17
Penicillium aethiopicum viridicatumtoxin cluster
5 4 3 2 A B C 6 8 9 10 D 14 1
Aspergillus fumigatus fumicycline cluster
7 3 1 4 8 C E
1. Zn(2)-Cys(6) zinc finger protein (VrtR2/fccR); 2. methyltransferase (VrtF); 3. metallo-beta-lactamase (VrtG/fccB); 4. FAD-dependent oxidoreductase (VrtH/fccC); 5. Zn(2)-Cys(6) zinc finger protein (VrtR1); 6. AMP-binding enzyme (VrtB); 7. secretory lipase; 8. PKS (VrtA/fccA); 9. isopenicillin N synthase-like (VrtI); 10. beta-eliminating lyase (VrtJ); 11. Bul1 domain protein; 12. multicopper oxidase; 13. CorA-like divalent cation transporter; 14. MFS transporter (VrtL); 15. hypothetical protein; 16. sulfatase; 17. Zn(2)-Cys(6) zinc finger protein,; A. p450 (VrtE); B. trans-isoprenyl diphosphate synthase (VrtD); C. aromatic prenyltransferase (VrtC/fccD); D. p450 (VrtK); E. NAD dependent epimerase/dehydratase (fccE). Viridicatumtoxin and fumicycline gene clusters are underlined.
X M-PKS32 SAT KS AT PT PP
Metarhizium spp.
1 2 3 4 5 6 7 8 9 10 11 12 13 14 15 16
Aspergillus fumigatus trypacidin cluster
A 14 15 6 5 2 11 B 9 3 12 13 8
Aspergillus nidulans monodictyphenone cluster
6 C D E 5 14 15 12 13 7 F 11 9
1. hypothetical protein; 2. glutathione S-transferase (tpcF); 3. multicopper oxidase (tpcJ); 4. GMC oxidoreductase; 5. Zn(2)-Cys(6) zinc finger protein (tpcE/mdpE); 6. regulatory prorein; (tpcD/mdpA); 7. AMP-binding enzyme (mpdI); 8. methyltransferase (tpcM) ; 9. DUF4243 domain protein (tpcI/mpdL); 10. short chain dehydrogenase; 11. NADH(P)-binding domain protein (mpdK/tpcG); 12. DUF1772 domain protein (tpcL/mdpH); 13. EthD domain protein (tpcK/mdpH); 14. metallo-beta-lactamase (mdpF/tpcB); 15. PKS (mdpG/tpcC); 16. ABC transporter; A, O-methyltransferase (tpcA); B. O-methyltransferase (tpcH); C. scytalone dehydratase (mdpB); D. THN reductase-like (mdpC); E. oxidoreductase (mdpD); F. glutathione S-transferase (mdpJ). Trypacidin and monodictyphenone gene clusters are underlined.

B.G.G. Donzelli and S.B. Krasnoff, Figure 2 *(continued)*.

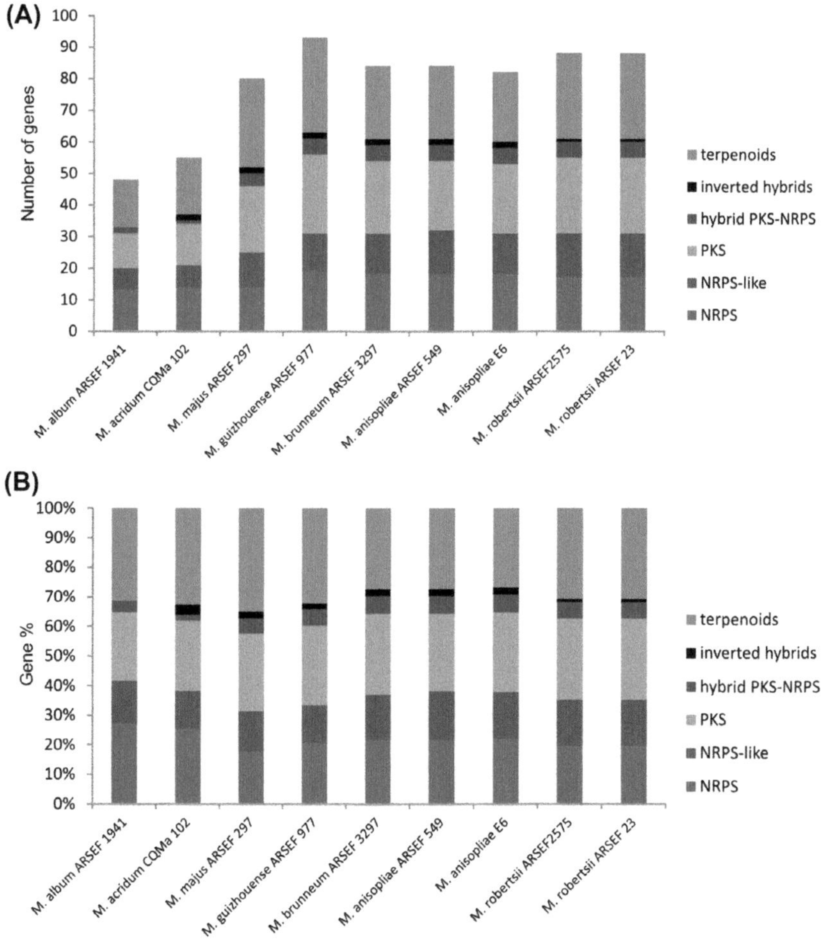

B.G.G. Donzelli and S.B. Krasnoff, Figure 3 Absolute (A) and relative (B) abundance of six classes of core secondary metabolite biosynthetic genes in nine *Metarhizium* species.

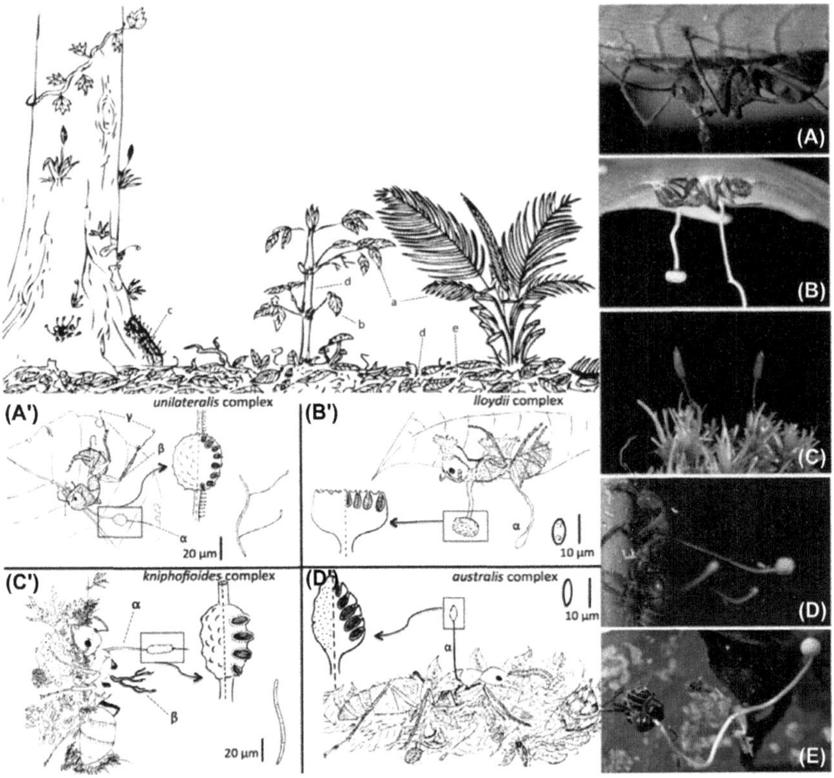

D.P. Hughes, *et al.*, Figure 1 ***Diversity of behavioral manipulation of ants by fungi in a tropical forest setting.*** Represented is some of the extensive taxonomic and functional diversity of *Ophiocordyceps* in ants. Top left shows a composite forest scene with six locations where cadavers are found: under leaves (a = *Ophiocordyceps unilateralis*), (b = *Ophiocordyceps lloydii*), tree bole/bark (c = *Ophiocordyceps kniphofioides*), leaf litter (d = *Ophiocordyceps australis, Ophiocordyceps myrmecophila, Ophiocordyceps irangensis, O. kniphofioides*), and stem (e = *O. australis* in Ghana). Each specific name represents a complex with more than one species. Line diagrams show the functional morphology (represented as a composite, all morphologies do not occur on one species). In (A′) *unilateralis* has one teleomorph (sexual stage) and three anamorphs (asexual, α–γ). The teleomorph (ascoma) shows the outside and inside where ascospores are produced and the ascospore is drawn with capilliconidia (secondary spores on hairs). In (B′) *lloydii* and (D′) *australis* only part spores are produced and they do not produce secondary structures and only one anamorph is found (asexual, α). In *lloydii* the ant is not biting but glued via hyphae from mouth (photo b). The complex (C′) *kniphofioides* (here on ant species *Dolichoderus bispinosus,* which is hidden from view in moss) is related to *unilateralis* and we can see two anamorphs (one does not occur on *Dolichoderus,* but only on *Cephalotes atratus)*. The photos A–E show host ants and position of death: A–E *Polyrhachis armata,* Thailand, *Camponotus atriceps*, Brazil, *Dolichoderus bispinosus*, Brazil, *Paltothyreus tarsatus*, Ghana, and *Polyrhachis robsoni*, Australia. Among these are two undescribed fungal species from the following complexes: *unilateralis* (A′) *and australis* (D′).

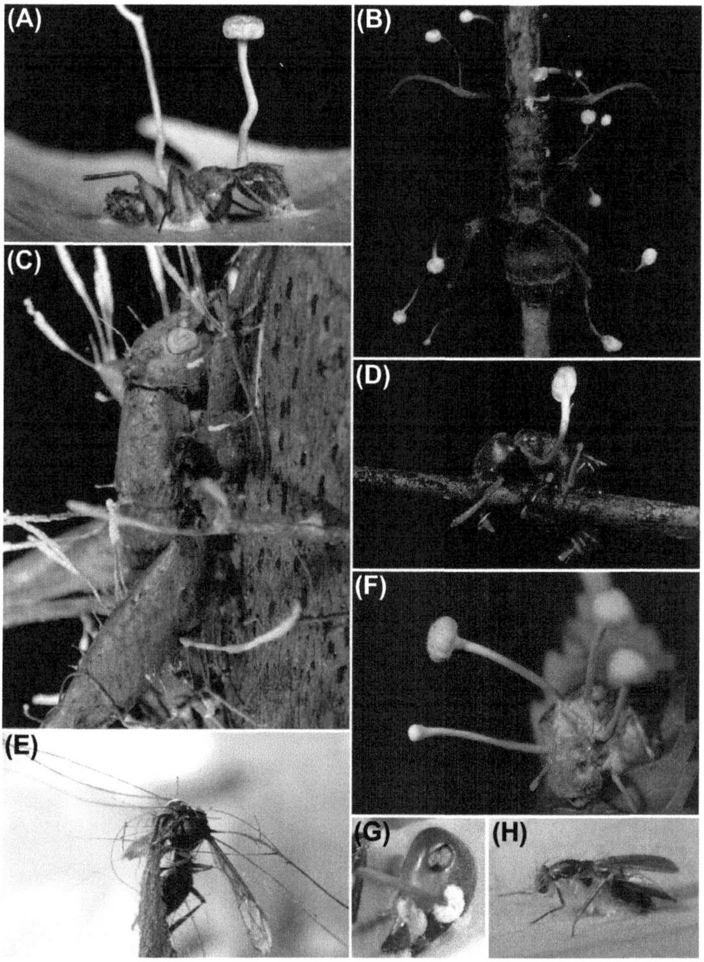

D.P. Hughes, *et al.*, Figure 2 ***Diversity of behavioral manipulation where insects are controlled to die attached to arboreal surfaces*** (A) *Ophiocordyceps lloydii* on *Camponotus atriceps* (Brazilian Amazon); (B) *Stilbella burmensis* on *Polyrhachis cf. militaris* (Ghana); (C) anamorphic Hypocreales on Orthopera (Atlantic rainforest, Brazil); (D) *Ophiocordyceps* sp. on *Pachycondyla impressa* (Brazilian Amazon); (E) *Hirsutella saussurei* (anamorph of *Ophiocordyceps humberti*, Atlantic rainforest in Brazil) on Polistinae wasp; (F) *Ophiocordyceps dipterigena s.l.* (Brazilian Amazon) on unidentified fly; (G) *Pandora formicae* on *Formica* ant (Finland); (H) *Ophiocordyceps dipterigena s.l.* early developmental stage (Atlantic Rainforest in Brazil) on unidentified fly.

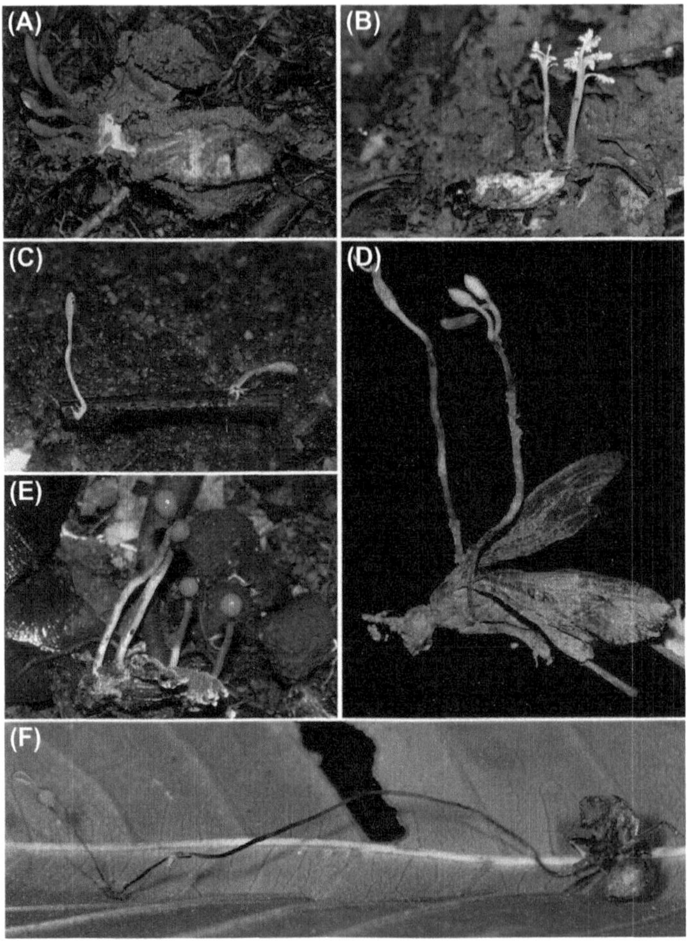

D.P. Hughes, *et al.*, Figure 3 ***Diversity of interactions where insects die in the soil where behavioral manipulation is not assumed to occur*** (A) *Ophiocordyceps* sp. on trapdoor spider (Brazilian Amazon); (B) *Isaria* sp. on Lepidoptera pupa (Brazilian Amazon); (C) *Ophiocordyceps cf. cardinalis* on Coleoptera larva (Brazilian amazon); (D) *Ophiocordyceps* sp. (*Neocordyceps* group) on wasp. (E) *Ophiocordyceps amazonica s.l.* on Orthoptera (Colombia); (F) *Ophiocordyceps australis s.l.* on *Polyrhachis* sp. (Ghana).

D.P. Hughes, *et al.*, Figure 4 ***Further examples of death on arboreal surfaces*** (A) *Entomophaga grylli* on a locust host, near the summit of a thistle plant, in the grasslands of Outer Mongolia (China). Note the legs clasping the stem and the creamy-white fungal bands bursting from the intersegmental sutures. All the thistles in the immediate area had infected locusts in the same position, mostly with two to three corpses per plant. (B) *Erynia* on an unknown dipteran host, attached to the underside of a shrub leaf by rhizoids, Atlantic rainforest, Mata do Paraíso, Minas Gerais, Brazil. A white spore halo is beginning to form on the leaf surface. (C) *Gibellula* sp. nov. on a huntsman spider (*Caayguara* sp., Sparassidae), underside of shrub leaf, Mata do Paraíso. The asexual stage is nonsynnematal and reduced to sporing heads on the fore legs, while the abdomen bears an abundance of flask-shaped perithecia with prominent necks. The host genus was erected in 2010 and the spider is a fast-moving predator living in tree bark. (D) *Gibellula*, of the *leiopus* group on a spider, on the underside of shrub leaf, Reserva Ducke, near Manaus, Amazonas, Brazil. (E) *Gibellula* sp. nov. on an unknown ghost spider (Anyphaenidae), underside of shrub leaf, Mata do Paraíso. Note the abundance of lilac-colored synnemata arising from the mycelial-covered abdomen. (F) *Gibellula* sp. nov. on a ghost spider (*Iguarima sensoria*, Anyphaenidae), underside of shrub leaf, Mata do Paraíso; showing similar synnematal production to above. This is a fast-moving spider in the forest understory and rests in silken retreats (sacs) in the litter, tree bark, and vegetation.

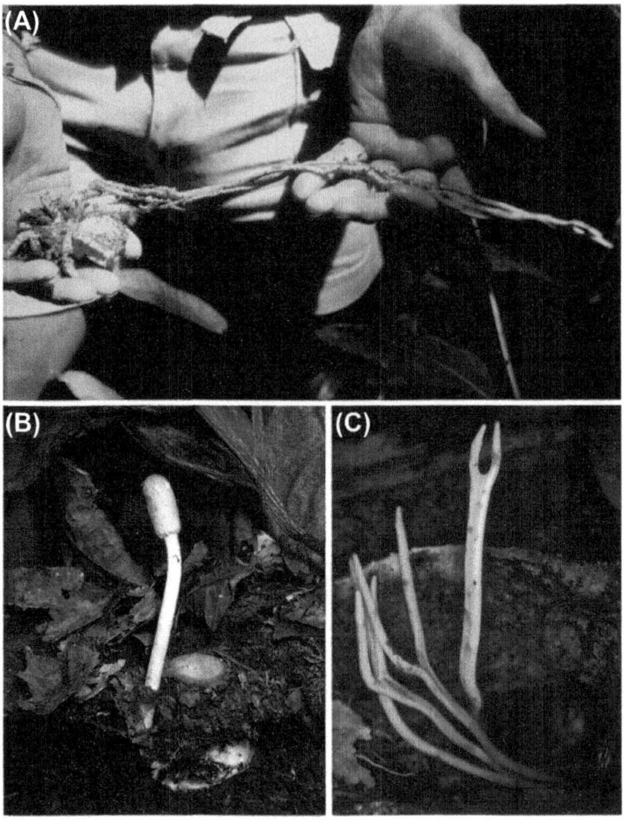

D.P. Hughes, *et al.*, Figure 5 ***Infection of ground dwelling spiders.*** (A) The entire fungus (*Ophiocordyceps caloceroides*) extricated from the deep burrow, showing the mycelial-covered but easily recognizable bird-eating mygalomorph host. (B) *Cordyceps* sp. emerging from the burrow of a trapdoor spider (Ctenizidae), Rio Negro, Amazonian Brazil. The fungal clava has pushed open the trapdoor (right) and formed a yellow perithecial-bearing head (ascostroma), with the host enveloped in a white mycelial mat (C). Clavae or ascomata of *O. caloceroides* (Hypocreales) emerging from the burrow of a *Mygale* spider (Ctenizidae), forest litter, Rio Napo, Amazonian Ecuador. The paler-colored apex contains embedded perithecia and the structures resemble the *Calocera* mushroom genus.

Edwards Brothers Malloy
Ann Arbor MI. USA
April 29, 2016